Composite Materials

Summarizing the recent advances in high strain rate testing, this book discusses techniques for designing, executing, analyzing and interpreting the results of experiments involving the dynamic behavior of multifunctional materials such as metals, polymers, fiber-reinforced polymers, hybrid laminates and so forth. The book also discusses analytical and numerical modeling of materials under high-velocity impact loading and other environmental conditions. Recent advances in characterization techniques such as digital image correlation and computed tomography for high strain rate applications are included.

Features:

- Presents exclusive material on high-rate properties of fiber-reinforced composites
- Provides numerical techniques on the analysis and enriched data on the high strain rate behavior of materials
- Explores cutting-edge techniques and experimental guidelines for an array of different materials subjected to high strain rate loading
- Explains a clear understanding of material behavior at various strain rates
- Reviews mechanical responses of different materials at high strain rates

This book is aimed at researchers and professionals in mechanical, materials and aerospace engineering.

Composite Materials

High Strain Rate Studies

Edited by
R. Velmurugan, Dong Ruan and S. Gurusideswar

CRC CRC Press
Taylor & Francis Group
Boca Raton London New York

CRC Press is an imprint of the
Taylor & Francis Group, an **informa** business

First edition published 2024
by CRC Press
6000 Broken Sound Parkway NW, Suite 300, Boca Raton, FL 33487-2742

and by CRC Press
4 Park Square, Milton Park, Abingdon, Oxon, OX14 4RN

CRC Press is an imprint of Taylor & Francis Group, LLC

Library of Congress Cataloging-in-Publication Data

Names: Velmurugan, R. (Professor of aerospace engineering), editor. | Ruan, Dong, editor. | Gurusideswar, S. (Professor of aerospace engineering), editor. Title: Composite materials : high strain rate studies / edited by R. Velmurugan, Dong Ruan, S. Gurusideswar. Description: Boca Raton : CRC Press, 2024. | Includes bibliographical references and index. | Identifiers: LCCN 2023024846 (print) | LCCN 2023024847 (ebook) | ISBN 9781032402932 (hardback) | ISBN 9781032402949 (paperback) | ISBN 9781003352358 (ebook) Subjects: LCSH: Composite materials--Testing--Case studies. | Strains and stresses--Case studies. Classification: LCC TA418.9.C6 C576 2024 (print) | LCC TA418.9.C6 (ebook) | DDC 620.1/18--dc23/eng/20230918 LC record available at https://lccn.loc.gov/2023024846LC ebook record available at https://lccn.loc.gov/2023024847

ISBN: 9781032402932 (hbk)
ISBN: 9781032402949 (pbk)
ISBN: 9781003352358 (ebk)

DOI: 10.1201/9781003352358

Typeset in Times
by Deanta Global Publishing Services, Chennai, India

Contents

Preface

R. Velmurugan and Dong Ruan

The dynamic behavior of materials is important, as most systems, whether created by nature or by humans, are subjected to dynamic loads in many of their applications. Few systems are subjected to quasi-static loads only.

There are natural calamities such as earthquakes, wind storms, rains, ocean waves, floods and ice falls which occur on large scales. Apart from these natural calamities, there are situations in our daily lives that involve dynamic loading, which includes the dropping of toys and cell phones, tools dropping on a shop floor and many more. There are accidents that happen in a short time globally which lead to loss of life, damage to vehicles, shattering of infrastructure and so forth.

The end users of the book are mostly postgraduate students, engineers, designers and researchers in the field of aerospace, civil, mechanical, chemical and ocean engineering, biotechnology and many more. Since materials used for different applications are plenty and their behavior also varies with the type of loads, the subject is most important for material scientists.

The editors of the book are highly accomplished researchers working as professors of highly reputed institutions. R. Velmurugan is a Senior Professor in the Department of Aerospace Engineering at the Indian Institute of Technology Madras, a premier institute in India. Dong Ruan is a Professor and Head of the Department of Mechanical and Product Design Engineering (MPDE) at Swinburne University of Technology, Australia, and a highly accomplished researcher and lecturer. S. Gurusideswar is working as an Assistant Professor at SRM Institute of Science and Technology, a highly-ranked university in India, and has very deep knowledge of the subject related to the published book.

Acknowledgments

The book *Composite Materials: High Strain Rate Studies* is part of the research program supported by SPARC (Scheme for the Promotion of Academic and Research Collaboration) from the Ministry of Human Resource and Development, Government of India, through the Indo-Australia collaborative research scheme. The authors are grateful to the Ministry of Human Resource and Development, India, for its support to make this book in its present form.

The area of research is highly focused and important from the design point of view. There are researchers in India and abroad who have contributed book chapters. The editors are grateful to all these contributors.

There are many professional reviewers who have reviewed the articles and the editors express their sincere thanks to the reviewers.

The editors are acknowledging the support received from the Indian Institute of Technology Madras, Chennai, India, Swinburne University of Technology, Australia, and SRM Institute of Science and Technology, Kattankulathur, India.

Finally, we are thankful to CRC Press who have kindly agreed to publish the chapters in book form. We are extremely grateful to Dr. Gagandeep Singh, who has been the source of inspiration from the beginning.

About the Editors

R. Velmurugan is a Senior Professor in the Department of Aerospace Engineering, Indian Institute of Technology Madras, Chennai. His areas of research include composite materials, nano materials, finite element analysis, structural crashworthiness and impact mechanics. He has completed many consultancies and sponsored projects from many DRDO labs, ISRO centers, government agencies and private industries. Dr. Velmurugan has published more than 300 papers in journals and conferences and guided many students in PhD, MS and MTech degrees. Dr. Velmurugan has a Google citation of more than 5,000 with an h-index of 40, an i-10 index of 116 and a Scopus h-index of 34.

Dong Ruan is a Chair Professor in the Department of Mechanical and Product Design Engineering, Swinburne University, Melbourne, Australia. Her research interests include additive manufacturing of continuous fiber-reinforced composite materials and structures, characterization of the mechanical properties of various materials at high strain rates and evaluation of the mechanical response of structures (such as multi-layered panels and tubes) under dynamic loadings. She has published over 240 academic papers in top international journals and prestigious international conferences. Dr. Ruan has secured over $20 million in research grants from the Australian Research Council (ARC), Defence Materials and Technology Centre (DMTC), Cooperative Research Centre for Advanced Automotive Technology (AutoCRC), CAST CRC and Rail Manufacturing CRC, among others. She has supervised more than 20 PhD students. Dr. Ruan has a Google citation of more than 6,400 with an h-index of 42 and an i-10 index of 111.

S. Gurusideswar is an Assistant Professor in the Department of Aerospace Engineering, SRM Institute of Science and Technology Madras, Kattankulathur, India. Prior to this, he worked as a postdoctoral fellow in the Department of Mechanical Engineering at the Indian Institute of Technology Bombay, Mumbai. During his postdoc tenure, Dr. Gurusideswar worked on dynamic characterization of ultra-high-performance concrete (UHPC) material, which is developed for blast and impact-resistant structures. His areas of research include composite structures, computer aided design, experimental mechanics, high strain rate characterization and digital image correlation (DIC) technique. Dr. Gurusideswar has published more than 20 academic papers in journals and conferences.

Contributors

Prakash A.
Indian Institute of Information
  Technology, Design and Manufacturing
  Kancheepuram, India

Nenshol Jayant Anand
Visvesvaraya National Institute of Technology,
  Nagpur, India

Singh A. P.
Dr. B. R. Ambedkar National Institute of
  Technology, Jalandhar, India

Bhonge A. S.
Visvesvaraya National Institute of Technology,
  Nagpur, India

Siva Prasad A. V. S.
Indian Institute of Information Technology,
  Design and Manufacturing, Kancheepuram,
  India

Naresh Bhatnagar
Indian Institute of Technology Delhi,
  New Delhi, India

Lakshmana Rao C.
Indian Institute of Technology Madras,
  Chennai, India

Tanusree Chakraborty
Indian Institute of Technology Delhi,
  New Delhi, India

Purnashis Chakraborty
Indian Institute of Technology Delhi,
  New Delhi, India

Anoop Chawla
Indian Institute of Technology Delhi,
  New Delhi, India

Ramdas Chennamsetti
Research and Development Establishment
  (Engineers Defence Research and
  Development Organisation, Pune, India

Naresh V. Datla
Indian Institute of Technology Delhi,
  New Delhi, India

Venkatesh M. Deshpande
Indian Institute of Technology Delhi,
  New Delhi, India

Devendra K. Dubey
Indian Institute of Technology Delhi,
  New Delhi, India

Kavita Ganorkar
Indian Institute of Technology Delhi,
  New Delhi, India

Navya Gara
Indian Institute of Technology Madras,
  Chennai, India

Sagar Ghatke
Defence Institute of Advanced Technology,
  Pune, India

Rajendra Gupta
Research and Development Establishment
  (Engineers), Defence Research and
  Development Organisation, Pune, India

Mahapatra I.
Indian Institute of Technology Madras,
  Chennai, India

Shanideo N. Jadhav
Indian Institute of Technology Bombay,
  Mumbai, India

Krishna Jonnalagadda
Indian Institute of Technology Bombay,
  Mumbai, India

Akshaya Gomathi K.
Indian Institute of Technology Hyderabad,
  Hyderabad, India

Kanny K.
Durban University of Technology, Durban,
  South Africa

Naresh K.
University of Southern California, Los Angeles, CA

Shankar K.
Indian Institute of Technology Madras, Chennai, India

Senthil K.
Dr. B. R. Ambedkar National Institute of Technology, Jalandhar, India

Singh K. K.
Indian Institute of Technology, Dhanbad, India

Kartikeya Kartikeya
Indian Institute of Technology Delhi, New Delhi, India

Aman Kumar
National Institute of Technology Hamirpur, Hamirpur, India

Ankit Kumar
Tata Steel Limited, Jamshedpur, India

Manoj Kumar
Dr. B. R. Ambedkar National Institute of Technology, Jalandhar, India

Sanjay Kumar
Delhi Technological University, New Delhi, India

Vimal Kumar
National Institute of Technology Hamirpur, Hamirpur, India

Nidhi Kumari
Dr. B. R. Ambedkar National Institute of Technology, Jalandhar, India

Iqbal M. A.
Indian Institute of Technology Roorkee, Roorkee, India

Goel M. D.
Visvesvaraya National Institute of Technology, Nagpur, India

Saravanan M. K.
Indian Institute of Information Technology, Design and Manufacturing, Kancheepuram, India

Puneet Mahajan
Indian Institute of Technology Delhi, New Delhi, India

Mahapatra I.
Indian Institute of Technology Madras, Chennai, India

Pabitra Maji
Indian Institute of Technology Kharagpur, Kharagpur, India

Ankit Malik
Indian Institute of Technology Delhi, New Delhi, India

Mayand Malik
Indian Institute of Technology Mandi, Mandi, India

Ashish Mohan
Research and Development Establishment (Engineers), Defence Research and Development Organisation, Pune, India

Sudipto Mukherjee
Indian Institute of Technology Delhi, New Delhi, India

Raguraman Munusamy
Indian Institute of Information Technology, Design and Manufacturing, Kancheepuram, India

Sirdesai N. N.
Visvesvaraya National Institute of Technology, Nagpur, India

Prakash Nanthagopalan
Indian Institute of Technology Bombay, Mumbai, India

Dhruv Narayan
Indian Institute of Technology Delhi, New Delhi, India

Sunil Nimje
Defence Institute of Advance Technology, Pune, India

Anoop Kumar Pandouria
Indian Institute of Technology Delhi, New Delhi, India

Velmurugan R.
Indian Institute of Technology Madras,
   Chennai, India

Abhishek Raj
Tata Steel Limited, Jamshedpur

Amirtham Rajagopal
Indian Institute of Technology Hyderabad,
   Hyderabad, India

Jayaganthan R.
Indian Institute of Technology Madras,
   Chennai, India

Dong Ruan
Swinburne University of Technology,
   Melbourne, Australia

Gurusideswar S.
Indian Institute of Technology Madras,
   Chennai and SRM Institute of Science and
   Technology,
Kattankulathur, India

Manojkumar S.
Indian Institute of Technology Madras,
   Chennai, India

Rupali S.
Dr. B. R. Ambedkar National Institute of
   Technology, Jalandhar, India

Nishant K. Sahu
Indian Institute of Technology Delhi, New
   Delhi, India

Rohit Sankrityayan
Indian Institute of Technology Delhi, New
   Delhi, India

Prateek Saxena
Indian Institute of Technology Mandi, Mandi,
   India

Ankush P. Sharma
Indian Institute of Technology Madras,
   Chennai, India

Payal Shirbhate
Visvesvaraya National Institute of Technology,
   Nagpur, India

Ruchir Shrivastava
Indian Institute of Technology, Dhanbad, India

Bhrigu Nath Singh
Indian Institute of Technology Kharagpur,
   Kharagpur, India

Makhan Singh
Indian Institute of Technology Delhi, New
   Delhi, India

Pundan Kumar Singh
Tata Steel Limited, Jamshedpur and Indian
   Institute of Technology Madras, Chennai,
   India

Joseph Solomon
Indian Institute of Technology Delhi, New
   Delhi, India

Ankush Thakur
Dr. B. R. Ambedkar National Institute of
   Technology, Jalandhar, India

Vikrant Tiwari
Indian Institute of Technology Delhi, New
   Delhi, India

Brahmadathan V. B.
Indian Institute of Technology Madras,
   Chennai, India

Rahul K. Verma
 Tata Steel Limited, Jamshedpur, India

Shivani Verma
Visvesvaraya National Institute of Technology,
   Nagpur, India

Introduction

R. Velmurugan and Dong Ruan

COMPOSITE MATERIALS: HIGH STRAIN RATE STUDIES

The book on high strain rate studies of materials, especially composite materials, is of prime importance as modern structures, either mechanical or civil, are eventually subjected to dynamic loads, for which the dynamic behavior of the materials is important. Conventional design considers only the quasi-static properties of the material. However, the dynamic behavior of most materials is entirely different and hence the rate-dependent behavior of the materials is needed.

Due to the change in the lifestyle of modern society, transportation is a very important part of human life and the vehicles which are moving on roads, rails, waterways and in the air are subjected to velocities ranging from 1 mm/min to thousands of m/s and the corresponding strain rates vary between 0.001/s and 10,000/s. The dynamic loading is mostly due to impact loads, blasts, explosives, wind storms, rains and earthquakes. Suitable materials are to be considered for different load cases. The book considers topics which are relevant to transport vehicles and civil structures.

The book considers the rate-dependent properties of the composite materials, sandwich and concrete structures for different strain rates. There are some chapters that cover the high-rate behavior of metallic materials also, since the designer needs the behavior of the materials at high strain rates. Some chapters cover topics related to the experimental facilities used for measuring the high strain rate behavior of the materials using the universal testing machine for strain rates up to 10^{-3}/s, drop mass setup for strain rates up to 1,000/s and Hopkinson bar experiments for high strain rates.

The experiments need a good capturing system and hence there are some chapters that describe the details of strain gauge systems and digital image correlation (DIC) systems. Most of the structural components are subjected to compression, tension and shear loads, and the behavior of the materials for all the load cases is discussed in the book.

Since rate-dependent behavior is important and needs to be studied from the basics, there are some chapters that cover the basics of high stain rate studies. Experimental studies related to high rate loading are sometimes difficult as it can be challenging to have facilities for strain-dependent and impact studies; hence numerical studies are highly useful. There are some chapters which cover numerical studies, which are very useful tools for getting the material properties through numerical modeling, and verification of these results with experimental results. Since the stress-strain relation for high dynamic loads is different from the stress-strain relations through quasi-static loading, the quasi-static relations are no longer valid. Hence the topics that cover the different strain rate models that can predict the dynamic behavior of the materials are important and are presented in the book.

Finally, the rate-dependent properties are important for impact loading, blast and explosive loading; hence there are some chapters that cover the behavior of the material for such dynamic loading conditions. The materials covered in the studies are mostly lightweight composite materials and there are some topics that cover different materials like sandwich structures, concrete structures and metallic materials.

Overall, the topics covered in the book will be useful for graduate students, researchers and designers who are involved in the design of structures for automobiles, spacecraft, tunnels and concrete buildings.

1 Strain Rate Studies on Metallic and Non-Metallic Materials for Tensile and Compressive Behaviour Under Impact Loading
A Review

Navya Gara, R. Jayaganthan and R. Velmurugan

1.1 INTRODUCTION

The metallic and non-metallic materials that are utilized extensively for the components of structures in aerospace, defence and civilian applications would be subjected to various loads [1–3]. These loads could be in the form of static (viz. dead loads, live or moving loads) and dynamic loads (viz. wind, earthquake, collision impact), which may or may not cause deformation within the deformability limits of the structures for their structural safety and reliability [4]. The component materials behave based on the exposure of the type of loads, the intensity of the load acting, the point of application of the load along with the method of production and machinability of the component [5]. This consequently results in the severity of the destruction of the components based on the stresses and strains developed in the materials [6]. The strains developed in the materials could vary with respect to the time of application of the load and are predominantly seen with dynamic loading in the materials [7]. Furthermore, the strain rate sensitivity of metals is comparatively the least while composites and polymers are highly strain rate sensitive in any velocity regime considered [8, 9]. Thus, the strain rate behaviour in the material is the state of the art to be understood in all the applications for the various materials used.

Metals and their alloys such as aluminium and high-strength steels are enormously used in different fields such as aerospace, defence, automobile and civil engineering structures [10–15]. The ease of the metals for their machinability, weldability and availability has made these materials more attractive for different load-bearing applications [16]. The large plastic deformation behaviour of the metals gives a warning of the damage, unlike brittle materials, and thus are the most preferred materials for many static and dynamic loadings [17]. Particularly, during the collision impact of the structures, the metals undergo severe plastic deformation owing to changes in the strains and strain rates, largely affecting the thermal behaviour. This phenomenon is known as shock-induced plasticity [18]. This subsequently changes the failure mechanisms of the material when subjected to such high strain rates. The microstructural morphology also undergoes changes in the formation of the secondary phase precipitates and changes in the dislocation mechanisms at the grain boundaries, thereby varying the mechanical properties such as strength and ductility [19]. Literature [20]

DOI: 10.1201/9781003352358-1

suggests that adiabatic shear bands (ASB) formation in a few aluminium alloys was also observed, thus weakening the material characteristics for its intended use. Thus, the necessity of high strain rate studies on metals is a must.

Fibre-reinforced polymer (FRP) composites are predominantly preferred for their high strength, stiffness and lightweight properties [21]. By and large, glass, carbon and Kevlar fibres are combined with the epoxy to obtain the FRP composites along with the filler materials which are extensively used in day-to-day life in almost all engineering applications [22]. The composites initially carry the loads with ease until they attain their load-carrying capacity, after which composite softening occurs with further damage in the composites such as delamination, fibre breakage, cracking of matrix, etc. [23]. Due to the dependency of the mechanical properties on loading, especially under dynamic impact loading, a rate-dependent deformation and the consequent rate-dependent failure mechanism are observed. This strain rate sensitivity is observed for all the epoxy-based composites [24]. Thus, a thorough understanding of the change in the behaviour is necessary for the development of impact-resistant composites and better high strain resistant composites.

Polymers are nowadays invariably used in many structural applications for their flexibility in changing mechanical properties [25, 26]. They are used in their raw form or along with the composites to attribute to the better properties for their intended use [27–29]. In recent times, their application has increased tremendously in fields such as retrofitting of civil engineering structures [30], bio-medical applications in prosthetics [31], aircraft components in aerospace and defence applications [32], bumpers and other automobile applications [33], sensors in electronic and electrical engineering [34], etc. Recent advances in additive manufacturing using fused filament fabrication (FFF) has gained sudden attention for its use even with the prototypes prepared using these polymers [35, 36]. The parameters such as glass transition temperature (T_g) and melting temperature (M_p) play an important role in the printing of the material through the extrusion process. Thus, a thorough understanding of the relation to the temperature, pressure and strain rate variation for their corresponding change in microstructural and chemical composition is essential [37–39]. Rudimentary studies were carried out by Kholsky, primarily in the strain rate range of 10^{-4} to 10^4/second [40]. It was observed that the mechanical properties were time-dependent and the properties such as tensile modulus and yield strength had a significant variance with change in temperature too [41]. Furthermore, the phase change from a glassy state to a rubbery one that further transforms into plastic and finally changes to a brittle nature is an imperative state change in many polymers [42]. Thus, polymers of such type are represented by the visco-elastic equations and generally with the spring-dash pot mechanisms [43–45]. However, the distinction between the different polymers especially in forms such as amorphous or semi-crystalline would have a change in the glass transition temperature (T_g). This T_g is obtained as a temperature where the onset of the polymer takes place, converting it into the rubbery flowy state [46–48]. This in turn aids in developing the constitutive models. Thus, a thorough understanding and their parametric correlation are mandatory for polymeric materials.

There is abundant literature on the behaviour of these materials at high strain rates and their studies revealed the best utilization of the material for the desired application. However, a consolidated review is proposed in the present study to understand the limitations and shortcomings with respect to the different materials. Furthermore, a clear distinction between the methodologies and their failure mechanisms is also made based on the velocity that the structures are subjected to. The experimental setup at different regimes is discussed along with the suitability of the material for the particular strain rate regime. The strain rate influence on the mechanical properties of all the materials is analyzed in detail in the following sections.

1.2 EXPERIMENTAL PROCEDURES AND MATERIALS

Characterization of materials such as metals, composites and polymers for their mechanical behaviour based on their suitability in their realistic applications is indispensable. The setting up of the experiments for the materials, especially in the case of composites and polymers, is arduous due to the multi-phase characteristics of their constituent material chemical composition. Literature [5, 49–51] suggests that over the past decades, the upgradation of the equipment to counter the physical challenges in setting up at such high velocities in realistic applications was feasible only by classifying the velocities. Furthermore, the deformability of the material such as metals being ductile, composites being comparatively brittle and polymers being visco-elastic also plays a significant role in the velocity classification [52–54]. Different equipment was set up to attain the classified velocities with a loading capacity, and the corresponding load-displacement graphs are obtained [55]. These velocities were further depicted in the form of strain rates to avoid the specimen and material influence during its real-time applications [56]. Although a clear constraint exists for the range of loads applied, an intuition of the behaviour of the different metallic and non-metallic materials is a requisite [57–60] to avoid any miscalculations of the stresses and the deformation behaviour of the materials. Figure 1.1 shows the strain rates and the corresponding experimental apparatus.

1.2.1 QUASI-STATIC STRAIN RATE REGIME

Based on the various load-carrying capacities, the load cell of the versatile servo-hydraulic Universal Testing Machine (UTM) is utilized for testing [61]. The choice of the load cell depends on the different materials to be tested which in turn depicts the precision of the load-displacement graph obtained [62]. These machines are extensively used to understand the stress-strain curves for the metals which are sub-sized to the comparatively larger gauge-length-sized specimen made of the composite material [63, 64]. Additionally, the homogenous, isotropic metallic behaviour of the fibre strand in composites along with the bulkier composite laminates with different orientations and their heterogeneous anisotropic mechanical behaviour are a variety of classes of materials estimated for such strain rate range [65–67]. Furthermore, there are several advancements in the UTM to measure the micron specimens such as thin films, molecular materials at micro scales, etc., along with the sophisticated electronic controls in the system that ease the procedure for testing and acquisition of the test results [68, 69]. Strain rate studies of the order of 10^{-4} to 10^{-1}/second are possible with these UTM test equipment.

A Dynamic Mechanical Analyzer (DMA) is used to estimate the loss modulus, storage modulus and the tan delta values that direct to the stiffness and the damping properties of the polymer in the quasi-static regime [70, 71]. The displacements are given in an oscillating pattern (generally sinusoidal in nature) onto the specimen with varying temperatures or frequencies. Tensile, flexure and compressive loads could be applied to the specimen for the desired temperature range and amplitude. The phase difference obtained from the force-displacement curves is used to estimate the parameters of the polymers. In general, to understand the strain rates, the frequencies are converted to the strain rates [72] using the formula as in Eqn. (1.1).

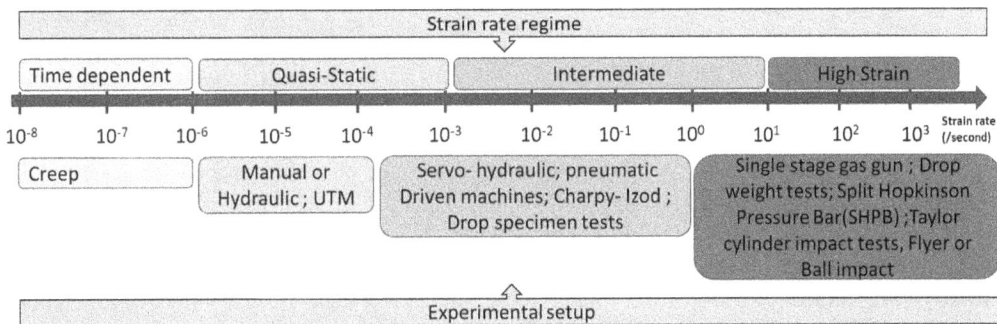

FIGURE 1.1 Magnitude of strain rate regimes and their corresponding equipment in strain rate regimes.

$$\dot{\varepsilon} = \frac{\Delta \varepsilon}{\Delta t} = \frac{\varepsilon_0}{1/4f} = 4f\varepsilon_0 \qquad (1.1)$$

Where ε_0 is the strain amplitude and f is the oscillation frequency.

1.2.2 INTERMEDIATE STRAIN RATE REGIME

The strain rate range from 10^0 to 10^2/second is challenging due to the effect of the interference between the natural frequencies of the equipment and those of the material tested [73]. With the increase in the strain rate, the energy absorption characteristics along with the mechanical properties such as strength, toughness and ductility play a crucial role in estimating the failure mechanism of the materials [74–78]. These failure mechanisms observed for all the materials may be different but generally are in the form of fracture; therefore, the toughness and the material impact resistance could be determined at velocities higher than the quasi-static tests. The commonly used intermediate strain rate equipment such as Izod [79], Charpy [80], drop mass tests [81], wheel systems [82] and very long Hopkinson bars [83] are utilized in this strain rate regime.

1.2.3 HIGH STRAIN RATE REGIME

These strain rates are replicated from the physical significance of events such as bird strikes and debris impact on aerospace components [84] or the shocks generated due to cyclones and earthquakes with higher intensity on civil engineering structures [4]. Furthermore, the generation of vibrations at such high strain rates in different engineering applications such as automobiles, workshop floors, etc., is also significantly observed in everyday life. Usually, strain rates ranging between 10^3 to 10^5/second are widely considered high strain rates. The split Hopkinson pressure bar (SHPB) [85], Taylor cylindrical rod [86] and plane impact tests [87] are some of the apparatus utilized for testing in this regime. However, the SHPB or Kolsky bar apparatus is widely used for its robust setup for material characterization. Almost all kinds of materials and structures can be easily tested for their high-strain properties by an accurate selection of the specimen geometry and dimensions to account for the inertia effects [88]. Moreover, the compatibility between the SHPB rods and the specimen plays a significant role in determining the flow stress and the corresponding failure in the material tested. The one-dimensional wave propagation principle in SHPB promotes an understanding of material deformation characteristics and the subsequent microstructure evolution after the impact [89]. The loss of energy to the surroundings in the form of heat after attaining equilibrium in the specimen also aids in the interpretation of the thermal behaviour of the material at such high strain rates [90].

1.3 METALLIC MATERIALS

A meticulous understanding of the metals such as aluminium and high-strength steels that are utilized extensively in aerospace, defence and automobile industries needs to be designed with utmost attention in terms of safety [84]. The mechanical properties vary due to the strain rate sensitivity in metals, specifically aluminium alloys, owing to the strain hardening of the metal after the yielding of the material. The change in the internal microstructure is because of the dislocation mechanisms, phase transitions and texture changes along with planar orientations due to various strain rates that the material is subjected to [91]. A substantial study was made on the strain rate sensitivity of aluminium alloys such as Al2xxx, Al 6xxx and Al7xxx alloys, and their corresponding strength and ductility were analyzed [92–95]. In the quasi-static regime in the strain rate range of 10^{-3} to 0.1/second, studies revealed that the tensile behaviour of the Al2024 alloy for the various strain rates considered was strain rate insensitive. Furthermore, the temperature did not influence any amount of the change in this alloy for the quasi-static range [96]. Nevertheless, the strain rate sensitivity (SRS) was predominantly seen when the Al2024 alloy was subjected to compression in this regime.

Furthermore, the temperature effect was also evident in the compressive testing. For the Al6061 alloy, the compressive behaviour has shown the strain rate sensitivity by an increase in the flow stress with an increase in strain rates [97, 98]. However, there was no change in the ductility of the alloy in the strain rate range of 0.01 to 10/second as shown in Figure 1.2. The effect of the heat treatments on these alloys also have shown the SRS and it was concluded that the peak-aged alloys have shown better SRS and highest stresses than the naturally aged, under-aged and over-aged alloys, particularly in the high strain regime under compressive behaviour [98]. Jung et al. [99] studied the effect of the naturally aged Al7075 alloy on the SRS and reported that the alloy has shown increasing flow stresses with an increase in strain rate. However, the fracture patterns were sporadic for the various strain rates considered. Thus, all the aluminium alloys have shown similar behaviour in quasi-static and intermediate strain rate regimes; however, a clear distinction in the flow stresses was observed.

The high strain rate studies using the split Hopkinson pressure bar (SHPB) for its compressive behaviour are utilized for estimation of the flow stress properties at very high velocities [100]. At the high strain rate loading or the dynamic strain rates, the inertia effect along with the SRS was predominantly observed for all the testing types – compression and tensile in metals. Thus, the specimen to be chosen needs to be smaller in dimensions to attain the dynamic equilibrium moderately [88, 89, 101]. The physical observations are generally observed in terms of dimensional changes such as a decrease in the longitudinal direction and an increase in the lateral dimensions. Furthermore, at such strain rates, shear fractures at the principal planes, i.e. at an angle of 45°, are also predominantly observed in Al7075 alloy compared to the other alloys. This depends on the alloy plastic deformation range and the extent to which the dislocation mechanisms at the grain boundaries are ceased [102]. The strengthening mechanisms generally depicted in terms of the angle boundaries also play a crucial role at such strain rates. The microstructure of Al2014 alloys attains the low angle grain boundaries (LAGB) due to its high energy state which requires more force to break the bonds, and thus strength attained is maximum with a large amount of plastically strained material [103]. Additionally, the microstructural influence on these aluminium alloys depicts the formation of the adiabatic shear bands (ASB) at higher strain rates. Nonetheless, there is no compulsion for the formation of ASB due to the factors such as the amount of impact velocity, the dimensions of the specimen and the extrusion process adapted for the alloy [104–107]. Figure 1.2 shows the stress-strain variation in aluminium at various strain rates.

The intensive use of high-strength steels for shipbuilding and construction activities demands a thorough estimation of the failure mechanisms at different strain rates. Studies on very high-strength steel (S960QL) were carried out by Cadoni et al. [108–110] in a strain rate range of 200 to 950/second and in a temperature range of 20 to 900° C. It was reported that with the increase in the strain rate, the steel exhibited noticeable improvement in mechanical properties such as strength and ductility while for an increase in temperature, a drastic decrease in properties was observed. Different high-strength steels such as DP780, DP980 and TRIP780 were compared for their tensile properties in the strain rate range of 0.1 to 500/second by Kim et al. [111]. Their studies have shown a prominent strain rate dependency on the tensile behaviour with a key understanding of the TRIP steel having a higher strain hardening rate, longer deformations and lower fracture strains than the DP steel. The high-strength steels under compression had shown an increase in yield strength due to the dislocation dependencies with strain rates in a quasi-static regime [112]. Additionally, the effect of the instant thermal degradation in the material and the phase changes in the precipitates play a crucial role in the strength and ductility parameters. Figure 1.3 shows the stress-strain curves for various strain rates in high-strength steels.

1.4 NON-METALLIC MATERIALS

1.4.1 Composites

Composite materials are cost-effective, easy to handle and provide a high strength-to-weight ratio as compared to metals and thus are emerging in all fields of engineering and medical applications

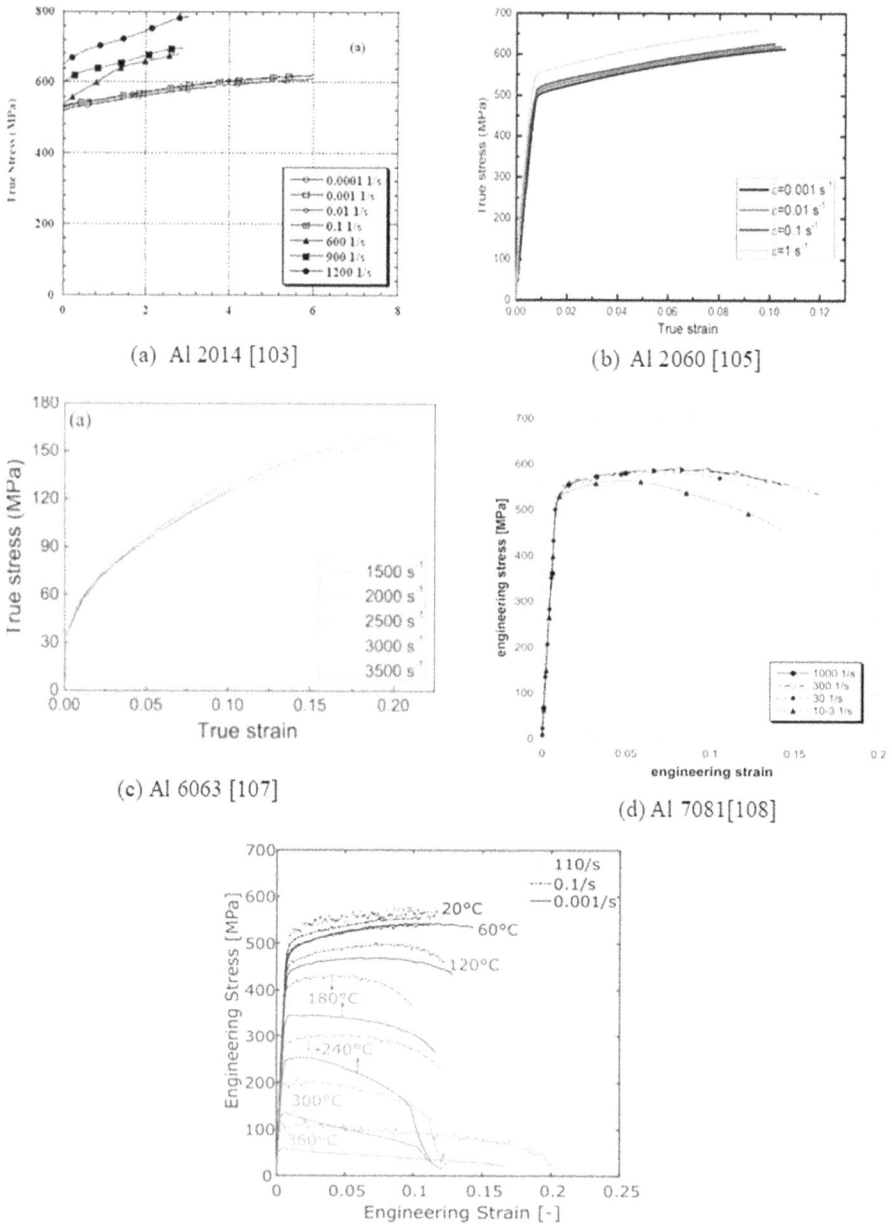

(a) Al 2014 [103]

(b) Al 2060 [105]

(c) Al 6063 [107]

(d) Al 7081[108]

FIGURE 1.2   Stress-strain curves for different aluminium alloys for their tensile behaviour at different strain rates.

[113, 114]. The mechanical properties of the composite material are modified by varying the number of layers of the fibre material used, the percentage of epoxy to the fibre content, the orientation of the fibres, the order of the fibre layers, the curing temperature of the thermoset used, void contents of layup and the efficiency in preparing the layups – manual or automated [77, 115–119]. Nowadays, immense use of environment-friendly bio-fibre composites such as jute fibres, coconut shell husk, wheat husk, etc., is also introduced in the composites to decrease the void content and increase the performance of the composite material [120–123]. Depending upon the orientation of the fibre mats and their areal weight (defined in terms of grams per square meter – GSM), the strength and the

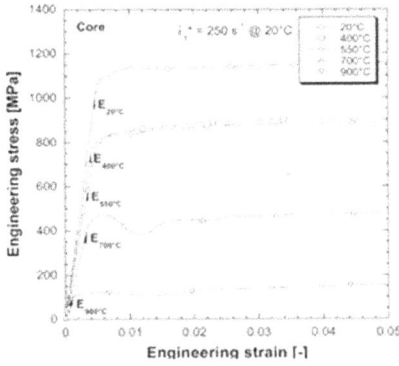

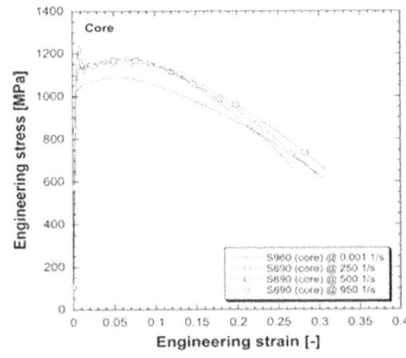

(a) High strength steels- S960QL [109]

(b) High strength steels- S960QL [109]

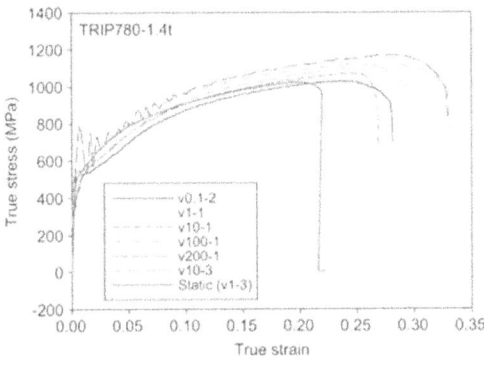

(c) TRIP steels [111]

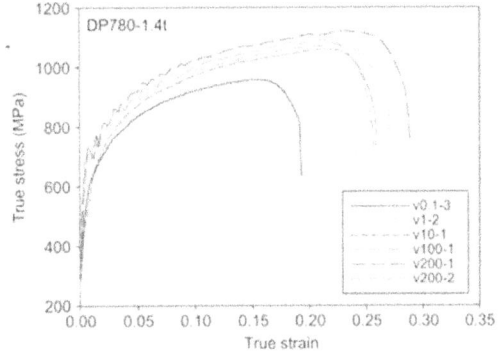

(d) DP steels [111]

FIGURE 1.3 Stress-strain curves for different strength steels at different strain rates.

failure strains vary in composites. Furthermore, fibre-reinforced polymers (FRPs) prepared with their constitutive material made of glass, carbon and Kevlar fibres and with an adequate proportion of the thermosets such as epoxy are widely used [124]. The harmonious existence between the fibres and thermoset is very much desirable and thus, various thermosets are developed to attain the desired properties for their co-existence with the fibres. For a better understanding of the appropriate thermoset, strain rate studies were carried out by Gilat et al. [114] on epoxy for its ductile response at low strain rates and brittle behaviour at high strain rates. Similar studies on other thermosets were carried out by other researchers on other thermosets [125–128]. Thus, composites are highly strain rate sensitive due to the inconsistent methodology of the preparation of the composite and polymeric components.

Glass fibre-reinforced polymer (GFRP) composites have better non-conductivity of electricity, are invulnerable to many chemical elements, etc. Researchers [103, 119, 129] suggested that GFRP/epoxy has shown an increase in ultimate tensile strength and modulus with increasing strain rates for unidirectional composites while the tensile strengths were independent for any angle ply layup composites. The plastic strain hardening was observed for the higher angle layup than the conventional 0° layup. However, the tensile strain decreased with an increase in the strain rates for all the layup composites. GFRP/polyester has shown an increase in tensile and compressive strength with an increase in strain rates [130, 131]. Nevertheless, all the GFRPs with thermoset resins have shown an increase in the tensile strength and tensile modulus and a decrease in failure strain with an increase in the strain rate.

Due to the requirement of electrical conductivity in many electrical and thermal applications, carbon fibre-reinforced polymer (CFRP) composites are utilized. The strain rate effect in the quasi-static and intermediate strain rates is relatively low as compared to the GFRP composites [87, 132]. However, the failure strength increased proportionately with the increase in strain rates. The strain rate dependency was predominant after the strain rate of 20/s until which there was no effect observed for the unidirectional composites [133]. The tensile strength, Young's modulus and failure strain increased with an increase in strain rates. Furthermore, the increase in the ply layup angle decreased the tensile strength of composites. The effect of gross misalignment of the plies due to the defects in the uni-directional fibre is studied in [134]. The compressive strength of the CFRP composites is affected which could be controlled to some extent by consolidation. However, the quality of consolidation also affects the compressive strength of the composites and it was observed that a poorer consolidation resulted in lesser strength and vice versa [132].

Kevlar composites are extensively used for defence and aerospace applications due to their excellent mechanical properties [77, 135, 136]. The anisotropic behaviour of the fibres could be due to the high tensile-to-shear modulus ratio. Kevlar fabric has a very good tensile strength in the longitudinal direction; however, this couldn't be expected from the Kevlar composite using the thermosets due to the variation from the rule of mixtures owing to the brittleness in the thermosets. Many researchers found that the Kevlar fibre composites have good tensile-to-compressive strength and thus, rigorous research has been carried out on the compressive strength of the composite [137–140]. The predominant failure in the composites was due to the kinking of fibres, smooth surface debonding and internal fibre failure which occurs when subjected to compressive load [138, 140]. The interfacial interaction and the shear laminar forces are responsible for the behaviour of the laminate. This further depends on the number of layers and the total thickness of the composite prepared. The delamination of the composites included the fibre bending, tearing and consequently shearing between the fibres [141]. Although the tensile properties are good, laminate failure after reaching the tensile strength due to kinking and fibre pull-outs was predominant. Literature suggests that research in the direction of further improving the mechanical properties was done by inducing fillers such as jute [122], carbon nano-fillers [37, 115], etc., or by the coating of the fabrics [142]. Furthermore, the introduction of shear thickening fluid (STF) also has enabled the composite to have good energy absorption and better impact-resistant properties [143–145]. Dyneema is also a composite that is equally good at resisting impact loads and has high energy absorption characteristics [143]. The strain rate sensitivity studies for different composites [130, 133, 143, 145] have reported that the effect was most prominent in that of GFRP as compared to other composites, as noticed in Figure 1.4.

Hybrid composites have gained popularity due to the feasibility of improving the mechanical properties by utilizing the best properties of the individual materials utilized for their intended use [135, 136]. When GFRP is integrated with other materials such as carbon, Kevlar and other polymeric materials such as PMMA, PP, PC, etc., there is a visible improvement in the mechanical properties. Literature [130, 142] suggests that the hybrid material is not limited, but extensive studies were made on the different amalgamation of properties. Aswani et al. [137] studied the effect of Kevlar/basalt plies at a dynamic strain rate range of 2,800 to 5,841/s and found that the strength, modulus and toughness increased with an increase in strain rates. Ankush et al. [141] carried out experiments on fibre metal laminates (FML) and reported that the placement of the metals with the fibre laminates has a significant effect on the energy absorption characteristics, impact resistance and delamination mechanisms that vary with the variation of the arrangement of each layer. Thus, studies in this direction are required for a feasible utilization of metal plasticity with the flexibility of composites.

1.4.2 POLYMERS

Polymers, particularly thermoplastic polymers, have gained attention due to low glass transition temperature, long durability, cheap availability and easy handling and manufacturing. With the

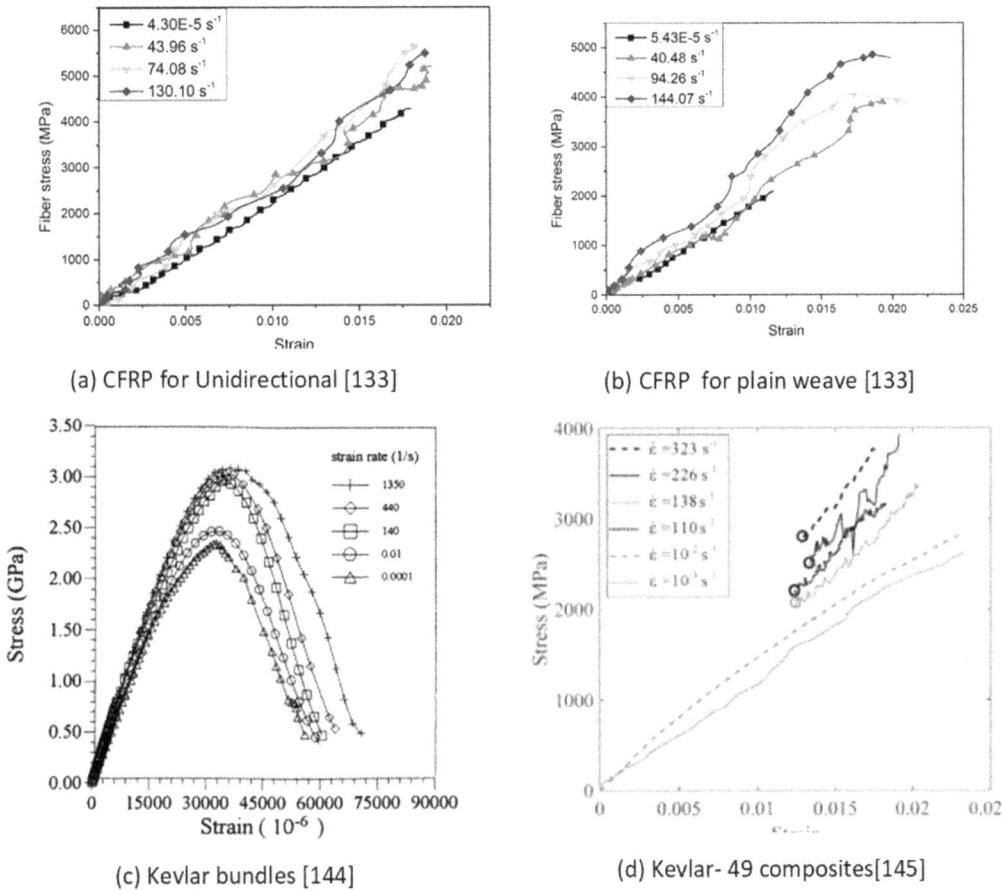

(a) CFRP for Unidirectional [133]

(b) CFRP for plain weave [133]

(c) Kevlar bundles [144]

(d) Kevlar- 49 composites[145]

FIGURE 1.4 Stress-strain curves for various strain rates for different fibre-reinforced polymer composites.

emergence of 3D printing or additive manufacturing, these polymers gained attention in various fields like the packing industry, bio-medical applications and also generation of the prototypes for various direct applications [146–149]. Polymers such as polypropylene (PP), polylactic acid (PLA), acrylonitrile butadiene styrene (ABS), polyamide 6 (PA 6), polyetheretherketone (PEEK) and polyethylene terephthalate glycol (PET G) are some of the current widely used polymeric materials using this additive manufacturing technique [150–153]. A wide range of strain rate studies was made on thermoplastic polymeric specimens using fused filament fabrication (FFF) in the quasi-static, intermediate and dynamic regimes. The mechanical behaviour of each polymer varies due to the organic chemical bonds that exist within the material and their corresponding thermal properties of polymeric material [154]. The mechanical properties are further influenced by the 3D printing measures such as the temperature of the nozzle and bed maintained; angle, feed rate and percentage density of infills, building angles and their orientations, layer thickness and the cooling rate mechanisms of the additive manufacturing process [146, 155–157]. In addition to these, the presence of voids in the form of defects during the printing of the specimen is predominantly observed in such specimens. These aid in the material deterioration due to the reaction with the environment containing moisture, and therefore swelling occurs that leads to the further delinking of the bonds resulting in degradation [161].

The modes of failure and the extent of deformation of various polymers at different regimes are different. In the quasi-static regime, polymers under tension have shown an increase in tensile strength with a decrease in the failure strain. Furthermore, the linear dependency was observed with an increase in the hydrostatic confining pressure. Additionally, with the increase in the strain rate, deformation in relation to the isothermal changes and consequently in the adiabatic properties was also observed in the strain rate range of 0.01 to 1/second [147, 153]. The material response due to plastic deformation aids with the mechanical work due to temperature rise and is observed in the intermediate strain rate range as reported by [154]. At lower and moderate strain rates, the polymeric material underwent brittle failure while at higher strain rates the material undergoes plastic deformation due to the strain hardening phenomenon [157]. Furthermore, the polymers showed more deformations within the elastic limit with the energy for the macromolecular chains suddenly released, thus mapping to smoother fracture surfaces. However, the specimens when subjected to higher strain rates underwent rougher fracture surfaces for PETG and PP but relatively lesser roughness in polymers such as PLA and ABS due to the deformation mechanism they are subjected to. Additionally, PLA is found to have higher strain rate dependency than ABS and PP, and it is even lower for PETG and PA6 [41].

The tensile tests conducted by [158] on the PMMA material at the cross-head velocities of 50, 100 and 500 mm/min suggested that the modulus and yield strength increased while the ductility decreased with an increase in velocity. The importance of temperature rise in specimens at the strain rate variation is studied by various authors [148–153] due to strain softening of the material after the yield point in the polymers. The high strain rate failure mode transitions for the PMMA and polycarbonate (PC) were reported by Chandar et al. [159]. At these strain rates, PEEK polymer was studied both experimentally and theoretically by [160], and it was reported that the elastic modulus and the back stress decreased with an increase in temperature. The creep flow for this polymer is significant above the glass transition temperature (T_g) and weak below it. In glassy amorphous polymers, the stress-strain curves depict an initial viscous behaviour of the materials which gradually becomes non-linear up to a peak value with an increase in strains. The strain softening and the consequent strain hardening in the materials take place after attaining these peak stress values. The dynamic compressive behaviour of ABS was studied by Peterson et al. [153] by conducting the tensile strain rate dependency at low strain rates and consecutively performing the split Hopkinson pressure bar (SHPB) compressive tests on ABS virgin material. It was reported that with an increase in the strain rates, the plastic deformation was evident with a sudden collapse of the stress levels after the failure. Further, a multi-stage collapse of the failure was observed unlike that in metals. The stress-strain curves for the different strain rate regimes from various literature are shown in Figure 1.5 for the various materials such as PLA, PEEK, ABS and PMMA.

Due to the innumerable material-based and printing parameters that affect the mechanical properties of the polymers depending on their mode of preparation, the temperature to which they are subjected and the strain rate sensitivity of the material in itself, constitutive models were proposed by several researchers [146, 149, 151]. The theoretical predictions with respect to the failure modes along with the significant thermal expansion and contraction characteristics of the polymers were first proposed by [165]. The strain rate sensitivity property was utilized in the numerical model for illustrating the mechanical properties as a function of temperature and strain rate by the Zhu-Wang-Tang (ZWT) model. A phenomenological constitutive model considering the transition between the small strains and large strains was proposed by Zhu et al. [166]. This developed model was good for both the tensile and the compressive behaviour of the polymeric material such as PEEK and PC for the lower and the higher strain rate regimes. The yield stress for polymers such as PMMA, PLA and PS was estimated using the Eyring flow model [160]. Models such as the finite strain elastic-viscoplastic model for large strain and temperature-dependent failure models on PLA and other photo-cured polymer materials [154, 158] were also developed. The Eyring model [151] was employed for the specimen to account for the defects generated during the printing of PLA and it was observed to have a low strain rate sensitivity. However, few researchers [149, 154, 161] have applied the continuum constitutive model considering the effects of strain rates and neglecting the voids induced during the printing. As the polymeric parent material in itself is strain rate and temperature dependent, more so when additively manufactured, the combined effect of such behaviour

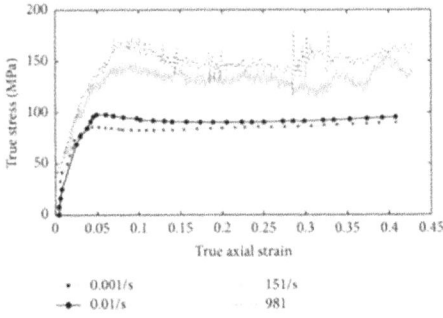

(a) PEEK [162]

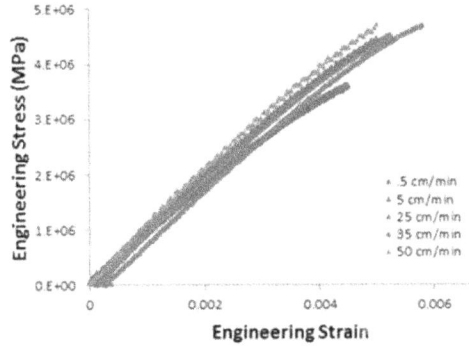

(b) Quasi static for ABS [153]

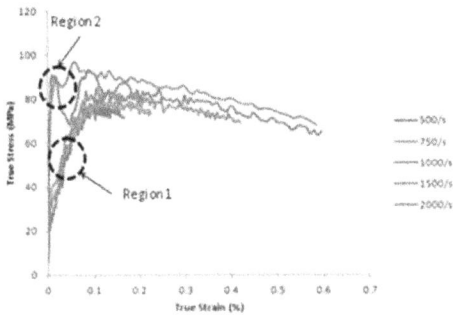

(c) dynamic strain rates for ABS [153]

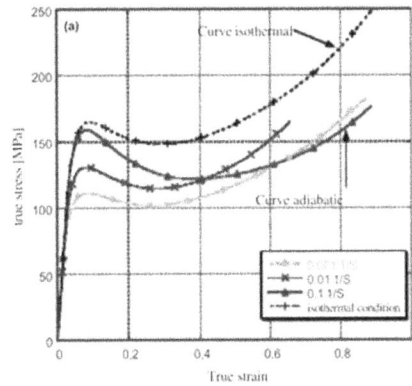

(d) PMMA [163]

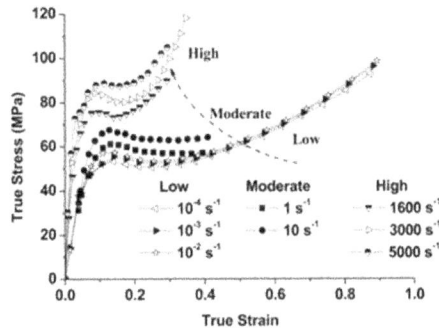

(e) PC [164]

FIGURE 1.5 Stress-strain curves for the different strain rates of (a) PEEK, (b–c) ABS, (d) PMMA and (e) PC. Sources: (a) [162], (b–c) [153], (d) [163], (e) [164].

estimation is quite challenging. Besides this, the experimental limitations imposed on the type of material being tested also led to the thought of theoretical models. For instance, Taylor testing on polymers was carried out way back in 1948 [166] by ejecting a long thin cylinder onto a rigid semi-infinite specimen. The impacted specimen is then calculated for its initial and final dimensions to find out the plastic deformation zone. However, due to the elastic properties of the polymers, the predicted results deviated from the experimental findings. The constitutive model developed by Hutchling et al. [167] to consider the complex behaviour of the material under the Taylor loading as a function of strain, strain rate and final strain is also obtained. The challenging task of considering the phase transitions in the polymeric materials which in turn need to be considered in the numerical or theoretical modelling requires more effort. Hence, constitutive models need to be developed in this direction.

1.5   SUMMARY

The design of a structure or components requires a thorough understanding of material behaviour when the structure could be subjected to various types of loadings. These loads consequently result in the variation of the strain rates of the materials. The strain rate sensitivity (SRS) of different materials such as metals, composites and polymers was studied. It was found that the SRS was high for the composites and polymers compared to the metals. Furthermore, the strain rate sensitivity was prominently visible for the higher strain rate regimes when compared to lower and moderate regimes. Additionally, at such strain rates, thermal effects become predominant and thus have severe consequences on the structures due to the notable changes in the microstructure. Furthermore, these studies help us to develop numerical constitutive models involving the behaviour of materials at such strain rates.

REFERENCES

1. A. P. Sharma, R. Velmurugan. 2020. "Uni-axial tensile response and failure of glass fiber reinforced titanium laminates." *Thin-Walled Structures* 154: 106859.
2. G. Navya, R. Velmurugan, R. Jayaganthan. 2018. "FE analysis of tensile behaviour of an aerospace alloy." Conference: NAFEMS, Bangalore.
3. R. W. Armstrong, S. M. Walley. 2008. "High strain rate properties of metals and alloys." *International Materials Reviews* 53(3): 105–128. DOI: 10.1179/174328008X277795.
4. G. Navya, P. Agarwal. 2016. "Seismic retrofitting of structures by steel bracings." *Procedia Engineering* 144: 1364–1372. DOI: 10.1016/j.proeng.2016.05.166.
5. C. Lakshmana Rao, K. R. Y. Simha, V. Narayanamurthy. 2016. "Applied impact mechanics." Wiley Online Book.
6. P. Agarwal, M. Shrikhande. 2011. *Earthquake Resistant Design of Structures*. PHI Learning Pvt. Ltd., -Technology & Engineering, 660 pages. ISBN-13978-8120328921
7. V. Kumar K. V. Kartik M. A. Iqbal. 2020. "Experimental and numerical investigation of reinforced concrete slabs under blast loading." *Engineering Structures*. DOI: 10.1016/j.engstruct.2019.110125.
8. G. Navya, R. Jayaganthan, R. Velmurugan, N. K. Gupta. 2022. "Finite element analysis of tensile behaviour of glass fibre composites under varying strain rates." *Thin-Walled Structures* 172: 108916.
9. N. A. Kazarinov V. A. Bratov N. F. Morozov Y. V. Petrov, V. V. Balandin M. A. Iqbal N. K. Gupta. 2020. "Experimental and numerical analysis of PMMA impact fracture." *International Journal of Impact Engineering*. DOI: 10.1016/j.ijimpeng.2020.103597.
10. K. Senthil R. Sharma, S. Rupali, A. Takhur M. A. Iqbal N. K. Gupta. 2021. "Evaluation of superior layer configuration of titanium Ti-6Al-4V and aluminium 2024-T3 against soft projectiles." *International Journal of Protective Structures*. DOI: 10.1177/20414196211035789.
11. S. K. Tak M. A. Iqbal. 2020. "Axial compression behaviour of thin-walled metallic tubes under quasi-static and dynamic loading." *Thin-Walled Structures*. DOI: 10.1016/j.tws.2020.107261.
12. R. D. Hussein H. T. Naeem, H. Atiyah, D. Ruan. 2022. "Lateral crushing of square aluminium tubes filled with different cores." *Materials Research* 25: e20220057.
13. R. Verma, S. K. Nath, R. Jayaganthan. 2018. "Effect of high strain rolling and multiaxial forging on tensile and fracture behaviour of ZE41 magnesium alloy." *Materials Today: Proceedings* 5(9): 17195–17202.
14. G. Tiwari, M. A. Iqbal P. K. Gupta. 2017. "Influence of target span and boundary conditions on ballistic limit of thin aluminum plate." *Procedia Engineering*. DOI: 10.1016/j.proeng.2016.12.054.
15. X. Xiang G. Lu Z. Li, D. Ruan. 2017. "Dynamic response of monolithic and sandwich structures subjected to impulsive and impact loadings." *Advances in Structural Engineering* 21(8): 1134–1147.

16. G. Navya, A. Joshi, R. Velmurugan, R. Jayaganthan, N. K. Gupta, E. V. Murashin. 2022. "Experimental and numerical simulation of mechanical behaviour of ultra fine-grained AA 2014 Al alloy." *Mechanics of Solids* 57(3): 590–596. ISSN No. 0025-6544. DOI: 10.3103/S0025654422030177.
17. E. N. Borodin, A. Gruzdkov, A. E. Mayer, N. S. Selyutin. 2018. "Physical nature of strain rate sensitivity of metals and alloys at high strain rates." *Journal of Physics: Conference Series* 991: 012012.
18. G. Li, Y. Wang, K. Wang, M. Xiang, J. Chen. 2019. "Shock induced plasticity and phase transition in single crystal lead by molecular dynamics simulations." *Journal of Applied Physics* 126: 075902. DOI: 10.1063/1.5097621.
19. N. Ma'at M. K. Mohd Nor, C. Sin Ho, N. Abdul Latif, A. E. Ismail, K.-A. Kamarudin, S. Jamian, M. N. Ibrahim Tamrin, M. K. Awang. 2020. "Effects of temperatures and strain rate on the mechanical behaviour of commercial aluminium alloy AA6061." *Journal of Advanced Research in Fluid Mechanics and Thermal Sciences* 54(1): 21–26. https://www.akademiabaru.com/submit/index.php/arfmts/article/view/2431
20. N. J. Edwards, W. Song, S. J. Cimpoeru, D. Ruan, G. Lu, N. Herzige. 2018. "Mechanical and microstructural properties of 2024-T351 aluminium using a hat-shaped specimen at high strain rates." *Materials Science and Engineering: A* 720(21): 203–213.
21. G. Navya, R. Jayaganthan, R. Velmurugan, N. K. Gupta. 2022. "Finite element analysis of tensile behaviour of glass fibre composites under varying strain rates." *Thin-Walled Structures* 172: 108916.
22. P. Nagasankar, S. Balasivanandha Prabhu, R. Velmurugan. 2014. "The effect of the strand diameter on the damping characteristics of fiber reinforced polymer matrix composites: Theoretical and experimental study." *International Journal of Mechanical Sciences* 89: 279–288.
23. A. P. Sharma, S. H. Khan, R. Velmurugan. 2019. "Effect of through thickness separation of fiber orientation on low velocity impact response of thin composite laminates." *Heliyon* 5(10): e02706.
24. S. Gurusideswar, R. Velmurugan, N. K. Gupta. 2017. "Study of rate dependent behaviour of glass/epoxy composites with nanofillers using non-contact strain measurement." *International Journal of Impact Engineering* 110: 324–337.
25. R. Abishera, R. Velmurugan, K. V. Nagendra Gopal 2018. "Shape memory behavior of cold-programmed carbon fiber reinforced CNT/epoxy composites." *Material Research Express, Iop Science* 5(8).
26. W. Li-li, Z. Xi-xiong, S. Shao-chiu, G. Su, B. He-sheng. 1991. "An impact dynamics investigation on some problems in bird strike on windshield of high-speed aircrafts." *Chinese Journal of Aeronautics* 12: 27–33.
27. J. Wang, Y. Xu, W. Zhang. 2014. "Finite element simulation of PMMA aircraft wind-shield against bird strike by using a rate and temperature dependent nonlinear viscoelastic constitutive model." *Composite Structures* 108: 21–30.
28. D. Garcia-Gonzalez, S. Garzon-Hernandez, A. Arias. 2018. "A new constitutive model for polymeric matrices: Application to biomedical materials." *Composites B* 139: 117–129.
29. T. Kanaya, T. Miyazaki, R. Inoue, K. Nishida. 2005. "Thermal expansion and contraction of polymer thin films." *Physica Status Solidi (B)* 242(3): 595–606.
30. S. N. Raman, T. Ngo, P. Mendis, T. Pham. 2012. "Elastomeric polymers for retrofitting of reinforced concrete structures against the explosive effects of blast." *Advances in Materials Science and Engineering*, Article ID 754142. DOI: 10.1155/ 2012/754142.
31. D. Klee, H. Hocker. 1999. "Polymers for biomedical applications: Improvement of the interface compatibility." *Biomedical Applications Polymer Blends*: 149 1–57.
32. W. W. Wright. 1991. "Polymers in aerospace applications." *Materials & Design* 12(4): 222–227.
33. S. A. Begum, A. V. Rane, K. Kanny. 2020. "Applications of compatibilized polymer blends in automobile industry." *Compatibilization of Polymer Blends Micro and Nano Scale Phase Morphologies, Interphase Characterization and Properties*: Vol 32, Issue 1: 563–593.
34. M. J. Tommalieh, A. M. Ismail, N. S. Awwad, H. A. Ibrahium, M. A. Youssef, A. A. Menazea. 2020. "Investigation of electrical conductivity of gold nanoparticles scattered in polyvinylidene fluoride/ polyvinyl chloride via laser ablation for electrical applications." *Journal of Electronic Materials* 49: 7603–7608.
35. P. Rezaeian, M. R. Ayatollahi, A. Nabavi-Kivi, S. M. Javad Razavi. 2022. "Effect of printing speed on tensile and fracture behavior of ABS specimens produced by fused deposition modeling." *Engineering Fracture Mechanics* 266: 108393.
36. W. M. H. Verbeeten, M. Lorenzo-Bañuelos, P. J. Arribas-Subiñas. 2020. "Anisotropic rate-dependent mechanical behavior of poly lactic acid processed by material extrusion additive manufacturing." *Additive Manufacturing* 31: 100968.

37. R. Abishera, R. Velmurugan, K. V. N. Gopal. 2018. "Free, partial, and fully constrained recovery analysis of cold-programmed shape memory epoxy/carbon nanotube nanocomposites." *Journal of Intelligent Material Systems and Structures*: 29(10): 1–13.

38. A. D. Mulliken, M. C. Boyce. 2006. "Mechanics of the rate-dependent elastic–plastic deformation of glassy polymers from low to high strain rates." *International Journal of Solids and Structures* 43(5): 1331–1356.

39. S. Kartikey, R. Boomurugan, R. Velmurugan. 2021. "Cold programming of epoxy-based shape memory polymer." *Structures* 29: 2082–2093.

40. S. Walley, J. Field, G. Swallowe, S. Mentha. 1985. "The response of various polymers to uniaxial compressive loading at very high rates of strain." *Journal de Physique Colloques* 46 (C5): C5-607-C5-616. DOI: 10.1051/jphyscol:1985578.

41. N. Vidakis, M. Petousis, E. Velidakis, M. Liebscher, V. Mechtcherine, L. Tzounis. 2020. "On the strain rate sensitivity of Fused Filament Fabrication (FFF) processed PLA, ABS, PETG, PA6, and PP thermoplastic polymers." *Polymers* 12(12): 2924. DOI: 10.3390/polym12122924.

42. X. Xiao, D. Kong, X. Qiu, W. Zhang, F. Zhang, L. Liu, Y. Liu, S. Zhang, Y. Hu, J. Leng. 2015. "Shape-memory polymers with adjustable high glass transition temperatures." *Macromolecules* 48(11): 3582–3589.

43. C. Briody, B. Duignan, S. Jerrams, J. Tiernan. 2012. "The implementation of a visco-hyperelastic numerical material model for simulating the behaviour of polymer foam materials." *Computational Materials Science* 64: 47–51.

44. Z. Liao, M. Hossain, X. Yao, M. Mehnert, P. Steinmann. 2020. "On thermo-viscoelastic experimental characterization and numerical modelling of VHB polymer." *International Journal of Non-Linear Mechanics* 118: 103263.

45. H. C. Booij, J. H. M. Palmen. 1982. "Some aspects of linear and nonlinear viscoelastic behaviour of polymer melts in shear." *Rheologica Acta* 21: 376–387.

46. J. Rieger. 2001. "The glass transition temperature Tg of polymers—Comparison of the values from differential thermal analysis (DTA, DSC) and dynamic mechanical measurements (torsion pendulum)." *Polymer Testing* 20(2): 199–204.

47. F. Nonque, A. Benlahoues, J. Audourenc, A. Sahut, R. Saint-Loup, P. Woisel, J. Potie. 2021. "Study on polymerization of bio-based isosorbide monomethacrylate for the formation of low-Tg and high-Tg sustainable polymers." *European Polymer* 160: 110799.

48. A. Schindler, M. Doedt, S. Gezgin, J. Menzel, S. Schmolzer. 2017. "Identification of polymers by means of DSC, TG, STA and computer-assisted database search." *Journal of Thermal Analysis and Calorimetry* 129: 833–842.

49. K. Houjou, K. Shimamoto, H. Akiyama, C. Sato. 2020. "Experimental investigations on the effect of a wide range of strain rates on mechanical properties of epoxy adhesives, and prediction of creep and impact strengths." *The Journal of Adhesion* 98(5). DOI: 10.1080/00218464.2020.1840368.

50. M. Peroni, G. Solomos, V. Pizzinato, M. Larcher. 2011. "Experimental investigation of high strain-rate behaviour of glass." *Applied Mechanics and Materials* 82: 63–68. DOI: 10.4028/www.scientific.net/AMM.82.63.

51. N. Khaire, G. Tiwari, M. A. Iqbal. 2021. "Energy absorption characteristics of sandwich shell structure against conical and hemispherical nose projectile." *Composite Structures*. DOI: 10.1016/j.compstruct.2020.113396.

52. C. R. Siviour, L. J. Jordan. 2016. "High strain rate mechanics of polymers: A review." *Journal of Dynamic Behavior Materials* 2: 15–32. DOI: 10.1007/s40870-016-0052-8.

53. G. C. Jacob, J. M. Starbuck, J. F. Fellers, S. Simunovic, R. G. Boeman. 2004. "Strain rate effects on the mechanical properties of polymer composite materials." *Journal of Applied Polymer Science* 94: 296–301.

54. L. Ma, F. Liu, D. Liu, Y. Liu. "Review of strain rate effects of fiber-reinforced polymer composites." *Polymers* 13: 2839. DOI: 10.3390/polym13172839.

55. M. Saleh, V. Luzin, M. A. Kariem, K. Thorogood, D. Ruan. 2020. "Experimental measurements of residual stress in ARMOX 500T and evaluation of the resultant ballistic performance." *Journal of Dynamic Behavior of Materials* 6(1): 78–95.

56. Q. Li, X. Jiang, T. Zeng, S. Xu. 2022. "Experimental investigation on strain rate effect of high-performance fiber reinforced cementitious composites subject to dynamic direct tensile loading." *Cement and Concrete Research* 157: 106825.

57. S. Tamrakar, R. Ganesh, S. Sockalingam, B. Z. Haque, J. W. Gillespie. 2018. "Experimental investigation of strain rate and temperature dependent response of an epoxy resin undergoing large deformation." *Journal of Dynamic Behavior of Materials* 4(1): 114–128. DOI: 10.1007/s40870-018-0144-8.

58. L. C. A. van Breemen, T. A. P. Engels, E. T. J. Klompen, D. J. A. Senden, L. E. Govaert. 2012. "Rate- and temperature-dependent strain softening in solid polymers." *Journal of Polymer Science Part B.* DOI: 10.1002/polb.23199.

59. H. Kweon, S. Choi, Y. Kim, K. Nam. 2006. "Development of a New UTM (Universal Testing Machine) system for the nano/micro in-process measurement." *International Journal of Modern Physics B* 20(25, 27): 4432–4438.

60. G. Navya, R. Jayaganthan, R. Velmurugan. 2021. "A study on the influence of johnson cook parameters on the flow behaviour of Al 2024 alloy against ballistic impact." 5th International Conference on Structural Integrity and Durability, ICSID, Procedia Engineering, Pre- Publish Press.

61. D. Kirkaldy. 1862. *Results of an experimental inquiry into the comparative tensile strength and other properties of various kinds of wrought-iron and steel.* Hamilton, Adams, & Co, London.

62. M. Singh. 2022. "Development of a portable Universal Testing Machine (UTM) compatible with 3D laser-confocal microscope for thin materials." *Advances in Industrial and Manufacturing Engineering* 4: 100069.

63. ASTM- E8/E8M − 13a. *Standard Test Methods for Tension Testing of Metallic Materials.*

64. ASTM- D368. *Standard Test Method for Tensile Properties of Plastics.*

65. N. Fantuzzi, P. Trovalusci, S. Dharasura. 2019. "Mechanical behavior of anisotropic composite materials as micropolar continua." *Frontiers of Material Science, Section- Mechanics of Materials.* DOI: 10.3389/fmats.2019.00059.

66. V. D. Azzi, S. W. Tsai. 1965. "Anisotropic strength of composites." *Experimental Mechanics* 5: 283–288. DOI: 10.1007/BF02326292.

67. S. Cai, B. Han, Y. Xu, E. Guo, B. Sun, Y. Zeng, H. Hou, S. Wu. 2022. "Anisotropic composition and mechanical behavior of a natural thin-walled composite: Eagle feather shaft." *Polymers* 14: 309. DOI: 10.3390/polym14020309.

68. E. Huerta, J. E. Corona, A. I. Oliva. 2010. "Universal testing machine for mechanical properties of thin materials." *Revista Mexicana De Fisica* 56(4): 317–322.

69. H. Kweon, S. Choi, Y. Kim, K. Nam. 2006. "Development of a New UTM (Universal Testing Machine) system for the nano/micro in-process measurement." *International Journal of Modern Physics B* 20(25): 4432–4438.

70. P. Sharma, U. Chauhan, Dr. S. Kumar and Dr. K. Sharma. 2018. "A review on dynamic rheology for polymers." *International Journal of Applied Engineering* 13(6): 363–368. Research ISSN 0973-4562.

71. G. Swaminathan, K. Shivakumar. 2008. "A re-examination of DMA testing of polymer matrix composites." *Journal of Reinforced Plastics and Composites* 28(8). DOI: 10.1177/0731684407087740.

72. K. Pae, S. Bhateja. 1975. "The effects of hydrostatic pressure on the mechanical behavior of polymers." *Journal of Macromolecular Science* 13(1): 1–75.

73. O. Ramon, S. Mizrahi, J. Miltz. 1994. "Merits and limitations of the drop and shock tests in evaluating the dynamic properties of plastic foams." *Polymer Engineering Science* 34(18): 1406–1410.

74. B. Song, B. Sanborn, J. D. Heister, R. L. Everett, T. L. Martinez, G. E. Groves, E. P. Johnson, D. J. Kenney, M. E. Knight, M. A. Spletzer, K. K. Haulenbeek, C. McConnell. 2019. "An apparatus for tensile characterization of materials within the upper intermediate strain rate regime." *Experimental Mechanics* 59: 941–951.

75. J. M. Logan, J. Handin. 1970. "Triaxial compression testing at intermediate strain rates." 12th U.S. Symposium on Rock Mechanics (USRMS).

76. S. Gurusideswar, N. Srinivasan, R. Velmurugan, N. K. Gupta. 2017. "Tensile response of epoxy and glass/epoxy composites at low and medium strain rate regimes." *Procedia Engineering* 173: 686–693.

77. A. Vasudevan, S. Senthil Kumaran, K. Naresh, R. Velmurugan. 2020. "Layer-wise damage prediction in carbon/Kevlar/S-glass/E-glass fibre reinforced epoxy hybrid composites under low-velocity impact loading using advanced 3D computed tomography." *International Journal of Crashworthiness* 25(1): 9–23.

78. R. Velmurugan, K. Naresh, K. Shankar. 2018. "Influence of fibre orientation and thickness on the response of CFRP composites subjected to high velocity impact loading." *Advances in Materials and Process Technologies* 4, Taylor & Francis.

79. U. O. Costa, L. F. C. Nascimento, J. M. Garcia, W. B. A. Bezerra, S. N. Monteiro. 2020. "Evaluation of Izod impact and bend properties of epoxy composites reinforced with mallow fibers." *Journal of Materials Research and Technology* 9(1): 373–382.

80. W. Hufenbach, F. MarquesIbraim, A. Langkampm R. Bohm, A. Hornig. 2008. "Charpy impact tests on composite structures − An experimental and numerical investigation." *Composites Science and Technology* 68(12): 2391–2400.

81. S. Lee, G. Swallowe. 2004. "Direct measurement of high rate stress–strain curves using instrumented falling weight and highspeed photography." *Imaging Science Journal* 52(4): 193–201.

82. P. Viot, F. Beani, J.-L. Lataillade. 2005. "Polymeric foam behavior under dynamic compressive loading." *Journal of Material Sciences* 40(22): 5829–5837.

83. B. Song, C. Syn, C. Grupido, W. Chen, W.-Y. Lu. 2008. "A long split Hopkinson pressure bar (LSHPB) for intermediate-rate characterization of soft materials." *Experimental Mechanics* 48(6): 809–815.

84. N. Gara, V. Ramachandran, J. Rengaswamy. 2021. "Analytical and FEM analyses of high-speed impact behaviour of Al 2024 alloy." *Aerospace* 8(281). ISSN NO. 2226–4310. DOI: 10.3390/aerospace8100281.

85. H. Zhao, G. Gary. 1996. "On the use of SHPB techniques to determine the dynamic behavior of materials in the range of small strains." *International Journal of Solids and Structures* 33(23): 3363–3375.

86. S. S. Adam, D. Mulliken, M. C. Boyce. 2007. "Mechanics of Taylor impact testing of polycarbonate." *International Journal of Solids and Structures* 44(7–8): 2381–2400.

87. A. Yoshimura, T. Nakao, S. Yashiro, N. Takeda. 2008. "Improvement on out-of-plane impact resistance of CFRP laminates due to through-the-thickness stitching." *Composites Part A: Applied Science and Manufacturing* 39(9): 1370–1379.

88. P. Pei, Z. Pei, Z. Tang. 2020. "Numerical and theoretical analysis of the inertia effects and interfacial friction in SHPB test systems." *Materials* 13(21): 4809. DOI: 10.3390/ma13214809.

89. N. N. Dioh, A. Ivankovic, P. S. Leevers, J. G. Williams. 1995. "Stress wave propagation effects in split Hopkinson pressure bar tests." *Proceedings: Mathematical and Physical Sciences* 449(1936): 187–204.

90. B. Davoodi, A. Gavrus, E. Ragneau. 2005. "A technique for measuring the dynamic behaviour of materials at elevated temperatures with a compressive SHPB." *WIT Transactions on Engineering Sciences* 51. ISSN 1743-3533.

91. P. E. Markovsky, J. Janiszewski, V. I. Bondarchuk, O. O. Stasyuk, D. G. Savvakin, M. A. Skoryk, K. Cieplak, P. Dziewit, S. V. Prikhodko. 2020. "Effect of strain rate on microstructure evolution and mechanical behavior of titanium-based materials." *Metals* 10: 1404. DOI: 10.3390/met10111404.

92. S. Kim, M. C. Jo, T. W. Park, J. Ham, S. S. Sohn, S. Lee. 2021. "Correlation of dynamic compressive properties, adiabatic shear banding, and ballistic performance of high-strength 2139 and 7056 aluminum alloys." *Materials Science & Engineering A* 804: 140757.

93. Y. Xiong, N. Li, H. Jiang, Z. Li, Z. Xu, L. Liu. 2019. "Microstructural evolutions of AA7055 aluminum alloy under dynamic and quasi-static compressions." *Acta Metallurgica Sinica (English Letters)* 27(2): 272–278. DOI: 10.1007/s40195-014-0041-7.

94. Y. Tuo, W. Yuanzhi, L. Anmin, T. Xu, L. Luoxing. 2019. "Deformation behavior and microscopic mechanism of as-extruded 6013-T4 aluminum alloy under dynamic shock loading." *Chinese Journal of Materials Research* 33(2). DOI: 10.11901/1005.3093.2018.504.

95. A. Azimi, G. M. Owolabi, H. Fallahdoost, N. Kumar, G. Warner. 2019. "High strain rate behavior of ultrafine grained AA2519 processed via multi axial cryogenic forging." *Metals* 9: 115. DOI: 10.3390/met9020115.

96. S. B. Bhimavarapu, A. K. Maheshwari, D. Bhargava, S. P. Narayan. 2011. "Compressive deformation behavior of Al 2024 alloy using 2D and 4D processing maps." *Journal of Materials Science* 46(9): 3191–3199. DOI: 10.1007/s10853-010-5203-z.

97. S. Ding, Q. Shi, G. Chen. 2021. "Flow stress of 6061 aluminum alloy at typical temperatures during friction stir welding based on hot compression tests." *Metals* 11: 804. DOI: 10.3390/ met11050804.

98. N. Ma, M. K. Mohd Nor, C. Sin Ho, N. Abdul Latif, Al Emran Ismail, K.-A. Kamarudin, S. Jamian, M. Norihan Ibrahim, M. Khairudin Awang. 2019. "Effects of temperatures and strain rate on the mechanical behaviour of commercial aluminium alloy AA6061." *Journal of Advanced Research in Fluid Mechanics and Thermal Sciences* 54(1): 21–26.

99. S.-H. Jung, G. Bae, M. Kim, J. Lee, J. Song, N. Park. 2022. "Effect of natural aging time on anisotropic plasticity and fracture limit of Al7075 alloy." *Materials Today Communications* 31: 103553.

100. A. T. Olasumboye, G. M. Owolabi, A. G. Odeshi, N. Yilmaz, A. Zeytinci. 2018. "Dynamic behaviour of AA2519-T8 aluminium alloy under high strain rate loading in compression." *Journal of Dynamic Behavior of Materials.* DOI: 10.1007/s40870-018-0145-7.

101. P. Guo, X. Liu, B. Zhu, W. Liu, L. Zhang. 2021. "The microstructure evolution and deformation mechanism in a casting AM80 magnesium alloy under ultra-high strain rate loading." *Journal of Magnesium and Alloys.* DOI: 10.1016/j.jma.2021.07.032.

102. X. Li, K. Pandya, N. Karathanasopoulos, C. C. Roth, D. Mohr. 2021. "Plasticity and fracture of aluminium 7075 at high strain rates and elevated temperatures." *EPJ Web of Conferences* 250: 05007. DOI: 10.1051/epjconf/2021250.

103. G. Prakash, N. K. Singh, N. K. Gupta. 2020. "Deformation behaviours of Al2014-T6 at different strain rates and temperatures." *Structures* 26: 193–203. DOI: 10.1016/j.istruc.2020.03.068.

104. Y. Nie, B. Claus, J. Gao, X. Zhai, N. Kedir, J. Chu, T. Sun, K. Fezza, W. W. Chen. 2019. "In situ observation of adiabatic shear band formation in aluminum alloys." DOI: 10.1007/s11340-019-00544-w.

105. A. Abd El-Aty, Y. Xu, S. H. Zhang, S. Ha, Y. Ma, D. Chen. 2019. "Impact of high strain rate deformation on the mechanical behavior, fracture mechanisms and anisotropic response of 2060 Al-Cu-Li alloy." *Journal of Advanced Research* 18: 19–37.

106. E. Scharifi, S. V. Sajadifar, G. Moeini, U. Weidig, S. Böhm, T. Niendorf, K. Steinhoff. 2020. "Dynamic tensile deformation of high strength aluminum alloys processed following novel thermomechanical treatment strategies." *Advanced Engineering Materials*. DOI: 10.1002/adem.202000193.

107. T. Ye, L. Li, P. Guo, G. Xiao, Z. Chen. 2016. "Effect of aging treatment on the microstructure and flow behavior of 6063 aluminum alloy compressed over a wide range of strain rate." *International Journal of Impact Engineering* 90: 72–80.

108. E. Cadoni, M. Dotta, D. Forni, H. Kaufmann. 2016. "Effects of strain rate on mechanical properties in tension of a commercial aluminium alloy used in armour applications." 21st European Conference on Fracture, ECF21, *Procedia Structural Integrity* 2: 986–993.

109. E. Cadoni, D. Forni. 2019. "Mechanical behaviour of a very-high strength steel (S960QL) under extreme conditions of high strain rates and elevated temperatures." *Fire Safety Journal* 109: 102869.

110. E. Cadoni, D. Forni. 2020. "Strain-rate effects on S690QL high strength steel under tensile loading." *Journal of Constructional Steel Research* 175: 106348.

111. J.-H. Kim, D. Kim, H. N. Han, F. Barlat, M. G. Lee. 2013. "Strain rate dependent tensile behavior of advanced high strength steels: Experiment and constitutive modelling." *Materials Science and Engineering: A* 559: 222–231.

112. A. Banerjee, R. Hossain, F. Pahlevani, Q. Zhu, V. Sahajwalla, B. Gangadhara Prusty. 2019. "Strain-rate-dependent deformation behaviour of high-carbon steel in compression: Mechanical and structural characterisation." *Journal of Materials Science* 54: 6594–6607.

113. W. Wang, Y. Ma, M. Yang, P. Jiang, F. Yuan, X. Wu. 2018. "Strain rate effect on tensile behavior for a high specific strength steel: From quasi-static to intermediate strain rates." *Metals* 8(1): 11. DOI: 10.3390/met8010011.

114. A. Gilat, R. K. Goldberg, G. D. Roberts. 2005. *Strain Rate Sensitivity of Epoxy Resin in Tensile and Shear Loading.* National Aeronautics and Space Administration, Glenn Research Centre.

115. P. Daniel, R. Velmurugan. 2018. "Analysis of the specific properties of glass microballoon-epoxy syntactic foams under tensile and flexural loads." *Materials Today: Proceedings* 5(9): 16956–16962.

116. K. Naresh, K. Shankar, R. Velmurugan. 2018. "Digital image processing and thermo-mechanical response of neat epoxy and different laminate orientations of fiber reinforced polymer composites for vibration isolation applications." *International Journal of Polymer Analysis and Characterization* 23(8): 684–709.

117. N. E. Mohamed, R. Velmurugan. 2019. "Effect of fiber orientations of composite panels under far-field pyroshock." *Polymer Composites* 40(1): 255–262.

118. A. P. Sharma, R. Velmurugan. 2020. "Uni-axial tensile response and failure of glass fiber reinforced titanium laminates." *Thin-Walled Structures* 154: 106859.

119. R. Velmurugan, S. Gurusideshwar. 2014. "Strain rate sensitivity of glass/epoxy composites with nano fillers." *Materials and Design* 60: 468–478.

120. A. Bourmaud, D. U. Shah, J. Beaugrand, H. N. Dhakal. 2020. "Property changes in plant fibres during the processing of bio-based composites." *Industrial Crops and Products* 154: 112705.

121. P. Pratim Das, V. Chaudhary. 2021. "Moving towards the era of bio fibre based polymer composites." *Cleaner Engineering and Technology* 4: 100182.

122. H. Chandekar, V. Chaudhari, S. Waigaonkar. 2020. "A review of jute fiber reinforced polymer composites." *Materials Today: Proceedings* 26, Part 2: 2079–2082.

123. D. O. Obada, L. S. Kuburi, M. Dauda, S. Umaru, D. Dodoo-Arhin, M. B. Balogun, I. Iliyasu, M. J. Iorpenda. 2020. "Effect of variation in frequencies on the viscoelastic properties of coir and coconut husk powder reinforced polymer composites." *Journal of King Saud University - Engineering Sciences* 32(2): 148–157.

124. S. Huo, P. Song, B. Yu, S. Ran, V. S. Chevali, L. Liu, Z. P. Fang, H. Wang. 2021. "Phosphorus-containing flame retardant epoxy thermosets: Recent advances and future perspectives." *Progress in Polymer Science* 114: 101366.

125. Y. Jin, Z. Lei, P. Taynton, S. Huang, W. Zhang. 2019. "Malleable and recyclable thermosets: The next generation of plastics." *Matter* 1(6): 1456–1493.

126. X. Ramis, X. Fernández-Francos, S. De la Flor, F. Ferrando, À. Serra. 2018. "Click-based dual-curing thermosets and their applications." *Thermosets -Structure, Properties, and Applications*, pp. 511–541.
127. D. Song, R. K. Gupta. 2012. "The use of thermosets in the building and construction industry." *Thermosets-Structure, Properties and Applications*, pp. 165–188.
128. J. P. Pascault, R. J. J. Williams. 2018. "Overview of thermosets: Present and future." *Thermosets (Second Edition)- Structure, Properties, and Applications*, pp. 3–34.
129. K. Naresh, K. Shankar, R. Velmurugan, N. K. Gupta. 2020. "High strain rate studies for different laminate configurations of bi-directional glass/epoxy and carbon/epoxy composites using DIC." *Structures* 27: 2451–2465.
130. Y. Ou, D. Zhu, H. Zhang, L. Huang, Y. Yao, G. Li, B. Mobasher. "Mechanical characterization of the tensile properties of Glass Fiber and Its Reinforced Polymer (GFRP) composite under varying strain rates and temperatures." *Polymers* 8(5): 196. DOI: 10.3390/polym8050196.
131. S. Amijima T. Fujii. 1980. In proceedings of the 3rd international conference on composite materials, ICCM III, A. R., ed.; Paris, France, pp. 399–413.
132. Y. Ou, D. Zhu, H. Zhang, Y. Yao, B. Mobasher, L. Huang. 2016. "Mechanical properties and failure characteristics of CFRP under intermediate strain rates and varying temperatures." *Composites B* 95: 123–136.
133. X. Zhang, Y. Shia, Z. X. Li. 2019. "Experimental study on the tensile behavior of unidirectional and plain weave CFRP laminates under different strain rates." *Composites Part B: Engineering* 164: 524–536.
134. P. Davidson, A. M. Waas. 2017. "The effects of defects on the compressive response of thick carbon composites: An experimental and computational study." *Composite Structures* 176: 582–596.
135. R. S. Sikarwar, R. Velmurugan, M. Vemuri. 2012. "Experimental and analytical study of high velocilty impact on Kevlar/epoxy composites plates." *Central European Journal of Engineering* 2(4): 638–630.
136. A. Vasudevan, S. Senthil Kumaraan, K. Naresh, R. Velmurugan, K. Shankar. 2018. "Advanced 3D and 2D damage assessment of low velocity impact response of glass and Kevlar fiber reinforced epoxy hybrid composites." *AMPT* 2018, VIT Chennai.
137. A. K. Bandaru, H. Chouhan, N. Bhatnagar. 2020. "High strain rate compression testing of intra-ply and inter-ply hybrid thermoplastic composites reinforced with Kevlar/basalt fibers." *Polymer Testing* 84: 106407.
138. J. H. Greenwood, P. G. Rose. 1974. "Compressive behaviour of Kevlar 49 fibres and composites." *Journal of Materials Science* 9: 1809–1814.
139. L. Zheng, K. Zhang, L. Liu, F. Xu. 2022. "Biomimetic architectured Kevlar/polyimide composites with ultra-light, superior anti-compressive and flame-retardant properties." *Composites Part B: Engineering* 230: 109485.
140. C. Audibert, A.-S. Andreani, E. Laine, J.-C. Grandidier. 2018. "Mechanical characterization and damage mechanism of a new flax-Kevlar hybrid/epoxy composite." *Composite Structures* 195: 126–135.
141. A. P. Sharma, R. Velmurugan, K. Shankar, S. K. Ha. 2021. "High-velocity impact response of titanium-based fiber metal laminates. Part I: Experimental investigations." *International Journal of Impact Engineering* 152: 103845.
142. S. Gowthaman, D. R. Sekhar. "Enhancing the interyarn friction properties of kevlar and glass fabrics through ZnO nanowire coating." *Journal of Composite Materials* 55(9). DOI: 10.1177/0021998320967416.
143. M. Chinnapandi, A. Katiyar, T. Nandi, V. Ramachandran. 2021 *"High Speed Impact Studies of Kevlar Fabric with and without STF"*. 10.1007/978-981-15-7711-6_74.
144. Y. Wang, Y. Xia. 1998. "The effects of strain rate on the mechanical behaviour of kevlar fibre bundles: An experimental and theoretical study." *Composites Part A: Applied Science and Manufacturing* 29(11): 1411–1415.
145. J. Zhou, U. Heisserer, P. W. Duke, P. T. Curtis, J. Morton, V. L. Tagarielli. 2021. "The sensitivity of the tensile properties of PMMA, Kevlar® and Dyneema® to temperature and strain rate." *Polymer* 225: 123781.
146. L. C. A. van Breemen, T. A. P. Engels, E. T. J. Klompen, D. J. A. Senden, L. E. Govaert. 2012. "Rate- and temperature-dependent strain softening in solid polymers." *Journal of Polymer Science, Part B* 50(24): 1757–1771.
147. F. Ullrich, S. Pal Veer Singh, S. McDonald, A. Krueger, J. Vera-Sorroche, A. Amirkhizi, D. Masato. 2022. "Effect of strain rate on the mechanical properties of polycarbonate processed by compression and injection molding." *Polymer Engineering & Science*. DOI: 10.1002/pen.25842.

148. N. Chikkanna, S. Krishnapillai, V. Ramachandran. 2022. "Static and dynamic flexural behaviour of printed polylactic acid with thermal annealing: Parametric optimisation and empirical modelling." *The International Journal of Advanced Manufacturing Technology* 119: 1179–1197.
149. A. Rahmati, M. Heidari-Rarani, L. Lessard. 2021. "A novel conservative failure model for the fused deposition modeling of polylactic acid specimens." *Additive Manufacturing* 48: 102460.
150. S. M. Mirkhalaf, M. Fagerstrom. 2019. "The mechanical behavior of polylactic acid (PLA) films: Fabrication, experiments and modelling." *Mechanics of Time-Dependent Materials* 25(2): 119–131.
151. D. Garcia-Gonzalez, R. Zaera, A. Arias. 2017. "A hyperelastic- thermo visco plastic constitutive model for semi-crystalline polymers: Application to PEEK under dynamic loading conditions." *International Journal of Plasticity* 88: 27–52.
152. A. Nabavi-Kivi, M. R. Ayatollahi, P. Rezaeian, S. M. J. Razavi. 2022. "Investigating the effect of printing speed and mode mixity on the fracture behavior of FDM-ABS specimens." *Theoritical Applied Fracture Mechanics* 118: 103223.
153. A. Peterson, E. Habtour, J. Riddick, M. Coatney, D. Bolling, G. Owolabi. 2015. "Dynamic evaluation of acrylonitrile butadiene styrene subjected to high-strain-rate compressive loads." Army Research Laboratory, Report No. ARL-TN-0648.
154. H. Wang, Y. Zhang, H. Zhou, Z. Huang. 2017. "Constitutive modelling of polycarbonate at low, moderate and high strain rates." *IOP Conf. Series: Materials Science and Engineering* 187: 012023. DOI: 10.1088/1757-899X/187/1/012023.
155. S. Bardiya, J. Jerald, V. Satheeshkumar. 2021. "The impact of process parameters on the tensile strength, flexural strength and the manufacturing time of fused filament fabricated (FFF) parts." *Materials Today Proceedings* (39): 1362–1366.
156. H. Gonabadi, A. Yadav, S. J. Bull. 2020. "The effect of processing parameters on the mechanical characteristics of PLA produced by a 3D FFF printer." *International Journal of Advanced Manufacturing Technology* 111(3–4): 695–709.
157. E. Jalon, T. Hoang, A. Rubio-Lopez, C. Santiuste. 2018. "Analysis of low-velocity impact on flax/PLA composites using a strain rate sensitive model." *Composite Structures* 202: 511–517.
158. M. Sepe. 2016. "Understanding strain-rate sensitivity in polymers." Plastic Technology. https://www.ptonline.com/articles/materials-understanding-strain-rate-sensitivity-in-polymers.
159. K. Ravi-Chandar, J. Lu, B. Yang, Z. Zhu. 2000. "Failure mode transitions in polymers under high strain rate loading." *International Journal of Fracture* 101: 33–72.
160. A. D. Drozdov, J. deClaville Christiansen. 2021. "Thermo-mechanical behavior of poly(ether ether ketone): Experiments and modeling." *Polymers* 13(11): 1779. DOI: 10.3390/polym13111779. PMID: 34071593; PMCID: PMC8199459.
161. A. Singh, R. M. Guedes, D. Paiva et al. 2020. "Experiment and modelling of the strain-rate-dependent response during in vitro degradation of PLA fibres." *Applied Sciences* 2: 177. DOI: 10.1007/s42452-020-1964-4.
162. Z. El-Qoubaa and R. Othman. 2015. "Tensile behavior of polyetheretherketone over a wide range of strain rates." *International Journal of Polymer Science* 2015, Article ID 275937. DOI: 10.1155/2015/275937.
163. M. Nasraoui, P. Forquin, L. Siad, A. Rusinek, E. Dossou. 2009. "Mechanical behaviour of PMMA: Influence of temperature and confining pressure." *DYMAT -9th International Conferences on the Mechanical and Physical Behaviour of Materials under Dynamic Loading*. DOI: 10.1051/dymat/2009154.
164. T. A. Engels, S. H. Sontjens, T. H. Smit, L. E. Govaert. 2010. "Time-dependent failure of amorphous polylactides in static loading conditions." *Journal of Material Science* 21(1): 89–97.
165. Z. Zhao, M. Lei, Q. Zhang, H.-S. Chen, D. Fang. 2020. "A general model for the temperature-dependent deformation and tensile failure of photo-cured polymers." *Extreme Mechanical Letters* 39: 100826.
166. G. I. Taylor. 1948. "The use of flat ended projectiles for determining yield stress- 1: Theoretical considerations." *Proceedings of the Royal Society London A* 194: 289–299.
167. I. Hutchings. 1978. "Estimation of yield stress in polymers at high strain-rates using GI Taylor's impact technique." *Journal of Mechanics and Physics of Solids* 26(5): 289–301.

2 Mechanical Properties of 3D Printed Materials at High Strain Rates

Mayand Malik and Prateek Saxena

2.1 INTRODUCTION

In contrast to subtractive and formative manufacturing techniques, additive manufacturing (AM) is defined by ISO/ASTM 52900 as the process of combining substances to create components from 3D model data, often layer by layer. Figure 2.1 displays different acronyms for AM.

According to the aforementioned definition, a process must meet the following requirements in order to be classified as AM:

a) It should involve combining different types of materials.
b) The data for the three-dimensional (3D) model should be the beginning point.
c) It should be manufactured using a layer-by-layer build-up technique.
d) Subtractive manufacturing techniques should not be used.
e) Formative manufacturing approaches should not be used.

By fusing the two edges of the metal together or by adding additional material to melt to form the joint, two metal pieces are joined together through welding. The method can be used to create structures of any size with greater flexibility. Although welding includes combining materials, it does not begin with a digital model or add materials one layer at a time. Therefore, a component created by welding simple metallic pieces (as opposed to material with a primitive shape, such as wire or powder) does not meet the criteria to be referred to as an AM process. But if building 3D components rather than combining two pieces or structural sections is required, welding can be expanded to the additive manufacturing process. One of the welding-based additive manufacturing (WAAM) technologies used to construct big metallic objects is the wire-based AM process. Electricity is used by WAAM as the heating source and welding wire as the raw material. Three-dimensional (3D) structures can be created with five-axis CNC milling by removing material from a workpiece with cutting tools. Similar to additive manufacturing (AM), milling begins with a 3D digital model file that is cut along an axis. Nevertheless, since the technology requires material removal utilizing a top-to-bottom approach rather than material addition, it is not considered an AM process. A further limitation on component size is the size of the workpiece, whereas AM may produce components irrespective of the size of raw materials.

Other non-conventional techniques of machining, like electrical discharge machining (EDM), electrochemical and ultrasonic machining, etc., also don't meet the requirements to be referred to as AM processes. Using centrifugal force throughout the casting process, centrifugal casting is an approach that is utilized to create thin-walled cylinders of metals, glass, and concrete. Although the method can be used to create pipes of any diameter and does not require the removal of any components, it does not entail layer-by-layer connecting of the materials. As a result, it cannot be referred to as an AM process. By transferring a pattern from a layer of an energy-definable polymer (photoresist)

 DOI: 10.1201/9781003352358-2

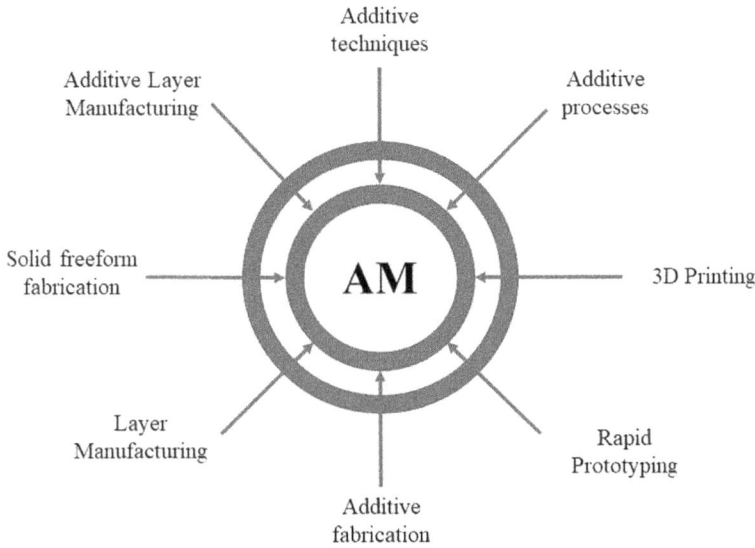

FIGURE 2.1 Acronyms for AM.

to a substrate, a design is created via optical lithography. Since it includes layer-by-layer chemical etching and proximity to a source of energy, optical lithography, a layer-by-layer manufacturing technology, is not regarded as an additive manufacturing process. By injecting molten material into a mold, the method of injection molding is utilized to create components. Injection molding isn't a technique of layer-by-layer material addition, despite the fact that the process requires solidifying the molten material without the removal of the material. This is not an AM procedure as a result.

Hydro-forming is a method that employs formative manufacturing principles to form metals into the appropriate geometry utilizing high-pressure fluid. Despite the fact that the method doesn't really require material removal, it is not an AM process. Laser shaping shapes the sheet metal pieces by making use of laser-induced thermal distortion/stresses. The laser beam's path determines how the sheet will take on the appropriate shape. The procedure does not require material removal; hence it is not an additive manufacturing procedure. The process neither involves a process of joining nor a layer-by-layer construction process, thus the reason. The approach behind the procedure, formative manufacturing, goes against one of the requirements for an additive manufacturing process. One of the earliest manufacturing processes, investment casting or "lost-wax casting," involves pouring molten metal into a disposable ceramic mold made by scraping away a wax pattern. Although there is no material removal involved, the procedure does not adhere to the layer-by-layer combining of formless material to construct parts using 3D model data. Material is shaped with concentrated compressive force during the forging process, which is a formative step. The method is not considered additive manufacturing (AM) since it does not entail layer-by-layer combining of formless material or direct component manufacture from 3D model data. The method uses formative methodology but does not require the removal of any materials.

2.2 CLASSIFICATION OF ADDITIVE MANUFACTURING TECHNIQUES

According to ISO/ASTM 52900, AM processes are divided into the following seven categories (see Figure 2.2):

a) *Vat photo-polymerization (VP)*: VP is a type of additive manufacturing in which light-activated polymerization is used to selectively cure liquid photopolymers stored in a vat (container).

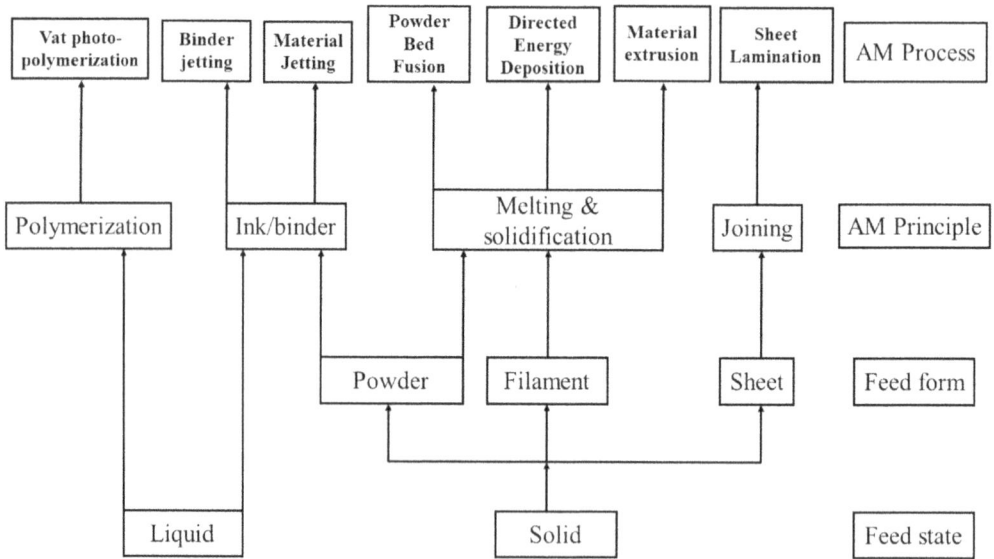

FIGURE 2.2 Categories of AM processes.

b) *Binder jetting (BJ)*: BJ is a technique of additive manufacturing that includes joining powdered materials together by carefully applying a liquid bonding agent.

c) *Material jetting (MJ)*: MJ is a type of additive manufacturing that includes placing building material droplets in specific locations.

d) *Sheet lamination (SL)*: SL is a method of additive manufacturing in which material sheets
are joined together to create an object.

e) *Material extrusion (ME)*: ME is a type of additive manufacturing in which material is
dispersed selectively through an aperture or nozzle.

f) *Powder bed fusion (PBF)*: PBF is a type of additive manufacturing in which specific powder bed regions are fused using thermal energy.

g) *Directed energy deposition (DED)*: DED is a type of additive manufacturing that includes
melting materials by fusion while they are layered using directed thermal energy.

2.3 MATERIALS USED IN 3D PRINTING

The three main categories of additive manufacturing materials are polymers, metals, and ceramics.
Materials are chosen based on the method or application.

2.3.1 POLYMER

a) *Polylactic acid (PLA)*: It is the most widely used polymer in additive manufacturing that
uses material extrusion. An abundant source such as corn starch, sugar cane, tapioca roots,
potato starch, etc. is used to create this biodegradable thermoplastic polymer. It is an eco-
friendly polymer that prioritizes appearance and durability over toughness. PLA melts
within a temperature ranging from 180–220° C, and for its deposition, the extrusion head
temperature is often fixed at 190–230° C. Additionally, the substance can be hydrolyzed or
thermally processed into its basic monomers. As a result, the produced monomers can be
cleaned up and used again to make more materials. Due to its low warping, PLA is simple
to deposit with additive manufacturing.

b) *Acrylonitrile butadiene styrene (ABS)*: Made of the polymer acrylonitrile, butadiene, and styrene, ABS is a thermoplastic. It is created by polymerizing acrylonitrile (20%) and styrene (55%), in the presence of polybutadiene (25%). Due to its superior material qualities, it is also among the most widely used polymers in material extrusion additive manufacturing. The material's recommended extrusion temperature is between 230 and 260° C, and its melting point is not far from 200° C. The material has good heat deflection resistance, stiffness, tensile strength, and impact resistance. Butadiene boosts the toughness, acrylonitrile provides wear and chemical resistance, and styrene gives the material exceptional workability and gloss.

c) *Polycarbonate (PC)*: Polycarbonate is a material that is transparent and impact resistant. It is a robust, light polymer that can be used in industrial settings. Bisphenol A and phosgene are poly-condensed to produce PC. Due to the difficulty of the material adhering to the build plate, a heated plate heated to a temperature between 80 and 120° C is employed as the build plate. For deposition, a temperature extrusion in the 260–310° C range is frequently utilized. There are many potential uses because the material is heat resistant and can endure a wide range of temperatures from –150 to 140° C.

d) *Acrylonitrile butadiene styrene/polycarbonate (ABS/PC) blend*: The best qualities of ABS and PC are combined in ABS/PC blends. High impact strength, high toughness, high heat resistance, and optical transparency are among PC's main benefits. Its poor processability and chemical resistance are a drawback, though. The aforementioned restriction can be overcome by adding chemicals to PC, however doing so decreases the material's hardness, which is undesirable for many engineering purposes. The chemical resistance and processability of ABS are both strong. It is also less expensive than PC.

e) *Polyetherimide (PEI)*: It has excellent flame retardancy, high dimensional stability, excellent defiance to chemicals (like hydrocarbons, alcohols, and halogenated solvents), good electrical properties, and high strength and rigidity at high temperatures.

f) *Polyethylene terephthalate glycol (PETG)*: PETG is a thermoplastic material with excellent strength, toughness, and chemical resistance. By combining polyethylene terephthalate and glycol, this co-polymer is created, enhancing the useful qualities of PET through the glycosylation process. The primary justification for employing PETG in additive manufacturing is that polyethylene terephthalate degrades and weakens under excessive heat, making it unsuitable for the process. However, the addition of glycol increases its resistance to oxidizing agents' corrosion and wear. The bed temperature should ideally be kept between 75° C and 80° C while the material extrusion temperature utilized for deposition is between 250 and 260° C.

g) *Acrylics*: Depending mostly on the acrylic co-polymer, acrylic materials have excellent optical clarity, outstanding rigidity, and excellent toughness. Additionally, they have excellent electrical qualities, strong chemical and thermal resistance, and biocompatibility. The substance is transparent and colorless, with a light transmission rate of around 90%.

h) *Epoxies*: Epoxies are thermoset resins with properties such as chemical resistance, mechanical strength, low shrinkage, high-temperature resistance, and insulation. The substance can stick to a wide range of substrates. Epoxy resins are less harmful and inexpensive. The exothermic nature of the curing process is the main restriction. The material is used to create lighter components for musical instruments, bicycle frames, racing cars, and railroads.

i) *Polyamide*: It is a thermoplastic polymer with low density, good chemical and oil resistance, wear resistance, thermal stability, and impact qualities, commonly referred to as nylon. The material's exceptional characteristics increase its use across a range of industrial industries, particularly the automobile industry. The substance can come in a variety of forms, including carbon fiber, glass-filled, and aluminum-filled. Glass-filled materials are made of polyamide powder that has been mixed with glass particles, and they have higher temperature tolerance (up to 110° C). The substance is typically employed in functional testing that includes heavy thermal stresses. Additionally, this substance exhibits strong tensile strength, low specific weight, and good stiffness.

j) *Polystyrene*: Polystyrene is an amorphous polymer that has a low cost and a hard physical makeup. There are two varieties of it: general-purpose polystyrene and high-impact polystyrene. High-impact polystyrene is created by combining rubber/butadiene co-polymer, which increases its toughness and impact strength, with general-purpose polystyrene, which is an unfilled polystyrene with a sparkling appearance. The main problems with polystyrene are its flammability, inadequate chemical resistance, and inadequate thermal stability. A strong material used in shipping containers and other applications requiring great impact resistance is high-impact polystyrene. High-impact polystyrene must be used with adequate ventilation since it releases vapors while being deposited.

k) *Polypropylene*: It is a translucent, off-white polymer with good toughness, fatigue resistance, and ductility. The bed temperature needs to be between 85° C and 100° C in an enclosed space, and deposition is done at processing temperatures between 220° C and 250° C with activated part cooling. Chemical resistance against bases and acids, high elasticity, hardness, and fatigue resistance are some of polypropylene's characteristics that draw applications from a variety of industries.

l) *Polyether ether ketone (PEEK)*: A polymer with good temperature tolerance up to 260° C is polyether ether ketone (PEEK). This draws interest for use in numerous pieces of high-temperature machinery. The substance possesses excellent temperature resistance, great corrosion resistance, and a low moisture absorption rate. PEEK is one of the primary materials used in the medical, oil and gas, aerospace, and automotive industries.

m) *Polyurethane*: Linear polymers with carbamate groups are known as polyurethanes ($-NHCO_2$). A diisocyanate and a polyol undergo a chemical process to create these groups. For high-quality deposition, the bed temperature must be between 45° C and 60° C, and the processing temperature must be between 225° C and 245° C.

2.3.2 METALS

a) *Aluminum alloys*: Due to their light weight, aluminum alloys are frequently used in the aerospace and automotive industries. For additive manufacturing, the most popular aluminum alloys are AlSi10Mg, AlSi7Mg0.6, AlSi9Cu3, and AlSi12.

b) *Cobalt chrome alloy*: The cobalt-chrome-molybdenum-based alloy known as cobalt chrome alloy (CoCr28M06) has good mechanical qualities in terms of strength, hardness, and great biocompatibility. The material also has great temperature, wear, corrosion, and fatigue resistance. Additionally, the material has a high degree of ductility and thermal fatigue resistance. Its processing by traditional machining is limited by its high strength and hardness.

c) *Precious alloys*: Despite the high cost, additive manufacturing makes it possible to process valuable materials like gold, silver, and platinum. Silver, namely 925 sterling silver, is frequently used in jewelry, bandages, and implants. Additionally, 950 platinum is the type of platinum that is most frequently utilized for jewelry applications.

d) *Nickel superalloy*: These materials are frequently employed in engineering applications where harsh duty conditions exist. The nickel superalloys Hastelloy-X, Inconel 625, Inconel 718, and Inconel 939 are frequently used in additive manufacturing.

e) *Iron alloys*: For a variety of applications in the aerospace, automotive, and forging industries, iron alloys are the most often utilized alloys in additive manufacturing. Iron-based alloys such as stainless steel 316L, 15-5-PH, 17-4PH, tool steels, Invar, and maraging steel are some of the most widely used alloys. The material stainless steel 316L is renowned for its high ductility, excellent drastic-temperature capabilities, and outstanding hardness and corrosion defiance.

f) *Titanium and its alloys*: Due to the material's low specific weight and biocompatibility, titanium and its alloys are mostly used in the engineering and medical sectors. Ti-6Al-4V is the most widely utilized titanium alloy in additive manufacturing because of its superior mechanical qualities, resistance to corrosion, low specific weight, and excellent bio-adhesion. Its use in aerospace, motorsports, and medicinal implants is attracted by this.

g) *Copper and its alloys*: In applications needing good electrical and thermal conductivity, copper and its alloys are used. CuCrZr is an alloy of copper that has strong electrical conductivity, thermal conductivity, and mechanical qualities. Pure copper is manufactured via additive manufacturing for a variety of applications, including heat exchangers and electronic components. It is used in induction coils, heat exchangers, and components of rocket engines. A bronze alloy with high elongation, medium hardness, good wear resistance, and resistance to air corrosion is CuSn10. It typically serves as a casing for heat exchangers and equipment utilized in marine systems.

There are many challenges associated with copper 3D printing due to its high reflectivity and high thermal conductivity. As printing copper requires melting copper powder in any of the printing processes like the electron beam method or selective laser melting, heat gets dissipated from the melt area, causing problems in its printing. There is also one problem of cracking of copper parts due to high thermal gradient; this can be eliminated by preheating the build plate so that the temperature can remain the same throughout the copper part.

2.3.3 Ceramics

Laser-based additive manufacturing processes can be used to directly produce ceramic components. Alumina and zirconia are the ceramic materials that are processed the most frequently.

Due to its superior hardness, strength, high-temperature defiance, high corrosion defiance, low thermal conductivity, and excellent chemical inertness, alumina is used in a number of engineering applications, including automotive parts, aerospace parts, electronic devices, machine tools, etc.

Due to its outstanding qualities, including high wear resistance, high strength, and high fracture toughness, zirconia, sometimes known as "ceramic steel," is widely utilized in the automotive, aerospace, and biomedical sectors. Zirconia is more durable than other ceramics because of the martensitic conversion of metastable t-ZrO_2 to stable m-ZrO_2. To keep the high-temperature metastable phases, such as c-ZrO_2 and t-ZrO_2, stable at normal temperature, dopants such as MgO, Y_2O_3, CeO_2, CaO, etc. Binder jetting uses sand as the material for the molds utilized in quick tooling applications. Zircon (zirconium silicate), chromite (iron chromium oxide), and silica sand are examples of naturally occurring and synthetic sands that are frequently used.

2.4 SOME ADVANCED MATERIALS USED IN 3D PRINTING

a) *Smart materials*: Additive manufacturing of smart materials has recently drawn a lot of interest. Smart materials are those that have the capacity to detect and react to environmental cues or stimuli, which can include mechanical, chemical, electrical, magnetic, or electrical impulses. Advanced structural materials or responsive materials are used in the fabrication of smart materials in additive manufacturing. Metamaterials, which are composite materials that have been carefully or purposely created, are an example of advanced structural materials. Their characteristics are not a result of the chemical components and are not easily apparent in nature. The artificial internal periodic structures affect the characteristics. Metamaterials are used in the creation of superlenses, antennae, sensors, etc.

b) *Bulk metallic glass*: Unlike crystalline materials, which have ordered atoms, bulk metallic glass has random atoms. As a result, this class of material has outstanding tensile strength, hardness, and toughness. For the fabrication of bulk metallic glass, numerous alloys based on Zr, Co, Cu, Pi, Pd, Ti, Ni, or Fe have been successfully produced. The high cooling rate needed for the manufacturing of bulk metallic glass, however, poses the biggest problem. This is so that crystallization, which is not necessary for glass formation, can occur during the cooling from a high temperature to room temperature at a slower rate.

c) *Functionally graded materials*: These materials differ gradually, at least along one direction, in their composition, components, or microstructures (such as grain size, texture, porosity, etc.). This will cause the material's functional properties to change. The most frequent uses of functionally graded materials in engineering applications involve the merging of metallurgically incompatible materials as well as the joining of materials with significant differences in their thermophysical properties.

d) *Metal matrix composites*: Materials with outstanding mechanical and physical characteristics are known as metal matrix composites. They are created by mixing reinforcement materials into a metallic matrix or spreading them across it. Metal is used as a matrix material because of its ductility and toughness, and ceramic is used as a reinforcement material to give metal matrix composites its qualities (high modulus and strength). Titanium, magnesium, aluminum, copper, and nickel alloys are the most popular matrix materials, whereas oxides, nitrides, and carbides are employed as reinforcing materials (e.g., tungsten carbide, titanium boride, alumina, etc.).

e) *Fiber-reinforced composites*: Due to their high ratio of strength to weight, fiber-reinforced composites have attracted the interest of both industry and researchers. They have applications in the automotive, sporting, and construction industries. The most popular additive manufacturing processes for creating fiber-reinforced composite components include powder bed fusion, sheet lamination, photo-polymerization, and material extrusion. The most widely used method for creating fiber-reinforced composites, with sizes ranging from nano to continuous fiber form, is material extrusion. Carbon, glass, graphene, carbon nanotube, and other fiber materials are among those being studied in this field. Thermoplastics, such as polypropylene, PLA, nylon, etc., are the most widely used matrix materials.

2.5 MECHANICAL PROPERTIES OF 3D PRINTED PARTS

Mechanical properties that can be evaluated after 3D printing a part are material hardness; yield tensile strength; break tensile strength; yield elongation; break elongation; flexural strength and modulus; Izod impact strength; Charpy impact strength; and hardness. More specifically, methods to measure these properties are defined below:

a) *Material hardness*: A 3D printed material's hardness is often assessed using the Shore durometer or the Rockwell scale. For instance, while Ultimaker ABS has a hardness of 97 on the Shore durometer, Ultimaker CPE+ has a hardness of 111 on the Rockwell scale. Both tests assess a material's susceptibility to indentation in order to determine how hard it is. A material is resilient to shape deformation when a compressive force is exerted the harder it is.

b) *Tensile strength at yield*: The amount of stress a material can endure before permanently deforming is known as its tensile strength at yield. Imagine a rubber tube that has been stretched out as an illustration of this. The tube would revert to its original shape if the pressure applied to it was released. The tube would eventually distort to the point that it would no longer be able to return to its former shape if the force on it kept rising. Megapascals (MPa) are the unit of measurement for a material's tensile strength at yield.

c) *Tensile strength at break*: The amount of force that may be applied to an object before it breaks is referred to as the tensile strength at break. Returning to our rubber tube case, if you applied more pressure, the material would deform past the point at which it would break because the molecules would not be able to resist the external pressures. The rubber tube's tensile strength at break rating is then determined by the amount of force necessary to rupture it.

d) *Elongation at yield*: Elongation at yield gauges a material's resistance to deformation before it deforms irreversibly. A material's elongation at yield rating is expressed as a percentage.

e) *Elongation at break*: The ratio of a material's original length to its enhanced length at its breaking point is known as elongation at break, also known as tensile elongation at break. In other words, this is a measurement of how long a material can withstand a change in shape before breaking.

f) *Flexural strength*: A measurement used to describe a material's propensity to bend, also known as bend strength. It is a ratio of flexural deformation's stress to strain. It is common practice to use this measurement with materials that considerably distort but do not shatter.

2.6 STRAIN RATE

The variation in strain/deformation of a material with reference to time is known as strain rate in the field of materials science.

The degree to which the distances between neighboring parcels of the material vary over time is measured by the strain rate at a certain location inside the material. It includes the rate of expansion or contraction of the material (expansion rate) as well as the rate of progressive shearing deformation without changing the volume of the material. If these distances remain constant, then the strain rate is zero, which occurs when all particles in a particular area are rotating or displacing with constant speed and in the same direction, the same as that section of the medium if it were a rigid body.

2.6.1 Effect of Strain Rate on Pure Polymer Printed Parts

Isotropic materials having the same mechanical behavior and material characteristics in every single direction have traditionally been utilized and studied extensively in mechanical engineering design and education because they are often simpler to understand, obtain, and produce. This is partly because a lot of materials used in engineering, such as metals and thermoset polymers, may be roughly described as isotropic. However, printers frequently produce components that are modeled in CAD software as solid objects, but when the parts are manufactured, they have infill patterns that swap out the solid internal volume for a more resource-efficient, structural lattice known as infill. These infill geometries are produced by either open-source or commercial slicing algorithms, and they are used to produce the G-code, or numerical code, required for manufacturing with consumer-grade 3D printers. Different infill patterns that can be used in printing any FFF part are lightning, lines, zig-zag, grid, triangles, tri-hexagon, cubic, cubic subdivision, octet, quarter cubic, gyroid, concentric, cross, cross 3D, etc.

a) *PLA*: The tensile strain rate significantly affects a material's characteristics, infill pattern, and density during mechanical testing. To extract and compare the mechanical characteristics of several parts that are produced using the Fused Filament Fabrication (FFF) technique with various infill geometries and infill geometry origins, monotonic tensile tests to fracture can be carried out using an Instron tensile testing machine. It is important to define the different infill geometries and orientations in a system of coordination associated with typical tensile test specimens in order to test them. To that aim, a set of axes, such as the length, breadth, and thickness axes, must be determined for all 3D printed parts. The final tensile strengths of printed specimens with hexagonal infill are substantially higher than those of rectilinear specimens. To establish a strain rate that reasonably approximates a static load, some components can be stressed at a higher strain rate.

The strength-to-mass ratio improves because of the utilization of infill. However, the ratios of strength by mass of the components with smaller infill percentages are lower. Additionally, parts having hexagonal infill are stronger than those having rectilinear infill, and their strength is more stable with respect to orientation. When strained at a higher strain rate, plastic stiffens. The optimum geometry for more dynamic loads might be different from the best geometry for relatively static loads because of the complicated nature of the infill geometry, which may cause the geometries to behave differently than they do at slower strain rates [1].

At greater strain rates, the PLA printed parts' tensile stiffness increases, while deformation and failure show little to no change. The capacity of parts to absorb energy increases with the strain rate. The hexagonal infill pattern is overall the strongest when compared to the specimen printed with a triangle infill pattern, having a greater Young's modulus and a lower elongation at fracture [2]. After performing a number of tensile tests on both PLA filament and PLA printed parts, it can be concluded that the strongest specimens have a 45° raster orientation. According to the findings of fatigue tests, 90° specimens are unquestionably the least resistant to fatigue loadings, and the fatigue lives of 45° specimens and 0° components are extremely comparable. The filament testing (at higher strain rates where creep is not a factor) yields results that are consistent with those of the printed specimen. This aids in determining whether unsuccessful print attempts can be recycled into fresh filament for further printing [3].

A high-speed camera is needed to measure the deformations optically at high testing speeds/strain rates. In order to make deformations trackable, a white primer and a graphite sparkle pattern are often placed on any of the surfaces of the printed part. PLA's storage modulus changes very little between 37 and 41° C since the material is still considerably below its glass transition temperature (T_g). Glass transition temperature is the temperature at which an amorphous polymer transitions from a hard or glassy state to a softer, frequently rubbery or viscous state. In the case of the PLA filaments, the cold crystallization, which is common for PLA, can be implicated in the increased storage modulus in the proximity of 100° C. Higher values for the strain at break are caused by the material's pronounced yielding before failure at medium strain rates. The images of the broken samples and the scanning electron microscopy (SEM) images do not reveal any appreciable differences in the fracture behavior between the various testing speeds, despite the fact that the overall behavior changes. A primarily brittle failure is indicated by SEM fracture surfaces. There are no air gaps or weld lines to be seen, which shows strong inter- and intra-layer bonding [4].

b) *Elastomeric polyurethane (EPU)*: Although soft materials like polymers and their composites are frequently used in digital printing, there are few complete mechanical experimental characterizations of some well-known polymeric materials, including elastomeric polyurethane (EPU), at high strain in the literature. Soft polymers that can be 3D printed hold great potential for soft robotics. Experimental characterization of these materials' viscoelastic behavior, particularly relaxation and strain rate-dependent effects, is crucial for creating accurate and realistic mathematical models for such materials. To research these materials, traditional experiments like loading-unloading cyclic testing, stress softening tests to identify Mullins effects, single- and multi-step relaxation tests, etc. are typically thought of. The primary experiments stated above are essential for building a thorough understanding of the viscoelastic soft polymeric materials used in 3D printing at high strains. Due to their great elasticity and strong resilience to impact and abrasion, polyurethanes and their derivatives, such as polyurea, find extensive use in products like ski boots, gaskets, seals, and the interlayers of windshield glass [5]. A potential material for the transparent armor design concept is transparent elastomeric polyurethane. Additionally, the recently developed magneto-rheological elastomer, a rapidly expanding composite polymeric material utilized in soft robotics and vibration control, makes extensive use of elastomeric polyurethane as a significant candidate. When elastomeric polyurethane is additively manufactured by the UV-curable DLS process all essential characterization methods are utilized like stress-softening tests, quasi-static tests, cyclic tests, and relaxation tests [6]. A viscoelastic polymer, EPU40 exhibits a very strong strain rate-dependent behavior. The polymer used in 3D printing has a significant degree of stretchability, with fracture occurring at strains greater than 600%. Due to its high energy return, high elongation at break, and high tear strength, additively produced EPU is an appealing material option for a variety of applications. When foam-filled re-entrant honeycomb structures with base material of polyurethane are 3D printed, it is typically observed that slow recovery foam gives re-entrant honeycombs lateral stiffness to effectively resist lateral buckling. The foam-filled re-entrant honeycombs offer greater rigidity and bearing capacity than the hexagonal honeycombs filled with foam. Slow recovery foam-filled honeycombs' ability to absorb energy steadily grows as the loading rate rises, but the auxetic impact of slow recovery foam-filled re-entrant honeycombs slowly reduces [7]. Broad application opportunities are provided by the slow recovery foam-filled re-entrant honeycombs, which are excellent buffer composites for uniaxial crushing with high strain rates. The mechanical reaction of the EPU is more dramatically impacted by 3D printing variables such as layer height and nozzle temperature than by strain rate [8]. The modulus and strength of parts printed with PU tough resin and PU hard resin show a substantial strain-rate effect, which increases as the strain rate rises. Under quasi-static loading, the two materials demonstrate high ductility; but, under dynamic stress, they display brittleness. The materials show greater strength and plateau stress when compressed, and an apparent hardening is seen before breakage [9].

c) *Acrylonitrile butadiene styrene (ABS)*: The MPD constitutive model is used in the modeling of uniaxial tensile tests as well as in the compression testing of the cellular structures to model the ABS material. The ABS printed parts with spiral geometries are more crashworthy than the honeycomb design in terms of their capacity to absorb energy. The plastic deformation of the layer material caused by a higher applied force than the layer's ultimate tensile strength (UTS) is the primary cause of fracture in PLA 3D printed parts when the strain rate is kept constant at 10^{-1} sec^{-1}. Additionally, there are indications of layer delamination. The failure of the adhesion bonding between the layers occurs when shear is given as a load to the interface of the layers. The two most significant variables that affect the maximum ultimate tensile strength of printed parts are layer thickness and print strength. A linear rise in tensile strength is observed with increasing strain rate when PLA and tough PLA materials are examined. The yield strength and Young's modulus of the PLA material increase steadily as the strain rate rises. PLA doesn't demonstrate the predicted linear decrease of strain with increasing strain rate [10].

The modulus of toughness of the ABS part is significantly impacted by the raster angle. When compared to specimens constructed with the [0/90°] raster angle, specimens built with the [45/–45°] raster angle have the potential to improve the modulus of toughness by 200%. A significant toughness modulus of 2.044 MPa can be attained with a strain rate of 5 cm/min and [45/–45°] raster angle. The strain rate values have a significant impact on the resilience modulus. The modulus of resilience can be significantly improved at increased strain rates for any build style combination of parameters. Strain rate and interior fill pattern have a disproportionately large impact on the UTS and yield strength parameters. At a strain rate of 10 cm/min, a high UTS of 27.44 MPa and a yield strength of 23.79 MPa are attained [11].

d) *Polyether ether ketone (PEEK)*: The layer thickness has a significant impact on the tensile strength of PEEK objects printed using 3D technology while having less of an impact on the bending and compressive strengths. The PEEK components are subjected to pressure that is aligned in the axial direction during compressive tests, and the neighboring layers carry the shear stress as a result of the additive buildup. This can lead the single layers to slide past one another until the sample finally breaks. Because printed parts are prone to defects, including both extensive pores that are initially pushed out and inadequate inter-layer bonding, their compressive strength is much lower than that of injected materials [12].

Parts' compressive strength is strain rate-dependent and rises as the strain rate does. Up until the apex, the PEEK behaves abruptly and linearly elastically; after that, it yields and the fracture propagates linearly till fracture failure. As the strain rate and fracture strain increase, PEEK's positive strain rate sensitivity increases from 0.02 to 0.4 for strain rates ranging from 0.01 s^{-1} to 1 s^{-1}. With a high strain rate, the rupture can spend less time navigating low-resistance paths, increasing toughness [13].

2.6.2 Effect of Strain Rate on Metal and Alloy Printed Parts

It is becoming easier to conduct research into novel lightweight constructions that have high specific structural stiffness and the potential for improved energy absorption mechanisms thanks to additive manufacturing (AM) techniques used to build metallic components. For the high-fidelity constitutive models required for the predictive design of high-performance structures, it is essential to comprehend the high strain rate deformation and failure mechanisms of these AM materials. A detailed review of the mechanical properties of various metals and alloy parts printed via 3D printing is highlighted below:

a) *Ti-6Al-4V (LMD Ti64)*: The mechanical properties of laser metal deposited (LMD) metals are influenced by process variables such as laser beam power, scanning speed, particle size, and layer thickness. The LMD Ti64 alloy can be recognized from conventionally manufactured Ti64 alloy by its first internal flaws brought on by its unique layered deposition, such as lack of fusion flaws between layers, gas void, and keyhole-induced porosity. Process parameters such as scan speed, scan strategy, energy density, laser power, etc. also regulate these initial flaws. As the strain rates increase, the ductility of LMD parts declines. The relationship between fracture strain and logarithmic strain rate is linear. With an increase in the logarithmic strain rate, the fracture strain drops linearly. At a high strain rate of 5,000/s, the fracture surface is perpendicular to the axis direction. The fracture surface becomes oblique to the loading direction for low strain rates. Compared to commercial-grade Ti64 alloy, LMD Ti64 alloy is more susceptible to strain rates. In addition, severe processing flaws significantly reduce tensile strength at high strain rates [14].

By observing the microstructure with an SEM, the failure mechanism of LMD Ti64 alloy, which varies from different stress states and strain rates, can be seen. SEM results indicate that the outer portion of the fracture surface is a shear fracture and the middle region is full of dimples. The shear fracture region becomes bigger and the number of dimples in the central region shrinks as the strain rate increases, indicating that shearing is predominating. At significantly higher strain rates, the fracture surface exhibits more initial cracks than voids. High strain rates depend significantly on the initial defect geometry [15]. Dynamic mechanical parameters like yield strength, ultimate compression strength, and total strain may rise by 21%, 18%, and 261%, respectively, when the strain rate rises. The reduction in interlamellar spacing and the resulting Hall-Petch strengthening, the creation of low-angle grain boundaries, and the dislocation density are all factors in the improvement [16].

b) *Aluminum alloy*: The lightweight, high strength-to-weight ratio, corrosion resistance, and outstanding mechanical characteristics of aluminum alloys processed by SLM are drawing attention, as are the distinctive benefits provided by SLM, such as tool-less fabrication, geometric freedom, personalized design, and intricate shapes. AlSi10Mg and AlSi12 alloys are the two types of aluminum alloys utilized in SLM systems most frequently. The split Hopkinson Pressure Bar (SHPB) apparatus, which was created for high strain rate testing, is typically used to conduct dynamic high strain rate tests [17].

When the as-built AlSi12 parts are tested at high temperatures, a considerable drop in both quasi-static and dynamic yield strength and ultimate compressive strength is seen. This decrease in flow stress is a result of the components becoming softer as a result of the development and aggregation of Si-rich precipitates at high temperatures (200° C and 400° C). Additionally, as the pieces are heated to a lower temperature, both the quasi-static and the dynamic flow stress diminish. Furthermore, both heat-treated and as-built parts have a positive strain rate effect at higher temperatures. Due to strain hardening and dislocation density, high strain rate deformation generally results in an increase in strength [18]. Up until the maximum stress is reached at a certain strain rate, deformation is mostly controlled by strain hardening. With an increase in strain rate, yield stress decreases. However, when compared to any die-cast material, such as A360.0 alloy, the components made using additive manufacturing exhibit greater yield strengths, ultimate tensile strengths, and elongation to fracture [19].

c) *Stainless steel*: The most popular engineering material is still steel, and the steel grades appropriate for AM include 316L and 304L stainless steels, maraging steels, and precipitation-hardening stainless steels. Due to the exceptional qualities of this stainless-steel grade, the AM of 316L has drawn particular attention. In order to evaluate the dynamic compressive behavior of various auxetic lattices created by laser powder bed fusion (LPBF) using 316L powdered austenitic steel, a split Hopkinson pressure bar (SHPB) outfitted with a high-speed camera can be utilized. Given that the stress values do not change noticeably, the mechanical reactions of the 45° and 90° orientations to the high-rate loading are comparatively uniform. Additionally, for these two directions, the oscillatory structure of the stress-strain curve brought on by the high strain rate is more obvious. All printing orientations exhibit considerable strain rate sensitivity; however, the 90° orientation exhibits the strongest strain rate effect. However, compared to the other investigated printing directions, the specimens tested under quasi-static loading exhibit a notable decrease in the hardening rate, resulting in lower values of stress in the plastic region. This significant increase is primarily the result of this decrease in the hardening rate. The decreased coherence of the neighboring layers in the 3D-printed material, which are oriented in a 90-degree direction parallel to the direction of loading, may be the cause of this phenomenon, which only manifests itself during quasi-static stress [20].

The 17-4PH stainless steel's high strain rate compressive response is primarily influenced by residual porosity. Internal voids prevent the anticipated improvement in strength brought on by post-manufacturing heat treatment. At higher strain rates, HIP successfully eliminates void flaws in the component and causes a larger than 20% increase in flow stress [21].

2.6.3 EFFECT OF STRAIN RATE ON FIBER/POLYMER COMPOSITE PRINTED PARTS

Continuous fibers are increasingly used as reinforcement in 3D printed materials to enhance the mechanical qualities of printed components. Due to their widespread use in the production of traditional composites, synthetic fibers were most frequently used as continuous fibers to strengthen 3D printed materials in literature.

a) *Carbon fiber*: The dispersion of materials and the post-test evaluation of composite parts are carried out using a high-resolution electron microscope. Prior to testing, parts are given a gold sputtering treatment to improve conductivity. Carbon fiber mechanical constraints on polymer chain agility improve the composite's capacity to retain elastic energy [22].

For the parts printed using carbon-fiber-reinforced polyamide composite, because the half-wavelength falls as cell number grows and more folding and energy absorption occurs, specific energy absorption (SEA) rises with filling density. In comparison to other densities, structures with a medium filling density exhibit a better crushing force efficiency (CFE). The process of exterior hexagonal tubes deforming is impacted by different filling shapes. Due to their more efficient plastic deformation and energy absorption cracks, triangular and circular filled structures have greater SEA than hexagonal filled ones at the medium filling density. Under the dynamic impact condition, all structures' SEA is considerably reduced. This results from specimens not deforming enough when subjected to dynamic loading. The failure mode of thin-walled structures shifts from plastic deformation mode to brittle fracture mode as a result of the strain rate effect [23]. To assess the strain rate sensitivity of carbon/PA composite, uniaxial quasi-static and dynamic compression tests using a direct Split Hopkinson Bar configuration are carried out. The mechanical properties of carbon/PA are sensitive to strain rate changes, and the compression strength is greater than under static conditions. Even though the strain-to-failure trend typically declines with strain rate, the carbon/PA component prematurely fails above a specific rate, which lowers the maximum stress [24]. In a semi-logarithmic graph, the resistance typically rises roughly linearly with the strain rate. Once a specific amount of strain rate is exceeded, the part's performance suddenly declines in terms of resistance and strain before failure. Due to their identical deposition strategies, the strain-stress curves of the transverse and longitudinal sections have a similar trend. On the other hand, the trend is the opposite for the vertical specimens, which exhibit a strain-hardening behavior in the plastic zone. Fiber pull-out and matrix brittle fracture are the primary failure mechanisms at low temperatures or high strain rates [25].

b) *Ramie fiber*: There are three changing trends in the dynamic strengths of printed continuous ramie fiber-reinforced bio-composites (CRFRC) as the strain rate rises:
1. The dynamic strength grows as the strain rate increases.
2. The dynamic strength increases at first and then changes insignificantly as the strain rate increases.
3. The dynamic strength increases at first, then decreases significantly as the strain rate increases.

Be aware that different printing parameter combinations can result in varying strain rates at the point where the dynamic strengths began to decline. Similar patterns may be seen in the dynamic strengths of printed CRFRC under additional printing parameter combinations and strain rates.

The CRFRC's dynamic strength is most significantly impacted by layer thickness. Higher layer thickness greatly increases the void content of the printed specimen, which lowers the CRFRC's dynamic strength. Because its implications on void content are less significant than those of layer thickness, hatch spacing has weaker effects on dynamic strength than layer thickness does. Additionally, by reducing layer thickness and hatch spacing, the composite's fiber content rises, enhancing the printed specimen's dynamic strength. However, increasing the fiber content will not result in a constant improvement of the dynamic strength. This is because more fiber-matrix interfaces were introduced as the fiber content increased. Additionally, a printed composite's mechanical characteristics may be compromised by an excessive number of fiber-matrix connections. Additionally, as the strain rate rises, the CRFRC's dynamic strength first rises and then changes barely at all. The rapid adiabatic temperature rise within the material at high strain rates is thought to be the source of the matrix material softening, which is the "threshold effect" of strain rate [26].

2.6.4 Effect of Strain Rate on Metal/Polymer Composite Printed Parts

Complex components that were earlier difficult to build using conventional techniques can now be produced thanks to three-dimensional printing technology. The variety of polymer materials available for 3D printing is likewise growing. The comprehensive nature of the characteristics of metal powder-filled polymers, as opposed to those of single metals or polymers, has led to their widespread application in the domains of powder injection molding and polymer material modification. The mechanical characteristics of the materials can also be altered by altering the type of powder and/or filling quantity.

a) *Copper powder-filled PLA*: Typically, quasi-static tensile, compression, and Split Hopkinson Pressure Bar (SHPB) tests are carried out to clarify the fundamental mechanical properties of the materials in order to study the impact of copper powder on the mechanical properties and microstructure of polylactic composites. The PLA material's tensile curve's form shows a strong relationship with the rate at which tension is applied, suggesting that the material experiences the strain rate effect. The material exhibits properties that indicate a transition from tough to brittle as the strain rate rises, including an increase in strength, a decrease in toughness and ductility, and an increase in strength. The stress and strain become proportional before the yield point is reached, following Hooke's law. The stress-strain curve in the elastic stage is fitted to obtain the initial elastic modulus of the PLA material. At this point, the polymer is deforming due to general elasticity brought on by the expansion of internal bonds and the modification of the bond angle. Because the impact rod applies less energy to the specimen when the strain rate is low and the original space and cracks in the specimen gradually close over time, a longer compaction period occurs, and the proportion of the linear elastic compression stage decreases as the strain rate rises. The energy that the impact bar transfers to the part and the speed at which the initial crack inside the part closes both increase as the strain rate rises. The linear step thus becomes smaller. As a direct result, both the material's dynamic elastic modulus and the slope of the stress-strain curve linked to the linear stage increase [27].

2.7 CHARACTERIZATION OF 3D PRINTED PARTS

Components fabricated using additive manufacturing are evaluated to assess their suitability before being employed in a variety of applications. Geometry evaluation and material property evaluation are two possible methods of quality evaluation.

2.7.1 Geometric Assessment

Below are some of the common characterization methods used for geometric evaluation:

1) *Coordinate measuring machine*: A coordinate measuring machine is referred to as a CMM machine. It is a device that utilizes coordinate technology to compute the dimensions of a machine or tool parts. The X, Y, and Z axes' height, width, and depth are among the dimensions that can be measured. The complexity of the CMM machine will determine how precise a measurement can be made of the object.

2) *Profilometer*: An extremely sophisticated meteorological assessment tool called a profilometer is used to assess the roughness of a surface. The profilometer is exceptionally accurate as a quantification tool because it can be used to recognize and measure extremely minute surface characteristics.

3) *Confocal microscope*: It uses a scanning procedure using a laser. To create a 3D restoration, the method simply scans an item with a focused laser beam. A confocal microscope takes photographs one depth level at a time, in contrast to a traditional microscope, which can only view as deep as light can go through an object.

4) *Computed tomography*: It is an imaging technique that employs X-rays to produce images of part cross sections. It is also a method of medical imaging that is used to get precise pictures of the inside of the body.

5) *Atomic force microscopy (AFM)*: AFM is a very high-resolution instrument that may be used to view a sample's morphology and precisely analyze its mechanical characteristics with atomic precision.

6) *Scanning electron microscope*: A beam of electrons is projected at a particular spot and scanned over a surface by a scanning electron microscope (SEM) to produce a picture. The interaction between the beam's electrons and the specimen results in a variety of signals that can be used to learn more about the surface's topography and chemistry.

7) *Raman spectroscopy*: Infrared spectroscopy is often used in conjunction with Raman spectroscopy which is utilized for measuring surface roughness; it is also a prominent method for analyzing molecular structure. The Raman effect, which underlies Raman spectroscopy, is based on the dispersion, which includes inelastic (Raman) dispersion at various wavelengths caused by molecular vibrations as well as elastic dispersion at the same wavelength as the input light.

2.7.2 MATERIAL PROPERTY ASSESSMENT

Below are some of the common characterization methods used for material property and defect evaluation.

1) *Optical microscopy*: An enlarged image of small specimens that would otherwise be invisible to the naked eye is produced by an optical microscope, also known as a light microscope, using visible light and lenses.

2) *X-ray diffraction*: In materials science, X-ray diffraction analysis (XRD) is a method used to ascertain the crystal structures of a substance. When using XRD, a material is exposed to incoming X-rays, and the intensity and dispersion directions of the X-rays that escape the material are then measured.

3) *Scanning electron microscope*: A concentrated electron beam is utilized by a scanning electron microscope (SEM) to examine a sample's surface and produce a high image quality. Imaging from SEM can provide details about an object's surface chemistry and topography.

4) *Universal testing machine*: A variety of tests can be run on universal testing machines (UTMs), also known as universal testers, to determine the compressive and tensile strengths of various materials. Basic compressive and tensile testing, seal strength tests, bond strength tests, bend tests, puncture tests, and spring tests are just a few examples of tests that can be performed.

2.8 SUMMARY

Researchers have been baffled by the effects of mechanical stress, especially at dynamic strain rates, on diverse materials since those structures are inherently unstable. Roughly all the major materials are included in this chapter with effects on their mechanical properties at high strain rates. It can be concluded that more or less all the materials' properties are significantly affected by strain rates during mechanical testing of specimens that are fabricated using various 3D printing techniques like fused deposition modeling (FDM) mainly for polymer composites, selective laser sintering (SLS), and selective laser melting (SLM) for polymers and metals respectively. It is evident that strain rate is also one of the aspects that must be taken into account when analyzing a part produced using 3D printing to get a clear picture of how it can be used and in which applications the material can be employed.

At greater strain rates, the PLA printed parts' tensile stiffness increases, while deformation and failure show little to no change. The capacity of parts to absorb energy increases with the strain rate. Elastomeric polyurethane printed parts show greater strength and plateau stress when compressed, and an apparent hardening is seen before breakage at higher strain rates. The ABS printed parts with spiral geometries are more crashworthy than the honeycomb design in terms of their capacity to absorb energy when strained at higher rates. PEEK printed parts' compressive strength is strain rate-dependent and rises as the strain rate does. Up until the apex, the PEEK behaves abruptly and linearly elastically. Dynamic mechanical parameters of Ti-6Al-4V printed parts, like yield strength, ultimate compression strength, and total strain may rise by 21%, 18%, and 261%, respectively, when the strain rate rises. Due to strain hardening and dislocation density, high strain rate deformation generally results in an increase in strength. Up until the maximum stress is reached at a certain strain rate, deformation is mostly controlled by strain hardening in the case of aluminum alloy specimens. Stainless steel's high strain rate compressive response is primarily influenced by residual porosity. Internal voids prevent the improvement in strength brought on by post-manufacturing heat treatment. Once a specific amount of strain rate is exceeded, the carbon fiber printed part's performance suddenly declines in terms of resistance and strain before failure. Due to their identical deposition strategies, the strain-stress curves of the transverse and longitudinal sections have a similar trend.

REFERENCES

1. Daniel Farbman, Chris D. McCoy, "Materials testing of 3D printed ABS and PLA samples to guide mechanical design", *Proceedings of the ASME 2016 International Manufacturing Science and Engineering Conference MSEC2016*.
2. Shafahat Ali, Said Abdallah, Deepak Devjani, Joel John, "Effect of build parameters and strain rate on mechanical properties of 3D printed PLA using DIC and desirability function analysis", *Rapid Prototyping Journal*, July 2022.
3. Todd Letcher, "Material property testing of 3D-printed specimen in PLA on an entry-level 3D printer", *Proceedings of the ASME 2014 International Mechanical Engineering Congress & Exposition IMECE2014*.
4. Sandra Petersmann, Martin Spoerk, "Mechanical properties of polymeric implant materials produced by extrusion-based additive manufacturing", *Journal of the Mechanical Behavior of Biomedical Materials* 104 (2020).
5. Mokarram Hossain, Rukshan Navaratne, Djordje Perić, "3D printed elastomeric polyurethane: Viscoelastic experimental characterizations and constitutive modelling with nonlinear viscosity functions", *International Journal of Non-Linear Mechanics* 126 (2020) 103546.
6. Nabila Elmrabet, Petros Siegkas, "Dimensional considerations on the mechanical properties of 3D printed polymer parts", *Polymer Testing* 90 (2020) 106656.
7. Nectarios Vidakis, Markos Petousis, Apostolos Korlos, Emmanouil Velidakis, Nikolaos Mountakis, Chrisa Charou, Adrian Myftari, "Strain rate sensitivity of polycarbonate and thermoplastic polyurethane for various 3D printing temperatures and layer heights", *Polymers* 13 (2021) 2752. https://doi.org /10.3390/ polym13162752
8. Hui Chen Luo, Xin Ren, Yi Zhang, Xiangyu Zhang, "Mechanical properties of foam-filled hexagonal and re-entrant honeycombs under uniaxial compression", *Composite Structures* 280 (2022) 114922.
9. Yitong Wang, Xin Li, Yang Chen, Chao Zhang, "Strain rate dependent mechanical properties of 3D printed polymer materials using the DLP technique", *Additive Manufacturing* 47 (November 2021) 102368.
10. Michał Kucewicz, Paweł Baranowski, "Modelling and testing of 3D printed cellular structures under quasi-static and dynamic conditions", *Thin-Walled Structures* 145 (2019) 106385

11. Kemar Hibbert, Grant Warner, "The effects of build parameters and strain rate on the mechanical properties of FDM 3D-printed acrylonitrile butadiene styrene", *Open Journal of Organic Polymer Materials* 9 (2019) 1–27.
12. Wenzheng Wu, Peng Geng, "Influence of layer thickness and raster angle on the mechanical properties of 3D-printed PEEK and a comparative mechanical study between PEEK and ABS", *Materials* 8 (2015) 5834–5846. https://doi.org/10.3390/ma8095271
13. Sagar M. Baligidad, G. Chethan Kumar, "Investigation on strain rate sensitivity of 3D printed sPEEK-HAP/rGO composites", *Journal of Manufacturing Processes* 79 (2022) 789–802.
14. Raffaele Barbagallo, Simone Di Bella, Giuseppe Mirone, Guido La Rosa, "Study of the electron beam melting process parameters' influence on the tensile behavior of 3D printed Ti6Al4V ELI alloy in static and dynamic conditions", *Materials* 15 (2022) 4217. https://doi.org/10.3390/ ma1512421
15. Chang Peng, Peng-Hui Li, "Failure behavior of laser metal deposited additive manufacturing Ti-6Al-4V: Effects of stress state and initial defects", *3D Printing and Additive Manufacturing* (2022).
16. Reza Alaghmandfard, Dharmendra Chalasani, "Activated slip and twin systems in electron beam melted Ti-6Al-4V subjected to elevated and high strain rate dynamic deformations", *Materials Characterization* 172 (2021) 110866.
17. Carter Baxtera, Edward Cyr, "Constitutive models for the dynamic behaviour of direct metal laser sintered AlSi10Mg_200C under high strain rate shock loading", *Materials Science & Engineering A* 731 (2018) 296–308.
18. Panneer Ponnusamy, S. H. Masood, Dong Ruan, "High strain rate dynamic behaviour of AlSi12 alloy processed by selective laser melting", *The International Journal of Advanced Manufacturing Technology* 97 (2018) 1023–1035. https://doi.org/10.1007/s00170-018-1873-5
19. C. K. Baxter, E. D. Cyr, "Mechanical behaviour and constitutive modeling of AlSi10Mg-200°C additively manufactured through direct metal laser sintering", *Proceedings of the 16th International Aluminum Alloys Conference* (ICAA16) 2018.
20. Michaela Neuhäuserová, Petr Koudelka, Tomas Fíla, Jan Falta, Vaclav Rada, Jan Šleichrt, Petr Zlámal, Anja Mauko, Ondrej Jirousek, "Strain rate-dependent compressive properties of bulk cylindrical 3D-printed samples from 316l stainless steel", *Materials* 15 (2022) 941. https://doi.org/10.3390/ ma15030941
21. Brandon McWilliamsa, Brahmananda Pramanik, "High strain rate compressive deformation behavior of an additively manufactured stainless steel", *Additive Manufacturing* 24 (2018) 432–439s.
22. Ibrahim M. Alarifi, "Investigation of the dynamic mechanical analysis and mechanical response of 3D printed nylon carbon fiber composites with different build orientation", *Polymer Composites*. https://doi .org/10.1002/pc.26838
23. Kui Wang, Yisen Liu, "On crashworthiness behaviors of 3D printed multi-cell filled thin-walled structures", *Engineering Structures* 254 (2022) 113907.
24. Mattia Utzeri, Emanuele Farotti, "High strain rate compression behaviour of 3D printed CarbonPA", *Journal of Materials Research*. https://doi.org/10.1557/s43578-021-00248-9
25. M. Mohammadizadeh, A. Imeri, "3D printed fiber reinforced polymer composites – Structural analysis", *Composites Part B* 175 (2019) 107112.
26. Ruijun Cai, Hao Lin, "Investigation on dynamic strength of 3D-printed continuous ramie fiber reinforced biocomposites at various strain rates using machine learning methods", *Polymer Composites*. https://doi.org/10.1002/pc.26816
27. Qing Ji, Zhijun Wang Jianya Yi Xuezhi Tang "Mechanical properties and a constitutive model of 3D-printed copper powder-filled PLA material", *Polymers* 13 (2021) 3605. https://doi.org/10.3390/ polym13203605

3 Strain Rate Studies of Polymer and Fibre-Reinforced Polymer Nanocomposites

S. Gurusideswar, Dong Ruan and R. Velmurugan

3.1 INTRODUCTION

Traditionally, polymers have been added with natural or synthetic fillers in order to improve their mechanical, thermal and electrical properties and also to reduce the cost. They are extensively used in the aerospace, automotive, defence and electronics industries. It is proven that the dispersion of nano-sized fillers with a larger aspect ratio in polymers leads to a dramatic improvement in mechanical properties. Most of the thermoset resins are inherently brittle, which restricts their performance to many structural applications. In general, polymers are incorporated with micro fillers, such as calcium carbonate, glass beads, mica and talc in order to enhance their performance. However, it is often reported that the addition of these fillers has certain drawbacks such as an increase in weight, brittleness and opacity [1]. It is also reported that the properties of those materials can be customized by changing the weight fraction, shape and size of the fillers. A further enhancement in performance can be achieved by adding fillers in the nanometre range, which have a high aspect ratio [2]. The addition of fillers in polymers for which at least one dimension of the dispersed particles is in the nanometre range (< 100 nm) is known as nanocomposites. Nanocomposites are considered one of the classes of nanomaterials, where nanofillers are dispersed in the matrix phase. The nanocomposites can be broadly classified based on the dimensions of the nanoparticles: i) iso-dimensional nanoparticles, ii) nanotubes and iii) nanolayers. If at least one of the dimensions is in the nanometre range, in the form of layers, sheets, laminas and/or shells dispersed in the polymer matrix, this is traditionally called polymer-layered nanocomposites. In recent decades, researchers [3–9] have found potential improvements in the properties and performances of fibre-reinforced polymer matrix materials with the addition of nano-scale fillers.

The mechanical properties of most of the polymers and polymer matrix composites are sensitive to strain rates, and in many instances, they are subjected to dynamic loadings, which require prior knowledge of dynamic mechanical properties to prevent catastrophic failure during their service. In practical scenarios, the composite structures undergo high-velocity impact loadings. A few examples are collisions, crash landings, a rigid body impact onto a structure, a bird impact on jet engine compressor rotating blades, automotive vehicle components, satellite solar panels, ship hull structures, shock loads, bomb blasts, etc. In order to understand the dynamic behaviour of polymers and their nanocomposite fibre-reinforced polymers, the current work focuses on the strain rate effects (low and high). Also, the current literature on low strain rate and high strain rate studies is discussed and presented in detail. In general, six different types of testing systems are employed to study the strain rate effects of a material. They are the i) conventional screw drive load frame (< 0.1 s^{-1}), ii) servo-hydraulic system (0.1–100 s^{-1}), iii) drop mass test setup (100–1,000 s^{-1}), iv) split Hopkinson pressure bar (100–10^4 s^{-1}), v) expanded ring (10^4 s^{-1}) and vi) flyer plate (> 10^5 s^{-1}). In the current study, an attempt is made to review the effects of both low and high strain rates on polymer nanocomposites and fibre-reinforced polymer nanocomposites.

DOI: 10.1201/9781003352358-3

3.2 LOW STRAIN RATE STUDIES OF POLYMERS

The mechanical properties of most of the polymers and their nanocomposites are sensitive to strain rates. Many researchers [10–20] have focused on the behaviour of composites at high strain rates. However, the behaviour of composites at low strain rates is rarely reported. Guo and Li [14] studied the compressive behaviour of epoxy/SiO$_2$ nanocomposites at various loading rates (10^{-4}–10^4 s^{-1}) and found that the compressive strength of the nanocomposites is higher than that of pure epoxy at high strain rates. Bao and Tjong [21] found that polypropylene/carbon nanotubes nanocomposites were strain rate (3.3×10^{-5}–3.3×10^{-1} s^{-1}) and temperature dependent. Lim et al. [22] investigated the effect of loading rate (10^{-4}–10^{-1} s^{-1}) and temperature on nylon 6/organoclay nanocomposites and found that its tensile yield strength is sensitive to both temperature and strain rate. Gurusideswar and Velmurugan [23] investigated the effect of nano-sized clay and strain rate on the tensile properties of neat epoxy and its nanocomposites. They observed that the mechanical properties of epoxy nanocomposites are rate sensitive, even at a low range of strain rate. When the strain rate increases from 0.0001 to 0.1 s^{-1}, the longitudinal ultimate strength increases by 15%, whereas the increase in Young's modulus is 13% for a neat epoxy system. They observed that the tensile strength is decreased for epoxy/clay nanocomposites as the strain rate increases, which could be due to the presence of nanoclay and its significant effect on mechanical properties.

3.3 LOW STRAIN RATE STUDIES OF FIBRE-REINFORCED POLYMERS

Okoli and Smith [24] performed tensile and shear tests on glass/epoxy composite at increasing rates of strain and showed that the logarithm of the strain rate on the tensile properties could be regarded as linear. Okoli [25] studied the low strain rate effects (0.01–2.72 s^{-1}) of woven-type glass/epoxy composites on tensile properties and their failure modes. He also proposed a semi-empirical relationship for predicting tensile modulus using micromechanics and experimental data. Berezhnytskyi and Panasyuk [26] studied the effect of strain rate and temperature on the mechanical properties of polymeric composite materials and proposed new empirical formulae for the strain rate and temperature dependence of the composites. Kallimanis and Kontou [27] studied the tensile behaviour of glass fibre polymer composites at various strain rates (8.33×10^{-5}, 3.33×10^{-4} and 1.66×10^{-3} s^{-1}) and explained the strain rate–dependent behaviour through a scaling law. Saniee et al. [28] investigated the rate dependency of glass/epoxy composites at low strain rates and observed that the longitudinal strength increased by 24.7% when the strain rate increased from 0.0001 to 0.11 s^{-1}. Sun et al. [29] investigated the low strain rate responses (2×10^{-5}–2×10^{-2} s^{-1}) of UD fibre composites and found that the ultimate strength increases with the increase in strain rate. They also observed that the macro-failure mode was changed from ductile to brittle fracture mode with increasing strain rate. Shokrieh and Omidi [30] studied the effects of quasi-static and intermediate strain rates (0.001–100 s^{-1}) for unidirectional glass polymeric composites under uniaxial loading and found a significant increase in tensile properties with increasing strain rate. Shadlou et al. [31] studied the effect of strain rate (0.01–10 s^{-1}) on the mechanical behaviour of epoxy/graphene nanoplatelets (GNPs) nanocomposites and showed that the yield strength and Young's modulus of epoxy and its nanocomposites increased at higher strain rates. They also compared the experimental results with some available models. Gurusideswar and Velmurugan [23] investigated the effect of nano-sized clay and strain rate on the tensile properties of glass/epoxy nanocomposites. They observed that the mechanical properties of glass/epoxy nanocomposites are rate sensitive, even at a low range of strain rate. When the strain rate increases from 0.0001 to 0.1 s^{-1}, the longitudinal ultimate strength increases by 23% whereas the increase in Young's modulus is 19% for glass/epoxy composite. From the visual inspection, it is observed that the failed specimens show significant changes in the fracture surface with an increased strain rate. From the brief literature review, it is observed that the mechanical properties of the polymers and their composites are rate-sensitive, even at low strain rates.

3.4 HIGH STRAIN RATE STUDIES OF POLYMERS

Many researchers [1–9] have devoted themselves to the field of "nanocomposites," mainly focusing on manufacturing, characterization, fracture mechanism, wear resistance and so on. In actual scenarios, polymeric materials and their composites are subjected to dynamic loading and high strain rate deformation in various important applications, such as aircraft and automotive components and marine structures. Also, the simulation of composite structures under high strain rate deformation requires a clear identification of the strain rate effect on the material behaviour. Hence, high strain rate studies are important, which is not reported in the above-mentioned literature. Very little literature reports the dynamic mechanical responses of this kind of nanocomposites due to the difficulty of high strain rate testing and data interpretation. It is noted that the viscoelastic nature of polymers exhibits significant rate dependence in their stress-strain responses. The effect of strain rate on the mechanical properties of epoxy nanocomposites was rarely studied, and the data is very limited [14]. Chen et al. [10] modified the split Hopkinson tension bar (SHTB) to determine the dynamic stress-strain responses (2.5×10^{-3}–1.2×10^{3} s^{-1}) of epoxy resin and polymethyl methacrylate (PMMA) in tension and compression loadings. They observed that the dynamic stress-strain behaviour under tension differs significantly from their dynamic compressive responses for both polymers. Also, they observed that the tensile specimens failed in a brittle manner during dynamic testing with a reduction in failure strain, in contrast to quasi-static testing where specimens failed in a ductile manner. Gilat et al. [11] studied the mechanical response of two different epoxy resins at different strain rates of 5×10^{-5}, 2, and 450–700 s^{-1} in tensile and shear loadings and found that the maximum stress is about the same in the intermediate and high strain rates tests and lower in the low rate tests. Evora et al. [12] employed a direct ultrasonication technique for producing polyester/TiO$_2$ nanocomposites and observed that the presence of TiO$_2$ nanoparticles had a significant effect on quasi-static fracture toughness and dynamic modulus and no marked effect on ultimate strength at high strain rate (2000 s^{-1}). Roland et al. [13] studied the effects of stain rates (0.06–573 s^{-1}) on tensile properties for an elastomeric polyurea using a drop weight test instrument and found an increase in stiffness and failure stress and a decrease in failure strain with increasing strain rate. Guo et al. [14] studied the quasi-static and dynamic compression behaviour of SiO$_2$/epoxy nanocomposites at different strain rates (10^{-4}–10^{4} s^{-1}) using a desktop split Hopkinson pressure bar (SHPB) and found that the nanocomposites are sensitive to loading rate and nanoparticle dispersion. Zebarjad et al. [15] studied the effect of nano-sized calcium carbonate and strain rate (0.1 s^{-1}) on the tensile properties of HDPE and concluded that the strain rate sensitivity of HDPE decreased with the addition of nanofillers. Xiao [16] carried out the dynamic tensile test at strain rates of 4 and 400 s^{-1} using a servo-hydraulic machine on four different polymers and validated the dynamic tensile tests by evaluating the condition of dynamic stress equilibrium using the SHPB criterion. Fu et al. [17] performed dynamic tensile tests (1750 s^{-1}) on polycarbonate using a split Hopkinson tension bar (SHTB) system and found that the tensile behaviour of polycarbonate is dependent on the strain rate. Raisch et al. [18] discussed the modification of the clamps of a servo-hydraulic tensile testing machine to achieve a strain rate of 670 s^{-1} and observed that both modulus and yield stress increased logarithmically with the strain rate. Cao et al. [19] studied the effect of strain rate on the tensile behaviour of polycarbonate over a wide range of strain rates (0.001 s^{-1}–1,700 s^{-1}) using the servo-hydraulic machine, a moderate strain testing apparatus and split Hopkinson tension bar and observed that the material is highly sensitive to strain rate. They also proposed a viscoelastic constitutive model to describe the stress-strain response of polycarbonate over a wide range of strain rates. Gurusideswar et al. [20] carried out dynamic tensile tests on epoxy/clay nanocomposites using a drop mass setup equipped with in-house fabricated specimen fixture assembly and high-speed camera. From the results, it is observed that the tensile strength increased by 41%, whereas the tensile modulus increased by 77% for the neat epoxy system when the strain rate increases from 0.008 to 445 s^{-1}. They observed a similar kind trend for epoxy/clay nanocomposites which showed a significant increase in elastic modulus and tensile strength and a clear reduction in failure strain under dynamic loading. From the microscopic observations of the fracture surface, it is noted that the fracture surface becomes rougher as the strain rate increases for epoxy/clay nanocomposites.

3.5   HIGH STRAIN RATE STUDIES OF FIBRE-REINFORCED POLYMERS

Composites are generally rate sensitive. Many researchers reported that glass/epoxy composites have shown an increase in tensile modulus and strength as the strain rate increases [32–34]. Several techniques were developed to study the rate sensitivity of composites for a wide range of strain rates such as conventional loading frame, servo-hydraulic testing machine, drop mass setup and split Hopkinson pressure bar (SHPB) technique. Daniel et al. [35] developed a method called the expanding ring technique for testing and characterizing graphite/epoxy composites at strain rates in the regime of 100–500 s^{-1}. Kawata et al. [36] introduced the SHPB technique for tensile testing of composites at high strain rates. Harding and Welsh [37] and Staab and Gilat [38] modified the standard SHPB for dynamic testing of composites. Hayes and Adams [39] studied the strain rate effects of glass/epoxy and graphite/epoxy composites using a pendulum impactor. Hamouda and Hashmi [40] discussed several techniques used for obtaining the mechanical behaviour of composite materials under impact loading at high strain rates. Jacob et al. [41] reviewed the strain rate dependence of mechanical properties of composite materials. Majzoobi et al. [42] achieved strain rates up to 10,000 s^{-1} using flying wedge apparatus for testing composites. In literature, many studies employed SHPB for high strain rates, which works in a high strain rate range and is expensive. It was found that the experimental techniques to determine tensile properties at medium-range strain rates of 1–100 s^{-1} are not well established [16]. The conventional servo-hydraulic machine is restricted to lower strain rates (< 10 s^{-1}), due to its inertial effects on the load cell and grips. The drop mass test setup is inexpensive and it can accommodate different specimen geometries and strain rates. Lifshitz [43] studied the tensile strength under dynamic loading of angle-ply glass/epoxy composites using an instrumented drop weight apparatus and failure stresses were found to be 20–30% higher than the static values; however, failure strain and modulus were the same for static and dynamic loadings. In this procedure, stress oscillations on the load cell signal obscured the details of the process and limited the maximum loading rate. Groves et al. [44] studied the high strain rate effects between 0.0001 s^{-1} and 2,660 s^{-1} for carbon fibre-reinforced polymer composites and found an unexpected exponential-like increase in strength and modulus beyond strain rates of 10 s^{-1} due to high-intensity stress waves. They also observed changes in fracture propagation patterns. The setup is made for compression loading and is limited to 10 s^{-1} due to high-intensity stress waves. In the current work, a drop mass tower with a specimen fixture is designed and fabricated for tensile loading and can accommodate strain rates up to 500 s^{-1}. Barre and Chotard [45] studied the dynamic response (10^{-1}–10^{1} s^{-1}) of glass phenolic/polyester composites and found that the modulus and strength tend to increase with strain rate. It was reported that the use of falling weight tup leads to vibration waves, which are superimposed on the load curve. Okoli [46] conducted tensile, shear and three-point bending tests to measure the energy absorbed by the failure of a material with an instrumented impact tester on glass/epoxy composites at increasing strain rates and found an increase in tensile, shear and flexural energy of 17, 5.9 and 8.5%, respectively, for the strain rates from 0.0106 s^{-1} to 2.72 s^{-1} per decade of increase in the log of strain rate. It was suggested that the inertial response of test systems obscure data obtained at high strain rates causing inaccuracies in the analysis. The instrumented impact tester is restricted to a velocity of 4 m/s for flexural impact. The strain rate effects on tensile properties in the range of low strain rates (0.00017–0.00830 s^{-1}) and medium strain rates (0.1–20 s^{-1}) were carried out using a low-speed tensile testing machine and servo-hydraulic machine. Pardo et al. [47] studied the tensile behaviour of glass/polyester composites at different strain rates and fibre orientations using the hydraulic testing machine and found an increase in tensile properties. A maximum of 20 m/s (approximately 100 s^{-1}) was reported. Shokrieh and Omidi [30] studied the dynamic response of glass/epoxy composites at different strain rates using a servo-hydraulic testing machine equipped with a special jig and fixture. They found an increase of 52% in tensile strength; 12% in tensile modulus; 10% in failure strain;

and 53% in absorbed failure energy as the strain rate increased from 0.001 to 100 s^{-1}. Though servo-hydraulic testing apparatus with a jig and fixture can accommodate medium strain rates, it restricts a strain rate to a maximum of 160 s^{-1} and is also expensive. Also, it is reported that servo-hydraulic equipment suffers from system ringing (noise) effects. The present study provides a cost-effective solution to tensile testing and can accommodate a wide range of intermediate strain rates. Brown et al. [48] studied the effects of strain rate (10^{-3}–10^{2} s^{-1}) on the tensile, compression and shear properties of glass/polypropylene composites using a drop weight tower and found an increase in tension and compression properties but a decrease in shear properties with increasing strain rate. The modified instrumental falling weight drop tower using a specially designed fixture was employed to get an intermediate strain rate of 70 s^{-1} on a dynamic tensile study of glass/polypropylene composites. The current study focuses on dynamic tensile studies on epoxy and glass/epoxy composites with nanofillers, which are required for aircraft and automobile structural applications. Li et al. [49] studied the compressive and tensile behaviour of carbon composites using a drop weight impact tester with a large impacting mass and achieved a constant strain rate by employing a shaper material. An affordable testing technique utilizing a drop weight impact tester was proposed for characterizing carbon composites at low strain rates. Perogamvros et al. [50] employed a modified drop tower to achieve medium strain rates (1–200 s^{-1}) and validated the experimental results using an explicit FE code. It was suggested that the drop tower apparatus with DIC optical devices is suitable for medium strain rate tensile testing. Gurusideswar et al. [51] carried out high strain rate studies (0.001 to 445 s^{-1}) on glass/epoxy/clay nanocomposites, using a drop mass test setup equipped with an in-house specimen fixture and a high-speed CMOS camera. From the quasi-static and dynamic experiment results, it is observed that the tensile behaviour of glass/epoxy/clay nanocomposites is dependent on the strain rate. They found that the tensile strength increased by 67%, whereas the increase in modulus is 106% for neat glass/epoxy composites when the strain rate increases from 0.001 to 445 s^{-1}. From the experimental results, they observed that there is a significant increase in elastic modulus and tensile strength and a clear reduction in failure strain under dynamic loading of glass/epoxy/clay nanocomposites. From the microscopic observation, it is observed that the fracture surface becomes rougher as the strain rate and clay loading increase.

3.6   CONCLUSION

Most of the literature focuses on thermoplastic materials. Hence, the current study is carried out to focus on thermoset polymers, which are high-end materials, and to study the effect of various nanofillers and strain rates on the mechanical properties of thermoset polymers. From the brief literature review, it is observed that the mechanical properties of the polymers and their nanocomposites are rate sensitive, even at low strain rates. Many researchers reported that fibre-reinforced polymers have shown an increase in mechanical properties as the strain rate increases. Several techniques were discussed to study the rate sensitivity of composites for a wide range of strain rates such as the conventional loading frame, servo-hydraulic testing machine, drop mass setup and split Hopkinson pressure bar (SHPB) techniques. From the comprehensive summary of published investigations, it is observed that studies on intermediate strain rate regions are rarely reported.

REFERENCES

1. S. Pavlidou and C. D. Papaspyrides, "A review on polymer-layered silicate nanocomposites," *Prog. Polym. Sci.*, vol. 33, no. 12, pp. 1119–1198, 2008.
2. S. C. Tjong, "Structural and mechanical properties of polymer nanocomposites," *Mater. Sci. Eng. R Reports*, vol. 53, no. 3–4, pp. 73–197, Aug. 2006.
3. A. A. Azeez, K. Y. Rhee, S. J. Park, and D. Hui, "Epoxy clay nanocomposites – Processing, properties and applications: A review," *Compos. Part B Eng.*, vol. 45, no. 1, pp. 308–320, 2013.
4. M. Kotal and A. K. Bhowmick, "Polymer nanocomposites from modified clays: Recent advances and challenges," *Prog. Polym. Sci.*, vol. 51, pp. 127–187, Dec. 2015.

5. A. Haque, M. Shamsuzzoha, F. Hussain, and D. Dean, "S2-Glass/Epoxy polymer nanocomposites: Manufacturing, structures, thermal and mechanical properties," *J. Compos. Mater.*, vol. 37, no. 20, pp. 1821–1837, Oct. 2003.

6. X. Kornmann, M. Rees, Y. Thomann, A. Necola, M. Barbezat, and R. Thomann, "Epoxy-layered silicate nanocomposites as matrix in glass fibre-reinforced composites," *Compos. Sci. Technol.*, vol. 65, no. 14, pp. 2259–2268, 2005.

7. R. Velmurugan and T. P. Mohan, "Epoxy-clay nanocomposites and hybrids: Synthesis and characterization," *J. Reinf. Plast. Compos.*, vol. 28, no. 1, pp. 17–37, Jul. 2008.

8. W. Daud, H. E. N. Bersee, S. J. Picken, and A. Beukers, "Layered silicates nanocomposite matrix for improved fibre reinforced composites properties," *Compos. Sci. Technol.*, vol. 69, no. 14, pp. 2285–2292, Nov. 2009.

9. K. Kanny and T. P. Mohan, "Resin infusion analysis of nanoclay filled glass fibre laminates," *Compos. Part B Eng.*, vol. 58, pp. 328–334, 2014.

10. W. Chen, F. Lu, and M. Cheng, "Tension and compression tests of two polymers under quasi-static and dynamic loading," *Polym. Test.*, vol. 21, no. 2, pp. 113–121, Jan. 2002.

11. A. Gilat, R. K. Goldberg, and G. D. Roberts, "Strain rate sensitivity of epoxy resin in tensile and shear loading," *NASA*, Mar. 2005.

12. V. Evora and A. Shukla, "Fabrication, characterization, and dynamic behaviour of polyester/TiO2 nanocomposites," *Mater. Sci. Eng. A*, vol. 361, no. 1–2, pp. 358–366, Nov. 2003.

13. C. M. Roland, J. N. Twigg, Y. Vu, and P. H. Mott, "High strain rate mechanical behaviour of polyurea," *Polymer (Guildf).*, vol. 48, pp. 574–578, 2007.

14. Y. Guo and Y. Li, "Quasi-static/dynamic response of SiO_2–epoxy nanocomposites," *Mater. Sci. Eng. A*, vol. 458, no. 1–2, pp. 330–335, Jun. 2007.

15. S. M. Zebarjad and S. A. Sajjadi, "On the strain rate sensitivity of HDPE/CaCO3 nanocomposites," *Mater. Sci. Eng. A*, vol. 475, no. 1–2, pp. 365–367, Feb. 2008.

16. X. Xiao, "Dynamic tensile testing of plastic materials," *Polym. Test.*, vol. 27, no. 2, pp. 164–178, Apr. 2008.

17. S. Fu, Y. Wang, and Y. Wang, "Tension testing of polycarbonate at high strain rates," *Polym. Test.*, vol. 28, no. 7, pp. 724–729, 2009.

18. S. R. Raisch and B. Möginger, "High rate tensile tests – Measuring equipment and evaluation," *Polym. Test.*, vol. 29, no. 2, pp. 265–272, 2010.

19. K. Cao, X. Ma, B. Zhang, Y. Wang, and Y. Wang, "Tensile behaviour of polycarbonate over a wide range of strain rates," *Mater. Sci. Eng. A*, vol. 527, no. 16–17, pp. 4056–4061, 2010.

20. S. Gurusideswar, R. Velmurugan, and N. K. Gupta, "High strain rate sensitivity of epoxy/clay nanocomposites using non-contact strain measurement," *Polymer*, vol. 86, pp. 197–207, Mar 2016.

21. S. P. Bao and S. C. Tjong, "Mechanical behaviours of polypropylene/carbon nanotube nanocomposites: The effects of loading rate and temperature," *Mater. Sci. Eng. A*, vol. 485, no. 1–2, pp. 508–516, Jun. 2008.

22. S.-H. Lim, Z.-Z. Yu, and Y.-W. Mai, "Effects of loading rate and temperature on tensile yielding and deformation mechanisms of nylon 6-based nanocomposites," *Compos. Sci. Technol.*, vol. 70, no. 13, pp. 1994–2002, Nov. 2010.

23. S. Gurusideswar and R. Velmurugan, "Strain rate sensitivity of glass/epoxy composites with nanofillers," *Mater. and Des.*, vol. 60, pp. 468–478, Aug 2014.

24. O. I. Okoli and G. Smith, "High strain rate characterization of a glass/epoxy composite," *J. Composites Technol. Res.*, vol. 22, no. 1, pp. 3–11, 2000.

25. O. I. Okoli, "An approach for obtaining the young's modulus in woven glass/epoxy reinforced composites," *J. Reinf. Plast. Compos.*, vol. 20, no. 14–15, pp. 1358–1368, Sep. 2001.

26. L. T. Berezhnyts'kyi and V. E. Panasyuk, "Effect of the strain rate and temperature on the physicomechanical properties of glass-fibre-reinforced plastics," *Mater. Sci.*, vol. 37, no. 1, pp. 53–58, 2001.

27. A. Kallimanis and E. Kontou, "Tensile strain-rate response of polymeric fibre composites," *Polym. Compos.*, vol. 26, no. 5, pp. 572–579, Oct. 2005.

28. F. Fereshteh-Saniee, G. H. Majzoobi, and M. Bahrami, "An experimental study on the behaviour of glass–epoxy composite at low strain rates," *J. Mater. Process. Technol.*, vol. 162–163, pp. 39–45, May 2005.

29. L. Sun, Y. Jia, F. Ma, S. Sun, and C. C. Han, "Mechanical behaviour and failure mode of unidirectional fibre composites at low strain rate level," *J. Compos. Mater.*, vol. 43, no. 22, pp. 2623–2637, 2009.

30. M. M. Shokrieh and M. J. Omidi, "Tension behaviour of unidirectional glass/epoxy composites under different strain rates," *Compos. Struct.*, vol. 88, no. 4, pp. 595–601, May 2009.

31. S. Shadlou, B. Ahmadi-Moghadam, and F. Taheri, "The effect of strain-rate on the tensile and compressive behaviour of graphene reinforced epoxy/nanocomposites," *Mater. Des.*, vol. 59, pp. 439–447, Mar. 2014.

32. J. M. Lifshitz and A. Rotem, "Longitudinal tensile failure of unidirectional fibrous composites," *J. Mater. Sci.*, vol. 7, no. 8, pp. 861–869, 1973.

33. A. E. Armenàkas and C. A. Sciammarella, "Response of glass-fibre-reinforced epoxy specimens to high rates of tensile loading," *Exp. Mech.*, vol. 13, no. 10, pp. 433–440, Oct. 1973.

34. R. G. Davies and C. L. Magee, "The effect of strain-rate upon the tensile deformation of materials," *J. Eng. Mater. Technol.*, vol. 97, no. 2, p. 151, Apr. 1975.

35. I. M. Daniel, R. H. LaBedz, and T. Liber, "New method for testing composites at very high strain rates," *Exp. Mech.*, vol. 21, no. 2, pp. 71–77, Feb. 1981.

36. K. Kawata, A. Hondo, S. Hashimoto, N. Takeda, and H. L. Chung, "Dynamic behaviour anaysis of composite materials," *Proc. Japan-US Conference on Compos. Mater.*, pp. 2–11, 1981.

37. J. Harding and L. M. Welsh, "A tensile testing technique for fibre-reinforced composites at impact rates of strain," *J. Mater. Sci.*, vol. 18, no. 6, pp. 1810–1826, Jun. 1983.

38. G. H. Staab and A. Gilat, "High strain rate response of angle-ply glass/epoxy laminates," *J. Compos. Mater.*, vol. 29, no. 10, pp. 1308–1320, Jul. 1995.

39. S. V. Hayes and D. F. Adams, "Rate sensitive tensile impact properties of fully and partially loaded unidirectional composites," *Journal of Testing and Evaluation*, vol. 10, no. 2, pp. 61–68, 1982.

40. A. M. Hamouda and M. S. Hashmi, "Testing of composite materials at high rates of strain: Advances and challenges," *J. Mater. Process. Technol.*, vol. 77, no. 1–3, pp. 327–336, May 1998.

41. G. C. Jacob, J. M. Starbuck, J. F. Fellers, S. Simunovic, and R. G. Boeman, "Strain rate effects on the mechanical properties of polymer composite materials," *J. Appl. Polym. Sci.*, vol. 94, no. 1, pp. 296–301, Sep. 2004.

42. G. H. Majzoobi, F. F. Saniee, and M. Bahrami, "A tensile impact apparatus for characterization of fibrous composites at high strain rates," *J. Mater. Process. Technol.*, vol. 162–163, pp. 76–82, May 2005.

43. J. M. Lifshitz, "Impact strength of angle ply fibre reinforced materials," *J. Compos. Mater.*, vol. 10, no. 1, pp. 92–101, Jan. 1976.

44. R. Groves and H. S. Rate, "High strain rate effects for composite materials," *Compos. Mater. Test. Des.*, vol. 11, pp. 162–176, 2011.

45. S. Barre and T. Chotard, "Comparative study of strain rate effects on mechanical properties of glass fibre-reinforced thermoset matrix composite," *Composites Part A: Applied Science and Manufacturing* vol. 27, no. 12, pp. 1169–1181, 1996.

46. O. I. Okoli, "The effects of strain rate and failure modes on the failure energy of fibre reinforced composites," *Compos. Struct.*, vol. 54, no. 2–3, pp. 299–303, Nov. 2001.

47. S. Pardo, D. Baptiste, F. Décobert, J. Fitoussi, and R. Joannic, "Tensile dynamic behaviour of a quasi-unidirectonal E-glass/polyester composite," *Compos. Sci. Technol.*, vol. 62, no. 4, pp. 579–584, 2002.

48. K. A. Brown, R. Brooks, and N. A. Warrior, "The static and high strain rate behaviour of a commingled E-glass/polypropylene woven fabric composite," *Compos. Sci. Technol.*, vol. 70, no. 2, pp. 272–283, 2010.

49. G. Li and D. Liu, "Low strain rate testing based on drop weight impact tester," *Exp. Tech.*, vol. 39, no. 5, pp. 30–35, 2015.

50. N. Perogamvros, T. Mitropoulos, and G. Lampeas, "Drop tower adaptation for medium strain rate tensile testing," *Exp. Mech.*, vol. 56, no. 3, pp. 419–436, 2016.

51. S. Gurusideswar, R. Velmurugan, and N. K. Gupta, "Study of rate dependent behavior of glass/epoxy composites with nanofillers using non-contact strain measurement," *International Journal of Impact Engineering*, vol. 110, pp. 324–337, Dec 2017.

4 High Strain Rate Testing Using Drop Mass Tower and Non-Contact Strain Measurement Techniques

S. Gurusideswar and R. Velmurugan

4.1 INTRODUCTION

In general, engineering materials are subjected to high-velocity dynamic loadings, which require prior knowledge of dynamic mechanical properties to prevent catastrophic failure at high loading rates. Hence, it is essential to understand the behaviour of these materials at high strain rates. From the literature studies [1–7], it is found that several experimental techniques are being employed to determine the behaviour of materials at high strain rates. The Kolsky bar, also known as the split Hopkinson pressure bar (SHPB) [1], is used to characterize the mechanical behaviour of materials under high strain rate loading. The setup consists of two long pressure bars: incident and transmission and a projectile made of high-strength material. The specimen is sandwiched between the bars and a projectile is impacted at one end of the input bar using a gas gun. Upon impact, a stress wave is generated in the incident bar and travels towards the specimen. After loading the specimen, a part of the wave is transmitted to the transmission bar and the remaining part is reflected to the incident bar depending on the impedance mismatch at the specimen-bar interface. Split Hopkinson apparatus can cover strain rates in the region of 100–10,000 s^{-1}. However, it cannot be used in the medium strain rate range (1–100 s^{-1}). Servo-hydraulic machines can cover medium strain rates; however, they are not affordable due to their high cost and also they may lead to a system ringing phenomenon. From the literature studies [8], it is found that the experimental techniques to achieve medium strain rates in the range of 1–100 s^{-1} are not well established and also to the best of our knowledge, literature regarding medium strain rate studies is very limited. In order to overcome these challenges, a drop mass test setup is proposed and developed for this work, which fills the gap between conventional testing machines (< 10 s^{-1}) and the split Hopkinson pressure bar technique (> 1,000 s^{-1}). The proposed setup is a very simple and direct conversion of impact to tension of specimen. By suitably modifying the same setup, it can be used for compression testing at high strain rates. Strain rates can be varied by varying the drop height of mass and by varying pressure in gas guns. The current work emphasizes the design and development of a drop mass test setup and DIC system for achieving medium to high strain rates. In this chapter, the technical specifications of the drop mass tower, load sensor and DIC setup are discussed in detail. The results of several trial tests using mild steel and aluminium materials are presented in this section to know the repeatability of the load sensor, and it is found that the DAQ system works well for the proposed high strain rate studies of epoxy and glass/epoxy and its clay nanocomposites. The significance of the DIC technique and its applications are discussed. The effect of DIC parameters, such as focal length, light source, speckle pattern and subset size for obtaining accurate results are studied and optimized to the current application for obtaining strains during high strain rate testing.

DOI: 10.1201/9781003352358-4

4.2   HIGH STRAIN RATE TEST SETUP

In general, six different types of testing systems are employed to study the strain rate effects of a material. They are

- Conventional screw drive load frame (< 0.1 s^{-1})
- Servo-hydraulic system (0.1–100 s^{-1})
- Drop mass test setup (100–$1,000$ s^{-1})
- Split Hopkinson pressure bar (100–10^4 s^{-1})
- Expanded ring (10^4 s^{-1})
- Flyer plate ($> 10^5$ s^{-1})

A drop mass test apparatus is chosen to cover a medium strain rate range for the current work. The drop mass test setup comprises the following equipment:

- Drop mass tower
- Specimen fixture
- Load sensor
- Data acquisition system
- High-speed camera
- Digital image correlation system
- Computer system

The drop mass test setup has the following merits:

- Strain rates range from 10 to 1,000 s^{-1}
- Inexpensive
- Allows different specimen geometries
- Allows easy variation of strain rate

4.2.1   Drop Mass Tower

The drop mass tower consists of the following components as shown in Figure 4.1.

- Base plate
- Guide rods
- Elevator
- Control unit

The guide rods, made up of induction-hardened chrome alloy (CK 45) with a tensile strength of 630 MPa, are used to guide the elevator or impactor unit. The impactor unit, which holds the electromagnet and the drop mass, has a linear bearing assembly to enable smooth sliding motion along the two guide rods. The entire drop mass unit can be lifted by an electric motor to a required height. The drop tower is designed for low-energy and low-velocity applications and it can be used to achieve a maximum velocity of 8.5 m/s using 25 kg drop mass, which is computed from where the mass can be dropped.

4.2.2   Specimen Fixture

Figure 4.2 shows a typical photograph of a specimen fixture. It has two grips, where the specimen is clamped at each end between steel grips. The top grip is directly bolted through the load cell to the

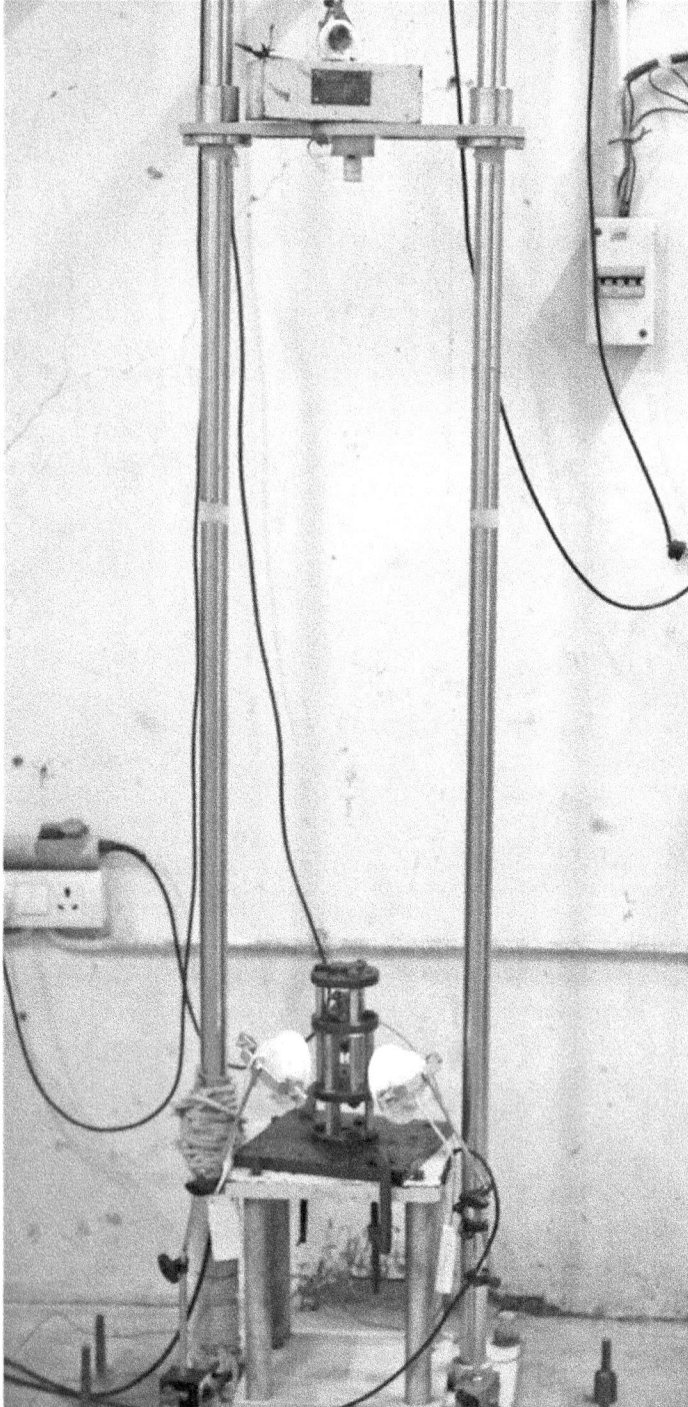

FIGURE 4.1   Photograph of drop mass tower with specimen fixture. Source: https://s100.copyright.com/
MyAccount/viewLicenseDetails?ref=e44b63f2-59d7-49bb-8cb4-1a963029ef77.

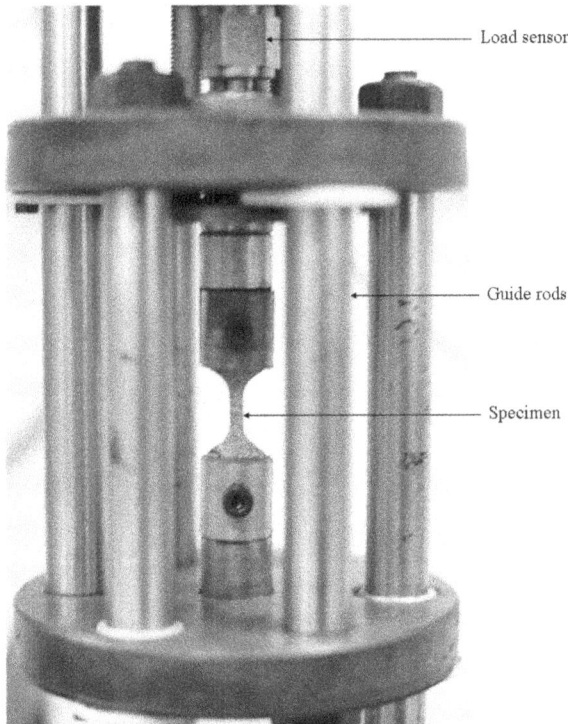

FIGURE 4.2 Photograph of in-house designed specimen fixture assembly. Source: https://s100.copyright .com/MyAccount/viewLicenseDetails?ref=e44b63f2-59d7-49bb-8cb4-1a963029ef77.

fixed carriage. A moving carriage is supported by the lower grip and is guided by three steel rods. The drop tower striker imparts a load on the moving carriage, which loads the specimen in tension through the lower grip as it travels downward.

4.2.3 DATA ACQUISITION SYSTEM

To get reliable stress data, a load sensor is required to acquire load data in a high-speed environment. In the current work, an integrated circuit piezoelectric (ICP) type load sensor was employed and it is noted that quartz-type load sensors are recommended for dynamic force applications. However, it cannot be used for static applications. Figure 4.3 shows a representative model of a 208C04 load sensor.

When the force is applied to the sensor, the quartz crystals generate an electrostatic charge proportional to the applied load. Then the output is collected through the electrodes, which are sandwiched between the crystals and routed to an external charge amplifier or converted to a low-impedance voltage signal within the sensor.

ICP sensors are powered by a separate constant current source, but many data acquisition systems now provide constant current power to the sensors. ICP sensors have an inbuilt microelectronic amplifier (MOSFET) to convert the high-impedance charge output into a low-impedance voltage signal for acquiring the data. Table 4.1 illustrates the technical specification of the load sensor. A maximum load of 2,224 N can be achieved with the given load sensor.

Several trial runs were carried out to know the repeatability of the load sensor measurements. The trial tests were carried out using a plastic hammer (impact loading) suggested an average load value of 1,700 N. The moving carriage of the specimen fixture was dropped freely and made to hit the load sensor to know its effect. The load values were consistent and the average load value was

FIGURE 4.3 A representative model of a load sensor.

TABLE 4.1
Specifications of ICP Load Sensor

Sensitivity	1,124 mV/kN
Measurement range in compression loading	4.448 kN
Measurement range in tension loading	2.224 kN
Maximum static force in compression	26.69 kN
Maximum static force in tension	2.224 kN
Temperature range	−54 to 121° C
Excitation voltage	20 to 30 VDC
Constant current excitation	2 to 20 mA
Output impedance	> 100 Ω
Output bias voltage	8 to 14 VDC
Output polarity in compression loading	Positive
Electrical connector	10–32 coaxial jack
Mounting torque	226 N-cm

Source: [9].

3,000 N. A mass of 0.5 kg was dropped from a height of 0.5 m on the specimen fixture. The load values were consistent (5,750, 5,500 and 5,600 N) and an average load of 5,620 N was observed. A mass of 1 kg was dropped from 0.5 m height on the specimen fixture and a load value of 9,500 N was observed. It is noted that the load values were saturated and the capacity of the load sensor is exceeded. A dummy specimen of mild steel was tested at different heights with two different drop masses of 0.5 and 1 kg and the results are tabulated in Table 4.2. A dummy specimen of aluminium was also tested at different heights and the results are tabulated in Table 4.3.

4.2.4 HIGH-SPEED CAMERA

A Phantom® V611 was employed to capture images at high speed. It has a widescreen 1,280 × 800 CMOS sensor, which enables it to capture moving targets. A maximum of 6,242 frames per second can be achieved at its full resolution, whereas a maximum of 680,000 fps can be achieved at a reduced resolution (128 × 8 pixels), and also it is possible to achieve 1,000,000 fps using the "fast option."

TABLE 4.2
Load Data of Mild Steel at Different Drop Mass Heights

Masskg	Height m	Rate s⁻¹	Trail 1 N	Trail 2 N	Trail 3 N	Trail 4 N	Trial 5 N	Trial 6 N	Average N
0.5	0.5	348	1,990.9	2,507.1	2,298.1	2,503.4	3,895.4	–	2,257.2
	1	443	3,441.3	3,219.5	3,147.3	2,869.5	3,354.2	3,396.3	2,838.7
	1.5	603	3,688.1	3,695.4	3,895.4	3,613.1	3,359.8	3,570.7	3,203.6
1	0.5	348	3,085.4	4,263.8	4,035.9	3,738.9	3,836.8	4,092.6	3,842.2
	1	443	5,812.2	4,560.5	4,841.3	5,287.8	4,811.4	4,497.1	4,968.4
	1.5	603	3,917.2	3,458.9	–	–	–	–	–

TABLE 4.3
Load Data of Aluminium Specimens with Different Widths at 1 m Drop Mass Height

Width	Sample	First peak load N	Second peak load N	Strength MPa
3 mm	Sample 1	400	2,900	130
	Sample 2	400	3,000	130
	Sample 3	450	3,000	150
5 mm	Sample 1	750	-	150
	Sample 2	1000	-	-
	Sample 3	750	2,200	150
	Sample 3	950	150	190

Table 4.4 shows the technical specifications of the V611 high-speed camera [10] and Table 4.5 illustrates the maximum frames per second, which can be gained from the high-speed camera at different resolutions.

4.3 DIGITAL IMAGE CORRELATION (DIC) TECHNIQUE

In literature [11–14], the digital image correlation technique is denoted using different names such as non-contact strain measurement technique, digital speckle correlation method (DSCM), computer-aided speckle interferometry (CASI), electronic speckle photography (ESP) and texture correlation. The specific features of this technique are highlighted as follows:

- Measures material properties both in the elastic and plastic range
- Determines localized necking strain (it is practically difficult to find the strain at the position of necking using traditional devices)
- Full-field (strain distribution over the entire surface)
- Non-contact type (applicable for thermal environment)
- Good accuracy and reliability

DIC is a non-interferometric optical technique and it is considered a powerful and flexible tool to get reliable strain data. It is based on digital image processing and numerical computing and it was first developed by a group of researchers at the University of South Carolina in the 1980s. Sutton et al. [12] summarized a wide range of applications of the 2D DIC technique.

- Measurement of velocity fields in 2D seeded flows and rigid-body mechanics
- Measurement of strains in thin films

TABLE 4.4

Specifications of High-Speed Camera V611

Sensor specifications	CMOS sensor (monochrome)
	20 μm pixel size
	25.6 mm × 16.0 mm (sensor window)
	8- and 12-bit depth
Exposure	1 μs minimum exposure (standard), 300 ns optional
	Global electronic shutter
	Extreme dynamic range (EDR)
	Auto exposure
Triggering	Programmable trigger location
	Image-based auto-trigger
	Trigger form software
	Hardware BNC
Power	100–240 V AC, 220 W
Dimensions	29.2 × 14 × 12.7 cm (L × W × H)

Source: [10].

TABLE 4.5

Frame Rates at Various Resolutions

Resolution	Frames per second
1,280 × 800	6,242
1,280 × 720	6,933
512 ×512	20,978
256 × 256	66,997
128 × 128	183,250
128 × 64	330,469
128 × 8	1,000,000

Source: [10].

- Measurement of strain fields near stationary and growing crack tips
- Measurement of crack-tip opening displacement during crack growth
- Deformation studies in concrete during compressive loading
- Measurement of deformation from microscale to nanoscale (2D DIC coupled with LSCM, SEM, AFM and STM)

To measure deformation on specimens with a curved surface, 3D DIC was developed, which works on the principle of binocular stereovision, and also digital volume correlation was developed to study the internal deformation of solid objects [11].

4.3.1 Basic Principle of DIC Technique

At first, the area of interest, AOI or the calculation area should be specified in the reference image or undeformed image or the image before loading. The basic principle of 2D DIC for the measurement of displacement (strain) involves tracking the movement of points between the two recorded images before and after loading. To achieve this, a virtual grid of subsets of a selected size and shape consisting of pixel grey value distributions is superimposed on the artificially sprayed surface pattern. The purpose of choosing a subset rather than a pixel is to have a wider range of grey level intensities, which helps to distinguish it from other subsets and to identify it in a unique manner. The processing involves the calculation of the average grey scale intensity over the subset of the reference image and deformed image (after loading) and correlating them.

In general, the 2D DIC method comprises the following principles:

- Preparation of the specimen (speckle pattern)
- Recording the images of the specimen before and during loading till failure
- Post-processing the recorded images using a computer program (image correlation algorithm)

The speckle pattern is a random grey intensity distribution, which deforms together with the specimen surface and provides information about the deformation in a specimen.

Some of the points that are to be considered while setting up the experiment setup include:

- The specimen must be flat and kept parallel to the CCD sensor
- Out-of-plane motion should be avoided for accurate determination of in-plane displacement
- The image should not have geometric distortion, which may lead to errors in the calculation of displacements

The DIC test setup comprises the following:

- Test specimen with speckle pattern
- CCD camera
- Illumination light source (LED)
- Computer system to record images
- DIC software for post-processing

4.3.2 DIC PARAMETERS

It is noted that the strain measurement accuracy of 2D DIC relies more on the quality of imaging, perfection of loading systems and the selection of post-processing parameters [11]. Many parameters are to be considered for obtaining accurate DIC results. Some of the parameters include:

- Image resolution (pixels)
- Sample dimension
- Distance between sample and camera
- Focal length of the lens
- Light source (uniformity)
- Speckle pattern (size and density)
- Type of algorithm
- Subset size and step size
- Grey level interpolation, etc.

Image resolution should be as high as possible because the quality of the image is based on the resolution. However, it is noted that higher resolution consumes more memory.

The focal length of the lens highly depends upon the specimen dimension and testing condition. The distance between the specimen and the camera also depends upon the focal length of the lens (a higher focal length provides a longer distance between the specimen and the camera).

Two different cameras were used in the current research work. A low-speed CCD camera with 15 frames per second (fps) was employed for quasi-static testing and a high-speed CMOS camera with a maximum fps of 600,000 was employed for high strain rate tensile testing.

It is reported in the literature [13] a laser speckle pattern can be used for surface displacement measurement, which can be produced by illuminating the optically rough surface with the laser beam. Unfortunately, this creates a decorrelation effect when the specimen is subjected to excessive strain as well as out-of-plane displacement. Hence, a white-light speckle pattern technique using natural light illumination is employed in the current research work.

Initially, compact fluorescent lamps (CFLs) were employed to illuminate the specimen. As the CFLs work on alternating current, flickering occurs and interferes with image recording at high frame rates during testing. Hence, the CFLs were replaced by 30 W flood light LED lamps. Since these LED lamps were illuminating in a divergent manner and could not be used to focus the specimen, it led to poor contrast and image quality. Finally, 12 W concealed LED lamps were employed to illuminate the specimen. The concealed lamps were placed very close to the specimen. Since the distance between the specimen and LED lamps was reduced, an aberration in the image was observed due to a large reflection. The scatter was considerably reduced by employing filter papers.

A focal length of 180 mm was chosen using TAMRON AF 75–300 mm f4 LD macro lens and the camera was positioned at 15 cm away from the specimen surface (minimum distance between the lens and specimen). The recorded image was not clear, and also large scatter in data was observed due to the larger distance between the lens and the specimen. Figure 4.4 shows the captured image using a 180 mm focal length lens. Hence, a 50 mm focal length lens was employed and the camera was kept closer to the specimen surface. Figure 4.5 shows the strain history for a 180 mm focal length lens.

FIGURE 4.4 Magnified image captured from a high-speed camera using a 180 mm focal length lens.

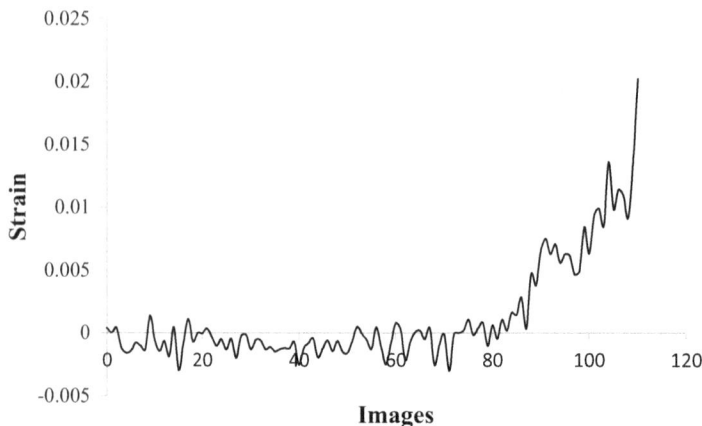

FIGURE 4.5 Strain history of an epoxy specimen using a 180 mm focal length lens.

The strain measurement of DIC is strongly related to the quality of the speckle pattern on the specimen surface. The speckle pattern can be made by

- Manual dots using a pen marker
- Spraying black and/or white paints

A speckle pattern comprises a considerable quantity of black speckles with different shapes and sizes on white background. It is noted that the effectiveness of the speckle pattern can be explained by the number of pixels per black speckle. A good speckle pattern must have i) small black speckles with ten pixels, ii) medium black speckles with 20 pixels and iii) large black speckles with 30 pixels. Figure 4.6 shows the quality of different speckle patterns.

The speckle pattern (black on white) made by spray paint was very fine and was not successful due to the lower resolution of the high-speed camera (Phantom V611) 1 MP. Hence, the speckle pattern was made manually using a pen marker on the white-painted surface of the specimen. The modified speckle pattern resulted in a better contrast required for the high-speed camera. Figure 4.7 shows the speckle pattern made from spray paint and Figure 4.8 shows the speckle pattern using a pen marker. Koerber et al. [15] used a fine speckle pattern (black on white) for quasi-static test specimens and coarse patterns for dynamic test specimens by considering the different image resolutions of the respective cameras for their research work on high strain rate characterization.

From the literature review, it is understood that the most important parameter affecting the accuracy of the result is the subset size [11]. The subset size indicates the minimum local gauge length being used to track the displacements between the reference and deformed subsets. So, care should be taken while choosing the subset size, which may vary from several pixels to even more than a hundred pixels for the analysis. In general, a large subset size provides distinct intensity patterns and large approximation, whereas a small subset size can be approximated in a better way by first-order or second-order subset shape functions. Since there is no systematic procedure, the selection of subset size depends upon operator experience and judgement.

The reliability of results and the selection of subset size were justified by the VIC 2D software confidence level (error value). It is noted that the quality of the speckle pattern plays a major role in the selection of subset size [11]. Subset size should be chosen based on the speckle pattern generated to get reliable and satisfactory data.

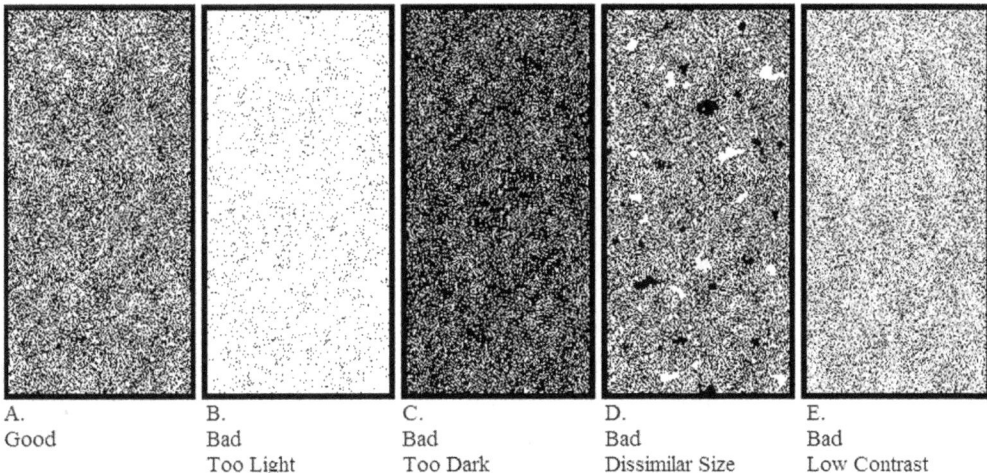

A.   B.   C.   D.   E.
Good   Bad   Bad   Bad   Bad
            Too Light   Too Dark   Dissimilar Size   Low Contrast

FIGURE 4.6   Different qualities of speckle pattern.

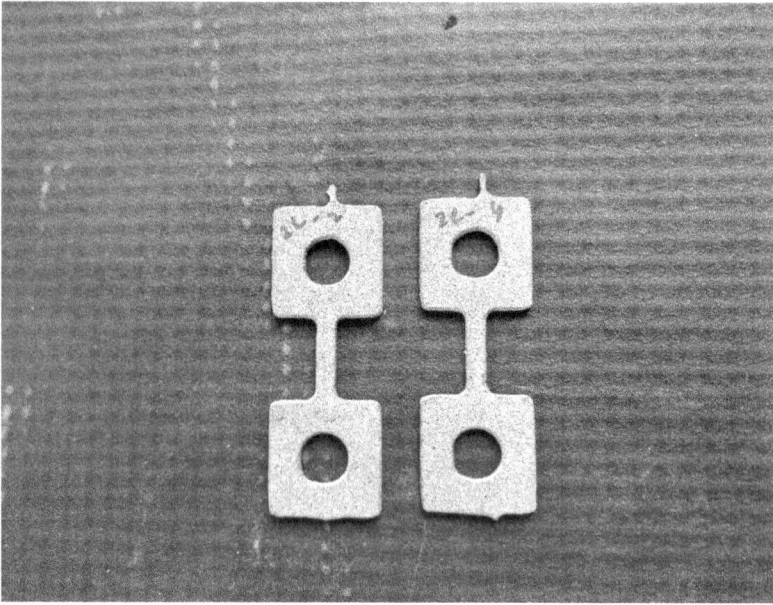

FIGURE 4.7 Specimen with spray paint speckle pattern.

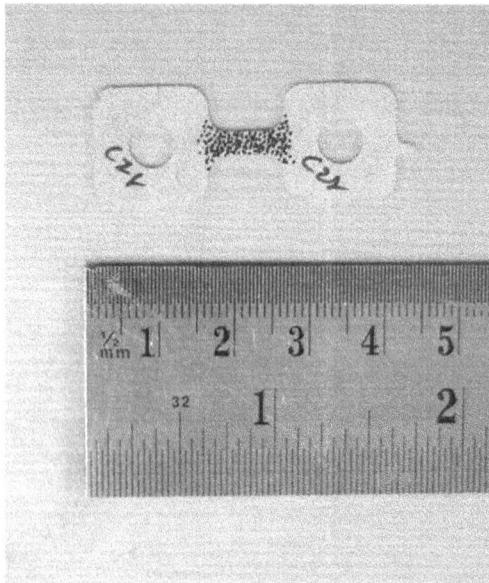

FIGURE 4.8 Specimen with manual speckle pattern. Source: https://s100.copyright.com/MyAccount/view-LicenseDetails?ref=e44b63f2-59d7-49bb-8cb4-1a963029ef77.

The selection of a subset size of 9 for the manual speckle pattern gave a high confidence level of 0.045. It is found that the increase in subset size results in a decrease in confidence level. At one particular value of subset size, the confidence level reached convergence (Figure 4.9). The optimal subset size for the selected speckle pattern was in the range between 19 and 25, respectively. Jerabek et al. [16] used quadratic subsets of 25 pixels and a step of 19 pixels for the determination of strain on polypropylene material.

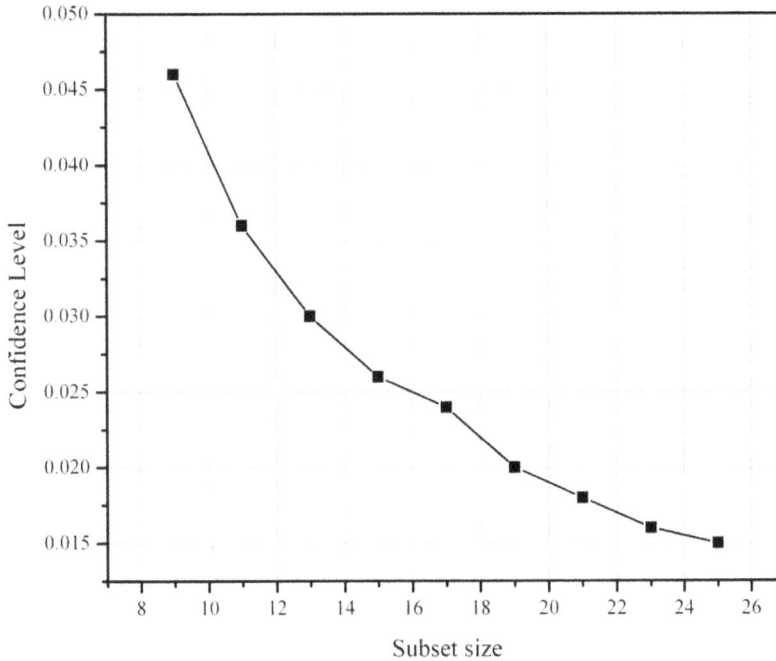

FIGURE 4.9   Effect of subset size for a given speckle pattern.

From trial runs, the following DIC parameters were optimized and chosen for high strain rate experiments to get accurate and reliable data using the 2D DIC technique.

- Guarantee parallelism between the CCD camera and the specimen surface
- 12 W concealed LED lamps
- 50 mm focal length lens
- 15 cm distance between the camera and the specimen surface
- Manual speckle pattern using pen marker to obtain high contrast
- 128 × 128 pixels to achieve 100,000 fps
- 19–25 subset size range

4.4   HIGH STRAIN RATE EXPERIMENTAL RESULTS

Dynamic tensile studies were performed in a drop mass tower using an in-house fabricated fixture assembly and the tensile specimen was clamped using steel grips. The upper grip which houses the load cell was fastened to the fixed carriage and the moving carriage was supported by the lower grip using guide rods. As the drop mass imparts a load on the moving carriage, the lower grip traverses downward, which causes the specimen to be loaded in tension mode. The load was measured using a PCB 208C04 integrated circuit piezoelectric (ICP) sensor. A PHANTOM V611 high-speed camera with 1 MP resolution, coupled with a SIGMA 50 mm f/2.8 EX DG macro lens, was used for the dynamic experiments. It was positioned 15 cm away from the specimen surface. It is noted that for the PHANTOM V611 camera, the maximum frame rate depends on the area of interest. Due to the smaller area of interest, the image size could be reduced to a resolution of 128×128 pixel^2, which resulted in a higher frame rate of 100,000 fps. Two standard 12 W LED lamps on either side of the camera guaranteed an even illumination of the specimen surface. The acquisition rate of the camera was set to 100,000 frames per second (fps) with a shutter speed of 9.51 µs and an aperture of f/2.8.

Tensile studies were carried out on the drop mass tower from heights of 0.5, 0.75 and 1 m, producing theoretical strain rates of 315, 385 and 445 s^{-1}, respectively. A drop mass of 0.5 kg was used. Figure 4.10 shows the variation of nominal strain rate versus drop mass height.

The actual strain rate can be determined from strain histories. It is observed that the initial part of the strain–time curve is not truly indicative of the effective strain rate experienced by the specimen, and hence, actual strain rates were thus determined from the gradient of the strain–time curves [17, 18]. The average actual strain rates during the experiments at heights of 0.5, 0.75 and 1 m (3.1, 3.8 and 4.4 ms^{-1}) were 25, 41 and 53 s^{-1}, respectively. However, the actual strain rates are used in practice for the purpose of analysis. A significant increase in strain rate at about 75 µs could be due to local failure as the duration of impact is very short. An average decrease in strain to failure of 12%, 20% and 27% at 315, 385 and 415 s^{-1}, respectively is observed for epoxy/clay nanocomposites with respect to quasi-static loading.

The strain rate can also be calculated by the application of conservation of energy to a drop mass for predicting its impact velocity as given in Equation (4.1).

$$mgh = \frac{1}{2}mv^2 \tag{4.1}$$

Cancelling mass on both sides; final velocity doesn't depend upon the mass.

$$v = \sqrt{2gh} \tag{4.2}$$

where,

g = acceleration due to gravity and

h = drop mass height

In general, the nominal or theoretical strain rate can be calculated by dividing the gauge length of the specimen by the cross-head speed (here, it is drop mass velocity).

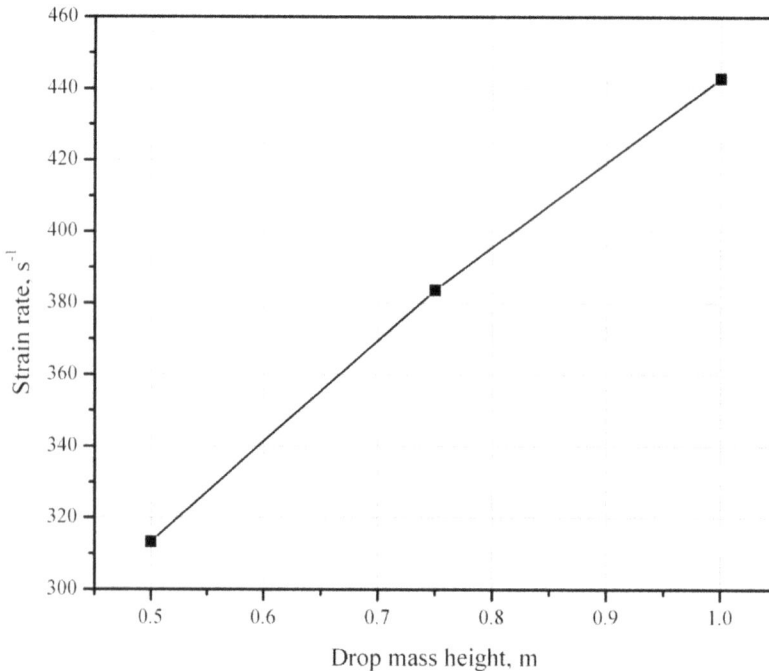

FIGURE 4.10 Variation of strain rate versus drop mass height.

$$\dot{\varepsilon} = \frac{v}{Gauge\,length} \tag{4.3}$$

In the present study, the nominal strain rates (315, 385 and 445 s^{-1}) are considered, which is only for understanding purposes [19]. But in a practical case, there would be a reduction in strain rate, which is calculated from the gradient of the strain-time history. Also, it is worth noting that the deformation velocity will reduce during the test, resulting in a decrease in strain rate. The actual strain rate (25 s^{-1}) calculated from the strain history varies 12 times from the nominal strain rate/initial velocity (315 s^{-1}). A detailed study based on this instantaneous strain rate can be considered as a direction for future work.

Figure 4.11a and b show load and strain histories for an epoxy specimen at dynamic loading (385 s^{-1}). The specimen failed at 550 N and the corresponding strain is 1.2% at a drop mass height of 0.75 m, which corresponds to the strain rate of 385 s^{-1}.

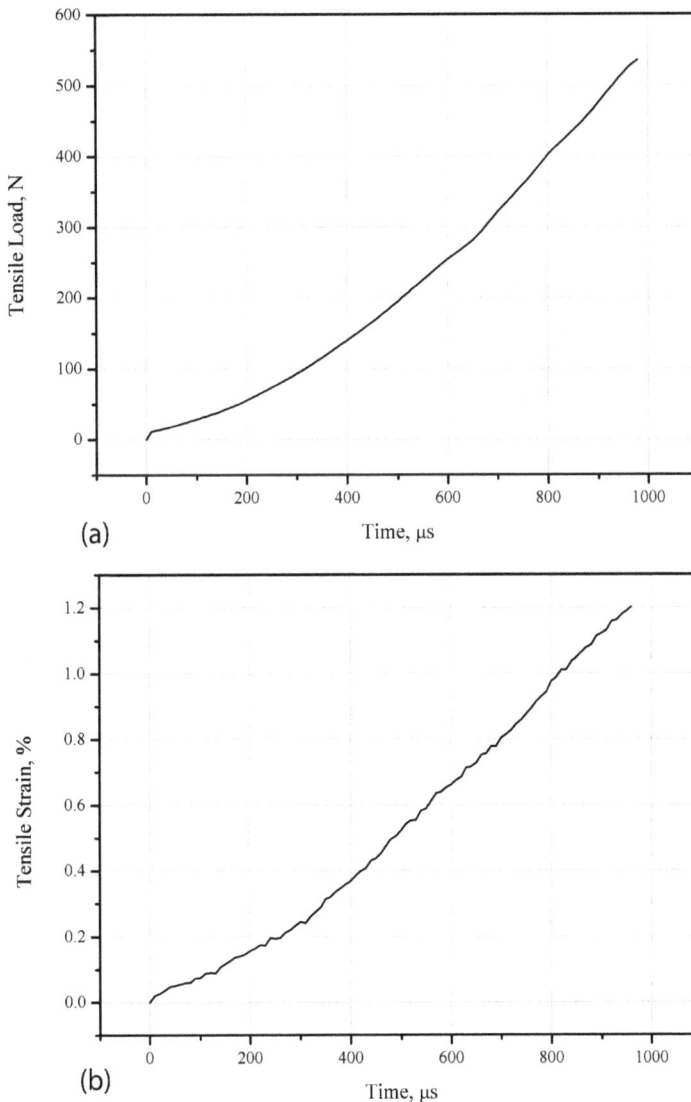

FIGURE 4.11 a) Load and b) strain histories of an epoxy specimen at a strain rate of 385 s^{-1}. Source: https://s100.copyright.com/MyAccount/viewLicenseDetails?ref=e44b63f2-59d7-49bb-8cb4-1a963029ef77.

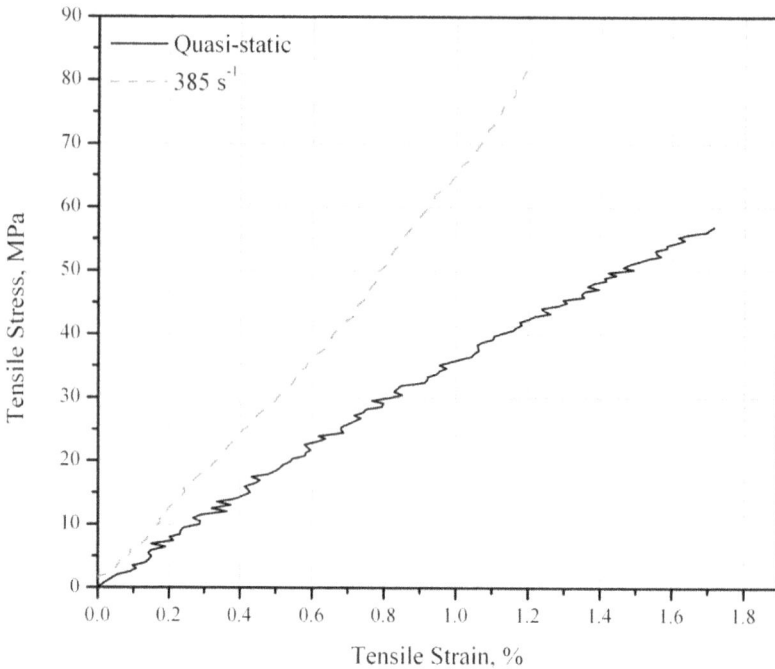

FIGURE 4.12   Stress-strain responses of an epoxy specimen. Source: https://s100.copyright.com/MyAccount /viewLicenseDetails?ref=e44b63f2-59d7-49bb-8cb4-1a963029ef77.

Figure 4.12 shows stress-strain responses for an epoxy specimen at quasi-static and dynamic loadings. At 0.75 m drop mass height corresponding to a strain rate of 385 s^{-1}, the stress-strain behaviour is different from quasi-static results. It showed increasing slope and strength with the increase in strain rate.

4.5   CONCLUSIONS

Dynamic tensile studies were conducted on epoxy specimens using a drop mass setup equipped with an in-house fabricated specimen fixture assembly and a high-speed camera. Data obtained from this method filled a gap between conventional testing machines and SHPB measurements. A non-contact strain measurement technique (DIC) using the high-speed camera was employed to capture the full-field strain measurement in the dynamic environment. The effects of the strain rate on the tensile behaviour of epoxy were investigated over a wide range of strain rates. Material behaviour at high strain rates obtained through experimental methods can be used for the validation of proposed material models and to focus on the development of test standards for the determination of dynamic mechanical properties.

REFERENCES

1. H. Kolsky, "An investigation of the mechanical properties of materials at very high rates of loading," *Proc. Phys. Soc., B*, vol. 62, pp. 676–700, 1949.
2. J. Harding, E. O. Wood, and J. D. Campbell, "Tensile testing of materials at impact rates of strain," *J Mech. Eng. Sci.*, vol. 2, pp. 88–96, 1960.
3. T. Nicholas, "Tensile testing of materials at high rates of strain," *Exp. Mech.*, vol. 21, pp. 177–188, 1981.
4. K. Ogawa, "Impact-tension compression test by using a split-hopkinson bar," *Exp. Mech.*, vol. 24, pp. 81–86, 1984.

5. A. Gilat and Y. H. Pao, "High-rate decremental-strain-rate test," *Exp. Mech.*, vol. 28, pp. 322–325, 1998.

6. A. Gilat, "Torsional Kolsky bar testing," *ASM Handbook*, vol. 8, pp. 505–515, 2000.

7. B. A. Gama, S. L. Lopatnikov, and J.W. Gillespie Jr., "Hopkinson bar experimental technique: A critical review," *Appl. Mech. Rev.*, vol. 57, no. 4, pp. 223–250, 2004.

8. Y. Guo and Y. Li, "Quasi-static/dynamic response of SiO_2–epoxy nanocomposites," *Mater. Sci. Eng. A*, vol. 458, no. 1–2, pp. 330–335, Jun. 2007.

9. PCB Piezotronics, Inc. Model 208C04, "Load sensor specifications," pcb.com, Feb. 2016. [Online] Available: http://www.pcb.com/products.aspx?m=208C04.

10. Vision Research, Inc. Phantom V611, "Camera specifications," *highspeedcameras.com*, Feb. 2016. [Online] Available: https://www.highspeedcameras.com/Products/v-Series-Cameras/v611.

11. B. Pan, K. Qian, H. Xie, and A. Asundi, "Two-dimensional digital image correlation for in-plane displacement and strain measurement: A review," *Meas. Sci. Technol.*, vol. 20, no. 6, pp. 062001, 2009.

12. P. K. Rastogi (Ed.): *Photomechanics, Topics Appl. Phys.* 77, 323–372 2000.

13. J. Brillaud and F. Lagattu, "Limits and possibilities of laser speckle and white-light image-correlation methods: Theory and experiments," *Appl. Opt.*, vol. 41, no. 31, p. 6603, Nov. 2002.

14. L. Hall, "Digital image correlation of flapping wings for micro-technologies," *U.S. Army Research Laboratory*, Jan. 2011.

15. H. Koerber, J. Xavier, and P. P. P. Camanho, "High strain rate characterisation of unidirectional carbon-epoxy IM7-8552 in transverse compression and in-plane shear using digital image correlation," *Mech. Mater.*, vol. 42, no. 11, pp. 1004–1019, Nov. 2010.

16. M. Jerabek, Z. Major, and R. W. Lang, "Strain determination of polymeric materials using digital image correlation," *Polym. Test.*, vol. 29, no. 3, pp. 407–416, May 2010.

17. H. M. Hsiao and I. M. Daniel, "Strain rate behaviour of composite materials, *Compos. Part B Eng.*, vol. 29, no. 5, pp. 521–533, Sep. 1998.

18. K. A. Brown, R. Brooks, and N. A. Warrior, "The static and high strain rate behaviour of a commingled E-glass/polypropylene woven fabric composite," *Compos. Sci. Technol.*, vol. 70, no. 2, pp. 272–283, 2010.

19. S. Gurusideswar, R. Velmurugan, and N. K. Gupta, "High strain rate sensitivity of epoxy/clay nanocomposites using non-contact strain measurement," *Polymer*, vol. 86, pp. 197–207, Mar 2016.

5 Effect of High Strain Rate on the Tensile Behaviour of 3D Printed ABS Polymer

I. Mahapatra, Dong Ruan, R. Velmurugan and R. Jayaganthan

5.1 INTRODUCTION

It is essential to understand the mechanical behaviour of polymers at different strain rates based on the required applications in designing mechanical and structural components. The present investigation aims to understand the tensile behaviour of 3D printed acrylonitrile butadiene styrene (ABS) at high strain rate loading ($> 10^2$ s^{-1}) for potential advantages in impact analysis. This interest arises because the maximum stress observed in an object before deformation or failure is directly related to strain rate, i.e., most polymers are usually strain rate sensitive [1]. One of the widely used polymeric materials in the fused deposition modelling (FDM) additive manufacturing (AM) method is acrylonitrile butadiene styrene (ABS) [2]. It is an important and commonly used engineering thermoplastic material for automotive, aerospace and electronics applications and household appliances [3]. One primary application of these polymers is in the overhead luggage storage compartment of aircraft. These compartments are subjected to quite a bit of strain in regular use, thus requiring highly durable materials. They also need to be light in weight such that their presence does not hinder the aerodynamics or operation of the aircraft. Polymers like ABS are excellent materials for these overhead storage bins [4].

In industries such as military, automotive, aerospace and medical devices, an understanding of the mechanical characteristics of polymers is necessary for precise and reliable material and product performance. These properties are dependent on the composition of the materials and are related to several factors such as temperature, pressure and strain rate (frequency) [5]. Viscoelastic and strain rate sensitive materials include polymers, composites and some metallic ones. These materials deform through different micro-mechanisms at high strain rates than they do at low strain rates. Therefore, at high strain rates, the use of quasistatic stress-strain data may not be precise and dependable in describing the material properties. The design of engineering components is inaccurate as a result of using such data in simulation through FEA. Thus, it is essential to study how polymers behave under various strain rates, temperatures, pressures and frequencies. High-speed crash analysis of automotive and aerospace structures, high-speed ballistic impacts and drop impacts of consumer durables and electronic devices are typical examples of applications with high strain rates [6].

According to Hibbert et al., 3D printed ABS samples showed a significant amount of toughness of 2.044 MPa at a strain rate of 5 cm/min and [45°/−45°] raster angle [7]. Additionally, strain rate dependency is crucial in determining the short-term behaviour of viscoelastic materials. It has an impact on long-term failures such as creep and fatigue [8]. Drop impact is one of the main reasons why plastic and electrical components fail. Using drop impact analysis, the component's brittleness is determined (Gosavi et al., 2021) [9]. However, sufficient literature is available on drop impact tests of composites prepared by compression moulding technique (Naresh et al., 2020), and less work is reported on impact analysis of 3D printed material [10]. In the present work illustrated in Figure 5.1, the effect of high strain rate studies of 3D printed polymer materials is carried out and implemented.

DOI: 10.1201/9781003352358-5

5.2 EXPERIMENTAL SETUP

5.2.1 SPECIMEN PREPARATION AND GEOMETRY

As shown in Figure 5.2, a 3D geometric model of a tensile test specimen with the specifications required in the ASTM D638-02a standard (specimen of 3.25 mm thickness) was first created using SolidWorks 2019. The central rectangular portion consists of a constant width of 3.18 mm and a gauge length of 9.53 mm. A hole of 4.5 mm diameter was provided on both ends for fixing the fixture to the test setup. A circular curve with a radius of 6.74 mm gradually increases from the centre to the ends of the tensile specimen. The 3D model was exported as a Standard Tessellation Language (STL) file to be further processed in the slicing tool used in the fused deposition modelling (FDM) process. Anisotropic characteristics, which alter based on print direction, are continually subjected to these 3D printed prototypes, which means the properties change according to print direction. According to the literature, tensile loading experiences more of an anisotropy impact than

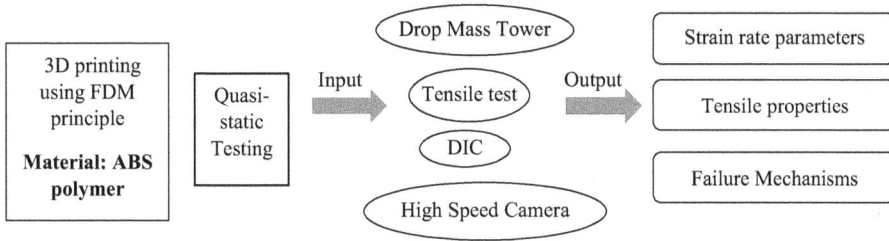

FIGURE 5.1 Schematic representation of the present work.

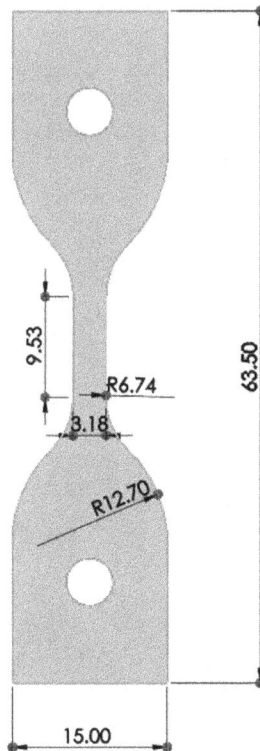

FIGURE 5.2 Strain rate specimen geometry.

compressive loading. The samples were printed in XY, XZ and YZ orientations to help in the study of the effect of anisotropy. The difference between the mechanical properties was also considered during the stress-strain analysis of quasi-static loading [10]. A total of three samples of the same design parameters were produced for a certain orientation. Flashforge Creator Pro 2, a desktop 3D printer based on fused filament fabrication (FFF), was utilized to fabricate each specimen. The material used was ABS polymer procured from Flashforge (China). For improved print quality, the manufacturer-recommended printing process parameters were used: a nozzle temperature of 230° C, printing bed temperature of 110° C, printing speed of 60 mm/s, travel speed of 80 mm/s and each layer height of 0.2 mm. In the slicing software Flashprint 5.3.3, the fill density was set to 100% in order to improve the sample's mechanical qualities [11].

5.2.2 Quasistatic Tensile Testing

Uniaxial tensile tests for quasi-static loading were conducted with an electromechanical universal testing machine (ZwickRoell Pvt. Ltd) as shown in Figure 5.3 with a load cell capacity of 50 kN. According to the ASTM D638 standard, each test was carried under displacement control load with a 1 mm/min strain rate. The corresponding displacement was recorded by applying the tensile load along the Z direction. The load and displacement data were recorded and converted into nominal stress and strain. Nominal strain values were determined by dividing displacement with the original length of the samples, and nominal stress values were calculated by dividing load with the cross-sectional area. To determine the tangent modulus along the Z direction, the slope of the stress-strain curve was plotted. The maximum stress that corresponds to the stress-strain curve's highest peak is known as tensile strength.

FIGURE 5.3   Uniaxial tensile test machine.

5.2.3 DIGITAL IMAGE CORRELATION (DIC)

The DIC analysis was carried out by capturing the deformation images using the high-speed camera at 6,000 fps. The initial series of images were captured at zero deformation, out of which one image is set as a reference image. Then, the next set of images was captured during the experiments to obtain full-field strain from the colour contrast of the pixels in the captured images. The quality of the speckle dots should be good, uniform and consistent enough to show a contrast between the pattern and the base substrate [12]. It wasn't necessary to paint the base substrate since it was already white. Due to the small size of the specimen, the black marker was used to manually create the speckle pattern for consistency and high quality. VIC-2D software was used to carry out the DIC evaluation.

5.2.4 DROP MASS SETUP

For dynamic testing, a drop mass impact machine (Figure 5.4) was used that produces strain rates between 10 and 600 s^{-1}. In order to get respective velocities of 2.21, 3.13, 4.43, 4.95 and 5.42 ms^{-1}, the mass of the falling bar was set at heights of 0.25, 0.5, 1, 1.25 and 1.5 m. These velocities were used to determine the nominal strain rates of 221, 313, 443, 495 and 542 s^{-1}. The ABS dog-bone-shaped specimen was tested at 221, 313, 443, 495 and 542 s^{-1} strain rates. The applied load was measured using a piezoelectric load cell with a 5 kN capacity, model PCB 208C04. By dividing the load data by the gauge region's cross-sectional area, the stress data were generated. A high-speed camera (Phantom V611) with a maximum resolution of 1,280 × 800 was used to take images. The data acquisition system NI PXI 1042 and Labview were used to obtain the load data [13].

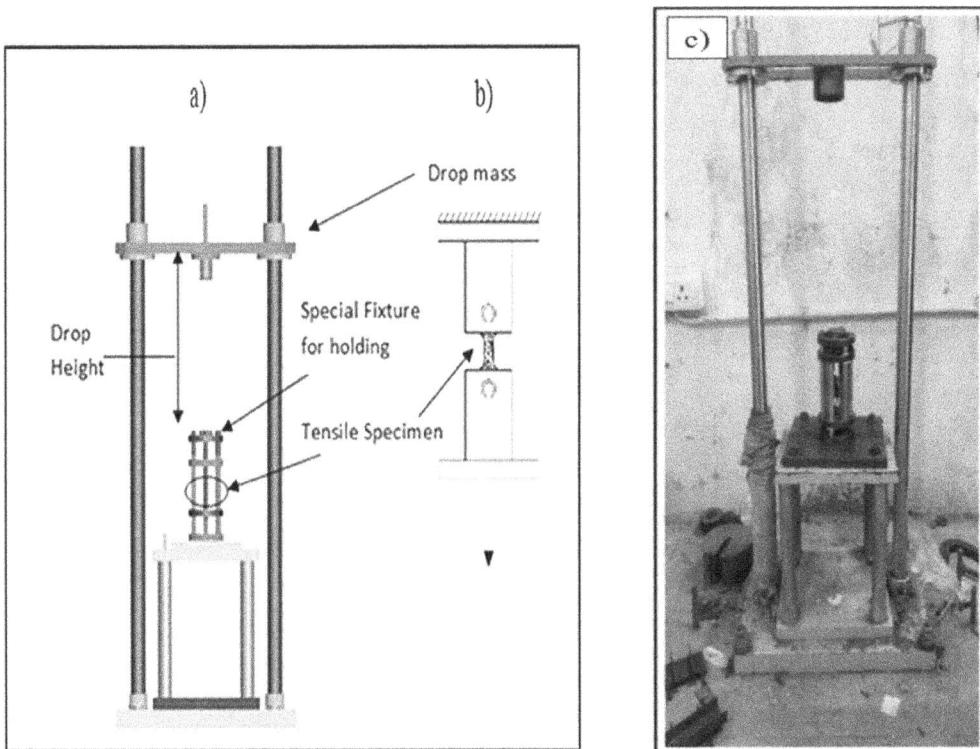

FIGURE 5.4 Schematic diagram of drop tower: (a) test setup model; (b) specimen fixture; (c) experimental setup. *Source: [14].*

5.3 RESULTS AND DISCUSSION

5.3.1 STRESS ANALYSIS

The specimens for each printing direction were tested at a crosshead speed of 1 mm/min for quasistatic testing. Three samples were prepared for each particular orientation, and the stress-strain curve was plotted by taking average values. The elastic modulus and tensile strength were calculated using the average of the curves as shown in Figure 5.5. The direction of printing affects the mechanical characteristics of the ABS dog bone samples. For quasi-static tensile testing, samples printed in XY, XZ and YZ directions showed tensile strength values of 51.2, 47.9 and 42.9 MPa, respectively. The mechanical characteristics of the samples were lowest in the YZ direction and highest in the XY direction. The average elastic modulus is 2,625.28 MPa, and the tensile strength is 51.2 MPa for the XY direction. Since for quasi-static testing, XY printing direction gave the highest properties, further dynamic analysis was continued by printing in XY direction only.

In dynamic studies, three tensile specimens for a particular strain rate are tested, and the average value of the results is displayed in the graph in Figure 5.6. The force values were obtained from National Instruments (NI PXI 1042). For accuracy, the sensor's force values are divided by the sensitivity value (1,124 mV/KN). The data obtained from Labview was correlated with the strain values obtained from the DIC analysis, which is explained further below. The stress-strain curves plotted at strain rates (221, 313, 443, 495, and 542 s^{-1}) are shown in Figure 5.6. The nominal strain rates are calculated from the formula

$$\varepsilon = \frac{V}{Lo} \tag{5.1}$$

where L_o is the gauge length of the tensile specimen (10 mm) for the dynamic loading and V is the impact velocity. The impact velocity is calculated by the following relation

$$V = \sqrt{2gh} \tag{5.2}$$

where g is the gravitational pull, and h is the height of the falling weight.

The stress-strain curves from multiple tests at varied strain rates are similar and partially overlap, depicting the effectiveness of the experimental approach. From Figure 5.6, it is observed that

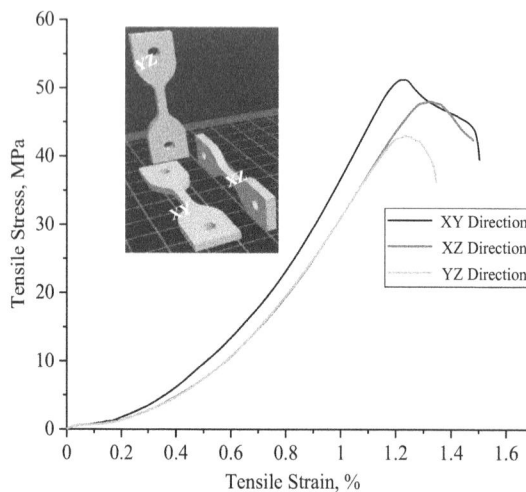

FIGURE 5.5   The tensile stress-strain curves of ABS samples printed in different directions (XY, YZ and ZX planes).

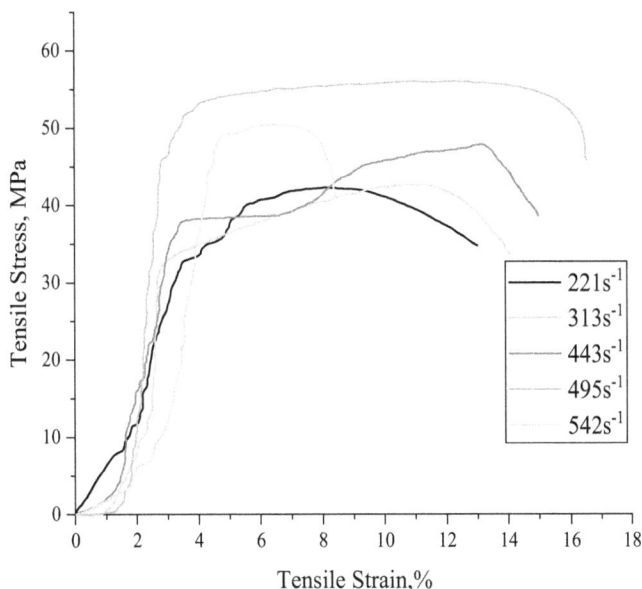

FIGURE 5.6   Stress-strain response of ABS samples at different strain rates.

as the strain rate increases from 221 to 542 s^{-1}, a stiffening mechanism occurs due to the increase in modulus and strength values. It is seen that the tensile yield strength value increases from 42.86 MPa at 221 s^{-1} to 57.08 MPa at 495 s^{-1}, indicating around a 33% increase in the tensile strength. However, the specimens tested at 313 and 542 s^{-1} show a decrease in strength and modulus values compared to the specimens tested at the previous strain rate.

The tensile stress-strain behaviour of quasistatic and dynamic loading differs for the 3D printed ABS material. This difference has also been observed for traditionally manufactured polymer materials. The initial stiffness of the 3D printed polymer material is practically negligible under quasi-static loading with a strain rate of 0.0016 s^{-1}. Stiffness rises under dynamic loading, and the slope of the stress-strain curve increases as the rate increases. Since the material reaction greatly depends on the strain rate, the stress-strain curve gets steeper as the strain rate rises. The mobility of molecular chain movements in a polymer, which is often low at high strain rates due to less relaxation time, can be linked to this trend. Given that material yielding happens after dynamic stress equilibrium, the slope of the stress-strain curve can be used to roughly estimate the dynamic tensile modulus. The material stiffness at the associated strain rate is indicated by the line slope, which is the tangent modulus. Figure 5.8 depicts the relationship between tangent modulus and strain rate.

The modulus has also increased from 1,477.5 MPa to 4,144.9 MPa, which increased by 1.8 times for the strain rate from 21 to 542 s^{-1}. Figure 5.7 illustrates the high-speed images immediately after the failure of the specimen at the respective strain rates. The damage mechanism is almost equivalent for all the tensile coupons at varying strain rates.

5.3.2 Strain Analysis

The specimen images before and after failure were captured by high-speed Phantom camera V611, and the image post-processing was carried out using Phantom CV 3.5 software. Specific measures were taken to synchronize the entire event of capturing images from the high-speed camera and simultaneously force values from the load cell to avoid overlapping data. The camera was kept at a distance of 50 cm with a resolution of 800 × 600, exposure time of 160 µs, and focal aperture of 2.8 cm. LED light was used and properly adjusted to get bright and clear images. These images were further processed in VIC-2D software for DIC analysis to obtain the strain contour plots. The subset

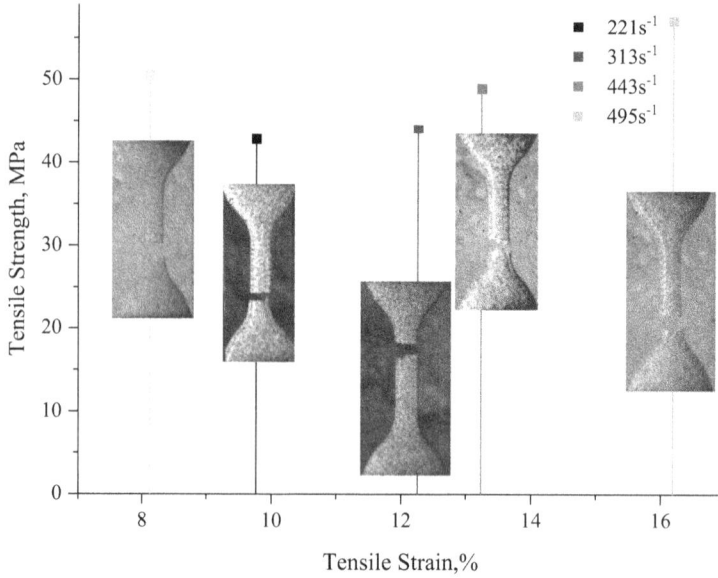

FIGURE 5.7 The effect of strain rate on the tensile strength of 3D printed ABS samples.

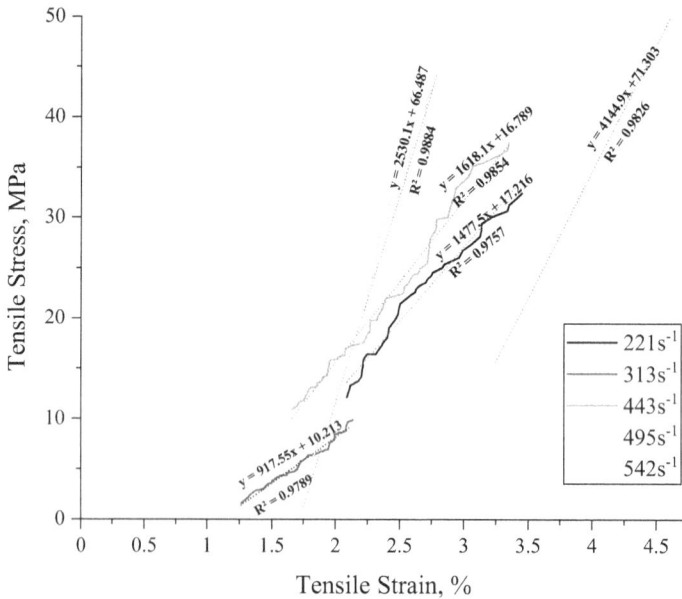

FIGURE 5.8 Relation of tangent modulus versus strain.

size was set as 17, and the step size is set as 1 to get reliable and accurate strain data. The colour maps were selected as a spectrum, and the strain contour plots are shown below at different time intervals at 0.25 m, 0.5 m, 1 m, 1.25 m and 1.5 m, respectively.

Figure 5.9 shows the strain contour plots of the additively manufactured ABS specimens at a strain rate of 221 s⁻¹. The maximum strain at t_o + 160 μs concerning the reference image is 1.85% which further increases to 5.35% at t_o + 480 μs, and the maximum strain of 9.2% is observed at t_o + 960 μs as shown in Figure 5.9c prior to fracture. Figures 5.10, 5.11, 5.12 and 5.13 illustrate that the percentage of failure strain is proportional to an increase in strain rate. The material shows mild ductile behaviour at dynamic loading due to higher testing velocities. For the strain rate of 542 s⁻¹, the maximum strain $t_o + 160$ μs for the reference image is 1.44% which further increases to 2.18% at $t_o + 320$ μs, and the maximum strain of 8% is observed at $t_o + 640$ μs as shown in Figure 5.13 prior to fracture. Therefore, failure occurs immediately at a strain rate of 542 s⁻¹ compared to the lower strain rates. Thus, a drop in failure strain for the strain rate increment from 495 to 542 s⁻¹ is observed.

For 495 s⁻¹, there is an increase in the strain of 1.7–13.7% before a fracture occurs. This abrupt increase in strain may be one of the explanations for the strain-hardening behaviour seen in the stress-strain curves [15]. For accurate real-time modelling of mechanical and structural analysis, design engineers would find the tensile characteristics, failure strain percentage and strain rate data acquired from this study to be helpful [16]. Two separate strain rate ranges can be used to categorize the behaviour of polymeric materials as a function of strain rate: the first includes strain rates between 200 and 450 s⁻¹, and the second includes the latter two strain rates of 450 and 600 s⁻¹.

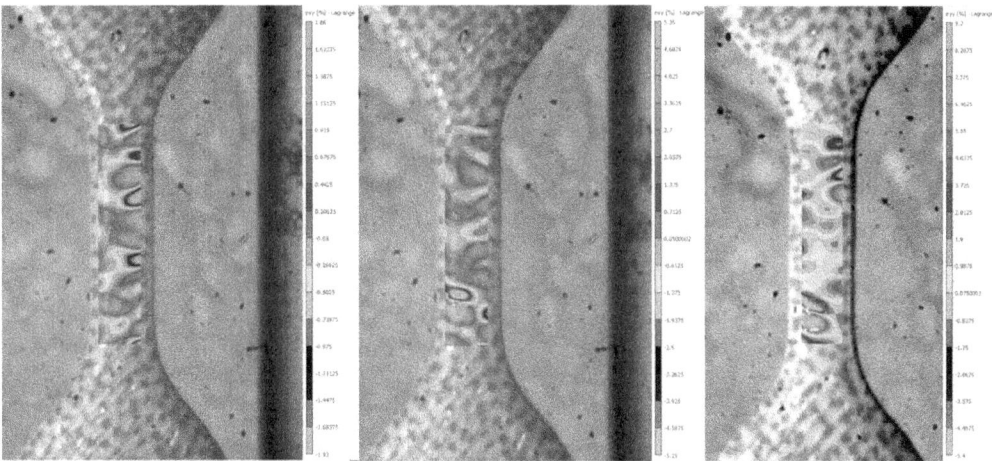

FIGURE 5.9 Strain contour plots of the ABS polymers at 0.25 m height: (a) $(\varepsilon yy)_{max} = 1.85\%$ at 160 μs; (b) $(\varepsilon yy)_{max} = 5.35\%$ at 480 μs; (c) $(\varepsilon yy)_{max} = 9.2\%$ at 960 μs.

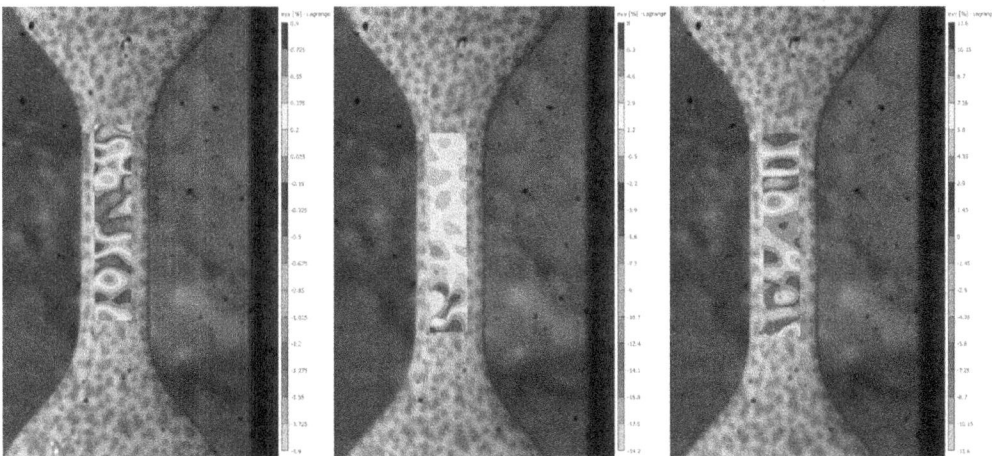

FIGURE 5.10 Strain contour plots of the ABS polymers at 0.5 m height: (a) $(\varepsilon yy)_{max} = 0.9\%$ at 160 μs; (b) $(\varepsilon yy)_{max} = 8\%$ at 480 μs; (c) $(\varepsilon yy)_{max} = 11.6\%$ at 800 μs.

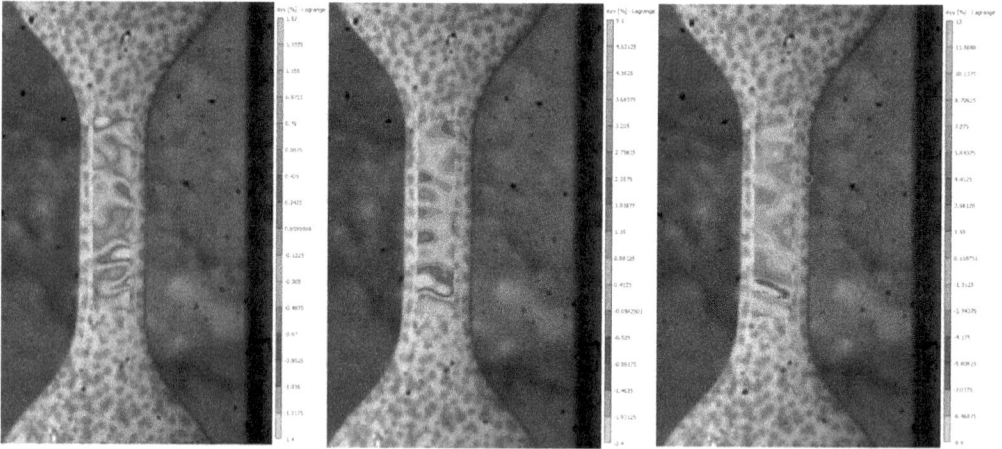

FIGURE 5.11 Strain contour plots of the ABS polymers at 1 m height: (a) $(\varepsilon yy)_{max} = 1.52\%$ at 160 μs; (b) $(\varepsilon yy)_{max} = 5.1\%$ at 320 μs; (c) $(\varepsilon yy)_{max} = 13\%$ at 960 μs.

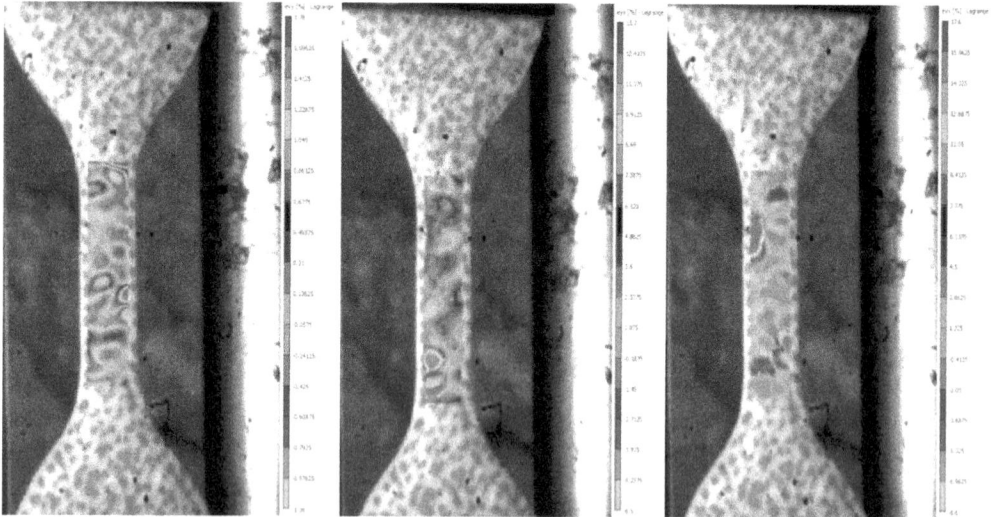

FIGURE 5.12 Strain contour plots of the ABS polymers at 1.25 m height: (a) $(\varepsilon yy)_{max} = 1.78\%$ at 160 μs; (b) $(\varepsilon yy)_{max} = 13.7\%$ at 480 μs; (c) $(\varepsilon yy)_{max} = 17.6\%$ at 1,120 μs.

5.4 CONCLUSION

This study emphasized understanding the increase in strain rates within the 200–600 s^{-1} range on the mechanical properties of FDM-based polymer under tensile loading. The following specific conclusions can be drawn based on this study:

1. There is an increase in strength and modulus values of high strain rate testing of the 3D printed polymer specimens as compared to the quasistatic loading.
2. There is a drastic change in the slope gradient between the quasistatic and dynamic stress-strain curves. A small amount of strain hardening is visible in the stress-strain curves for dynamic loading, which may be one of the reasons for the increase in strength and modulus values at higher strain rates.

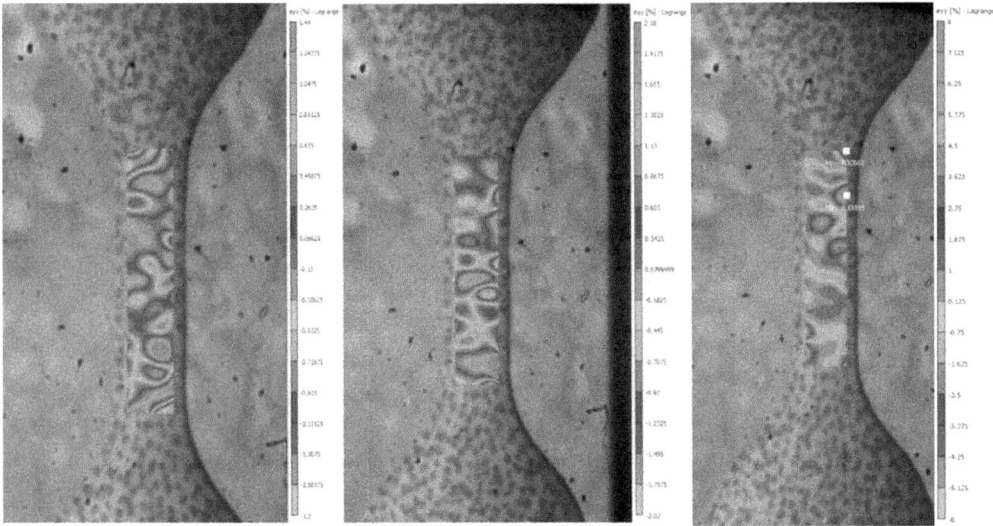

FIGURE 5.13 Strain contour plots of the ABS polymers at 1.5 m height: (a) $(\varepsilon yy)_{max}$ =1.44% at 160 μs; (b) $(\varepsilon yy)_{max}$ = 2.18% at 320 μs; (c) $(\varepsilon yy)_{max}$ = 8% at 640 μs.

3. The specimens showed tensile stress between 42.8 to 57.08 MPa prior to fracture for strain rates of 221 to 542 s^{-1}, respectively, as shown in Figure 4.2.
4. The corresponding yield point changed along with the maximum stress as the transition occurred from one strain range to the other. For a strain rate of 443 s^{-1}, the tensile modulus and the maximum stresses were 1,618.1 and 48.9 MPa, respectively. For the strain rate of 484 s^{-1}, we find an increase in tensile strength value by 16.7% and an increase in modulus by 1.56 times.
5. In the range of 450–600 s^{-1} strain rate, there was also a significant increase in stress, followed by strain hardening and then a drastic decrease in stress resulting from stress collapse.
6. The tensile strength properties were found to be strongly influenced by the increase in strain rate values. A very high strength of 57.08 MPa was observed for additively manufactured ABS polymer at a strain rate of 495 s^{-1}. However, it is seen that the failure strain rate also increases simultaneously.
7. It is found that generally, as the strain rate increases, the tensile strength values increase along with an increase in failure strain.

REFERENCES

1. Owolabi G., Peterson A., Habtour E., Riddick J., Coatney M., Olasumboye A. and Bolling D. 2016. 'Dynamic response of acrylonitrile butadiene styrene under impact loading'. *International Journal of Mechanical and Materials Engineering*, 11(1):1–8.
2. Vairis A., Petousis M., Vidakis N. and Savvakis K. 2016. 'On the strain rate sensitivity of abs and abs plus fused deposition modeling parts'. *International Journal of Mechanical and Materials Engineering*, 25(9):3558–3565.
3. Verbeeten W. M., Lorenzo-Bañuelos M., Saiz-Ortiz R. and González R. 2020. 'Strain-rate-dependent properties of short carbon fiber-reinforced acrylonitrile-butadiene-styrene using material extrusion additive manufacturing'. *Rapid Prototyping Journal*, 26(10):1701–1712.
4. Siviour C. R. and Jordan J. L. 2016. 'High strain rate mechanics of polymers: A review'. *Journal of Dynamic Behavior of Materials*, 2(1):15–32.
5. Peter S. and Woldesenbet E. 2008. 'Nanoclay syntactic foam composites—High strain rate properties'. *Materials Science and Engineering: A*, 494(1–2):179–187.

6. Vidakis N., Petousis M., Velidakis E., Liebscher M., Mechtcherine V. and Tzounis L. 2020. 'On the strain rate sensitivity of fused filament fabrication (FFF) processed pla, abs, petg, pa6, and pp thermoplastic polymers'. *Polymers*, 12(12):2924.

7. Hibbert K., Warner G., Brown C., Ajide O., Owolabi G. and Azimi A. 2019. 'The effects of build parameters and strain rate on the mechanical properties of FDM 3D-Printed acrylonitrile Butadiene Styrene'. *Open Journal of Organic Polymer Materials*, 9(1):1.

8. Naresh K., Shankar K., Velmurugan R. and Gupta, N. K. 2020. 'High strain rate studies for different laminate configurations of bi-directional glass/epoxy and carbon/epoxy composites using DIC'. *Structures October*, 27:2451–2465.

9. Gosavi A., Kulkarni A., Dama Y., Deshpande A. and Jogi B. 2022. 'Comparative analysis of drop impact resistance for different polymer based materials used for hearing aid casing'. *Materials Today: Proceedings*, 49:2433–2441.

10. Naresh K., Shankar K., Rao B. S. and Velmurugan R. 2016. 'Effect of high strain rate on glass/carbon/ hybrid fiber reinforced epoxy laminated composites'. *Composites Part B: Engineering*, 100:125–135.

11. Sharma D. and Hiremath S. S. 2022. 'Engineering the failure path with bird feather inspired novel cellular structures'. *Engineering Fracture Mechanics*, 264:108350.

12. Chikkanna N., Krishnapillai S. and Ramachandran V. 2022. 'Static and dynamic flexural behaviour of printed polylactic acid with thermal annealing: Parametric optimisation and empirical modelling'. *The International Journal of Advanced Manufacturing Technology*, 119(1):1179–1197.

13. Gurusideswar S., Velmurugan R. and Gupta N. K. 2016. 'High strain rate sensitivity of epoxy/clay nanocomposites using non-contact strain measurement'. *Polymer*, 86:197–207.

14. Anuse V. S., Shankar K., Velmurugan R. and Ha S. K. 2022. 'Compression-after-impact analysis of carbon fiber reinforced composite laminate with different ply orientation sequences'. *International Journal of Impact Engineering*, 167:104277.

15. Fan J., Fan X. and Chen A. 2017. 'Dynamic mechanical behaviour of polymer material'. In: Faris Yilmaz, editors. *Aspects of Polyurethanes*. IntechOpen, 3:193–212.

16. Navya G., Jayaganthan R., Velmurugan R. and Gupta N. K. 2022. 'Finite element analysis of tensile behaviour of glass fibre composites under varying strain rates'. *Thin-Walled Structures*, 172:108916.

6 Effects of Different Strain Rates on the Tensile Properties of Bi-Directional Glass/Epoxy, Carbon/Epoxy and Interply Hybrid Composites Using DIC

K. Naresh, K. Shankar and R. Velmurugan

6.1 INTRODUCTION

Glass fiber-reinforced polymer (GFRP) and carbon fiber-reinforced polymer (CFRP) composites are being widely used in the aerospace and automotive industries [1]. The former is owing to its higher failure strain and moderate tensile strength with low cost, while the latter is owing to its excellent tensile strength and modulus with lightweight [2]. These structures are subjected to extreme impact loading events [3] over a wider range of strain rates [4]. Hence, understanding the mechanical behavior of these materials under high strain rates is of great concern for the optimum design of composite structures [5]. However, understanding the strain rate effects is crucial in measuring the load-bearing capabilities of composites; for that, several trials and modeling aspects are required [6], because the properties of composites are directionally dependent and also vary with the nature of loading, i.e., the properties for compressive loading are different from tensile loading [7]. Earlier studies have focused on understanding the properties of different fiber orientations of unidirectional (UD) composites [8, 9] and along three principal directions (warp, weft, and thickness) of woven roving composites [10]. Besides, most of these studies were performed under high strain rate compressive loading, and fewer studies under high strain rate tensile loading, using the split Hopkinson pressure bar (SHPB) [11–13]. Very few researchers also performed high strain rate experiments with both tensile and compressive fixtures using the SHPB and compared the properties between tensile and compressive loadings [14, 15]. The mechanical properties of GFRP and CFRP composites under various strain rates are outlined briefly.

Jenq and Sheu [16] studied the compressive behavior of unidirectional glass/epoxy composites over the range of strain rates from quasi-static (3×10^{-4} s^{-1}) using an MTS tester to 104 s^{-1} using SHPB. The dynamic strength was found to be two times higher than the static strength and the dynamic modulus was 92% higher than the static modulus. In contrast to progressive failure, a structure observed in quasi-static specimens, the dynamic specimens were crushed into small pieces. Tasdemirci et al. [17] and Kara et al. [18] performed the quasi-static (0.001 to 0.01 s^{-1}) and high strain rate (1,000 s^{-1}) compression response of woven roving [$\pm 45°$]$_s$ glass/polyester composite in three different directions (longitudinal, transverse, and thickness directions). Using SHPB, high strain rate tests were conducted, and Yen-Caiazzo (Y-C) logarithmic equations were used for finding out the strain rate sensitivity of composites. The failure mechanisms such as damage initiation and progression and axial splitting of fibers were observed in situ using a high-speed camera. They

DOI: 10.1201/9781003352358-6

71

described that the strain rate sensitivity was higher in the in-plane direction as compared to the thickness direction.

Acharya et al. [11] used the power law regression equation for finding out the strain rate sensitivity of compressive strength of the glass/epoxy composite over two different ranges of strain rates: (i) from low to high (0.001–550 s^{-1}) and (ii) from high to very high strain rate regime (550–2,500 s^{-1}). The results revealed the compressive strength to be more rate sensitive in the latter regime than the low to high strain rate regime, which can be attributed to the mobility of molecular chain movement in the polymer matrix which is high at low strain rates due to more relaxation time. The higher the loading rate, the better could be the slope of stress-strain curves. As a result, the steeper stress-strain of the glass/epoxy composite was obtained at high strain rates.

Chen et al. [15] experimentally studied the strain rate effect of cross-ply carbon/epoxy composite using the Cowper-Symonds model over a wide range of strain rates (10^{-3}–1,400 s^{-1}) under compressive and tensile loadings. The quasi-static tests were conducted using a displacement-controlled (CCSS88010) electronic universal test machine of 10 kN capacity and the dynamic tests were conducted using the SHPB. They have reported that the failure mechanisms are different in both loadings. Fibers play an important role during tensile loading, while compressive loading depends on both matrix and interface properties, and also fibers are influenced by strain rate than that of the matrix. Therefore, the strain rate sensitivity was found to be higher in tension than that in compression. Luca et al. [19] performed experiments on different stacking sequences [(90)$_6$, (0/90/45/-45/0/90)$_S$ and (45/-45/0/90/45/-45)$_S$] of UD CFRP composite under tensile loading over the range of strain rates from quasi-static to high strain rate (100 and 200 s^{-1}) and investigated the effect of strain rate of composites theoretically with two constitutive models, namely the Cowper-Symonds model and Yen-Caiazzo (Y-C) equations. The Cowper-Symonds model was found as an efficient one that accurately predicts the strain rate effect of composites compared to Y-C equations.

Koerber et al. [9] studied compression properties of UD carbon/epoxy prepreg composite with different fiber orientations of off-axis (15°, 30°, 45°, 60°, and 75°) and on-axis (90°) laminates. They have performed quasi-static (4 × 10^{-4} s^{-1}) and high strain rate (90–350 s^{-1}) tests using the Instron instrument and SHPB, respectively. DIC software was used to measure the in-plane strain data. The 45° off-axis specimen showed higher failure strain. Gilat et al. [20] have demonstrated the importance of the DIC technique and compared DIC results with those of strain gauge results for both SHPB compression and tensile tests. Large deformations in most of the tensile tests were seen beyond the gauge length. The strain gauge measures strain at the location where the gauge was pasted, whereas DIC measures the full field strain data. Therefore, DIC results provided more information than strain gauge results in tensile tests, but in compression tests, both results match well. The advantages of using the DIC technique over other strain measuring techniques (strain gauge, extensometer, etc.) can be found in our recent articles [12, 21].

However, studies performed on different fiber orientations of bi-directional composites subjected to high strain rate tensile loading are limited and have not been understood clearly. Especially, the strain rate sensitivity of CFRP composite is not clear. Some researchers [22, 23] reported that CFRP composite was strain rate insensitive while others [15, 24, 25] reported that CFRP composite was less or more sensitive to the strain rate. Hence, a detailed study is needed for further confirmation of the rate sensitivity results of CFRP composite. In the present work, an in-house test facility, the drop mass test setup equipped with the tensile fixture, has been used to investigate the effect of strain rate which ranges from quasi-static (1.6 × 10^{-3} s^{-1}) to 542 s^{-1} on different fiber orientations of woven roving glass/epoxy and carbon/epoxy composites: (i) on-axis (0/90), (ii) and (iii) partial off-axis/on-axis (0/90/30/-60) and (0/90/45/-45), and (iv) off-axis (30/-60/60/-30). Also, the tensile properties were investigated on (0/90) hybrid (glass/carbon/epoxy) composite and compared these properties with (0/90) glass/epoxy and carbon/epoxy composites.

6.2 MATERIALS AND METHODS

6.2.1 MATERIALS AND SPECIMEN FABRICATION

Reinforcement materials such as woven roving mat (WRM), glass fiber and carbon fiber, and matrix materials such as epoxy resin [Araldite (LY 556)] and hardener [Aradur (HY-951)] were used in this study for manufacturing the composites. Fiber to resin ratio was maintained at 1:1 and resin to hardener ratio at 10:1, by weight.

Each laminate was made of four layers and each layer had 0.4 mm thickness. Figure 6.1 shows the different stacking sequences [(0/90), (0/90/30/-60), (0/90/45/-45), and (30/-60/60/-30)] of E-glass/epoxy, carbon/epoxy, and interply hybrid [(glass fabric layers)$_2$/(carbon fabric layers)$_2$]/epoxy composite laminates. All laminates were prepared using a compression molding technique with a uniform specimen size of 300 mm × 300 mm.

6.2.2 EXPERIMENTAL DETAILS

6.2.2.1 Quasi-Static Tensile Test

Quasi-static tests were carried out using an Instron 3367 test machine of capacity 30 kN at a crosshead speed of 5 mm/min which corresponds to the strain rate of 1.6×10^{-3} s^{-1}. The width, thickness, and gauge length of the specimen used were 6 mm × 2 mm × 50 mm, respectively. The full-field displacement fields were captured in situ by the DIC technique using a Point Grey Grasshopper 3 (Model – GS3-U3-50S5M-C), 5 MP, Mono CCD camera with a maximum resolution of 2,448 × 2,048 pixels^2. The maximum speed that can be achieved by this camera is 15 fps. Further, postprocessing of the in situ images was carried out to obtain the strain data using VIC-2D software.

6.2.2.2 High Strain Rate Tensile Test

High strain rate tensile tests were carried out at different heights (0.25, 0.50, 0.75, 1.00, 1.25, and 1.50 m) which correspond to different nominal strain rates (221, 313, 384, 443, 495, and

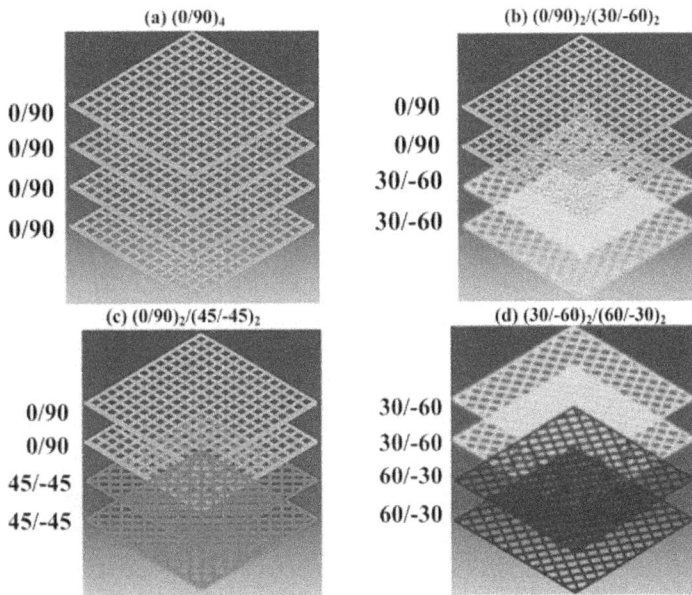

FIGURE 6.1 Schematic representation of different stacking sequences of four layer laminates: (a) (0/90)$_4$, (b) (0/90)$_2$/(30/-60)$_2$, (c) (0/90)2/(45/-45)$_2$, and (d) (30/-60)2/(60/-30)$_2$.

542 s^{-1}) using the drop mass test setup. Further details about the drop mass test setup used were discussed in our previous articles [5, 7, 12, 26]. The specimen geometry and tensile fixture used for the high strain rate tests are shown in Figure 6.2(a) and (b), respectively. The full-field displacement fields were captured in situ by the DIC technique using a Phantom (Model V611) high-speed camera, with a resolution of 128 × 128 pixels^2 and a speed of 1,00,000 fps. Further, post-processing of the in situ images was carried out to obtain the strain data using VIC-2D software.

6.3   RESULTS AND DISCUSSION

6.3.1   LOCAL AND GLOBAL STRAIN HISTORIES

The local strains were calculated on the lower portion of the specimen within the gauge section using VIC-2D software and compared with the global (full-field) strains. The local and global strains vs. time graphs for GFRP and CFRP composites of (0/90/30/-60) orientation for the strain rate ($\dot{\varepsilon}$) of 221 s^{-1} are shown in Figure 6.3(a) and (b), respectively. It was observed from these figures that the linear portion of local strain values was found to be slightly higher than the global strain values. The strains to failure values were observed to be lower in the CFRP composite than those in the GFRP composite.

6.3.2   EFFECT OF STRAIN RATE ON TENSILE PROPERTIES OF GLASS/EPOXY COMPOSITES

In this section, the tensile properties of glass/epoxy composite for four different laminate configurations have been presented. The change in properties (strength, strain, and modulus) with the increase in strain rate was studied. The tensile modulus was calculated from the initial slope of the stress-strain curve. Typical fractured specimens of glass/epoxy composite specimens for quasi-static and high strain rate loading are shown in Figures 6.4 and 6.5, respectively. These indicate that the failure has occurred in the gauge portion of the specimen.

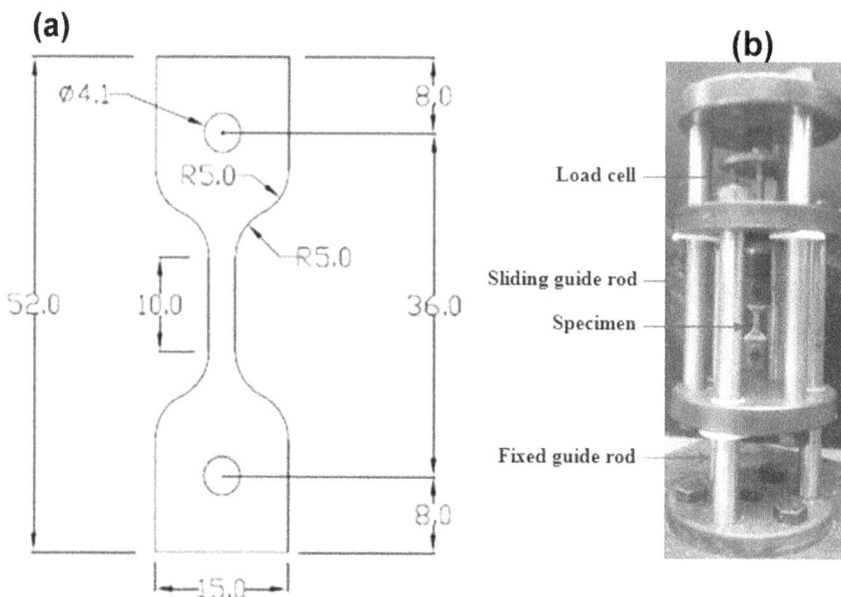

FIGURE 6.2   Specimen geometry (a) and tensile fixture (b), used for the high strain rate test (all dimensions are in mm).

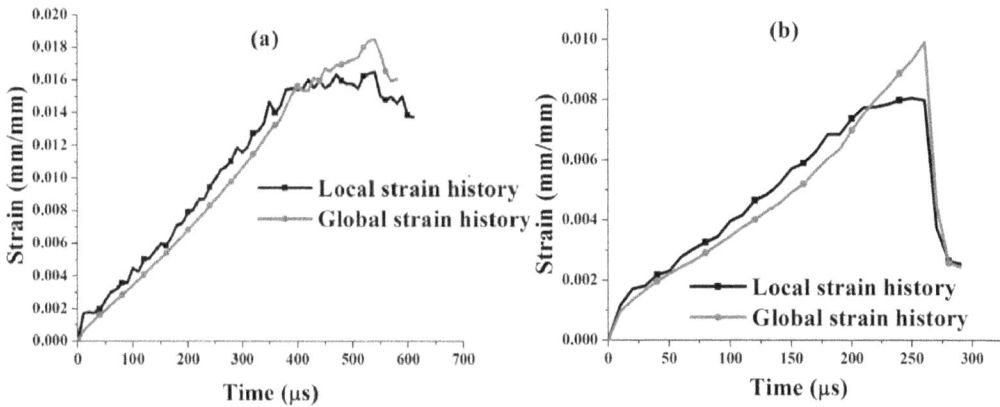

FIGURE 6.3 Local and global strain histories for (0/90/30/-60) laminate configurations at 221 s⁻¹: (a) GFRP and (b) CFRP.

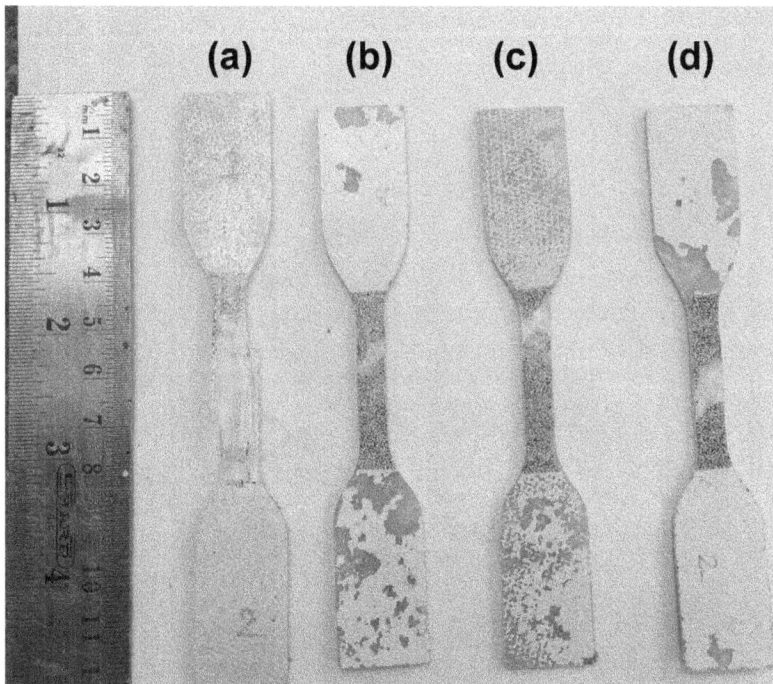

FIGURE 6.4 Fractured specimens of glass/epoxy composite subjected to quasi-static tensile loading: (a) (0/90), (b) (0/90/30/-60), (c) (0/90/45/-45), and (d) (30/-60/60/-30).

6.3.2.1 Strain Analysis of Glass/Epoxy Composite

Typical strain contour plots of (0/90), (0/90/30/-60), (0/90/45/-45), and (30/-60/60/-30) glass/epoxy composites under quasi-static loading are shown in Figures 6.6 and 6.7. Three images are presented for each laminate configuration. In Figures 6.6 and 6.7, images (a) represent the deformed specimen just before failure, the strain contour plot for the corresponding state of the specimen is shown in images (b), and images (c) represent the damaged specimen for their respective laminates. It is clear from images (c) that the failure has occurred along the fiber direction for all the laminate configurations.

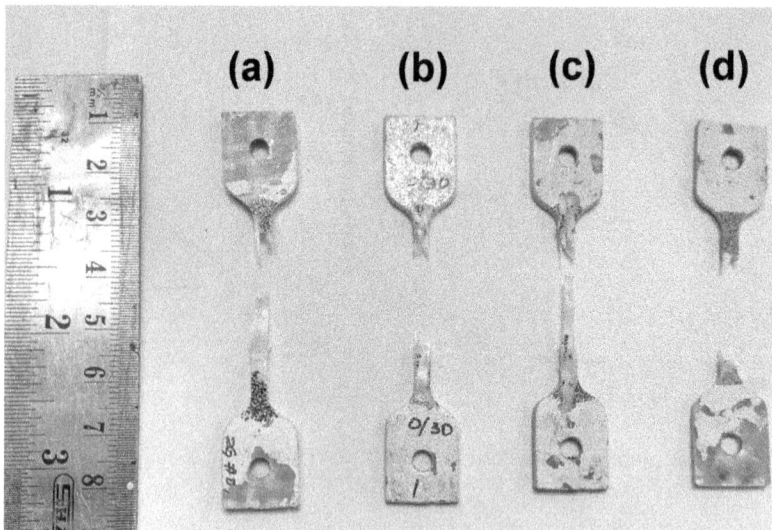

FIGURE 6.5 Fractured specimens of glass/epoxy composite in dynamic loading: (a) (0/90), (b) (0/90/30/-60), (c) (0/90/45/-45), and (d) (30/-60/60/-30).

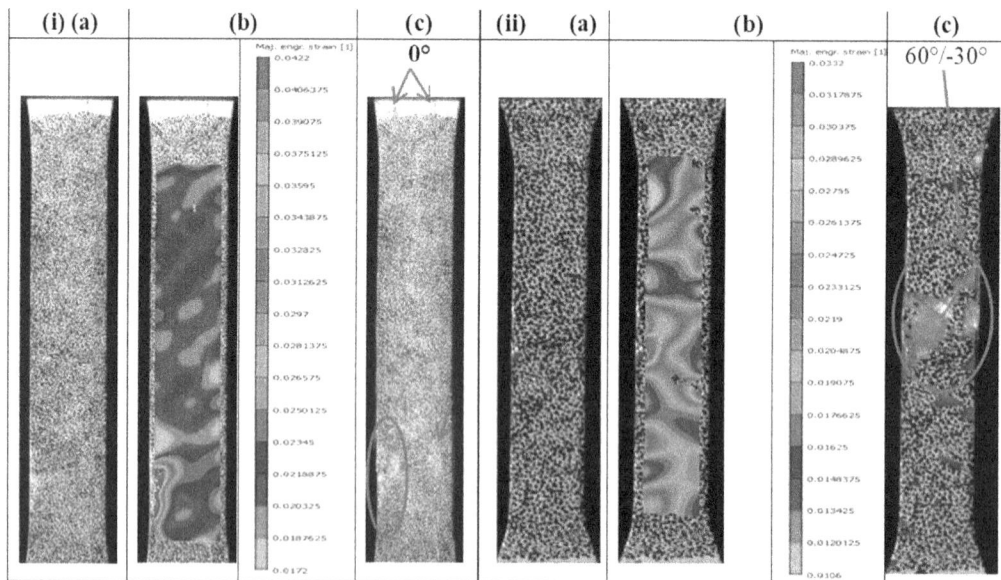

FIGURE 6.6 GFRP (i) (0/90) and (ii) (0/90/30/-60): (a) speckle movements just before failure, (b) strain component (ε_{yy}), and (c) damaged specimen.

Figures 6.6(i)(a–c) represent the DIC images for (0/90) laminate. Figures 6.6(i)(a–b) indicate the same image before and after its post-processing corresponding to the uniaxial tensile load of 4.33 kN. The failure strain (ε_{yy}) corresponding to the load was 4.22% [Figure 6.6(i)(b)]. Figure 6.6(i)(c) indicates the damaged specimen which reveals the fiber-dominated failure along the 0° direction. Similarly, Figures 6.6(ii)(a–b) indicate the speckle images before and after post-processing for (0/90/30/-60) laminate at a load of 2.71 kN. The corresponding failure strain (ε_{yy}) was 3.32%. The failure had occurred along the (60°/–30°) direction, as shown in Figure 6.6(ii)(c). Figures 6.7(i) (a–b) indicate the speckle images for the (0/90/45/-45) laminate at the load of 2.26 kN and the

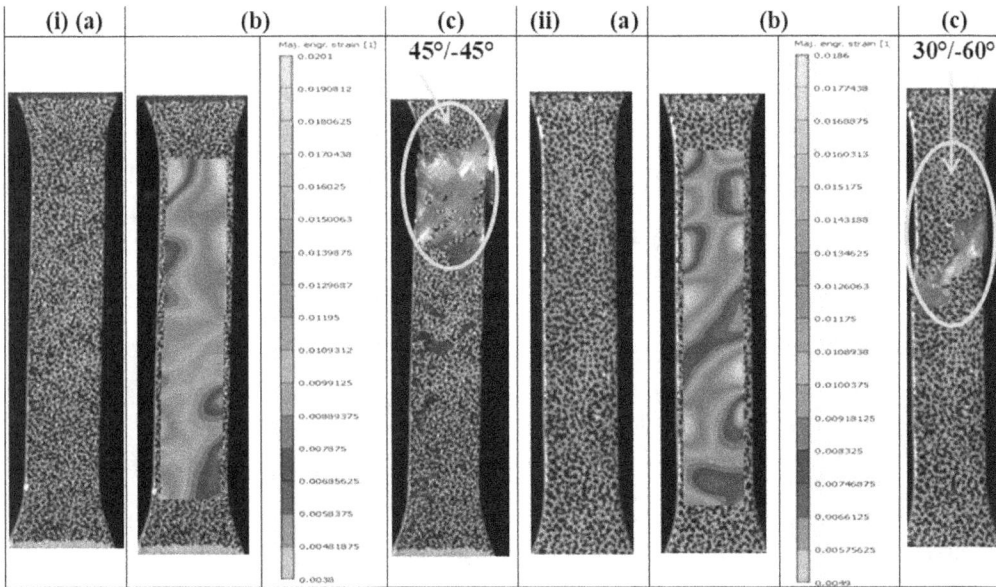

FIGURE 6.7 GFRP composite: (i) (0/90/45/-45) and (ii) (30/-60/60/-30): (a) speckle movements just before failure, (b) strain component (ε_{yy}), and (c) damaged specimen.

corresponding failure strain is 2%. Figure 6.7(i)(c) indicates that the direction of failure is along the (45°/–45°) layers. The load and failure strain values for (30/- 60/60/-30) laminate [Figures 6.7(ii) (a–b)] were 1.25 kN and 1.86%, respectively. Figure 6.7(ii)(c) shows the failure along the (30°/–60°) direction. Fewer fibers are present along the loading direction for (30/-60/60/-30) laminate. As a result, this off-axis laminate bears a low load and fails before other laminates.

Similarly, Figure 6.8(i–iv) indicate the recorded images during dynamic loading at the strain rate of 542 s^{-1} for glass/epoxy composite with lay-up sequences of (0/90), (0/90/30/-60), (0/90/45/-45), and (30/-60/60/-30), respectively. These images were captured using a Phantom (V611) high-speed camera.

In Figure 6.8, images (a) show the state just before failure, images (b) show the strain contour corresponding to the state in images (a), and images (c) show the failure along the respective fiber direction in all cases. The failure strain (ε_{yy}) values observed from Figure 6.8(i–iv)(b) were 2.23%, 0.99%, 1.20%, and 0.70 % for (0/90), (0/90/30/-60), (0/90/45/-45), and (30/-60/60/-30) laminate configurations, respectively. The corresponding load values for these laminate configurations were 3.93 kN, 2.42 kN, 1.85 kN, and 1.07 kN, respectively. These load values were significantly lower than the quasi-static values as the width is 6 mm for quasi-static specimens and 3 mm for high strain rate specimens.

6.3.2.2 Comparison of Tensile Properties for Different Laminate Configurations of GFRP

The variation in stress-strain curves with respect to strain rate for different laminate configurations of GFRP is seen in Figures 6.9(a–d). It was observed that the magnitude of tensile properties is higher for the (0/90) laminate [Figure 6.9(a)] whereas the (30/-60/60/-30) laminate [Figure 6.9(d)] has the lowest properties at all strain rates compared to the other laminates. However, the variation in tensile properties was found to be higher for the (0/90/45/-45) laminate [Figure 6.9(c)] than the other laminates, as the strain rate increases from quasi-static to 542 s^{-1}. The material had a brittle behavior at dynamic loading due to higher testing speeds. Therefore, failure took place immediately, resulting in an enhancement in strength and stiffness and a drop in failure strain for all lay-up configurations.

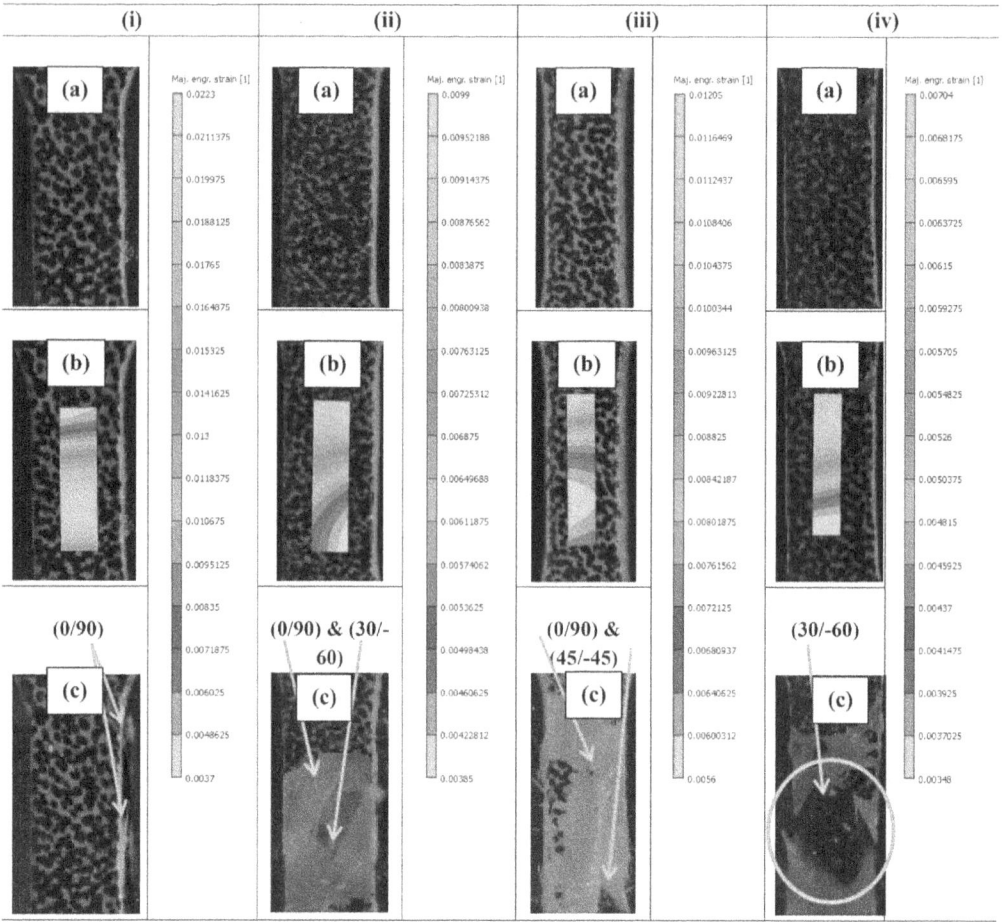

FIGURE 6.8 High-speed camera images for glass/epoxy composite of different fiber orientations at 542 s⁻¹: (i) (0/90), (ii) (0/90/30/−60), (iii) (0/90/45/−45), and (iv) (30/−60/60/−30).

6.3.3 Effect of Strain Rate Sensitivity on Tensile Properties of Glass/Epoxy Composite

The degree of strain rate sensitivity of composites varies with the type of fiber and matrix, the type and the direction of loading, etc. For a clear understanding of the strain rate dependency of composites, the Cowper-Symonds model [27] has been used in this study.

6.3.3.1 Determination of Strain Rate Parameters

The parameters for tensile strength (A and m_σ) and modulus (B and m_E) were obtained using the known experimental values by fitting Eqns. (6.1) and (6.2), respectively [12], which are given by power law equations.

$$K_\sigma = \frac{\sigma_d}{\sigma_0} - 1 = \left(\frac{1}{A}\right)^{m_\sigma} (\dot{\varepsilon})^{m_\sigma} \qquad (6.1)$$

$$K_E = \frac{E_d}{E_0} - 1 = \left(\frac{1}{B}\right)^{m_E} (\dot{\varepsilon})^{m_E} \qquad (6.2)$$

where σ_0 and σ_d are the reference (or) quasi-static strength and dynamic strength, respectively. E_0 and E_d are the quasi-static and dynamic moduli, respectively. $\dot{\varepsilon}$ is the strain rate which ranges

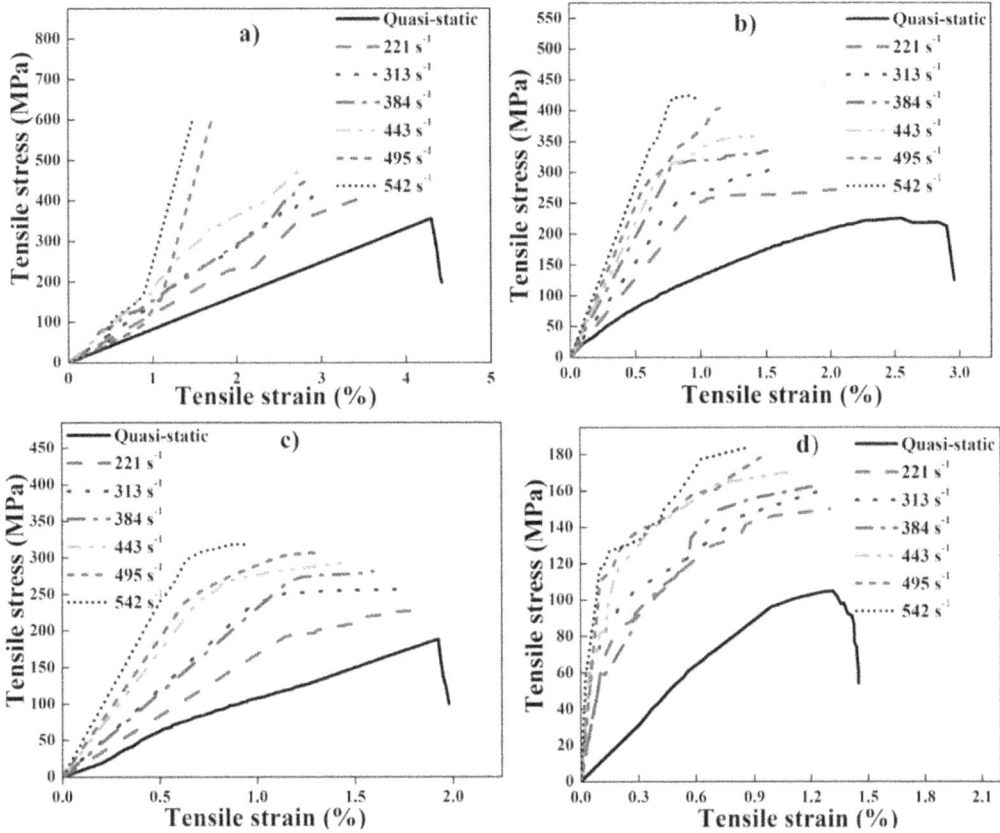

FIGURE 6.9 Effects of different strain rates on the tensile properties of GFRP: (a) (0/90), (b) (0/90/30/-60), (c) (0/90/45/-45), and (d) (30/-60/60/-30)

from 1.6×10^{-3} to 542 s^{-1}. A and B are material parameters for the strength and modulus, respectively. m_σ and m_E are the strain rate sensitivity estimating parameters for strength and modulus, respectively. A higher value of the coefficients of strain rate sensitivity (m) leads to a more strain rate effect in composites.

Figure 6.10 indicates the effect of strain rate on the tensile strength of glass/epoxy composite for different fiber orientations [(0/90), (0/90/30/-60), (0/90/45/-45), and (30/-60/60/- 30)]. R^2 (coefficient of determination) value gives the accuracy of experimental and curve fit values. R^2 values obtained from the figure were greater than 0.99 for all laminate configurations which confirms the accuracy of curve fit and experimental values.

The slope of Kσ vs. strain rate predicts the strain rate effect of composites. The higher the slope, the better could be the strain rate sensitivity. The (0/90) laminate has the least slope and the (0/90/45/-45) laminate has the highest slope compared with the other laminates. Therefore, the influence of strain rate is more for the (0/90/45/-45) laminate compared to the other laminates.

Similarly, Figure 6.11 indicates the effect of strain rate on the tensile modulus of glass/epoxy composite for different orientations. R^2 values obtained were greater than 0.99 for all laminate configurations. The correlation between experimental values and Cowper- Symonds equations was in good agreement. A similar trend was observed here like the strain rate effect on the tensile strength of glass/epoxy composite. The (0/90) laminate has the least slope and the (0/90/45/-45) laminate has the highest slope compared to the other laminates.

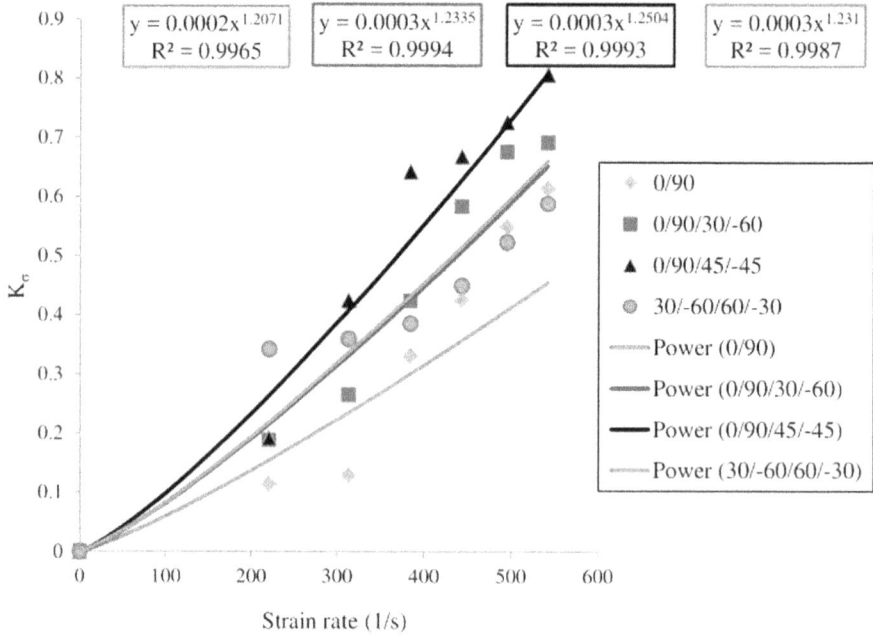

FIGURE 6.10 Strain rate effect on the tensile strength of different GFRP laminates.

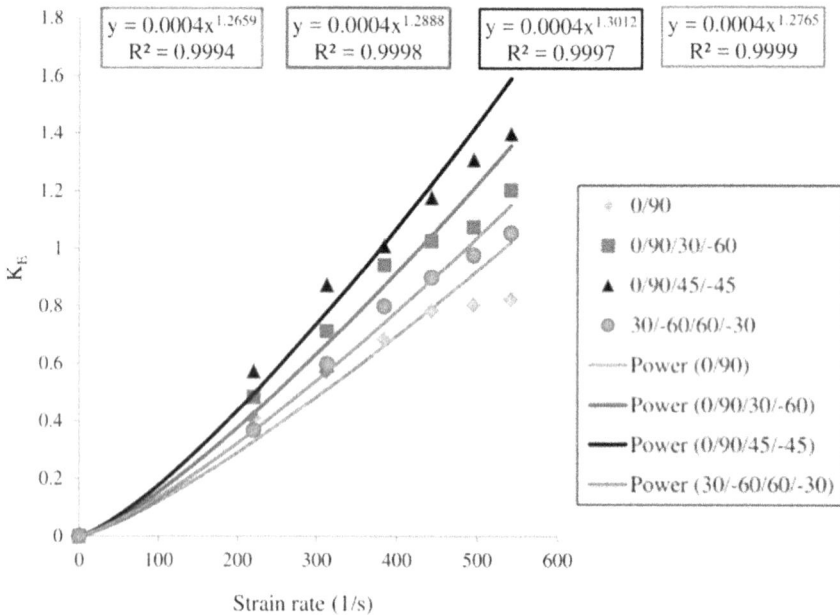

FIGURE 6.11 Strain rate effect on the tensile modulus of different GFRP laminates.

The parameters for tensile strength and modulus of GFRP composite for different laminate configurations are given in Table 6.1. It is worth observing that the (0/90/45/-45) laminate has higher and values. Hence, it can be concluded that the (0/90/45/-45) laminate exhibits more strain sensitivity than the other lay-up configurations. This can be attributed to (±45) lay-ups which have longer fibers along the diagonal. In general, the on-axis (0/90) specimens offer low strain rate sensitivity than the off-axis specimens [26].

TABLE 6.1
Strain Rate Parameters for GFRP Composite

Laminate	Strength parameters [Eqn. (6.1)]			Modulus parameters [Eqn. (6.2)]		
	A (s⁻¹)	m_σ	R^2	B (s⁻¹)	m_E	R^2
(0/90)	1,159.70	1.207	0.996	483.055	1.266	0.999
(0/90/30/-60)	717.798	1.233	0.999	432.619	1.289	0.999
(0/90/45/-45)	656.752	1.250	0.999	408.686	1.301	0.999
(30/-60/60/-30)	727.450	1.231	0.998	459.113	1.276	0.999

6.3.4 EFFECT OF STRAIN RATE ON TENSILE PROPERTIES OF CARBON/EPOXY COMPOSITE

Studies in prior literature show a lack of clarity in information related to the strain rate sensitivity of CFRP composite. Some studies report that CFRP composite is sensitive to the strain rate [15, 24, 25] but some conclude that CFRP is less or insensitive to the strain rate [22, 23].

Tensile properties (strength, failure strain, and modulus) of carbon/epoxy composite were studied for different strain rates. The dog bone geometry specimens (Figures 6.12 and 6.13) ensure that the failure has taken place in the gauge portion for both the quasi-static and dynamic loadings. The specimen fails quickly in dynamic loading as compared to quasi-static loading. As a result, global failure occurred in the former case (Figure 6.12) while localized failure occurred in the latter case (Figure 6.13).

6.3.4.1 Strain Analysis of Carbon/Epoxy Composite

CFRP composite of different fiber orientations was considered for the study. Typical strain contour plots of (0/90), (0/90/30/-60), (0/90/45/-45), and (30/-60/60/-30) carbon/epoxy composite under

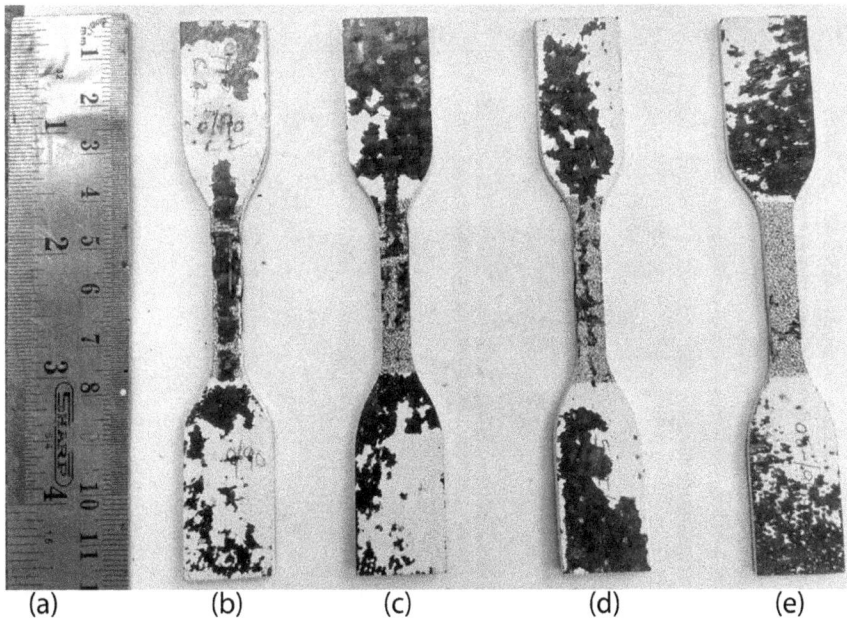

(a)   (b)   (c)   (d)   (e)

FIGURE 6.12 Fractured specimens of carbon/epoxy composite subjected to quasi-static tensile loading: (a) (0/90), (b) (0/90/30/-60), (c) (0/90/45/-45), and (d) (30/-60/60/-30).

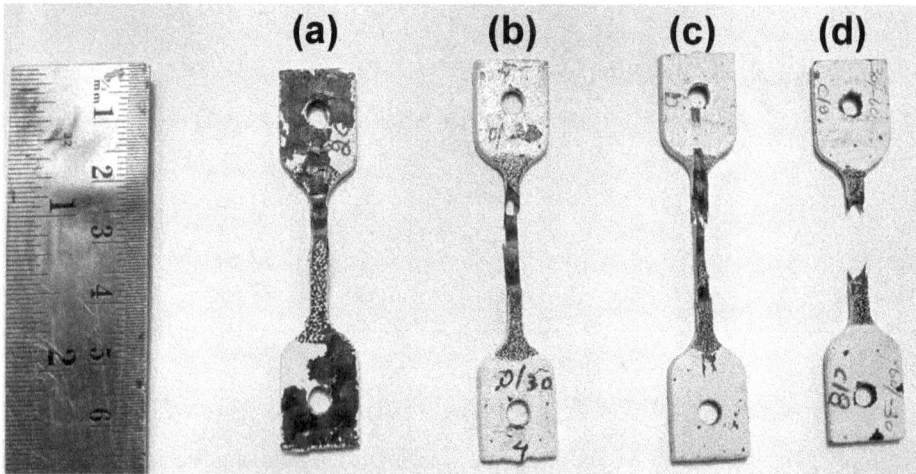

FIGURE 6.13 Fractured specimens of carbon/epoxy composite subjected to high strain rate tensile loading: (a) (0/90), (b) (0/90/30/-60), (c) (0/90/45/-45), and (d) (30/-60/60/-30).

quasi-static loading are shown in Figures 6.14 and 6.15. These figures show the direction of the failure of composite along the fiber direction for their respective laminate configurations.

Figures 6.14(i)(a–c) represent the recorded images of (0/90) laminate, just before and at the failure during the quasi-static loading. Figures 6.14 (i)(a–b) indicate the images before failure with and without strain contours. The strain contour plot [Figure 6.14(i)(b)] was obtained after post-processing the images using the VIC-2D software. The corresponding failure load was 6.12 kN. The failure strain (ε_{yy}) observed was 1.87% [Figure 6.14(i)(b)]. Figure 6.14(i)(c) indicates that the crack has started along the loading direction. Similarly, DIC images for (0/90/30/-60) laminate under quasi-static loading are shown in Figure 6.14(ii)(a–c). Figures 6.14(ii)(a–b) correspond to the load of 3.88 kN

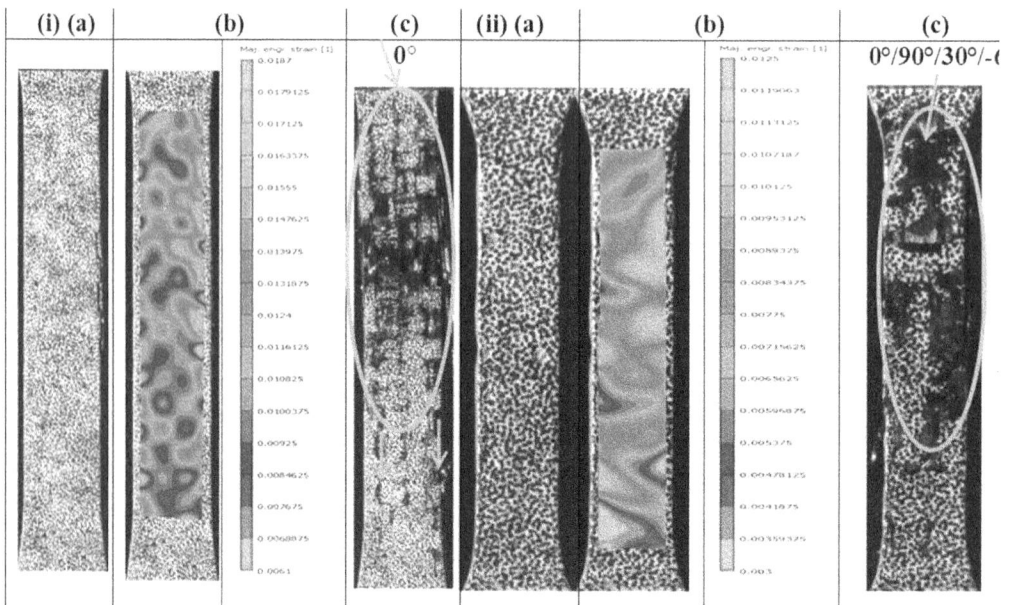

FIGURE 6.14 CFRP (i) (0/90) and (ii) (0/90/30/-60): (a) speckle movements just before failure, (b) strain component (ε_{yy}), and (c) damaged specimen.

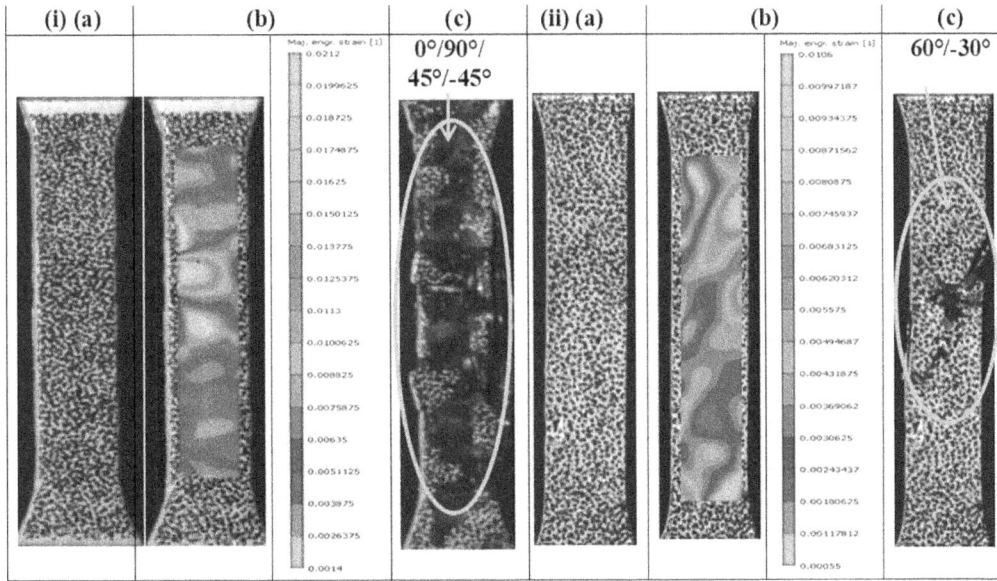

FIGURE 6.15 CFRP (i) (0/90/45/-45) and (ii) (30/-60/60/-30): (a) speckle movements just before failure, (b) strain component (ε_{yy}), and (c) damaged specimen.

and the failure strain (ε_{yy}) of 1.25%. The direction of the fracture angle is seen in Figure 6.14(ii)(c) indicates the failure midway between (0°/90°) and (60°/–30°) directions.

Figures 6.15(i)(a–b) show the DIC images just before the failure of the (0/90/45/-45) laminate. The failure strain seen in Figure 6.15(i)(a) is 2.12% for the corresponding load value of 3.44 kN. Figure 6.15(i)(c) indicates the failure is along both the (0°/90°) and (45°/–45°) directions. The corresponding load and failure strain values were 1.55 kN and 1.06%, respectively for (30/-60/60/-30) laminates [Figures 6.15(ii)(a–b)]. Figure 6.15(ii)(c) indicates the failure along the (60°/–30°) direction.

Similarly, Figures 6.16(i–iv) indicate the state just before failure as shown in images (a), strain contours as shown in images (b) at the corresponding state in images (a), and damaged specimen in images (c), for carbon/epoxy composite of different fiber orientations at 542 s⁻¹.

The failure strain (ε_{yy}) values for (0/90), (0/90/30/-60), (0/90/45/-45), and (30/-60/60/- 30) laminates were 0.90%, 0.94%, 0.99%, and 0.88%, respectively [Figures 6.16 (i–iv)(b)]. The corresponding load values for these laminate configurations were 3.35 kN, 2.69 kN, 2.46 kN, and 1.04 kN, respectively. In contrast to progressive failure observed in quasi-static loaded specimens [Figures 6.14(c) and 6.15(c)], the tested specimens at 542 s⁻¹ showed severe damage [Figures 6.16(i–iv)(c)]. These images also indicate failure along the fiber direction for their respective laminate configurations.

The stress-strain response of different laminate configurations of CFRP composite is seen in Figures 6.17(a–d). In Figure 6.17(a), it was observed that the properties were almost the same for quasi-static and high strain rate, which depicts less strain rate effect on tensile properties of (0/90) CFRP laminate. An increase in the slope of curves with the increase of strain rate was observed in (0/90/30/-60) laminate in Figure 6.17(b), though the magnitude of properties was lower than the (0/90) laminate. However, a significant increment in the slope of stress-strain curves was observed in (0/90/45/-45) laminate, as the strain rate increased. The magnitude of tensile properties obtained was lower in (0/90/45/-45) laminate than (0/90) and (0/90/30/-60) laminates, though these slopes were higher than the other laminates. In Figure 6.17(d), it was observed that the initial portion of the dynamic stress-strain curves closely resembles one another in (30/-60/60/-30) laminate, though these curves are steeper when compared to the static curve. This confirms the lower

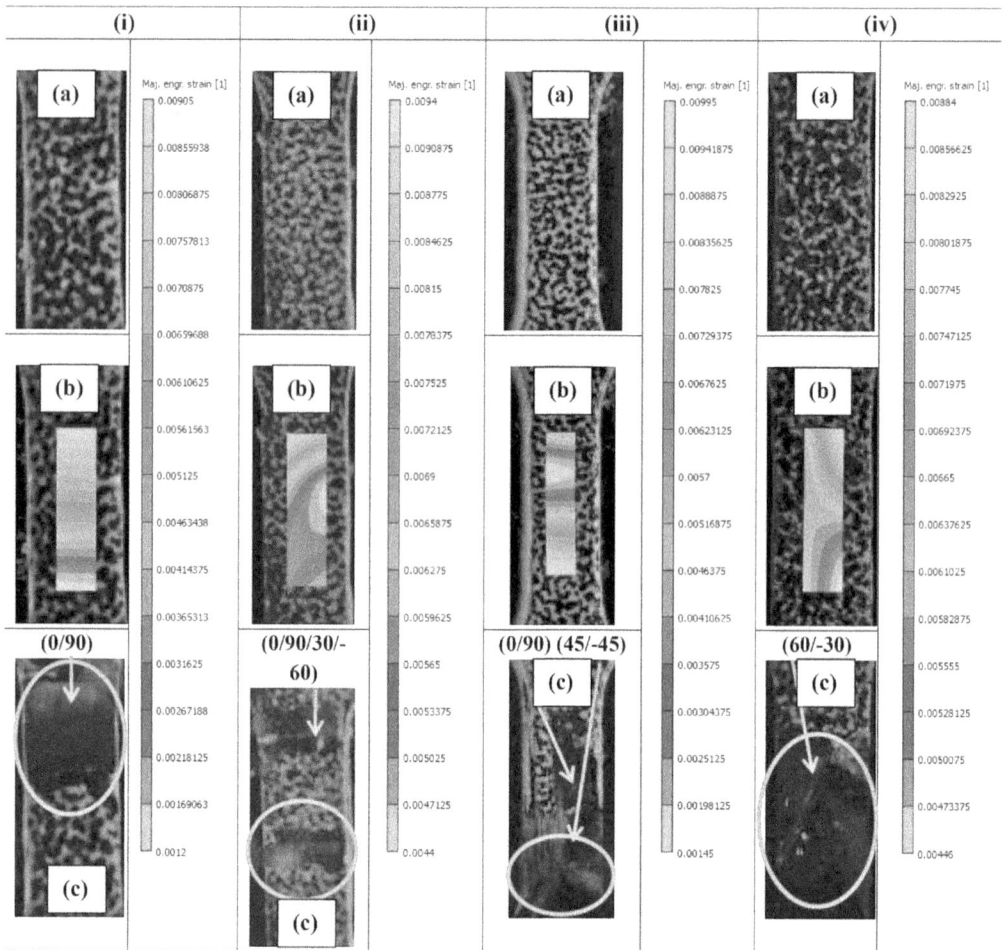

FIGURE 6.16 High-speed camera images for carbon/epoxy composite of different fiber orientations at 542 s^{-1}: (i) (0/90), (ii) (0/90/30/−60), (iii) (0/90/45/−45), and (iv) (30/−60/60/−30).

strain rate effect of off-axis woven laminate compared to partial on-axis/off-axis woven laminate (0/90/45/-45).

The fiber-matrix interfacial bonding increased with test speed, due to which the material had a tendency to behave in a brittle manner, leading to an enhancement of strength and stiffness and a reduction of failure strain with the increase of strain rate. However, the increment in Young's modulus was found to be lower compared to the strength and failure strain as the strain rate increased from quasi-static to 542 s^{-1}, for CFRP composite of all orientations except for (0/90) laminate. Both strength and Young's modulus of (0/90) laminate were less sensitive to the strain rate. Similar kinds of observations were found by Powell et al. [24] for the woven roving mat, twill weave (0/90) carbon/epoxy composite.

Moreover, the variations in tensile strength, strain, and Young's modulus with respect to strain rate were observed to be less for all four orientations of carbon/epoxy composite [Figures 6.17(a–d)], when compared to glass/epoxy composite [Figures 16.9(a–d)]. In the same lay-up sequences of GFRP and CFRP composite, the tensile strength and modulus were obtained to be higher for the CFRP composite whereas the percentage of elongation and strain rate sensitivity were obtained to be higher for the GFRP composite. As in the case of the GFRP composite, the (0/90) CFRP laminate also has higher tensile properties, and the (30/-60/60/-30) laminate has lower properties. This

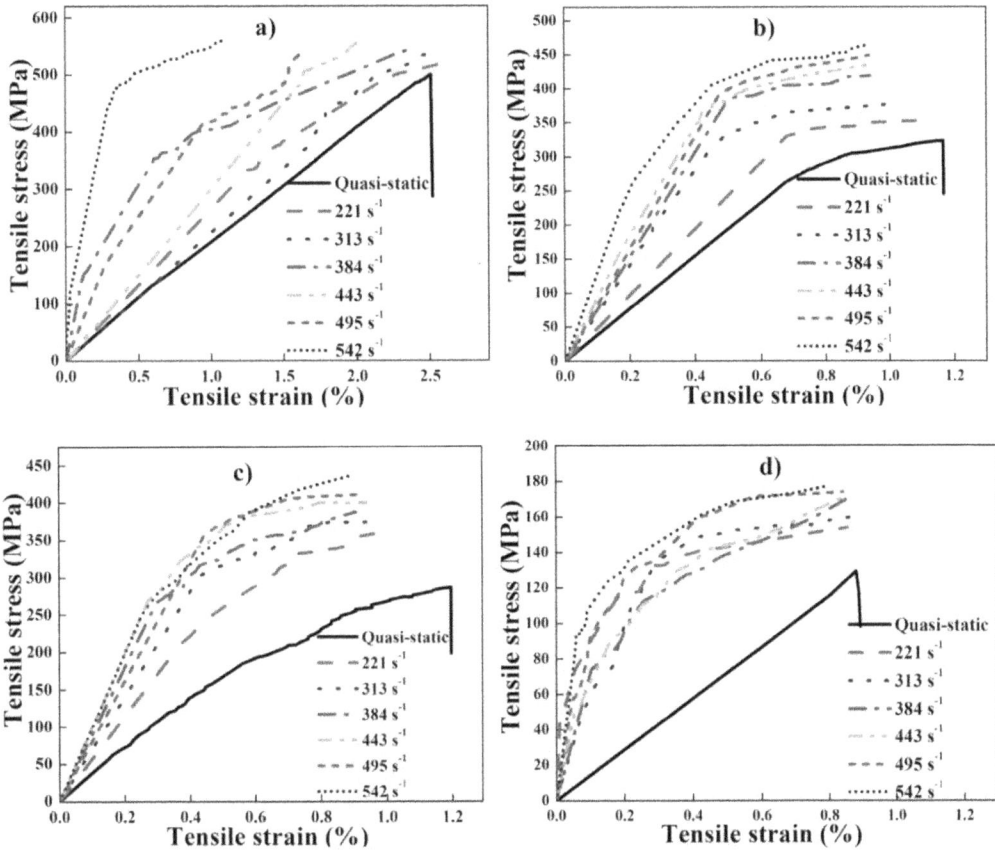

FIGURE 6.17   Stress-strain curves of carbon/epoxy composite: (a) (0/90), (b) (0/90/30/−60), (c) (0/90/45/−45), and (d) (30/−60/60/−30).

can be attributed to the length and amount of fibers present along the loading direction being more and the fibers are tightly aligned with respect to the warp and weft directions for the (0/90) laminate (on-axis). However, the length and amount of fibers present along the loading direction are less in the (30/-60/60/-30) laminate (off-axis) compared to the other laminate configurations.

6.3.5   EFFECT OF STRAIN RATE SENSITIVITY ON TENSILE PROPERTIES OF CARBON/EPOXY COMPOSITE

The degree of strain rate sensitivity of composites varies with the type of fiber and matrix, the type and the direction of loading, etc. For a clear understanding of the strain rate dependency of composites, the Cowper-Symonds model has been used in this study.

6.3.5.1   Determination of Strain Rate Parameters of Carbon/Epoxy Composite

The effect of strain rate on the tensile strength of CFRP composite for different fiber orientations [(0/90), (0/90/30/-60), (0/90/45/-45), and (30/-60/60/-30)] was investigated using the Cowper-Symonds model [Eqn. (6.1)] from the known experimental values; this is shown in Figure 6.18 and the parameters (A and m_σ) were determined. R^2 values obtained were greater than 0.99 for all four laminate configurations which confirms that the curve fit and experimental values are accurate. It is also observed from the figure that the slope of the K_σ vs. strain rate curve is very low for (0/90) laminate while the (0/90/45/-45) laminate has a higher slope than the other laminates. Hence,

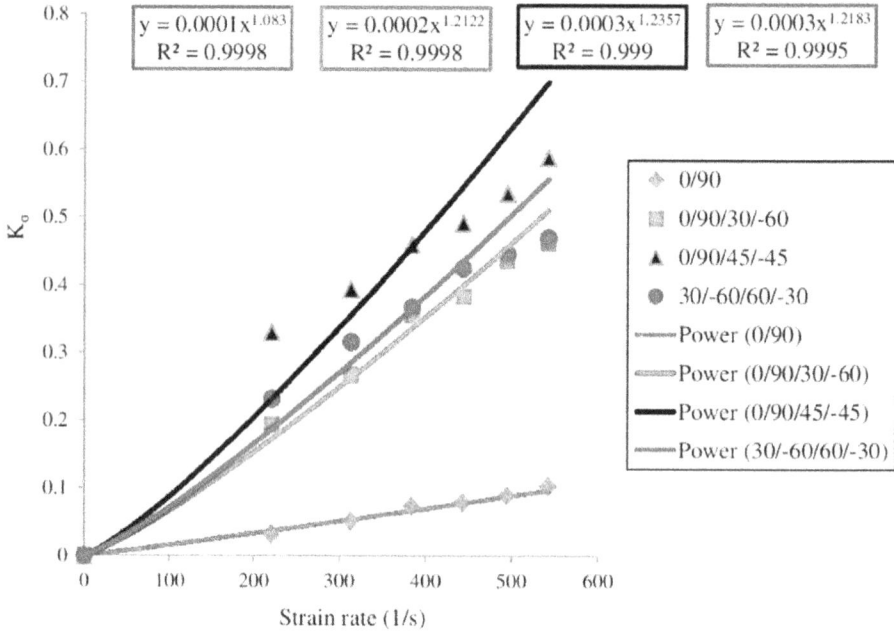

FIGURE 6.18 The effect of strain rate on the tensile strength of different CFRP laminates.

the strain rate sensitivity is higher for the (0/90/45/-45) laminate compared to the other laminate configurations.

Strain rate constants for the tensile modulus of CFRP composite for different fiber orientations were determined using Eqn. (6.2) from known experimental values and are shown in Figure 6.19. It is seen from the figure that R^2 values obtained were greater than 0.99 for all four laminate configurations [(0/90), (0/90/30/-60), (0/90/45/-45), and (30/-60/60/-30)]. Slopes indicate that the effect of strain rate on the tensile modulus was higher for (0/90/45/-45) laminate. This indicated that the (0/90/45/-45) laminate was more sensitive to strain rate than the laminates of other fiber orientations.

The strain rate parameters for strength and modulus are given in Table 6.2. It was observed from Table 6.2 that the strain rate sensitivity determination parameter () was lower for the tensile strength of the (0/90) carbon/epoxy composite. It revealed that there was no significant change in tensile strength for the (0/90) carbon/epoxy composite as the strain rate increased from 1.6×10^{-3} to 542 s^{-1}. In contrast to the (0/90) laminate, other laminates exhibited a significant increase in tensile strength and modulus for the same range of strain rates. This was due to the higher shear effect induced in the matrix for disorientated laminates compared to aligned fibrous laminates. It is clear that the strain rate sensitivity is higher for the (0/90/45/-45) laminate, which has [±45°] plies, than the other three laminates. The [±45°] plies are more sensitive to the strain rate [25, 28]. These results match well with the experimental results.

6.3.6 EFFECT OF STRAIN RATE ON TENSILE PROPERTIES OF HYBRID COMPOSITE

Epoxy-reinforced (0/90) interply hybrid fabrics (two glass and two carbon fabric layers) composite was considered for the study to obtain unambiguous conclusions on tensile properties over a wide range of strain rates from quasi-static to 542 s^{-1}. The fractured surface of quasi-static and high strain rate specimens are shown in Figures 6.20(a) and (b), respectively, which indicate that the direction of failure was along the loading direction.

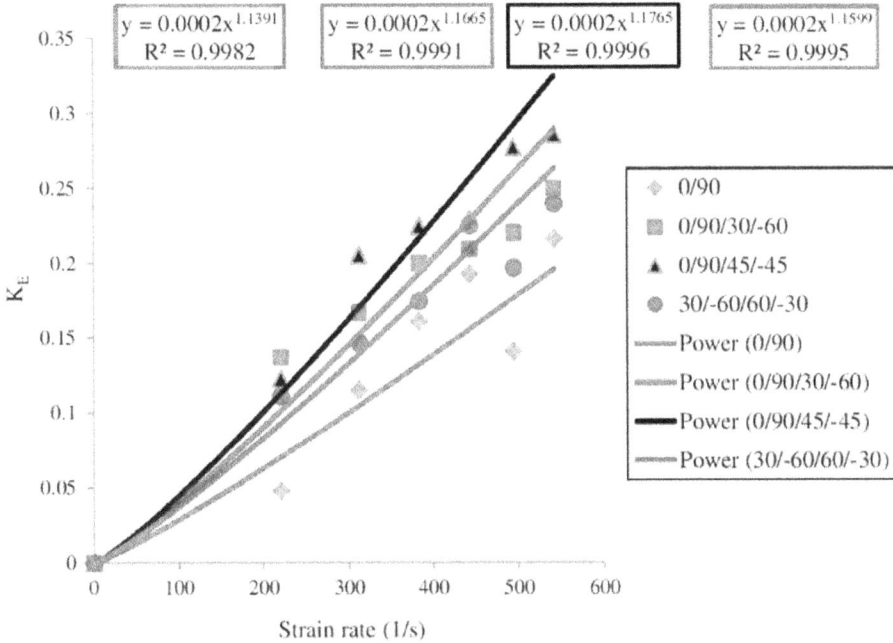

FIGURE 6.19   The effect of strain rate on the tensile modulus of different CFRP laminates.

TABLE 6.2
Strain Rate Parameters of CFRP Composite

Laminate	Strength parameters [Eqn. (6.1)]			Modulus parameters [Eqn. (6.2)]		
	A (s⁻¹)	m_σ	R^2	B (s⁻¹)	m_E	R^2
(0/90)	4,936.784	1.083	0.999	1,767.153	1.139	0.998
(0/90/30/-60)	1,119.275	1.212	0.999	1,477.89	1.167	0.999
(0/90/45/-45)	708.314	1.236	0.999	1,389.025	1.177	0.999
(30/-60/60/-30)	779.175	1.218	0.999	1,544.427	1.160	0.999

Typical strain contour plots of the hybrid (glass/carbon/epoxy) composite for the strain rates of 1.6×10^{-3} s⁻¹ (quasi-static) and 542 s⁻¹ are shown in Figures 6.21(a) and (b), respectively. These figures show a decrease in the failure strain from 3.76% to 1.285% with an increase in strain rate from 1.6×10^{-3} s⁻¹ to 542 s⁻¹. However, these values are higher than the (0/90) carbon/epoxy composite [Figure 6.9(a)] and lower than the (0/90) glass/epoxy composite [Figure 6.17(a)]. These failure strain values correspond to the load of 4.74 kN and 3.6 kN, respectively.

The tensile properties of hybrid (glass/carbon/epoxy) composite are given in Table 6.3. It was observed that the tensile strength increased from 393.7 MPa to 579 MPa (47% increase), as the strain rate increased from 1.6×10^{-3} to 542 s⁻¹. The modulus increased from 24.58 GPa to 32.85 GPa (33.6% increase), as against, the failure strain decreased from 3.51% to 1.60% (54.4% decrease) for the corresponding strain rates. As the specimen elongation happened rapidly in dynamic loading, this resulted in higher strength and modulus and lower failure strain compared to quasi-static loading. Details of the stress-strain response are shown in Figure 6.22.

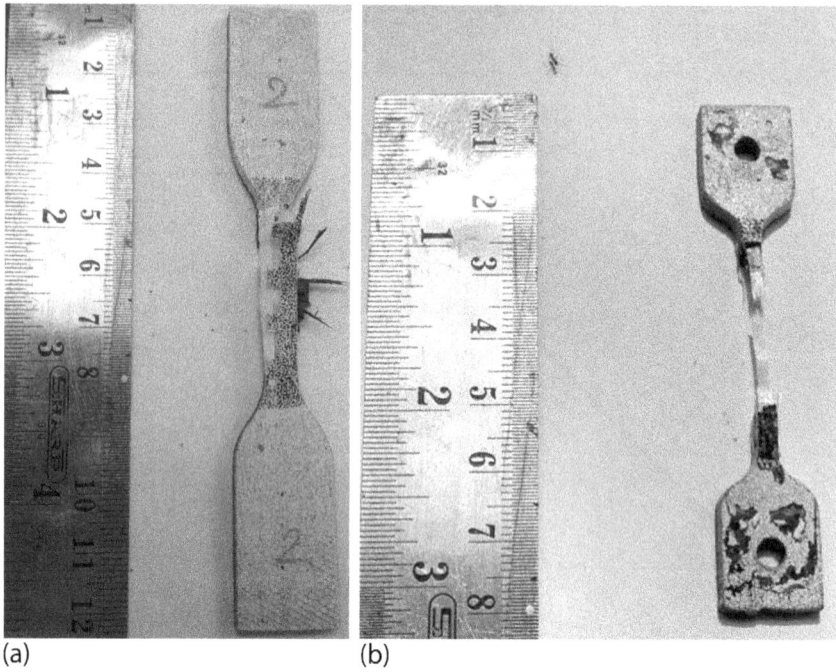

(a)   (b)

FIGURE 6.20 Typical fractured specimens of (0/90) hybrid (glass/carbon/epoxy) composite at different strain rates: (a) 1.6×10^{-3} s^{-1} and (b) 542 s^{-1}.

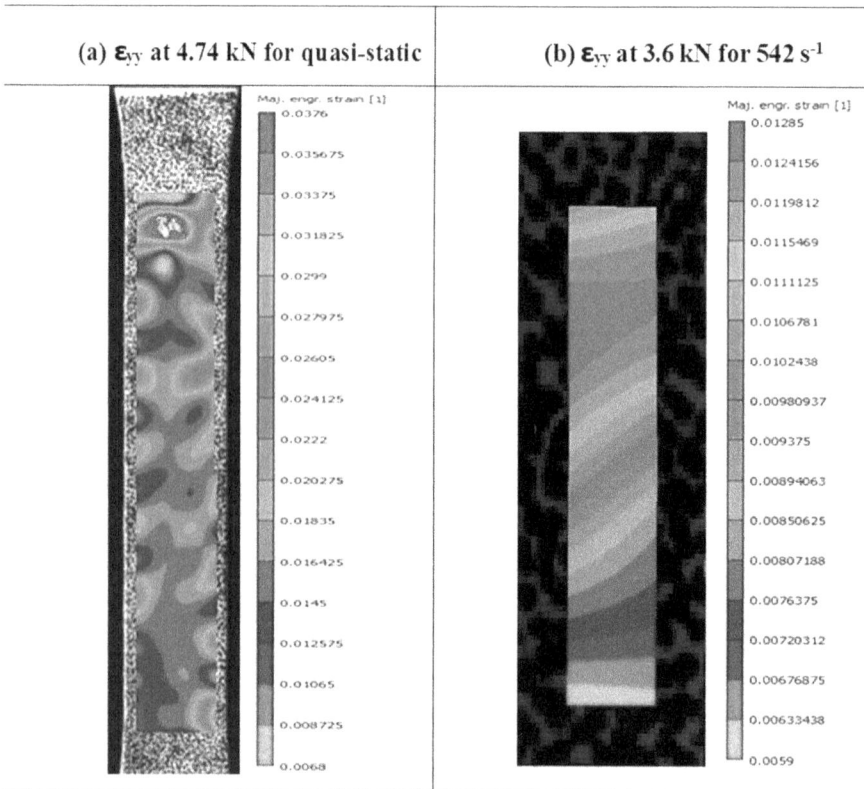

FIGURE 6.21 Strain contour plots of (0/90) hybrid (glass/carbon/epoxy) composite.

TABLE 6.3

Tensile Test Results of the (0/90) Hybrid (Glass/Carbon/Epoxy) Composite

Height (m)	Velocity (ms⁻¹)	Strain rate (s⁻¹)	Tensile stress (MPa)	Failure strain (%)	Tensile modulus (GPa)
Quasi-static	8.3×10^{-5}	1.6×10^{-3}	393.7 ± 14.1	3.51 ± 0.54	24.58 ± 1.69
0.25	2.21	221	428.3 ± 04.3	2.95 ± 0.81	27.61 ± 1.54
0.50	3.13	313	438.3 ± 11.1	2.54 ± 0.43	31.10 ± 1.63
0.75	3.84	384	461.2 ± 21.0	2.1 ± 0.82	32.65 ± 7.14
1.00	4.43	443	472.6 ± 20.9	1.81 ± 0.27	31.0 ± 4.11
1.25	4.95	495	530.6 ± 26.5	1.67 ± 0.48	32.21 ± 3.15
1.50	5.42	542	579.0 ± 29.0	1.60 ± 0.55	32.85 ± 5.20

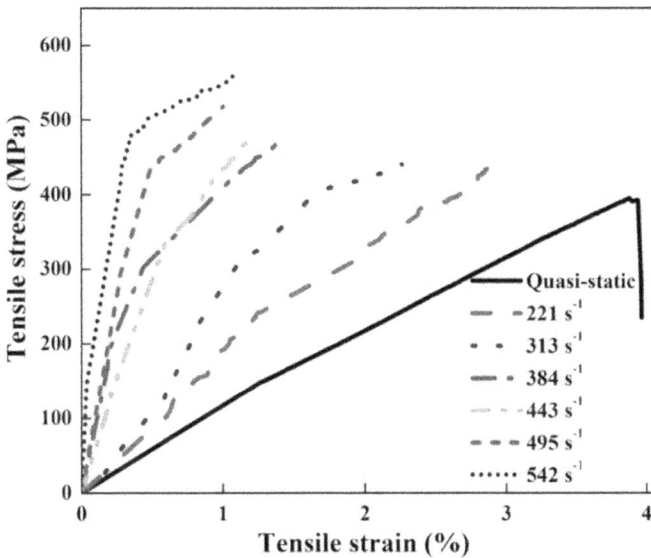

FIGURE 6.22   Stress-strain curves of the hybrid (glass/carbon/epoxy) composite

6.3.7   STRAIN RATE PARAMETERS OF HYBRID COMPOSITE

Strain rate constants for tensile strength and modulus of hybrid (glass/carbon/epoxy) composite were determined by using the Cowper-Symonds model [Eqns. (6.1) and (6.2)] from known experimental values and are given in Figure 6.23. R^2 (determination coefficient) values obtained were above 0.99, which confirms the accuracy of experimental and curve fit values. The strain rate parameters for the (0/90) hybrid composite are given in Table 6.4. These parameters indicate that the values are less in the hybrid composite than those of the (0/90) glass/epoxy composite (Table 6.1) and higher than those of the (0/90) carbon/epoxy composite (Table 6.2).

6.4   CONCLUSIONS

In this study, the effects of different strain rates on tensile properties for different fiber orientations of glass/epoxy, carbon/epoxy, and hybrid composites were investigated and the strain rate parameters were predicted using the Cowper-Symonds model. The important conclusions are summarized:

- In both GFRP and CFRP composites, the on-axis laminates exhibited higher tensile strength and modulus compared to the other laminate configurations. This is because, in on-axis laminates, fibers are aligned tightly with respect to the warp and weft directions and are longer along the loading axis.
- In both GFRP and CFRP composites, the on-axis and off-axis laminates are less sensitive to the strain rate compared to the partial on-axis/off-axis laminates, particularly the (0/90/45/−45) laminates possess significantly higher strain rate sensitivity. In the presence of (±45) lay-ups in (0/90/45/−45) laminates, the higher shear effect is induced which is the reason for having higher strain rate sensitivity than the other ply sequences.
- In the same lay-up configurations of GFRP and CFRP, the (0/90) CFRP laminate exhibited higher tensile strength and modulus at all strain rates compared to the other laminate configurations. However, the (0/90/45/-45) GFRP laminate exhibited a higher strain rate sensitivity compared to the other laminate configurations leading to the increase in strength (80.64%) and modulus (139.9%) for the increase of strain rate from 1.6×10^{-3} s^{-1} to 542 s^{-1}.
- Hybrid composite exhibited higher failure strain and strain rate sensitivity than the (0/90) carbon/epoxy laminate and higher tensile strength and modulus than the (0/90) glass/epoxy laminate. The best curve fit was obtained from the Cowper-Symonds model.
- The data presented in this study will be useful to develop strain-rate-dependent numerical models.

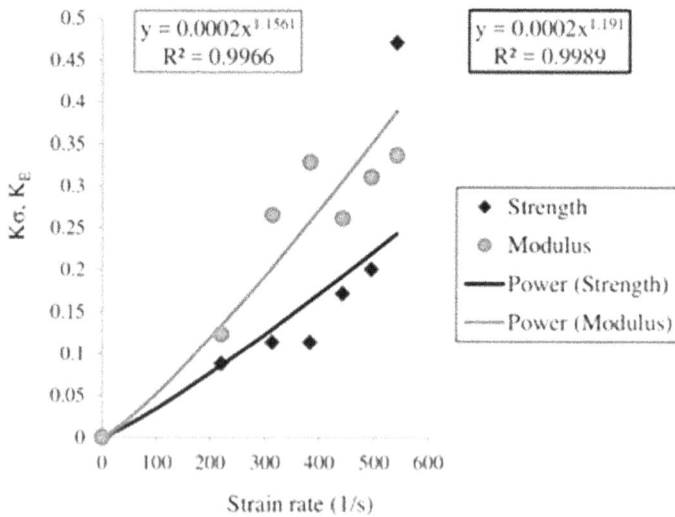

FIGURE 6.23   The effect of strain rate on the tensile strength and modulus of the (0/90) hybrid composite

TABLE 6.4
Strain Rate Parameters for the (0/90) Hybrid (Glass/Carbon/Epoxy) Composite

Strength parameters			Modulus parameters		
A (s^{-1})	m_σ	R^2	B (s^{-1})	m_E	R^2
1,583.158	1.156	0.996	1,275.758	1.191	0.998

REFERENCES

1. Perry JI, Walley SM. Measuring the effect of strain rate on deformation and damage in fibre-reinforced composites: A review. *Journal of Dynamic Behavior of Materials.* 2022;8(2):178–213.
2. Naresh K, Shankar K, Velmurugan R, Gupta NK. Statistical analysis of the tensile strength of GFRP, CFRP and hybrid composites. *Thin-Walled Structures.* 2018;126:150–61.
3. Balaganesan G, Velmurugan R, Srinivasan M, Gupta NK, Kanny K. Energy absorption and ballistic limit of nanocomposite laminates subjected to impact loading. *International Journal of Impact Engineering.* 2014;74:57–66.
4. Leite LFM, Leite BM, Reis VL, Alves da Silveira NN, Donadon MV. Strain rate effects on the intra-laminar fracture toughness of composite laminates subjected to tensile load. *Composite Structures.* 2018;201:455–67.
5. Naresh K, Shankar K, Rao BS, Velmurugan R. Effect of high strain rate on glass/carbon/hybrid fiber reinforced epoxy laminated composites. *Composites Part B: Engineering.* 2016;100:125–35.
6. Ma L, Liu F, Liu D, Liu Y. Review of strain rate effects of fiber-reinforced polymer composites. *Polymers (Basel).* 2021;13(17): 1–31
7. Gurusideswar S, Velmurugan R, Gupta NK. Study of rate dependent behavior of glass/epoxy composites with nanofillers using non-contact strain measurement. *International Journal of Impact Engineering.* 2017;110:324–37.
8. Schmack T, Filipe T, Deinzer G, Kassapoglou C, Walther F. Experimental and numerical investigation of the strain rate-dependent compression behaviour of a carbon-epoxy structure. *Composite Structures.* 2018;189:256–62.
9. Koerber H, Xavier J, Camanho PP. High strain rate characterisation of unidirectional carbon-epoxy IM7-8552 in transverse compression and in-plane shear using digital image correlation. *Mechanics of Materials.* 2010;42(11):1004–19.
10. Naik NK, Kavala VR. High strain rate behavior of woven fabric composites under compressive loading. *Materials Science and Engineering: A.* 2008;474(1–2):301–11.
11. Acharya S, Mondal DK, Ghosh KS, Mukhopadhyay AK. Mechanical behaviour of glass fibre reinforced composite at varying strain rates. *Materials Research Express.* 2017;4(3): 1–14.
12. Naresh K, Shankar K, Velmurugan R, Gupta NK. High strain rate studies for different laminate configurations of bi-directional glass/epoxy and carbon/epoxy composites using DIC. *Structures.* 2020;27:2451–65.
13. Pandya TS, Dave MJ, Street J, Blake C, Mitchell B. High strain rate response of bio-composites using split Hopkinson pressure bar and digital image correlation technique. *International Wood Products Journal.* 2019;10(1):22–30.
14. Li X, Yan Y, Guo L, Xu C. Effect of strain rate on the mechanical properties of carbon/epoxy composites under quasi-static and dynamic loadings. *Polymer Testing.* 2016;52:254–64.
15. Chen X, Li Y, Zhi Z, Guo Y, Ouyang N. The compressive and tensile behavior of a 0/90 C fiber woven composite at high strain rates. *Carbon.* 2013;61:97–104.
16. Jenq ST, Sheu SL. High strain rate compressional behavior of stitched and unstitched composite laminates with radial constraint. *Compos Struct.* 1993;25:427–38.
17. Tasdemirci A, Kara A, Turan AK, Tunusoglu G, Guden M, Hall IW. Experimental and numerical investigation of high strain rate mechanical behavior of a [0/45/90/ - 45] quadriaxial E-glass/polyester composite. *Procedia Engineering.* 2011;10:3068–73.
18. Kara A, Tasdemirci A, Guden M. Modeling quasi-static and high strain rate deformation and failure behavior of a (±45) symmetric E-glass/polyester composite under compressive loading. *Materials & Design.* 2013;49:566–74.
19. Luca AD, Di Caprio F, Caputo F, Lamanna G, Ignarra M. On the tensile behaviour of CF and CFRP materials under high strain rates. *Key Eng Mater.* 2017;754:111–4.
20. Gilat A, Schmidt TE, Walker AL. Full field strain measurement in compression and tensile split Hopkinson bar experiments. *Experimental Mechanics.* 2008;49(2):291–302.
21. Naresh K, Shankar K, Velmurugan R. Digital image processing and thermo-mechanical response of neat epoxy and different laminate orientations of fiber reinforced polymer composites for vibration isolation applications. *Int J Polym Anal.* 2018;23(8):684–709.
22. Harding J, Welsh LM. A tensile testing technique for fibre-reinforced composites at impact rates of strain. *Journal of Materials Science.* 1983;18:1810–26.

23. Kawata K, Hashimoto S, Sekino S, Takeda N. Macro- and micro-mechanics of high velocity brittleness and high-velocity ductility of solids. *Proceedings of the IUTAM Symposium on MMMHVDF*, Tokyo, Japan. 1985:1–26.

24. Powell LA, Luecke WE, Merzkirch M, Avery K, Foecke T. High strain rate mechanical characterization of carbon fiber reinforced polymer composites using digital image correlation. *SAE Int J Mater Manf*. 2017;10:138–46.

25. Gilat A, Goldberg RK, Roberts GD. Experimental study of strain-rate-dependent behavior of carbon/epoxy composite. *Compos Sci Technol*. 2002;62:1469–76.

26. Naresh K, Shankar K, Velmurugan R. Reliability analysis of tensile strengths using Weibull distribution in glass/epoxy and carbon/epoxy composites. *Composites Part B: Engineering*. 2018;133:129–44.

27. Cowper GR, Symonds PS. Strain hardening and strain rate effect in the impact loading of cantilever beams. Brown University, DTIC. 1957.

28. Sebaey TA, Costa J, Maimí P, Batista Y, Blanco N, Mayugo JA. Measurement of the in situ transverse tensile strength of composite plies by means of the real time monitoring of microcracking. *Composites Part B: Engineering*. 2014;65:40–6.

7 Rate-Dependent Cohesive Zone Modelling for Polyurea–Steel Interface

M. K. Saravanan, A. V. S. Siva Prasad
and Raguraman Munusamy

7.1 INTRODUCTION

It is necessary, in the defence industry, to protect structures from various conditions such as impacts, blasts and so on. Polymer is widely used to protect structures from both accidental and incidental failures. Polymer coatings on structures can withstand abrasion, impacts and extreme mechanical loads. Polyurea is a highly rate-sensitive polymer that plays an important role in the protection of critical buildings and defence vehicles, among other things.

Adhesive joints are now widely used to replace mechanical joints due to their ability to create a uniform stress distribution along the bonded area, resulting in higher stiffness, load transmission and fatigue resistance. It also reduces the structure's cost and weight. Steel coated with polyurea is tough, flexible and chemically resistant [1]. Traction separation law (TSL) with mode-I and mode-II behaviour is used to investigate the adhesive bonding between steel and polyurea.

So far, no work has been found that considers polyurea delamination from the substrate in the numerical modelling of the penetration response of polyurea-coated structures. All of the modelling approaches assume that the coating and the substrate are perfectly bonded. It is necessary to develop a method for modelling the delamination of polyurea from the substrate during an impact event using the cohesive zone method.

7.2 TRACTION SEPARATION LAW (TSL) FOR POLYUREA–STEEL INTERFACE

The double cantilever beam (DCB) test can be used to estimate the mode-I traction separation law (ASTM D5528-01) [2]. The end-notched flexure (ENF) test is used to estimate the mode-II traction separation law (ASTM D7905/D7905M-14) [3]. Figure 7.1 depicts the dimensions of the DCB and ENF tests.

Zhu et al. [1] reported experimental results of DCB and ENF on a polyurea–steel interface, which are shown in Figure 7.2 and Figure 7.3. The TSL between polyurea and steel is rate dependent, with some initial stress present before it begins to displace. As a result, appropriate cohesive zone modelling for various strain rates is required to capture the result. Sorensen [8] and Jin et al. [9] observed and explained a similar result between polyurethane and steel interfaces.

7.3 COHESIVE ZONE MODELLING

According to Zhu et al. [1], the traction separation law for each loading rate can be expressed by a quadratic polynomial equation.

DOI: 10.1201/9781003352358-7

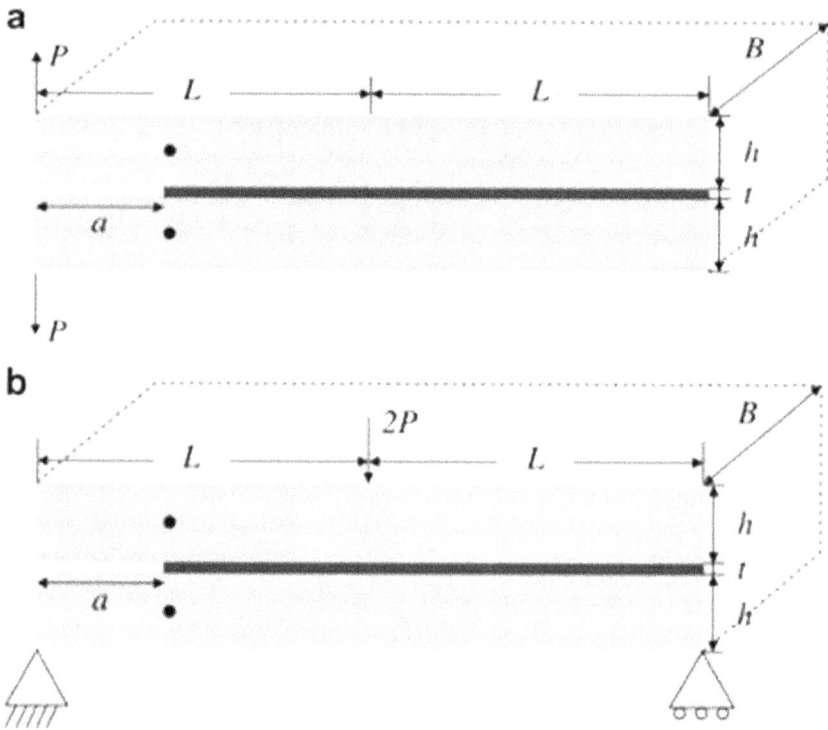

FIGURE 7.1 Schematics of (a) the DCB test configuration and specimen (L = 75 mm, B = 10 mm, and a range from 25 to 70 mm) and (b) the ENF test configuration and specimen (L = 75 mm, B = 10 mm, and a range from 15 to 30 mm). For both configurations, h = 4.76 mm and t = 0.7 mm. Sources: [1]; https://doi.org/10.1016/j.ijsolstr.2008.08.019 and http://myaccount.copyright.com.

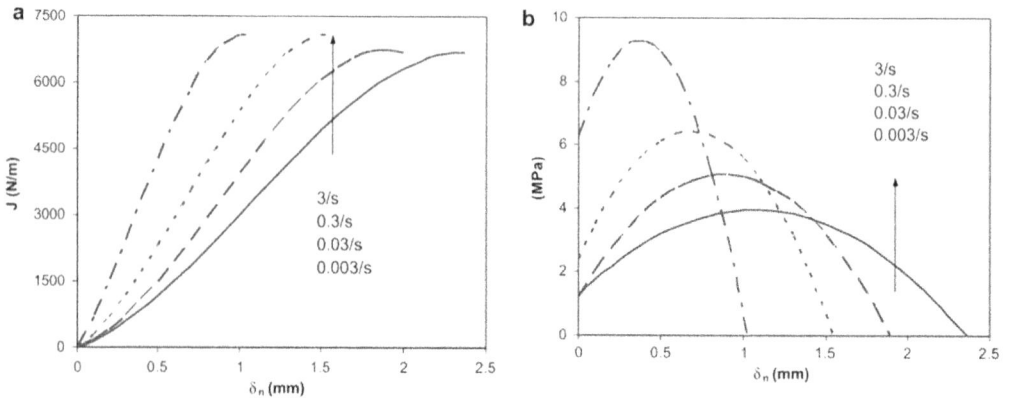

FIGURE 7.2 (a) J-δn curves; (b) traction-separation laws for opening mode fracture at different loading rates. δn in both the figures is the normal displacement. Sources: [1]; https://doi.org/10.1016/j.ijsolstr.2008.08.019 and http://myaccount.copyright.com.

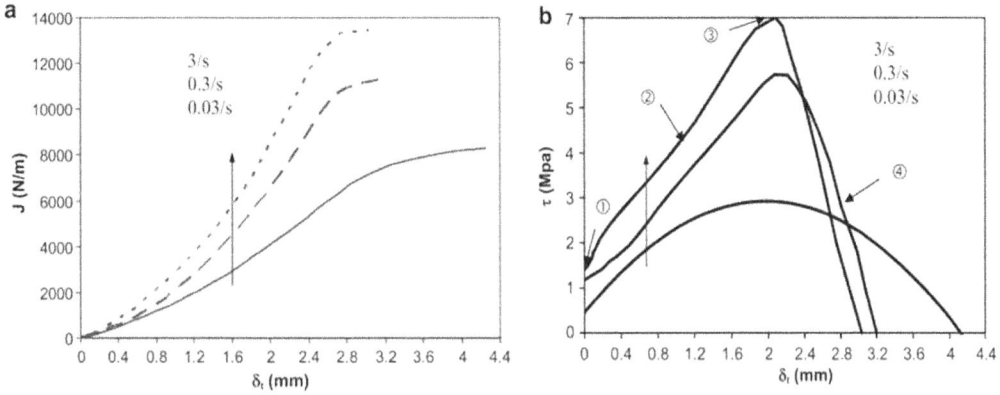

FIGURE 7.3 (a) J-δt curves; (b) traction-separation laws for shear fracture at different loading rates. δt in both the figures is the shear displacement. Sources: [1]; https://doi.org/10.1016/j.ijsolstr.2008.08.019 and http://myaccount.copyright.com.

$$\sigma = \sigma_c \left[1 - \frac{(\delta - \delta_0)^2}{(\delta_c - \delta_0)^2} \right] \tag{7.1}$$

where, σ_c is the peak stress, δ_c critical end opening displacement and δ_o is the end opening displacement at the peak stress. Table 7.1 lists the parameters for the quadratic traction separation law equation.

This model approximates the mode-I and II traction separation laws. Due to the presence of initial traction stress, the polynomial model and other existing models such as the exponential model and the bilinear model could not capture the experimental results. In Equations (7.2) and (7.3), a modified CZM with initial traction stress is proposed to predict TSL behaviour. Figure 7.4 shows how the developed cohesive zone model fits the experimental results with the least amount of deviation.

$$\sigma = \begin{cases} \sigma_0 \left[1 - \left(\dfrac{\delta}{\delta_0} \right)^{\frac{1}{2}} \right] + \sigma_c \left(\dfrac{\delta}{\delta_0} \right)^{\frac{1}{2}} & \text{if } \delta < \delta_0 \\[3ex] \sigma_c \left[1 - \dfrac{(\delta - \delta_0)^2}{(\delta_c - \delta_0)^2} \right] & \text{if } \delta < \delta_0 \end{cases} \tag{7.2}$$

$$\Gamma = \begin{cases} \sigma_0 \left(\delta - \dfrac{2\delta^{\frac{3}{2}}}{3\delta_0^{\frac{1}{2}}} \right) + \sigma_c \dfrac{2\delta^{\frac{3}{2}}}{3\delta_0^{\frac{1}{2}}} \infty & \text{if } \delta < \delta_0 \\[3ex] \dfrac{1}{3}\sigma_0\delta_0 + \dfrac{\sigma_c}{3} \left[3\delta - \delta_0 - \dfrac{(\delta - \delta_0)^3}{(\delta_c - \delta_0)^2} \right] & \text{if } \delta > \delta_0 \end{cases} \tag{7.3}$$

TABLE 7.1

Parameters for the Quadratic Traction-Separation Law Equation as a Function of Loading Rate and Mode

	Mode-I					Mode-II				
Rate (s⁻¹)	Γ_1 (KJ/m²)	σ_c (MPa)	δ_{n0} (mm)	δ_{nc} (mm)	σ_0 (MPa)	Γ_2 (KJ/m²)	τ_c (MPa)	δ_{t0} (mm)	δ_{tc} (mm)	τ_0 (MPa)
0.003	6.69	3.95	1.21	2.36	1.27	–	–	–	–	–
0.03	6.68	5.12	0.92	1.90	1.24	8.29	2.72	1.99	4.15	0.46
0.3	7.08	6.45	0.76	1.56	2.47	11.26	5.45	2.13	3.18	1.18
3	7.08	9.25	0.41	1.02	6.37	13.44	6.97	2.08	3.02	1.45

Source: [1].

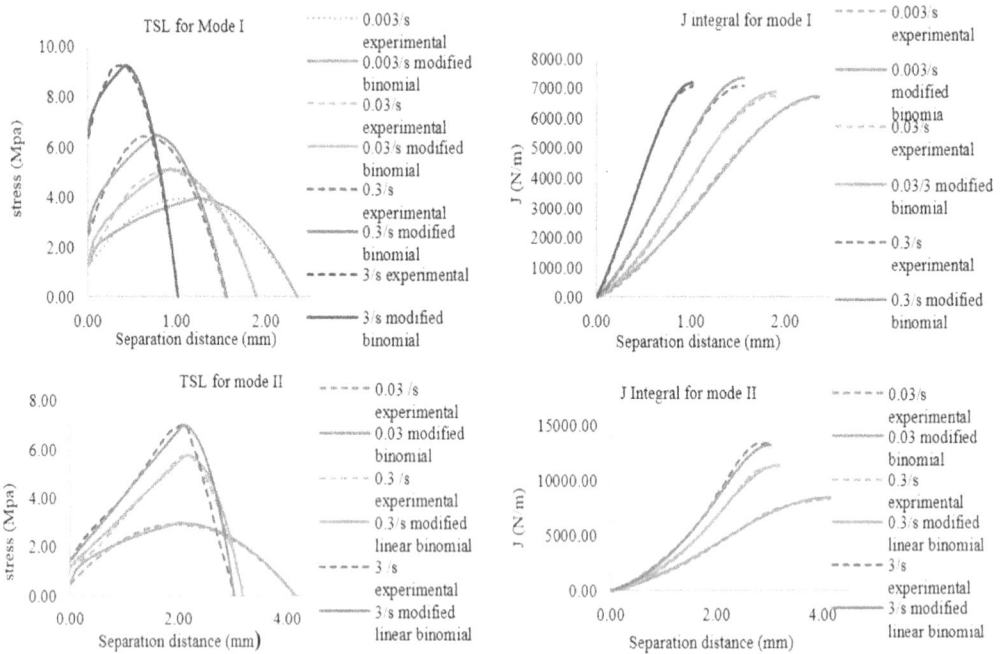

FIGURE 7.4 Comparing experimental data for mode-I and mode-II given by Zhu et al. [1] with the traction separation rule and J-integral of the modified cohesive zone model.

7.4 RATE-DEPENDENT COHESIVE ZONE MODELLING

There can be a drop in fracture energy as the strain rate rises in specific circumstances, which is reported by Karac et al. [11] and Blackman et al. [12]. The lower loading rates permit a more significant crack-tip deformation, which may result in improved toughness; toughness decreases with a rising strain rate.

In some cases, such as reported by Carlberger et al. [13], as the strain rate increased, the fracture energy significantly increased in the rubber-modified epoxy. The polyurea–steel interface also exhibited this behaviour for a mode-II ENF test.

In some cases, the fracture energy is constant below a lower limit strain rate; after a lower limiting strain rate, the fracture toughness grows a linear relation in a semi-logarithmic scale until a second boundary strain rate is reached. After reaching the second boundary stain rate, the fracture energy gets constant again, which is reported by Marzi et al. [6] and May et al. [10]. The polyurea–steel interface also exhibits the same behaviour in a mode-I DCB test.

For the polyurea–steel interface, it is important to create the model for the CZM parameters, in order to capture these rate-dependent CZM properties at various strain rates. For J_c and σ_c the models are adapted from Marzi et al. [6] and Borges et al. [7], which are the exponential and logarithmic equations. The same trend is developed for other CZM parameters which are listed in Table 7.2. The experimental findings given by Zhu et al. [1] are contrasted with the developed rate-dependent cohesive zone model in Figure 7.5. The rate-dependent cohesive zone model produces parameters that are within 5% of the experimental results.

7.5   NUMERICAL SIMULATION OF DCB IN LS-DYNA

Figure 7.1 shows the dimensions of the DCB according to the ASTM standard. Polyurea has a total thickness of 0.7 mm (0.6 mm polyurea and 0.5 mm thickness cohesive zone). The initial crack measures 50 mm in length. Steel is used as the adherent, and polyurea is used as the adhesive. Figure 7.6 depicts the DCB LS-Dyna model. The model is made up of a 1 mm hexahedron mesh. Steel and polyurea are meshed with a fully integrated S/R solid designed for elements. Cohesive zone elements are meshed with four-point cohesive elements.

TABLE 7.2
Equations Used to Define the Changes in Various Parameters as a Function of Strain Rate

S. no	Parameter	Type of equation	Equation
1	J_c	Exponential	$J_c = J_0 + \left(J_\infty - J_0\right)\exp\left(-\dfrac{\dot{\varepsilon}_i}{\dot{\varepsilon}}\right)$
2	σ_c	Logarithmic	$\sigma_c = \sigma_{ci} + \sigma_{c1}\left(ln\dfrac{\dot{\varepsilon}}{\dot{\varepsilon}_{min}}\right)^2$
3	δ_0	Logarithmic	$\delta_0 = \delta_{0i} + \delta_{01}$ $ln\dfrac{\dot{\varepsilon}}{\dot{\varepsilon}_{min}}$ For mode-I $\delta_0 = $ constant For mode-II
4	δ_c	Logarithmic	$\delta_c = \delta_{ci} + \delta_{c1}\ ln\dfrac{\dot{\varepsilon}}{\dot{\varepsilon}_{min}},$ For mode-I $\delta_c = \delta_{ci} + \delta_{c1}\left(ln\dfrac{\dot{\varepsilon}}{\dot{\varepsilon}_{min}}\right)^{1/2}$ For mode-II
5	σ_c	Exponential	$\sigma_0 = \sigma_{0i} + \left(\sigma_{0i} - \sigma_{0\infty}\right)\exp\left(-\dfrac{\dot{\varepsilon}_i}{\dot{\varepsilon}}\right)$

Sources: S. no 1 and 2: [6][7].

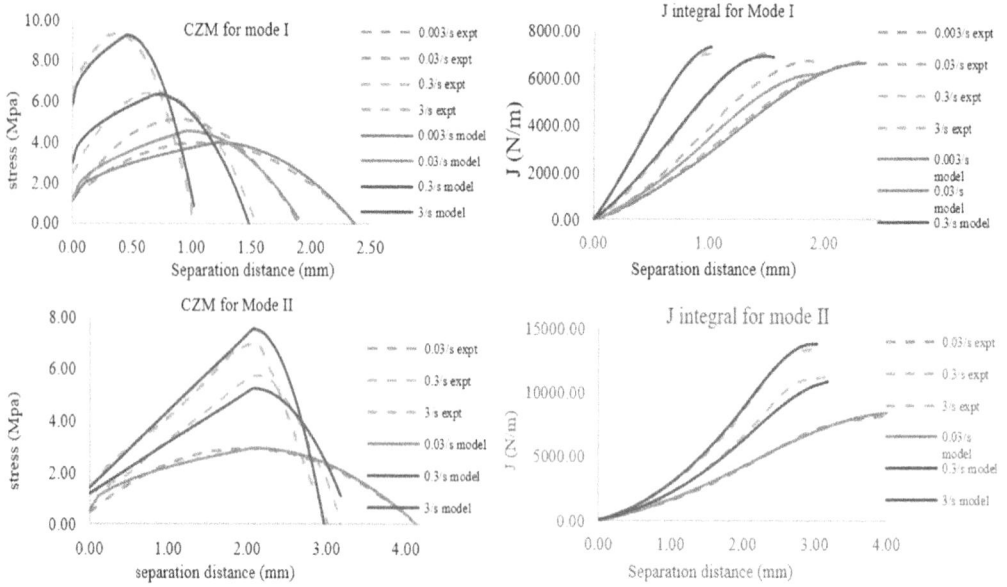

FIGURE 7.5 Comparison of developed rate-dependent cohesive zone modelling with experimental results from Zhu et al. [1] for (a) mode-I and (b) mode-II.

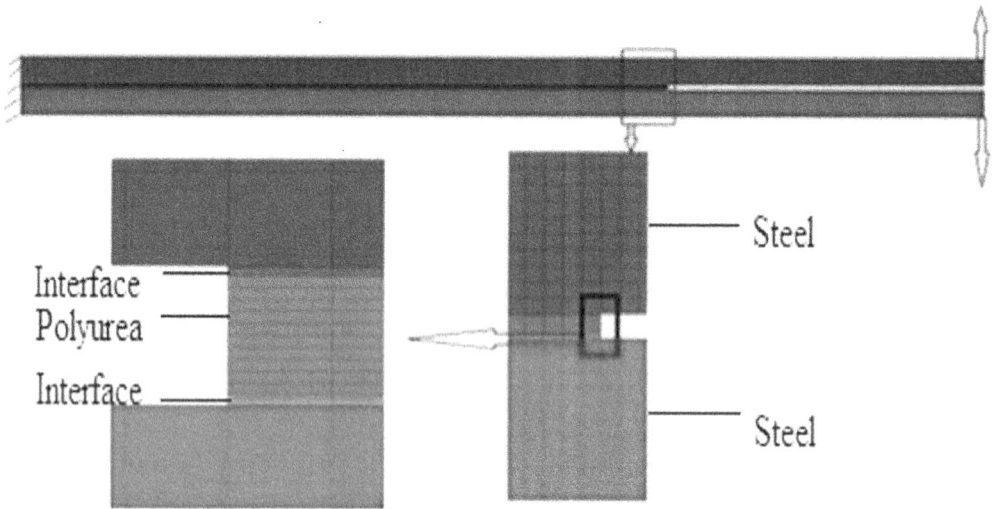

FIGURE 7.6 DCB specimen boundary conditions and details.

The boundary conditions are as follows: one end of the beam is fixed by constraining all displacements and moments, and the other end is subjected to a constant velocity upward for the top beam and downward for the bottom beam (1.05 m/s, 0.105 m/s and 0.0105 m/s for strain rates of 3 /s, 0.3 /s and 0.03 /s, respectively). For steel and polyurea, the elastic-plastic model is used, along with the failure strain [4]. Table 7.1 lists the material parameters for the cohesive zone model.

7.6 RESULTS AND DISCUSSIONS

Figure 7.7 depicts the obtained load-displacement plot. The compliance calibration is calculated by plotting the slope of log (C) versus log (a), where C is the compliance. For strain rates

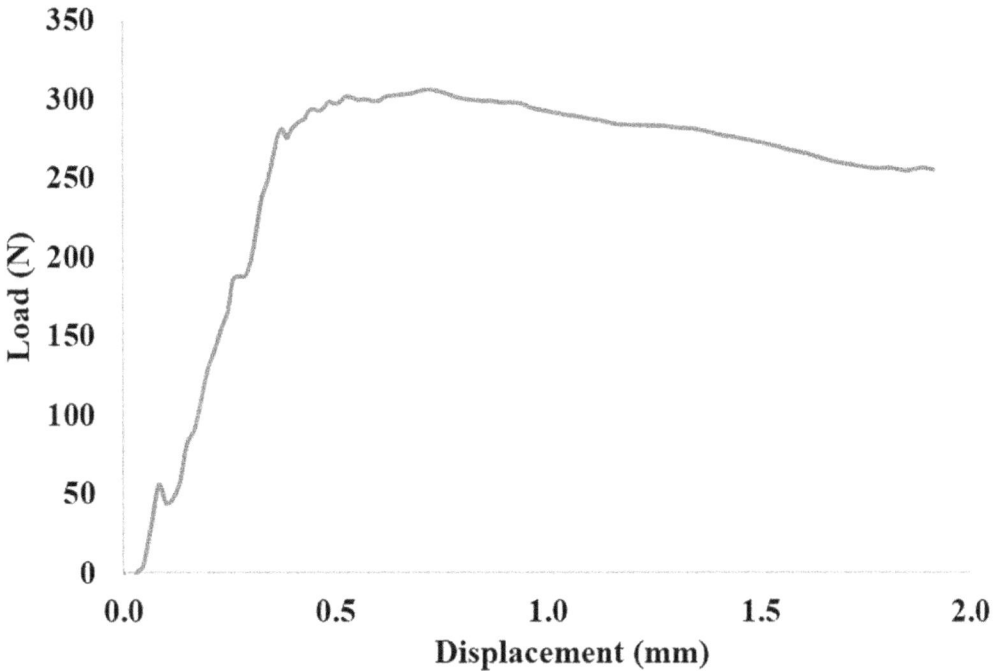

FIGURE 7.7 Load-displacement plot of steel–polyurea interface at the strain rate of 3/s from numerical simulation.

of 0.03 /s, 0.3 /s and 3 /s, the compliance calibration 'n' obtained from the plots is 3.39, 3.64 and 4.18, respectively.

ASTM D5528-01 specifies the mode-I energy release rate per unit width (J-integral) [2].

$$J = G_1 = \frac{nP\delta}{2ba} \tag{7.4}$$

where P is the applied load, δ is the open-end displacement and n is obtained from compliance calibration. For small displacement n = 3, b is the width of the specimen and a is the delamination length.

Figure 7.8 depicts the various phases of DCB simulation. The expansion and failure of the coherent parts are readily seen. The experimental, bilinear and rate-dependent cohesive zone models are contrasted with the J-integral from the numerical simulation for various stain rates in Figure 7.9. Since the numerical simulation uses a bilinear model for the cohesive zone, it correlates well with the bilinear model. However, the experimental results disagree with the numerical findings.

7.7 CONCLUSION

The existing cohesive zone model (CZM) is invalid since the steel–polyurea interface's delamination behaviour is significantly non-linear and rate dependent. As a result, the rate-dependent cohesive zone model is created and successfully fits the experimental data. The bilinear cohesive zone model that is currently in use has helped to create and verify the approach for the DCB test in LS-Dyna. The results also agreed well with the analytical findings of the bilinear CZM of J-integral. The developed model will be implemented for numerical simulation and validated with the developed CZM in the future.

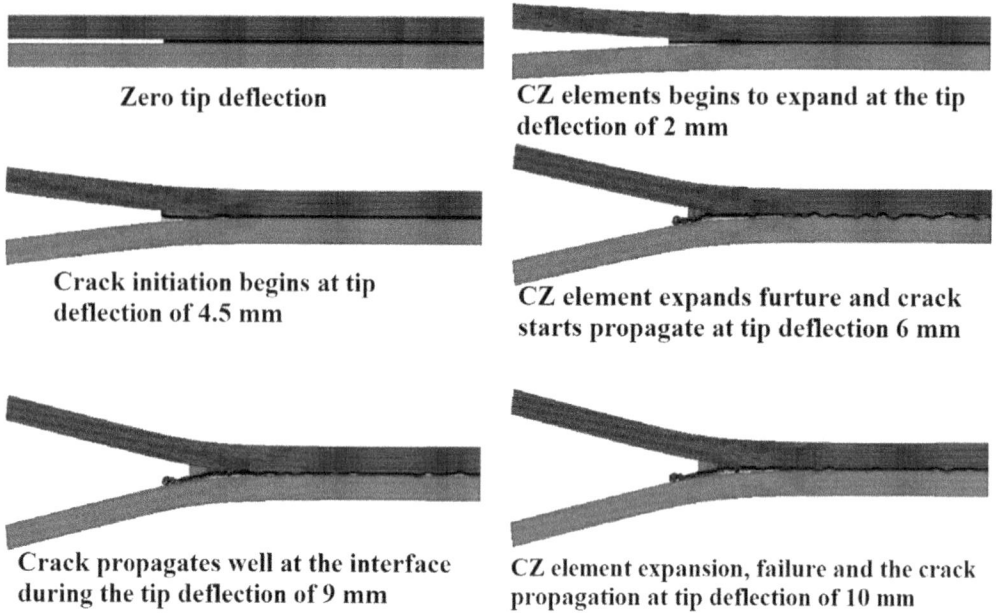

FIGURE 7.8 Expansion of cohesive elements for different tip deflection.

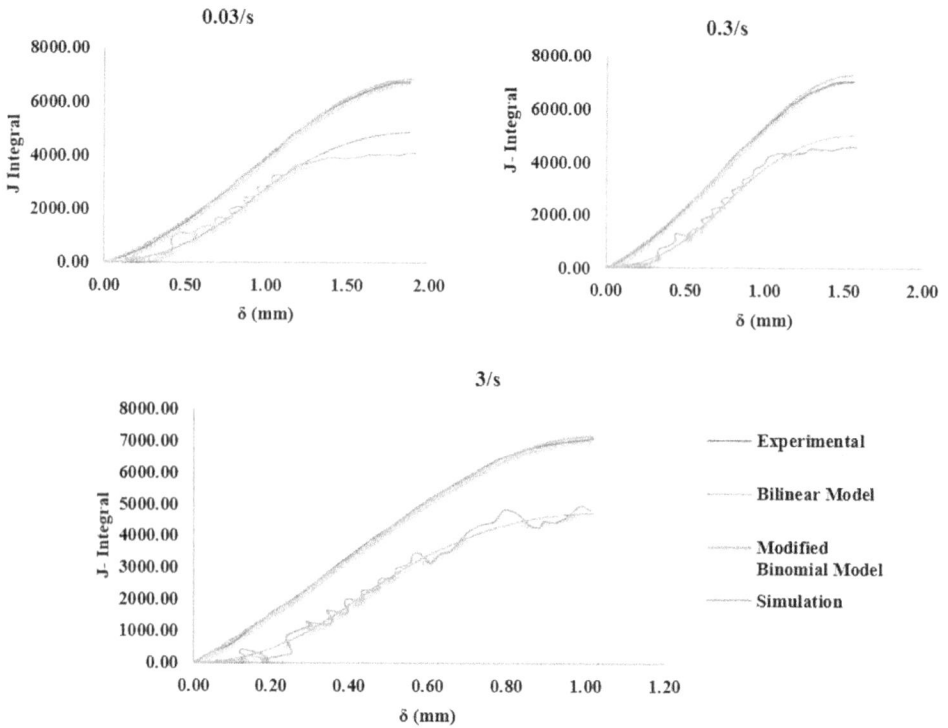

FIGURE 7.9 Comparison of experimental, bilinear and modified binomial models with the J-integral from numerical simulations at various strain rates.

REFERENCES

1. Zhu, Y., Liechti, K. M., & Ravi-Chandar, K. Direct extraction of rate-dependent traction–separation laws for polyurea/steel interfaces. *International Journal of Solids and Structures*, 46(1), 31–51, 2009.
2. ASTM D5528-13, Standard test method for mode I interlaminar fracture toughness of unidirectional fiber-reinforced polymer matrix composites, ASTM International, West Conshohocken, PA, www.astm.org, 2013.
3. ASTM D7905 / D7905M-19e1, Standard test method for determination of the mode II interlaminar fracture toughness of unidirectional fiber-reinforced polymer matrix composites, ASTM International, West Conshohocken, PA, www.astm.org, 2019.
4. Mohotti, D., Ali, M., Ngo, T., Lu, J., & Mendis, P. Strain rate dependent constitutive model for predicting the material behaviour of polyurea under high strain rate tensile loading. *Materials & Design*, 53, 830–837, 2014.
5. Xue, L., Mock, Jr, W., & Belytschko, T. Penetration of DH-36 steel plates with and without polyurea coating. *Mechanics of Materials*, 42(11), 981–1003, 2010.
6. Marzi, S., Hesebeck, O., Brede, M., & Kleiner, F. A rate-dependent cohesive zone model for adhesively bonded joints loaded in mode I. *Journal of Adhesion Science and Technology*, 23(6), 881–898, 2009.
7. Borges, C. S. P., Nunes, P. D. P., Akhavan-Safar, A., Marques, E. A. S., Carbas, R. J. C., Alfonso, L., & Silva, L. F. M. A strain rate dependent cohesive zone element for mode I modeling of the fracture behavior of adhesives. *Proceedings of the Institution of Mechanical Engineers, Part L: Journal of Materials: Design and Applications*, 234(4), 610–621, 2020.
8. Sørensen, B. F. Cohesive law and notch sensitivity of adhesive joints. *Acta Materialia*, 50(5), 1053–1061, 2002.
9. Jin, Z. H., & Sun, C. T. Cohesive fracture model based on necking. *International Journal of Fracture*, 134(2), 91–108, 2005.
10. May, M., Voß, H., & Hiermaier, S. Predictive modeling of damage and failure in adhesively bonded metallic joints using cohesive interface elements. *International Journal of Adhesion and Adhesives*, 49, 7–17, 2014.
11. Karac, A., Blackman, B. R. K., Cooper, V., Kinloch, A. J., Sanchez, S. R., Teo, W. S., & Ivankovic, A. Modelling the fracture behaviour of adhesively-bonded joints as a function of test rate. *Engineering Fracture Mechanics*, 78(6), 973–989, 2011.
12. Blackman, B. R. K., Kinloch, A. J., Sanchez, F. R., Teo, W. S., & Williams, J. G. The fracture behaviour of structural adhesives under high rates of testing. *Engineering Fracture Mechanics*, 76(18), 2868–2889, 2009.
13. Carlberger, T., Biel, A., & Stigh, U. Influence of temperature and strain rate on cohesive properties of a structural epoxy adhesive. *International Journal of Fracture*, 155(2), 155–166, 2009.

8 A Study on the Dynamic Behaviour of Multilayered Aluminium Alloy A5083 Due to Loading at High Strain Rates

Sanjay Kumar, Anoop Kumar Pandouria,
Purnashis Chakraborty and Vikrant Tiwari

8.1 INTRODUCTION

As we know, aluminium and its alloys have low density and high strength, so they are used for structural applications. AA5xxx alloy plates obtained through rolling operations are commonly used in the fabrication of engineering structures. Due to their corrosion resistance properties, the preferred application of AA5xxx alloys is the manufacturing of naval structures such as ship hulls and offshore topsides [1]. During the repair of any damaged structures, patches of the same or different materials are used. Patched surfaces, along with original structures known as multilayered structures, and their mechanical and impact properties are different than those of single-layered structures.

Many researchers have already performed experimental and simulation work related to the study of the variation of dynamic properties of aluminium and steel. However, Chiddister et al. [2], Lindholm et al. [3], Rosen et al. [4], Campbell et al. [5], and Wulf [6] found that the stress-strain behaviour of an aluminium alloy is affected by loading rates. They found that pure aluminium has moderate strain sensitivity. Tanka et al. [7] also showed that the flow stress of aluminium increases linearly with its logarithm up to a strain rate of 10^3 s^{-1} at room temperature; after that, it becomes nonlinear. However, Lindholm et al. [3] and Oosterkamp et al. [8] discovered that the strain sensitivity of aluminium alloys decreases with strain rates. Danish and Tiwari [9] studied the compressive behaviour of stacking aluminium alloys AA6063-T6 and IS 1570 alloys under high strain loading conditions.

The present chapter focused on the compressive behaviour of single and multilayered AA5083 aluminium alloy under a high strain rate. Also, the effect of strain rate on the maximum compressive strength of AA5083 alloy was investigated.

8.2 EXPERIMENTAL PROCEDURE

8.2.1 Split Hopkinson Pressure Bar

The experimental setup used for dynamic compressive experiments in this research article is the split Hopkinson pressure bar (SHPB). The mechanical properties at high strain rates of a material were first investigated by Kolsky in 1949 using a split Hopkinson pressure bar [10]. The schematic diagram of the experimental setup shown in Figure 8.1 consists of an air cylinder, striker, incident and transmission bar, data acquisition system, and computer. The strain gauges are pasted on the

DOI: 10.1201/9781003352358-8

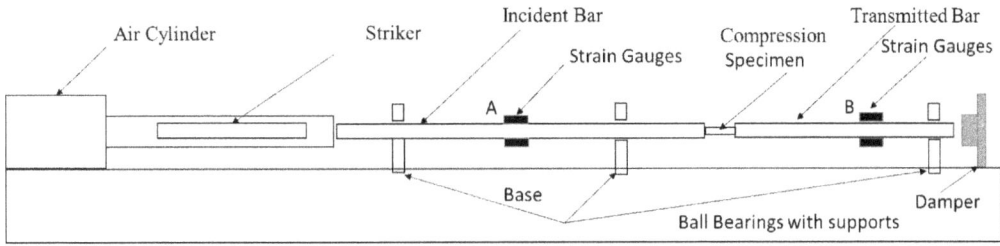

FIGURE 8.1 Schematic diagram of split Hopkinson pressure bar.

incident and transmitted bars for the recording of incident, reflected, and transmitted pulses. The strain gauges pasted on the incident bar record the incident and reflected pulse whereas the strain gauges on the transmitted bar measure the transmitted pulse. The length of the striker bar was 0.3 m, and the incident and transmitted bars used for experiments were equal lengths of 2 m each. The two strain gauges of resistance 120 Ω were used on the incident bar whereas an electrical circuit was used as a half Wheatstone bridge for measurement of incident strain pulses. A similar circuit was also used for the measurement of transmission strain pulses. The incident, reflected, and transmitted pulses were recorded using a signal conditioner and data acquisition system. The strain rates and strains were obtained using reflected strain, and stresses were obtained using transmitted strains. Both strain signals, reflected and transmitted, were substituted in the formula mentioned in Equations (8.1), (8.2), and (8.3) as mentioned in sections 8.2.2, and strains and stresses were obtained using MATLAB software.

8.2.2 MATERIALS AND METHODS

The material used for the experiments was AA5083, and its spectroscopy was performed to obtain the chemical composition of different elements shown in Table 8.1. From Table 8.1, it can be seen that it has magnesium as a significant composition that makes it a corrosion-resistant alloy.

After performing different experiments on SHPB, the strain rates are obtained by using Equation (8.1) expressed in terms of elastic wave speed, reflected strain, and specimen length.

$$\dot{\varepsilon} = \frac{2C_b \varepsilon_R(t)}{l_s} \tag{8.1}$$

Where C_b denotes elastic wave speed in the incident bar, $\varepsilon_R(t)$ is the reflected strain pulse and l_s denotes specimen length. Engineering strain is obtained by integrating strain rate and engineering stress is evaluated using transmitted strain pulses. Expressions for engineering strain and stress written in Equation (8.2) are used for calculating engineering strain and stress from experimental data.

$$Engineering\ strain,\ \varepsilon_s = \frac{2C_b}{l_s} \int \varepsilon_R(t) dt \quad and$$

TABLE 8.1

The Chemical Composition of AA5083 Alloy

Elements	Al	Cr	Cu	Fe	Mg	Mn	Si	Ti	Zn	Ni
Present weight percentage (wt%)	94.58	0.007	0.023	0.241	4.251	0.653	0.106	0.026	0.216	0.004

$$\textit{Engineering stress} \quad \sigma_s = \frac{A_b E_b}{A_s} \varepsilon_t(t) \tag{8.2}$$

The modulus of elasticity of the incident and transmitted bars is denoted by E_b. The areas of the cross-section of incident and transmitted bars are equal and represented by A_b, and specimen cross-sections are denoted with A_s. The transmitted strain is denoted by $\varepsilon_t(t)$. After getting engineering stress and strain, true stress (σ_T) and true strain (ε_T) were obtained by using Equation (8.3).

$$\textit{True strain,} \qquad \textit{and}$$

$$\textit{True stress,} \quad \sigma_T = \sigma_s(1 + \varepsilon_s) \tag{8.3}$$

The specimen used for experiments had a circular cross-section of 12 mm in diameter, but different lengths were considered to maintain different aspect ratios. Specimens of different lengths were cut from an aluminium rod of diameter 12 mm, and each specimen was faced with a facing tool on the lathe machine. To obtain three different aspect ratios (length to diameter ratio) of 1, 0.75, and 0.5, the length of specimens was cut to 12, 9, and 6 mm, respectively. The specimens of aspect ratios 1 and 0.75 were subdivided into two and three parts and their individual aspect ratios were obtained as 0.5 and 0.25, respectively [9]. The stacked specimen of length-to-diameter ratio 1 was made by stacking two specimens of aspect ratio 1/2. In a similar way, the stacked specimen of length-to-diameter ratio 0.75 was obtained by stacking three specimens of aspect ratio 1. To avoid the stress concentration during experiments in specimens due to irregular surfaces on faces, the rough surfaces were initially made smooth using sandpaper of grade 400 and then polished with a polishing machine. Using grease, commonly known as molybdenum disulphide, frictional effects were reduced at specimen interfaces with incident and transmitted bar surfaces. The nomenclature used to differentiate different specimens with different L/D ratios, number of partitions, and size of stacked elements is x-y-1/z (Table 8.2). Here x represents the L/D ratio, y is the number of partitions, and 1/z represents the fraction of the size of an element of the L/D ratio 1. For example, 1-2-1/3 represents that L/D is 1, the number of partitions is 2, and the size of the stacked element is 1/3 of the single specimen, whose L/D ratio is 1.

The photographs of specimens before the experiment without and with stackings are shown in Table 8.3, with scale and respective nomenclatures for stackings also shown with aspect ratios and number of partitions.

8.3   RESULTS AND DISCUSSION

The mechanical behaviour of AA5083 specimens under dynamic compressive loading conditions, different stacking sequences, and different length-to-diameter ratios were obtained using a split Hopkinson pressure bar (SHPB). Figure 8.2a represents the incident, reflected, transmitted, and reflected plus transmitted pulses. It can be seen that the incident pulse is approximately equal to the summation of reflected and transmitted pulses. The dynamic properties obtained from experiments will be valid if the specimen follows equilibrium conditions. For a specimen to be in dynamic

---

TABLE 8.2

Specimen Nomenclature According to Length-to-Diameter Ratio and Stacking Sequence

L/D ratio	Name of stacking sequences			
L/D = 1	R1-0-1/1	[ ]	R 1-1-1/2	[ ]
L/D = 0.75	R 0.75-0-1/1	[ ]	R 0.75-2-1/3	[ ]

TABLE 8.3

Photographs of Specimens and Their Stacking with Different Aspect Ratios

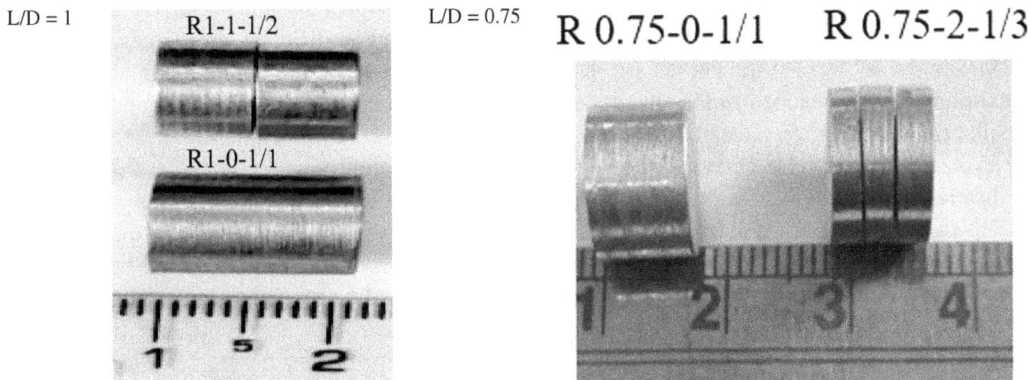

equilibrium, the forces acting on it towards the incident and transmitted sides of Hopkinson bars must be equal, as shown in Figure 8.2b. To avoid a stress concentration effect, the cross-sectional surfaces of specimens should be smooth and perpendicular to the longitudinal axis. The cross-section of specimens was made plane by using emery paper of 400, 600, and 800 starting from lower-grade numbers to higher grade. After that, it was polished using a polishing machine. The equilibrium conditions may also be affected due to friction, which should be minimized using some chemicals. Different chemicals may be used to reduce the friction between the surfaces of specimen contact and the incident and transmitted bars surfaces, but molybdenum disulphide grease is used commonly between specimen surfaces and rod ends to reduce friction effects because the true stress-strain curves obtained with experiments differ from the desired stress curve due to friction. In this experiment, molybdenum disulphide grease was used to reduce friction.

Figure 8.3a shows the true stress-strain diagram for AA5083 for aspect ratio 1 at strain rates of 1,126/s and 1,588/s without stacking of specimens, as well as the true stress-strain diagram for two stacked specimens of aspect ratio 1/2 each. Stacked specimens were used for dynamic compression experiments at two different strain rates, 1,160 and 1,617/s, and true stress-strain curves were obtained as shown in Figure 8.3a. From the figure, it can be seen that the maximum stress developed

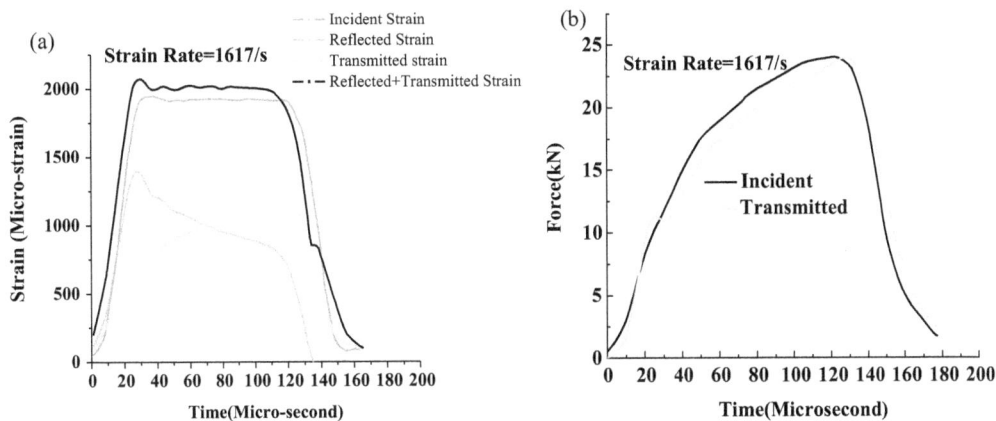

FIGURE 8.2 (a) Incident, reflected, transmitted, and reflected plus transmitted wave for l/d = 1. (b) Forces acting on specimen towards incident and transmitted sides for l/d = 1.

in the specimens with or without stacking is the same at strain rates of approximately 1,100 and 1,600/s, but strain induced at a strain rate of approximately 1,617/s is greater than that of 1,588/s. The true stress-strain diagram for AA5083 for aspect ratio 0.75 at strain rates of 1,376/s and 1,697/s, as well as three stacked specimens of aspect ratio 1/3 each, is shown in Figure 8.3b. Dynamic compression experiments were performed on stacked specimens at two different strain rates, 1,378 and 1,754/s, and true stress-strain curves are shown in Figure 8.3b. From the figure, it can be seen that maximum stress developed in the specimens with stacking at strain rates of 1,378/s and 1,754/s is smaller than those strain rates of 1,376/s and 1,697/s for non-stacked specimens.

Experiments were also carried out on specimens with different aspect ratios of 1, 0.75, and 0.5 at different strain rates corresponding to two different launch impact pressures of 1.5 and 2 bars, as shown in Figure 8.4. At 1.5 bar pressure and aspect ratios of 1, 0.75, and 0.5 of specimens, the corresponding maximum stresses induced are 231, 262, and 290 MPa. Similarly, for 2 bar pressure and aspect ratios of 1, 0.75, and 0.5 of specimens, the corresponding maximum stresses induced are 272, 282, and 348 MPa.

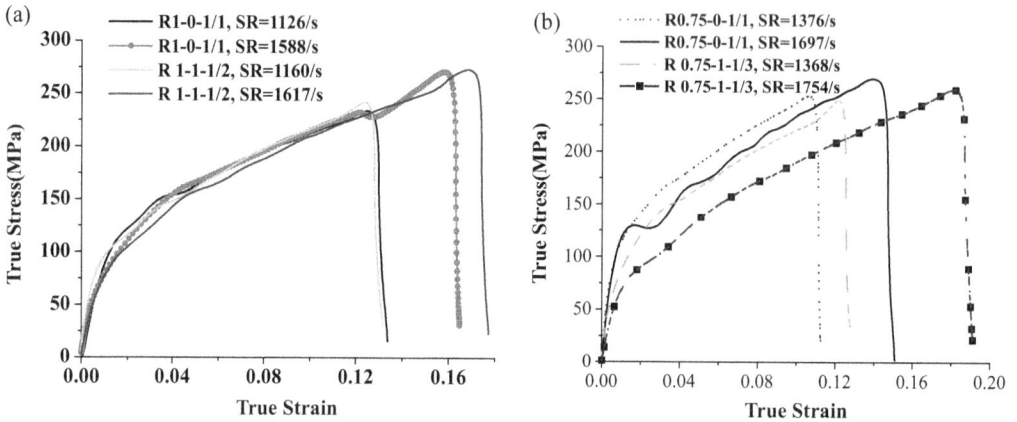

FIGURE 8.3   (a) True stress-strain diagram at different impact velocities and different stackings for L/D = 1. (b) True stress-strain diagram at different impact velocities and different stackings for L/D = 0.75.

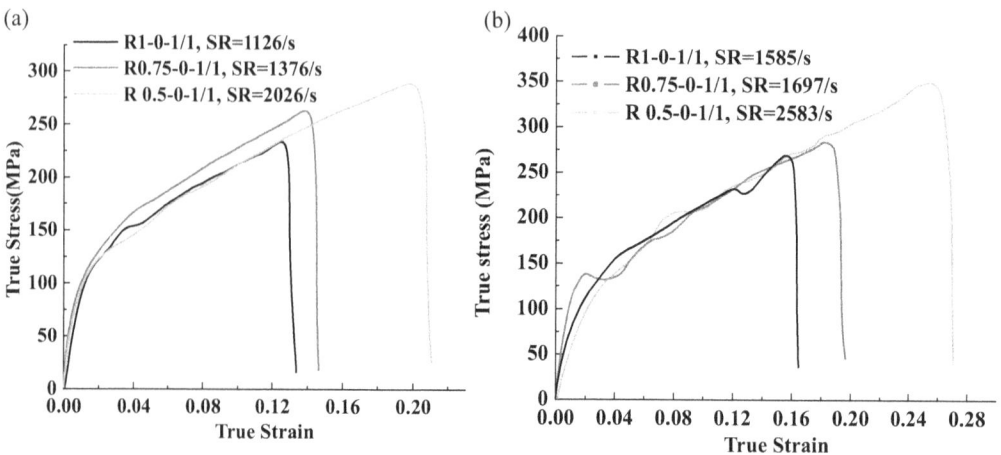

FIGURE 8.4   True stress-strain diagram for L/D = 1, 0.75, and 0.5. (a) Different strain rates corresponding to impact pressure 1.5 bar. (b) Different strain rates corresponding to impact pressure 2 bar.

8.4 CONCLUSION

The present chapter shows compression experiments on aluminium alloy AA5083 with and without stacking for different aspect ratios using a split Hopkinson pressure bar. Experimental results obtained showed that for aspect ratio 1 (L/D = 1), maximum stresses developed in specimens were the same for both types of specimens, either single or stacked. Also, the rise in maximum stresses was the same for both specimens with or without stacking. For specimens of aspect ratio 0.75 (L/D = 0.75), the maximum stress developed in a single specimen is more than that of stacked specimens. Also, true stress-strain curves for specimens of different aspect ratios (1, 0.75, and 0.5) without stacking were drawn for different levels of split Hopkinson pressure bar cylinder pressure.

REFERENCES

1. A. H. Clausen, T. Børvik, O. S. Hopperstad, and A. Benallal, "Flow and fracture characteristics of aluminium alloy AA5083-H116 as function of strain rate, temperature and triaxiality," *Mater. Sci. Eng. A*, vol. 364, no. 1–2, pp. 260–272, 2004.
2. J. L. Chiddister and L. E. Malvern, "Compression-impact testing of aluminum at elevated temperatures," *Exp. Mech.*, vol. 3, no. 4, pp. 81–90, 1963.
3. U. S. Lindholm, "Some experiments with the split Hopkinson pressure bar," *J. Mech. Phys. Solids*, vol. 12, no. 5, pp. 317–335, 1964.
4. A. Rosen and S. R. Bodner, "The influence ageing of strain on the flow rate and stress commercially-pure aluminium," *J. Mech. Phys. Solids*, vol. 15, no. 52, pp. 47–62, 1967.
5. B. J. D. Campbell and A. R. Dowling, "The behaviour of materials subjected to dynamic incremental shear loading," *J. Mech. Phys. Solids*, vol. 18, pp. 43–63, 1970.
6. G. L. Wulf, "The high strain rate compression of 7039 aluminium," *Int. J. Mech.*, vol. 20, no. 9, pp. 609–615, 1978.
7. K. Tanaka and T. Nojima, "Strain rate change tests of aluminum alloys under high strain rate," *Proceedings of The 19th Japan Congress on Materials Research*, pp. 48–51, 1975.
8. L. D. Oosterkamp, A. Ivankovic, and G. Venizelos, "High strain rate properties of selected aluminium alloys," *Mater. Sci. Eng. A*, vol. 278, pp. 225–235, 2000.
9. D. Iqbal and V. Tiwari, "Structural response of multilayered aluminum and steel specimens subjected to high strain rate loading conditions," *J. Theor. Appl. Mech.*, vol. 56, no. 4, pp. 1139–1151, 2018.
10. H. Kolsky, "An investigation of mechanical properties of materials at very high rates of loading," *Proceedings of the Physical Society, B*, vol. 62, no. 11, pp. 676–700, 1949.

9 High Strain Rate Tensile Testing of High-Strength Steel Using Split Hopkinson Pressure Bar

Makhan Singh, Kartikeya Kartikeya,
Dhruv Narayan and Naresh Bhatnagar

9.1 INTRODUCTION

High-strength steels are extensively explored for arresting the damage induced by impact loading. The high strength of low-alloy steels along with excellent deformability makes them a candidate choice for impact-absorbing applications. In order to provide protection against the threat levels of ballistic impact, the high-strength steels can prove to be the cheaper solution as compared to ceramic composite materials. The impact of a fast-moving projectile against any material leads to the rise of stress level from zero to thousands of MPa in a few milliseconds (Davies, 1956). Therefore, these types of events correspond to the high strain rate loading of the ballistic-resistant materials. The properties of these materials are also influenced by changes in strain rate which are different as compared to the quasi-static loading (Nicholas, 1981). The mechanical characterization of these materials at high strain rates, thus, becomes essential for design, verification, and simulation. The high strain rate regime lies between 10^2 and 10^4 s^{-1} in the strain rate range. The split Hopkinson pressure bar (SHPB) setup is used to perform high strain rate testing experimentally in various modes such as compression, tension, and torsion.

Bertram Hopkinson (1914) designed a setup in 1914 to understand the shape of the pressure-time pulse produced by the impact of bullets or detonation of explosives on a steel rod. This setup was later modified by H. Kolksy (1949) in 1949 by splitting the bar into two parts. Therefore the name of this experimental technique is the split Hopkinson pressure bar (SHPB) or Kolksy bar method. In this method, the specimen to be tested dynamically is placed between the two bars, namely the incident bar/input bar and transmission bar/output bar. The loading of the specimen under high strain tensile testing has also evolved over a period of time. Harding et al. (1960) generated a compression pulse in a tube surrounding a solid inner rod. A mechanical joint connects the tube and rod. When the compression pulse in the outer tube reaches the joint, which is a free end, it gets reflected through the solid inner rod as a tensile pulse. Lindholm and Yeakley (1968) designed the sample in the form of a complex hat shape that allowed the direct use of the compression wave to achieve a dynamic tensile wave. Harding and Welsh (1983) incorporated an instrumented input bar between specimens and an output bar for testing unidirectionally reinforced CFRP, thereby modifying the design of Hopkinson's tensile split bar. Staab and Gilat (1991) designed a direct-tension SHPB apparatus for storing the elastic energy in the incident bar and instantaneously releasing it to load the specimen at a high strain rate. Smerd et al. (2005) modified the SHPB apparatus in order to minimize the wave distortions at the interface between the specimen and the bar and to the specimen design in order to minimize specimen geometry effects. They identified the constitutive response and damage evolution in AA5754 and AA5182 aluminum alloy sheets at high strain rates using this modified SHPB. Song et al. (2009) improved the SHPB bar design by incorporating a highly precise optical table to achieve perfect alignment of the system. Pillow

DOI: 10.1201/9781003352358-9

blocks with Frelon-coated linear bearings were used in order to reduce friction between the bar and the supports. A Teflon-coated striker bar was used to minimize friction within the gun barrel. A laser alignment system was integrated for the perfect alignment of the bars. These modifications were employed in SHPB to test 6,061 aluminum specimens, and the results reported the precision of the system. Nie et al. (2018) designed an innovative approach utilizing the principle of electro-magnetism to generate stress pulses. An LC circuit was used to produce an eddy current in the copper coil which provides the necessary stress waves.

The existing mechanisms available in the literature for loading material under high strain rate testing were explored to develop an in-house split Hopkinson tension bar setup for high strain rate testing setup as shown in Figure 9.1. The incident and transmission bars are 20 mm diameter rods made from C-250 Maraging steel. The length of the incident bar is 5 m, while the transmission bar is 2.5 m long. The developed SHTB can generate stress waves of 1 ms duration for testing even ductile materials. The hydraulic jack stores the elastic energy in the incident bar until a brittle intermediate piece between the jack and the incident bar is broken in shear. The instantaneous release of the stored energy in the incident bar dynamically loads the specimen placed between the incident bar and transmission bar with the help of form-fit mounting, as shown in Figure 9.2. The strain time history is recorded by pasting one strain gauge over the incident bar to record the incident and reflected signals while a second strain gauge is pasted over the transmission bar to record the transmitted signals. The quarter Wheatstone bridge powered by a DC power source connects these strain gauges to the data acquisition system. The incident (ε_I), reflected (ε_R), and transmitted strain signals (ε_T) are used to compute the stress-strain response of the material using a one-dimensional wave propagation solution. The strain rate, strain, and stress are derived using the D'Alembert solution of wave equation as per Equations (9.1), (9.2), (9.3), and (9.4) respectively.

The strain rate in the specimen is given by,

$$\dot{\epsilon} = \frac{2c_b\left(\varepsilon_R\right)}{L_s} \tag{9.1}$$

FIGURE 9.1 Split Hopkinson bar setup.

(a)

(b)

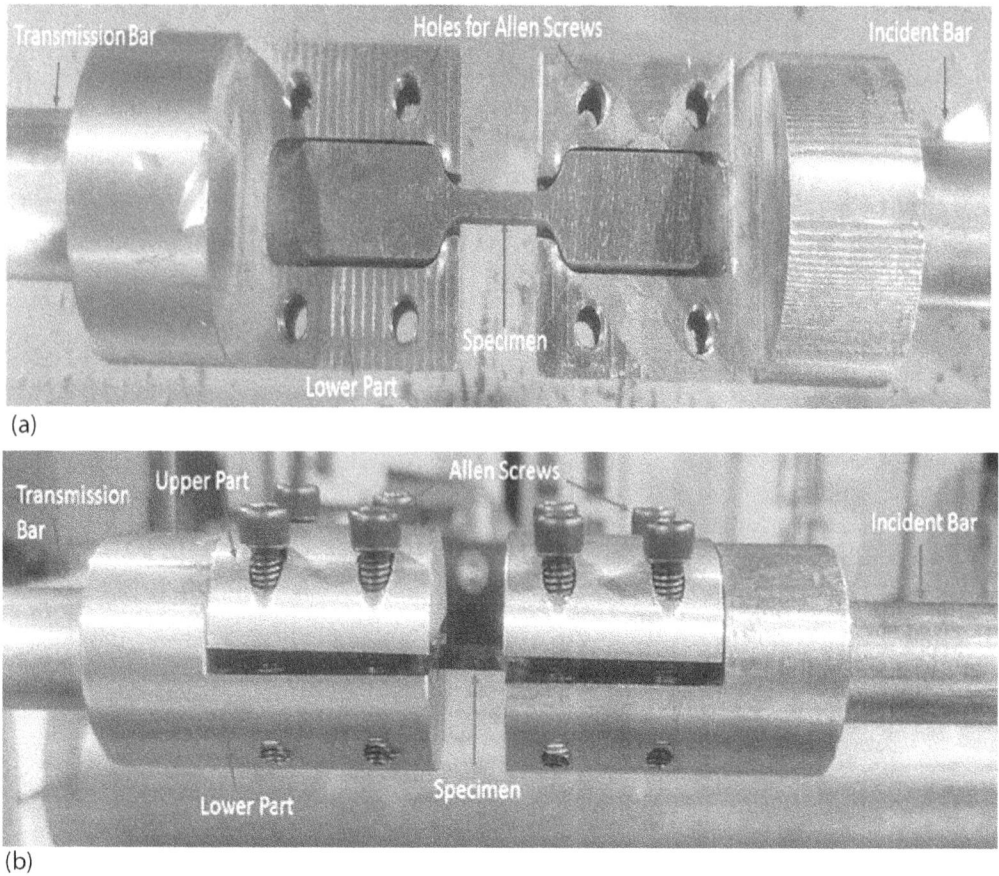

FIGURE 9.2 Form-fit mounting.

Strain in the specimen can be computed by integrating Equation (9.1) from 0 to t,

$$\varepsilon = \frac{2c_b}{L_s} \int_0^t \varepsilon_R dt \qquad (9.2)$$

Stress in the specimen can be derived by,

$$\sigma = \frac{F}{A_s} \qquad (9.3)$$

Where, A_s is the cross-section of the specimen.

Therefore, $\qquad\qquad\qquad \sigma = \frac{A_b E_b \varepsilon_T}{A_s} \qquad (9.4)$

Thus, Equations (9.1), (9.2), and (9.4) are used to compute the strain rate, strain, and stress, respectively, for the specimen under high strain rate testing.

9.2 MATERIALS

The material used in this study is high-strength steel. The chemical composition of the steel is shown in Table 9.1. The material is selected to investigate dynamic mechanical properties at high

TABLE 9.1
Chemical Composition of High-Strength Steel

C	Si	Mn	P	S	Cr	Mo	Ni	V	Cu	Fe
0.43	1.70	0.63	0.007	0.002	0.84	0.45	1.89	0.085	0.050	Bal.

strain rate tensile loading. The material is supplied by Starwire (India) Limited, Faridabad, in the form of plates of thickness 2.6 mm, 2.8 mm, and 3 mm. Four samples from each plate of different thicknesses are cut using an abrasive water jet cutting machine at 4,500 bar of water pressure. The tensile specimens for high strain rate testing are obtained as per the geometry configuration shown in Figure 9.3 (Antoun et al., 2017; Sasso et al., 2019). The waterjet cut specimens are shown in Figure 9.4. The physical dimensions of each specimen are also illustrated in Table 9.2.

9.3 EXPERIMENTAL RESULTS

Experiments are performed on SHTB by mounting the specimen between the incident bar and transmission bar using form-fit mounting. The high-speed camera with 1,184 × 416 resolution at 9,000 fps is used to record the deformation behavior of the material as shown in Figure 9.5. The foil

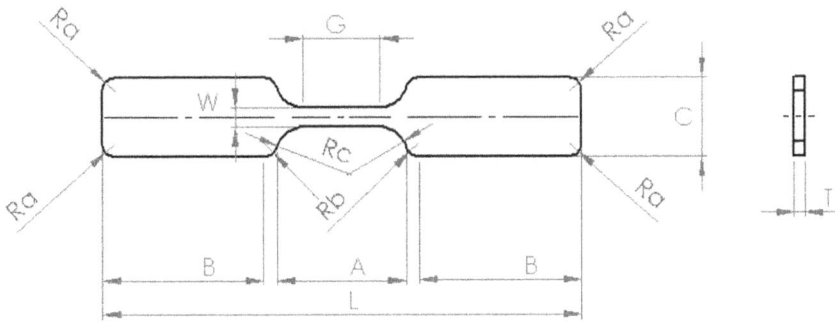

FIGURE 9.3 Specimen geometry configuration.

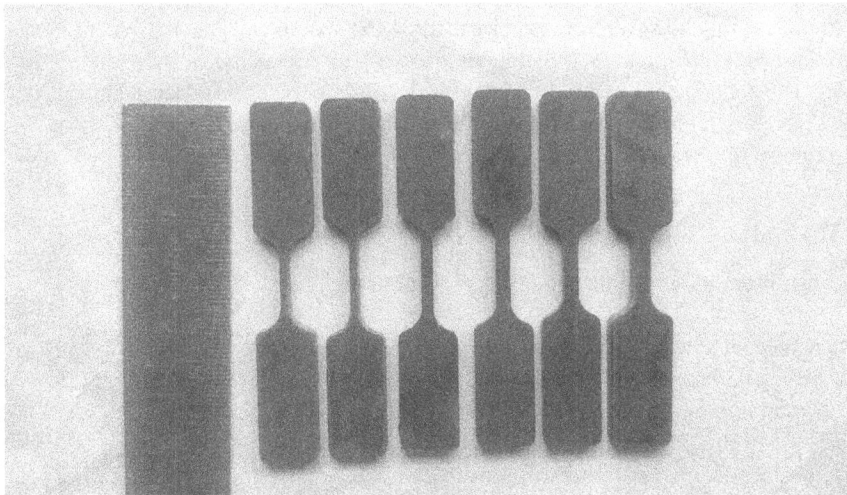

FIGURE 9.4 Tensile specimen extracted from the plate using an abrasive waterjet cutting machine.

TABLE 9.2
Physical Dimensions of the Samples

Sample no.	T (mm)	L (mm)	A (mm)	B (mm)	C (mm)	W (mm)	G (mm)	Ra (mm)	Rb (mm)	Rc (mm)
S-1	2.6	80	23.4	25.8	14	3.5	13.4	2.5	2.5	5
S-2	2.6	80	23.4	28.2	14	3	13.4	2.5	2.5	5
S-3	2.6	80	18.2	25.8	14	2.5	8.2	2.5	2.5	5
S-4	2.6	80	18.2	28.2	14	2	8.2	2.5	2.5	5
S-5	2.8	80	23.4	25.8	14	3.5	13.4	2.5	2.5	5
S-6	2.8	80	23.4	28.2	14	3	13.4	2.5	2.5	5
S-7	2.8	80	18.2	25.8	14	2.5	8.2	2.5	2.5	5
S-8	2.8	80	18.2	28.2	14	2	8.2	2.5	2.5	5
S-9	3	80	23.4	25.8	14	3.5	13.4	2.5	2.5	5
S-10	3	80	23.4	28.2	14	3	13.4	2.5	2.5	5
S-11	3	80	18.2	25.8	14	2.5	8.2	2.5	2.5	5
S-12	3	80	18.2	28.2	14	2	8.2	2.5	2.5	5

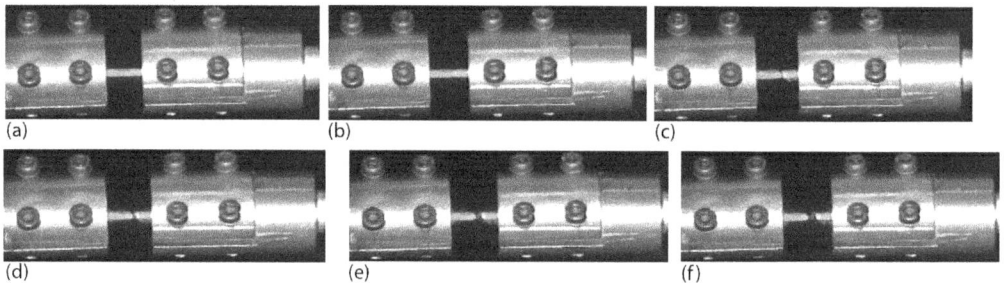

FIGURE 9.5 High-speed camera images at the onset of material failure.

strain gauge I measures the signals of the incident bar, while strain gauge II measures the signals of the transmitted bar. The measured signals are amplified and conditioned using a signal conditioner. The amplified signals are recorded by a National Instruments PXI-8115 data acquisition system at a rate of one million samples/sec. The incident, reflected, and transmitted signals are used to compute stress, strain, and strain rate experienced by the specimen. The analysis of measured data using Equations (9.1), (9.2), and (9.4) results in a stress-strain response of the material, as shown in Figure 9.7(a), (b), and (c). The summary of the results obtained for each specimen is also illustrated in Table 9.3. The failed material specimens from 2.6 mm, 2.8 mm, and 3 mm thickness plates after high strain rate testing are shown in Figure 9.6(a), (b), and (c) respectively.

9.4 DISCUSSION

The following observations are made after high strain rate loading of high-strength steel:

(i) Four tensile specimens from a 2.6 mm thick plate were tested in the strain rate range of 389–1,206 s^{-1}. The maximum flow stress of 2,432 MPa was achieved at a 1,206 s^{-1} strain rate with maximum strain of 0.149. These specimens show increasing flow stress with an increase in strain rate.

(ii) Four tensile specimens from a 2.8 mm thick plate were tested in the strain rate range of 669–1,441 s^{-1}. The maximum stress of 2,917 MPa is obtained for specimen S-8 at the strain rate of 1,441 s^{-1}.

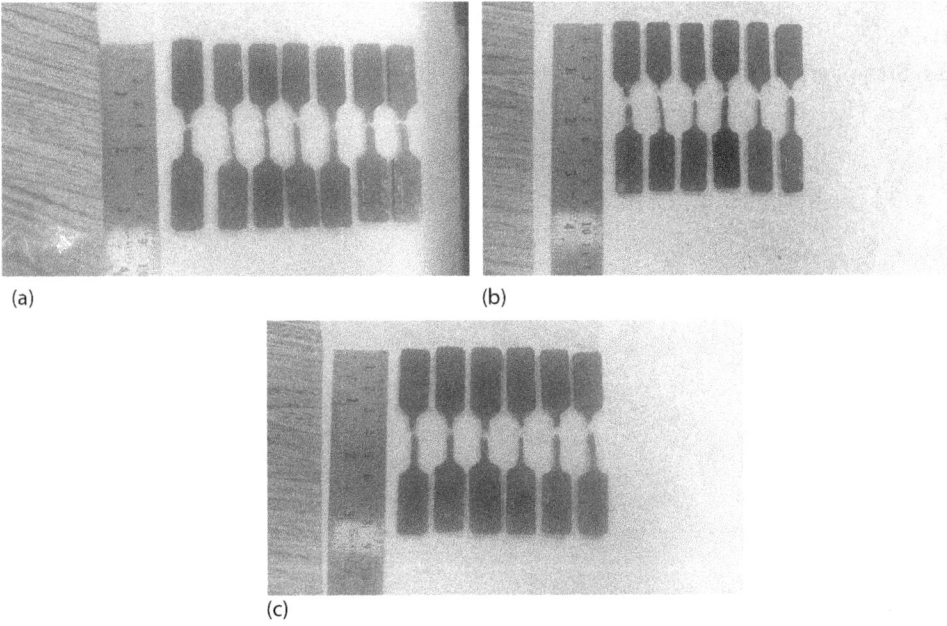

FIGURE 9.6 Failure of the specimen after high strain rate tensile testing.

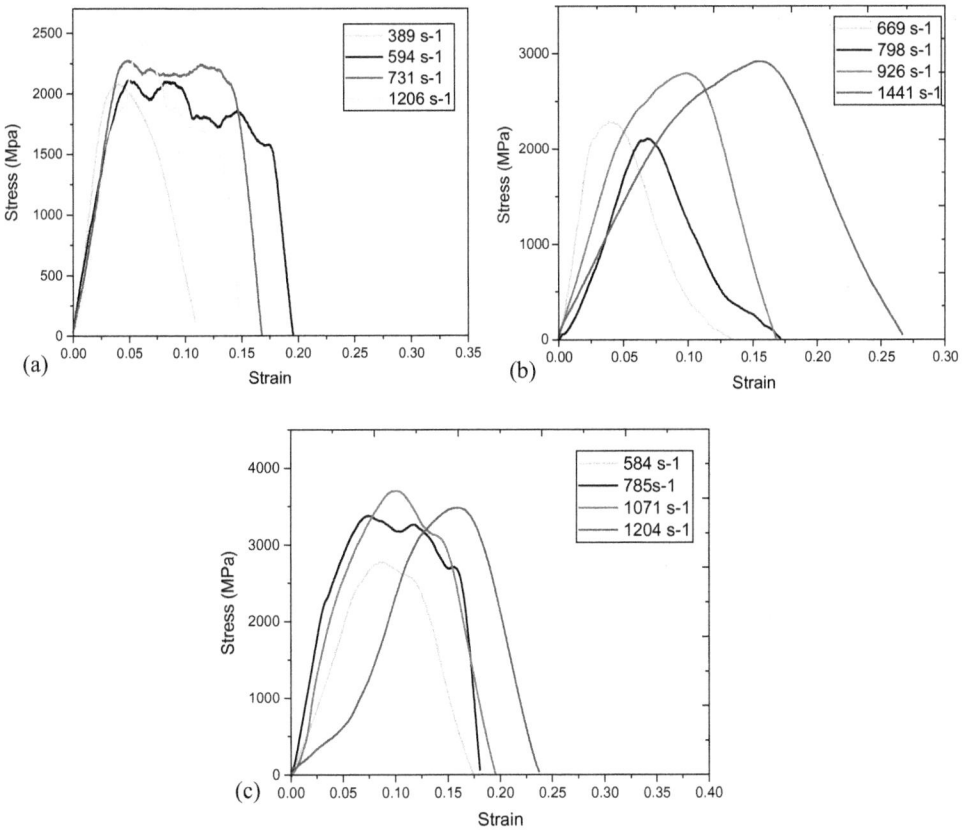

FIGURE 9.7 Stress-strain response of (a) 2.6 mm thickness, (b) 2.8 mm thickness, and (c) 3 mm thickness specimens at different strain rates.

TABLE 9.3
Stress, Strain, and Strain Rate of Specimens

S. no.	Specimen	Flow stress (MPa)	Strain	Strain rate (s^{-1})
1.	S-1	2,083	0.109	389
2.	S-2	2,111	0.194	594
3.	S-3	2,276	0.169	731
4.	S-4	2,432	0.149	1,206
5.	S-5	2,284	0.131	669
6.	S-6	2,110	0.159	798
7.	S-7	2,794	0.166	926
8.	S-8	2,917	0.26	1,441
9.	S-9	2,768	0.174	584
10.	S-10	3,381	0.179	785
11.	S-11	3,702	0.194	1,071
12.	S-12	3,482	0.236	1,204

(iii) Four tensile specimens from a 3 mm thick plate were tested in the strain rate range of 584–1,204 s^{-1}. The maximum stress of 3,702 MPa is obtained for specimen S-11 at the strain rate of 1,071 s^{-1}.

It is observed from the above discussion that the thickness of high-strength steel plates also plays an important role during high strain rate tensile loading. The dynamic strength of these materials increases significantly with an increase in thickness. Future work with high-strength steel involves the development of a constituent model based on quasi-static, high strain rate, and high-temperature testing results. The microstructure of the steel can also be studied before and after high strain rate testing to understand the criteria contributing towards an increase in strength with increasing strain rate and thickness of the specimen.

ACKNOWLEDGMENTS

The authors acknowledge the use of the Split Hopkinson Tensile Bar Facility at CoE – Personal Body Armor Lab of JATC-IIT Delhi. This facility provided the necessary tests for this fundamental study. The authors would also like to acknowledge Starwire India Ltd. for the high-strength steel sheets used for this study.

REFERENCES

Antoun, B., Arzoumanidis, A., Qi, H. J., Silberstein, M., Amirkhizi, A., & Furmanski, J. (2017). Identification of plastic behaviour of sheet metals in high strain rate tests. *2*, 203–209. https://doi.org/10.1007/978-3 -319-41543-7

Davies, R. M. (1956). Stress waves in solids. *British Journal of Applied Physics*, *7*, 203–209.

Harding, J., & Welsh, L. M. (1983). A tensile testing technique for fibre-reinforced composites at impact rates of strain. *Journal of Materials Science*, *18*(6), 1810–1826. https://doi.org/10.1007/BF00542078

Harding, J., Wood, E. O., & Campbell, J. D. (1960). Tensile testing of materials at impact rates of strain. *Journal of Mechanical Engineering Science*, *2*(2), 88–96. https://doi.org/10.1243/jmes_jour_1960_002 _016_02

Hopkinson, B. (1914). A method of measuring the pressure produced in the detonation of high, explosives or by the impact of bullets. *Philosophical Transactions of the Royal Society of London. Series A, Containing Papers of a Mathematical or Physical Character*, *213*(497–508), 437–456. https://doi.org/10.1098/rsta .1914.0010

Kolsky, H. (1949). An investigation of the mechanical properties of materials at very high rates of loading. In *Proceedings of the Physical Society. Section B. 62*(11), 676.

Lindholm, U. S., & Yeakley, L. M. (1968). High strain-rate testing: Tension and compression. *Experimental Mechanics, 8*(1), 1–9. https://doi.org/10.1007/bf02326244

Nicholas, T. (1981). Tensile testing of materials at high rates of strain. *Experimental Mechanics, 21*(5), 177–185. https://doi.org/10.1007/bf02326644

Nie, H., Suo, T., Wu, B., Li, Y., & Zhao, H. (2018). A versatile split Hopkinson pressure bar using electromagnetic loading. *International Journal of Impact Engineering.* https://doi.org/10.1016/j.ijimpeng.2018.02.002

Sasso, M., Mancini, E., Dhaliwal, G. S., Newaz, G. M., & Amodio, D. (2019). Investigation of the mechanical behavior of CARALL FML at high strain rate. *Composite Structures, 222*(January), 110922. https://doi.org/10.1016/j.compstruct.2019.110922

Smerd, R., Winkler, S., Salisbury, C., Worswick, M., Lloyd, D., & Finn, M. (2005). High strain rate tensile testing of automotive aluminum alloy sheet. *International Journal of Impact Engineering, 32*(1–4), 541–560. https://doi.org/10.1016/j.ijimpeng.2005.04.013

Song, B., Connelly, K., Korellis, J., Lu, W. Y., & Antoun, B. R. (2009). Improved Kolsky-bar design for mechanical characterization of materials at high strain rates. *Measurement Science and Technology, 20*(11). https://doi.org/10.1088/0957-0233/20/11/115701

Staab, G. H., & Gilat, A. (1991). A direct-tension split Hopkinson bar for high strain-rate testing. *Experimental Mechanics, 31*(3), 232–235. https://doi.org/10.1007/BF02326065

10 High Strain Rate Testing of Automotive Sheet Steel with Evaluation of a Double-Cell Crash Box

*Pundan Kumar Singh, Ankit Kumar, Abhishek Raj,
Rahul K. Verma and C. Lakshmana Rao*

10.1 INTRODUCTION

The design of crash-critical components of an automobile requires appropriate material data for simulation purposes. The use of strain-rate-dependent material behaviour has been studied well and understood to be crucial in the prediction of the crash behaviour of structures and components. However, the methodology and technique to arrive at the required stress-strain data which is useful for modelling purposes vary widely with respect to the type of machine, specimen used for testing, and data acquisition employed for load and strain measurement. At high strain rates (100/s and higher), the load data provided by machines is affected due to ringing, and appropriate measures like the use of a load strain gauge are used to determine the load response. The strain acquisition at all rates must be done by a non-contact approach like digital image correlation or by bonding strain gauges. The technique to utilize these external devices for reliable high strain rate data is crucial since it could affect the overall quality of the data and hence jeopardize the simulation results for a particular design.

Safe vehicle structural design is one of the important aspects of passenger safety which is challenged by tightening crash protocols [1] and progressive use of strong but lightweight (thin) sections for vehicle design. Crash energy management in a vehicle demands that the primary structure must be able to sustain and dissipate the kinetic energy without being excessively deformed. A certain amount of plastic deformation is essential to attenuate the crash loading; however, too much could be intrusive and fatal to the occupants onboard. Various structural members like the frontal crash box, b-pillar, sill and side impact beams are a few of the components which help the crashworthiness of the vehicle. The crash box specifically is located between the bumper and side-front rail of any vehicle and absorbs most of the kinetic energy in case of full frontal and offset frontal crash before the front rail structure absorbs and transfers the rest of the kinetic energy to the main structure. The idea is to dissipate the kinetic energy so that damage to the main cabin is minimized with less shock transferred to the occupant. This is not only achieved by the materials of the crash box but also depends significantly on the design. Many researchers have studied the designs, experiments, and simulation of energy-absorbing devices made of different materials and geometry [2, 3, 4, 5, 6, 7].

This work attempts to acquire the experimental data by high-speed tensile testing using a servo-hydraulic tensile testing machine. The experimental data consists of the determination of engineering stress and engineering strain curves at strain rates from quasistatic until 500/s, which is considered a suitable range for constitutive modelling in simulating the crash behaviour. Further, a

DOI: 10.1201/9781003352358-10

double-cell crash box design is attempted and the axial collapse behaviour of the double-cell crash box is studied at 35kmph and 50kmph using numerical simulations, thus highlighting the low-speed and high-speed behaviour of the section during crash loading. Various studies exist with respect to the energy absorption behaviour of thin-walled structures applicable to vehicle structures; however, most of them do not discuss the ease and cost of manufacturing. For example, for premium vehicles, there is no constraint with respect to either use of a particular material or the cost involved in the manufacturing process; however, for mass-manufactured vehicles, the designs are still predominantly based on steel and are cost-sensitive. Thus, it is important for any novel crash component design to take note of ease of manufacturing for easy implementation in a mass-manufactured vehicle. It is notable that double-cell or multicell crash boxes are manufactured using extrusion and are used on premium cars.

The same double-cell crash box prototype is also manufactured using conventional processes like sheet metal stamping and bending and is planned for further testing. This work thus demonstrates the methodology to generate high strain rate data, the design of a novel double-cell crash box, and the simulation and further prototyping of the crash box.

10.2 EXPERIMENTAL METHODOLOGY

10.2.1 MATERIAL, MICROSTRUCTURE, AND CHEMISTRY

Dual-phase steel (DP590), a high-strength steel grade having ferritic-martensitic microstructure, is chosen for this study. The microstructure is shown in Figure 10.1 and the chemical composition is shown in Table 10.1.

10.2.2 EXPERIMENTAL SET-UP AND SPECIMEN GEOMETRY

Although there are recommendations for conducting high strain rate experiments for sheet metal, the methodologies could vary from laboratory to laboratory with respect to machine (top vs bottom actuator), specimen geometry, and data acquisition system for load and strain, thereby affecting the quality of final stress-strain data. The earlier recommendations [8] have now materialized into a testing standard [9].

The flow behaviour of DP590-grade sheet steel of thickness 1.42 mm is evaluated in this study by conducting high-speed tensile tests in the range of 0.001/s–500/s. All the experiments are performed using a servo-hydraulic high-speed tensile testing machine (Zwick HTM 5020) except the

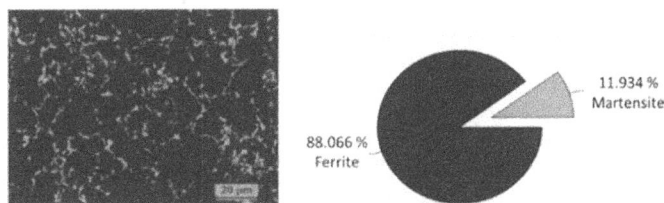

FIGURE 10.1 Microstructure and phase fraction of ferrite and martensite of DP590 grade.

TABLE 10.1
Chemical Composition of DP590 Steel

Element	C	Mn	Si	Ca	Cr	Ni	V	Nb	B	Al	N	S
Wt. %	0.079	1.855	0.476	0.0270	0.001	0.019	0.001	0.003	0.0001	0.0460	0.0045	0.002

experiments at 0.001/s strain rate which were conducted on Instron's tensile testing machine. Zwick's HTM5020 servo-hydraulic machine is equipped with an extremely stiff double-column load frame design to mitigate the impact effects during high-speed testing. The maximum load capacity of the machine is 50 kN and a wide testing speed selection is possible from a minimum of 0.001 m/s to a maximum of 20 m/s. The hydraulic energy is supplied by accumulators which are placed on top mounted actuators. The testing chamber is encased in a transparent housing which is locked once the hydraulic safety circuit is activated prior to any testing once the testing mode is activated.

This machine has a unique patented lightweight double rod actuator design with hydrostatic bearing to achieve the required acceleration in a short distance of actuator movement before the specimen is loaded for the desired experiment. The picture in Figure 10.2 shows the actuator, machine load cell, and mounted specimen with strain gauges. The force measurement is done by an inbuilt piezoelectric load cell, the data of which is free from any noise and oscillation for a low strain rate regime; however, at 100/s strain rate and beyond there is a considerable ringing in the machine and the load data from the machine may not be reliable for constitutive modelling purpose.

The desired strain rate is achieved by setting an appropriate actuator velocity and measuring the strain rate by using strain-time data from an elongation strain gauge. The range of strain rates is such that the experimental data generated will be useful for a crash box simulation study as shown subsequently. All the tests are conducted on a high strain rate tensile specimen geometry [10] as shown in Figure 10.3.

Different specimen geometry like the use of ASTM E8 standard geometry at a low strain rate is avoided to keep uniformity of the specimen with changing strain rates. The specimen geometry could have a profound effect on the quality of the experimental data and thus the specimen chosen is such that only minimal noise is produced both in the gauge area and grip area of the specimen. The gauge area and grip area of the specimen allow for pasting the elongation strain gauge and load strain gauge respectively. For both the strain gauges, with deformation, the change in resistance is converted to strain and load as described in the strain gauge user's guide [11]. The use of the right adhesives and proper methodology to paste gauges is important to conduct the rate testing without failed gauges or ending up with erroneous data. The methodology for strain gauging needed for good data acquisition is described in the next section.

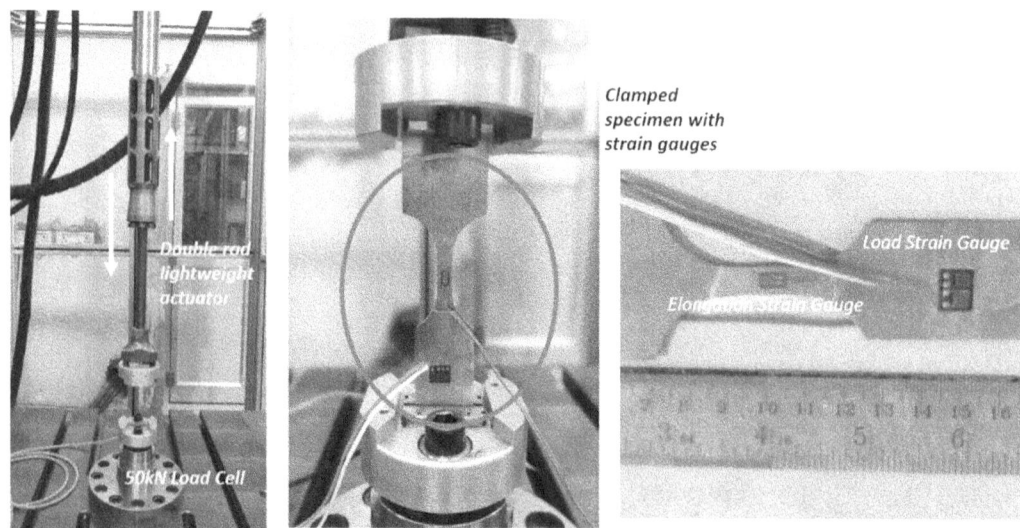

FIGURE 10.2 Double rod lightweight actuator to achieve the desired speed in a high strain rate test, and mounted specimen with load strain gauge and elongation strain gauge.

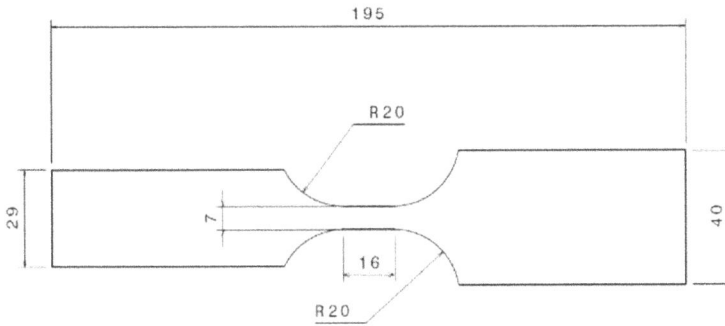

FIGURE 10.3 Tensile specimen geometry used for conducting the experiments (all dimensions in mm). Source: [10].

10.2.3 STRAIN GAUGES FOR ELONGATION AND LOAD DATA ACQUISITION

In this work, elongation data is determined using strain gauges at all strain rates. Load data from a machine is only applicable for strain rates under 100/s of strain rate. For strain rates beyond 100/s, an external strain gauge load cell is used which is pasted in a half-bridge connection on both sides of the specimen in the top grip area to acquire the load signal. The top grip area is mostly under elastic deformation, and stress wave reflections are minimal in this area. The load signature from load cell strain gauges is mostly free from any ringing. However utmost care must be taken to paste these gauges aligned and collinear with the centreline of the mounted specimen. It is notable that as the strain rate is increased there could be oscillation in the strain gauge data itself or the gauge itself could peel off due to stress waves. Specifications for both elongation strain gauge (YEFLA-2-1LJC) and strain gauge load cell (FCB-3-350-11-1LJB) for room temperature test conditions (23°C and 50% RH) are given in Table 10.2. The testing samples are cleaned using acetone and abrasive paper (120–180) to remove any debris (rust, paint, grease, etc.) and appropriate centreline marking is done for gauge installation. The strain gauges are pasted thereafter using prescribed post-yield adhesives (cyanoacrylate: CN-Y) which are cured at room temperature within 60–120 seconds with suitable thumb pressure. This specific adhesive is suitable for use on post-yield strain gauge, offers good bonding performance, and the degradation in peel resistance is minimum due to ageing. Both the strain gauges, i.e., load cell strain gauge and elongation strain gauge, including CNY bonding adhesive, are manufactured by Tokyo Measuring Instruments Lab (TML, Japan).

10.3 CONSTITUTIVE MODELLING

In this work, the strain-rate-dependent behaviour is modelled using model [12] given by Equation (10.1).

$$\sigma = \sigma_0 \left(1 + A ln \left(\frac{\dot{\varepsilon}}{\dot{\varepsilon}_0} \right) + K \sqrt{\frac{\dot{\varepsilon}}{\dot{\varepsilon}_0}} \right) + \left(B\varepsilon_p + C \left(1 - e^{-\beta \varepsilon_p} \right) \right) \left(1 - Gln \left(\frac{\dot{\varepsilon}}{\dot{\varepsilon}_0} \right) - H \sqrt{\frac{\dot{\varepsilon}}{\dot{\varepsilon}_0}} \right) \tag{10.1}$$

Where, σ_0, A, K, B, C, β, G, and H are the material constants which are determined using non-linear least squares minimization using Microsoft Excel® Solver. In the equation, $\dot{\varepsilon}$ and $\dot{\varepsilon}_0$ refer to strain rate and reference strain rate respectively. These parameters are used to generate true-stress vs effective plastic strain data for Abaqus input to validate experimental stress-strain data. The correlation of experimental stress-strain curves with tensile test simulation is shown in Figure 10.4.

The validated material model then can be used in crash box energy absorption simulation as demonstrated in section 10.5. The model development work for the high strain rate behaviour of different sheet steel for automotive applications is discussed in [11] and adopted here for simulation purposes.

TABLE10.2
Strain Gauge Specifications

Strain gauge parameters and type	Load cell strain gauge	High elongation strain gauge
Gauge length, mm	3	2
Gauge factor	2.11 (± 1%)	2.16 (± 2%)
Gauge resistance, Ω	350 ± 1.0	119.5 ± 0.5
Lead wire resistivity (Ω/m)	0.44	0.32

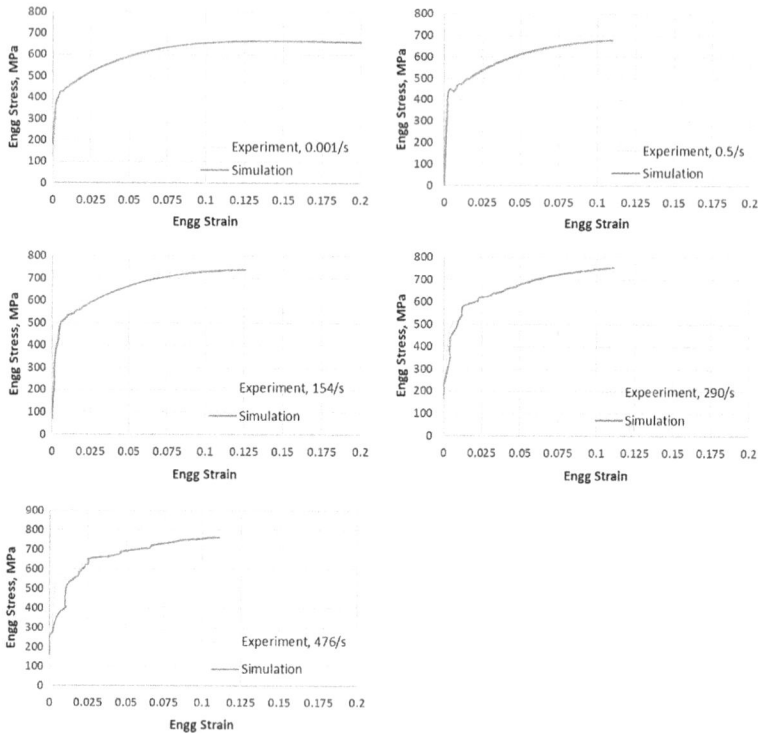

FIGURE 10.4 Tensile test simulation and correlation with experiments using rate-dependent constitutive model.

10.4 DOUBLE-CELL CRASH BOX DESIGN

A double-cell crash box design made of DP590 (1.42 mm) is proposed for enhanced energy absorption and could be useful as an automotive axial crash member. Multicell designs are normally extruded and are found in high-end premium cars. These designs provide better energy absorption compared to single-cell crash box designs. The cost of manufacturing for extrusion is high and thus this design is restricted to premium cars. However, steel as a material is significantly used in mass-produced cars, and a double-cell crash box design using stamping and bending could be a frugal and efficient choice for automotive manufacturers. The proposed double-cell crash box design discussed in this study is also patented [13]. The three design elements of the crash box are listed below:

- Equally spaced grooves and ribs on the walls to facilitate axial collapse achieved through stamping.

- S-shape cross-section achieved through bending.
- Spot welds on the broader side of the crash box where the S section ends.

It could be inferred from the foregoing design elements of the proposed crash box that a prototype could be realized using conventional manufacturing keeping the cost low. The geometry, dimensions, and cross-section of the proposed double-cell crash box and its prototype are shown in Figure 10.5.

The next section shows how the material model is derived based on high strain rate data generated for DP590 steel grade which could be coupled with the proposed crash box design to simulate various axial collapse situations at different speeds to gain insights into progressive folding and energy absorption processes.

10.5 ENERGY ABSORPTION STUDY UNDER AXIAL CRASH LOADING

An FE simulation model in Abaqus is developed to study the axial collapse behaviour under crash loading at 35kmph and 50kmph. The crash box is impacted using a loading plate weighing 250 kg. The schematic of the crash model is depicted in Figure 10.6.

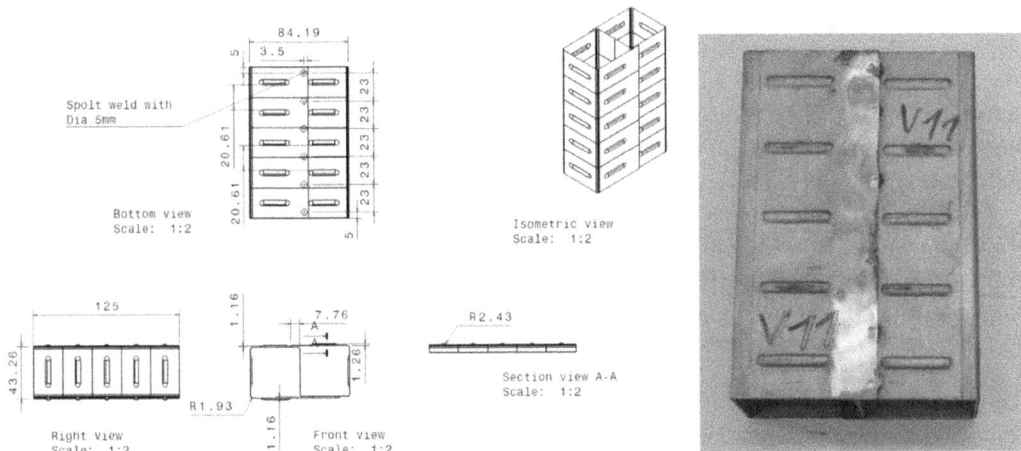

FIGURE 10.5 Double-cell crash box drawing showing specific design elements (grooves, S-shape cross-section and spot welds) and its prototype (all dimensions in mm).

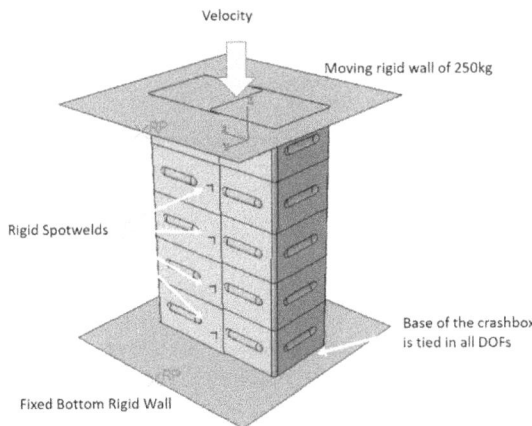

FIGURE 10.6 Schematic of crash box simulation model.

The crush performance study is carried out using FE analysis using Abaqus/Explicit. The crash box is discretized using four nodes of a reduced integration S4R shell element (~1 mm mesh size) which is computationally efficient. The general contact algorithm used in the simulation includes rigid plates at the end with the crash box and contact between folding sheet metals in the wall region during progressive crush. The coefficient of friction between contacting surfaces in the model is taken as 0.12.

10.5.1 Comparison of Performance Metrics of Crash Box

The axial collapse of tubular structures like a crash box has a unique force-deflection signature. The rise of the force-deflection curve is associated with sudden loading of the crash box and the maximum load achieved is termed as peak force. Buckling of the side wall leads to a drop in force, and with the progressive folding of the side walls, the load oscillates lower than peak force which could be termed as average force or steady-state crush force. The best-performing axial crash members have the maximum ratio of steady-state crush force to peak force, which is known as crush efficiency. Energy absorption can be calculated as the area under the crush force-deflection curve. The crush force-deflection response along with energy displacement for 9.73 m/s and 13.89 m/s are shown in Figure 10.7. The choice of speeds is based on the premise of understanding the low-speed and high-speed impact response of the crash box. The force and energy response of the crash box differs significantly beyond 30 mm of displacement. The steady-state force in the case of 13.89 m/s impact is approximately 12% higher than 9.73 m/s, signifying the effect of strain rate due to change in speed. Similarly, crush efficiency at 13.89 m/s is 12% higher compared to crush efficiency at 9.73 m/s. Energy absorption at high speed is 10% higher than at low speed. However, the peak force for both cases is almost similar.

The definition of performance metrics of the axial crash box follows the metrics used by Kohar et al. [14]. Table 10.3 shows the steady-state force, peak force, crush efficiency, and energy absorption of the crash box at 9.73 m/s and 13.89 m/s.

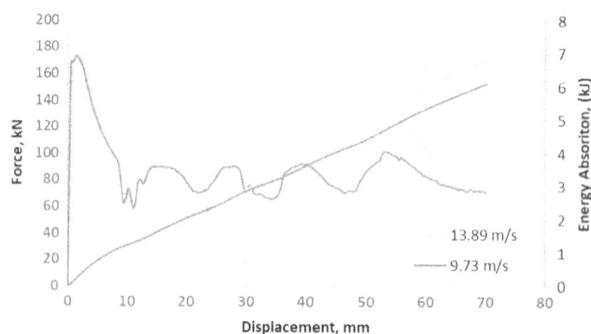

FIGURE. 10.7 Force-displacement and energy absorption-displacement curves (secondary axis) at 13.89 m/s and 9.73 m/s.

TABLE 10.3
Crash Performance Results

Velocity, m/s	Steady-state force (F_s), kN	Peak force (F_{peak}), kN	Crush efficiency, η	Energy absorption (kJ)
9.73	81.34	172.76	0.47	6.10
13.89	92.91	174	0.53	6.77

10.5.2 Deformation Mode of Crash Box at Low Speed (9.73 m/s) and High Speed (13.89 m/s)

Progressive deformation of the crash box under low-speed impact and high-speed impact is significantly different as observed in simulations. For the low-speed case, initial progressive deformation starts both from the top and bottom and later buckles at mid-height. For the high-speed case, progressive folding initiates from the bottom until the energy absorption process terminates due to asymmetric buckling later. Progressive axial crush profiles for both cases are shown in Figure 10.8. Stable collapse behaviour is essential wherein the crash box should deform without any unusual pattern. Despite the grooves which have been designed to facilitate the crush, the deformation pattern at maximum displacement, i.e., at 70 mm, shows a distinct difference. In the simulation, the welds are modelled as rigid beams and their location could influence the deformation mode too.

10.6 CONCLUSIONS AND FURTHER WORK

This work aimed at characterizing a steel grade (DP590, 1.42 mm) applicable for energy-absorbing devices like a crash box used in a passenger vehicle. High strain rate experimental procedure for tensile testing is elucidated wherein the use of strain gauges for elongation and load measurement purposes is explained. Experimental stress-strain data up to 500/s of strain rate is generated and further used in a rate-dependent constitutive material model for validating high-speed tensile test experimental results. A simulation study of double-cell crash box design at low speed and high speed is conducted to understand the crash performance. The proposed double-cell crash box design is patented and prototyped using conventional manufacturing methods. It is envisaged that the prototyped design will be evaluated for crash performance under impact/drop loading at 35kmph and 50kmph. Local failures like spot weld rupture may trigger an unwanted structural response, which could not be easily predicted in simulations. The second part of this study will evaluate the structural response of the double-cell crash box and the results will be utilized to further improve design and manufacturing.

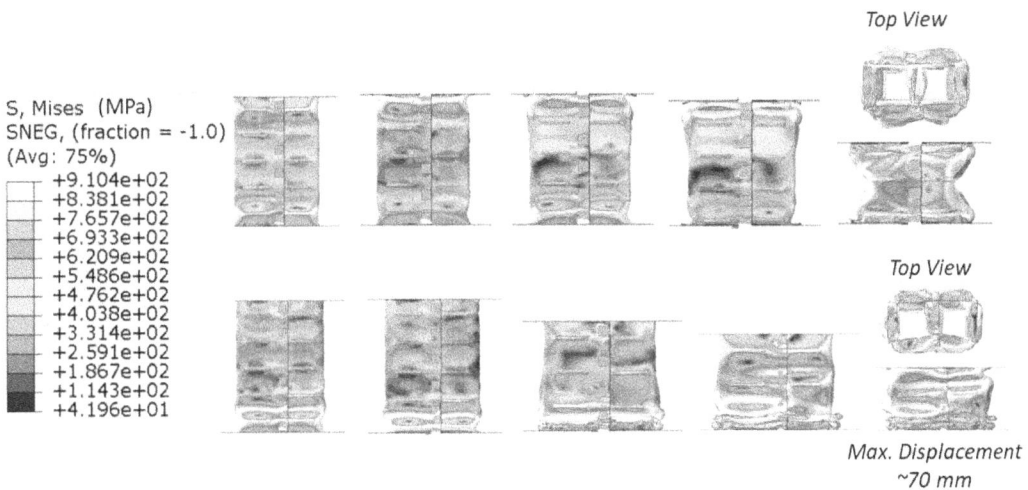

FIGURE 10.8 Progressive deformation of crash box at (a) low speed (9.73 m/s) and (b) high speed (13.89 m/s).

ACKNOWLEDGEMENTS

The authors are grateful for the support extended by R&D Tata Steel Ltd. for material and experimental facilities in support of this research work. We also thank our laboratory support staff Nitish Ranjan Goutam, Bhaskar, and Pavithran for supporting experimental work and CAD.

REFERENCES

1. https://www.euroncap.com/en/vehicle-safety/the-ratings-explained/adult-occupant-protection/frontal-impact/ Accessed 1st March 2022.
2. Olabi AG, Morris E, Hashmi MSJ. 2007. "Metallic tube type energy absorbers: A synopsis". *Thin-walled Structures* 45: 706–726. https://doi.org/10.1016/j.tws.2007.05.003
3. Alghamdi AAA. 2001. "Collapsible impact energy absorbers: An overview". *Thin-Walled Structures* 39 (2): 189–213. https://doi.org/10.1016/S0263-8231(00)00048-3
4. Langseth M, Hopperstad O, Berstad T. 1999. "Crashworthiness of aluminum extrusions: Validation of numerical simulation, effect of mass ratio and impact velocity". *International Journal of Impact Engineering* 22: 829–854. https://doi.org/10.1016/S0734-743X(98)00070-0
5. Hussain NN. 2015. "Automobile crash box design improvement using HyperStudy". India Altair Technology Conference, Bangalore.
6. Song J, Xu S, Liu S, Zou M. 2020. "Study on the crashworthiness of bio-inspired multi-cell tube under axial impact". *International Journal of Crashworthiness*. https://doi.org/10.1080/13588265.2020.1807686
7. Singh PK, Das A, Sivaprasad S, Biswas P, Verma RK, Chakrabarti D. 2017. "Energy absorption behaviour of different grades of steel sheets using a strain rate dependent constitutive model". *Thin-Walled Structures* 111: 9–18. https://doi.org/10.1016/j.tws.2016.11.005
8. High Strain Rate Experts Group. 2005. "Recommendations for dynamic tensile testing of sheet metals". International Iron and Steel Institute.
9. ISO 26203-2:2011. "Metallic materials - Tensile testing at high strain rates - Part 2: Servo-hydraulic and other test systems".
10. National Metallurgical Laboratory (India). Technical Report. (2015). "High Strain Rate Deformation of Automotive Grade of Sheet Steels".
11. *Strain Gauge Users Guide*. Tokyo Measuring Instruments Lab.
12. Paul SK, Raj A, Biswas P, Manikandan G, Verma RK. 2014. "Tensile flow behavior of ultra-low carbon, low carbon and micro alloyed steel sheets for auto application under low to intermediate strain rate". *Materials and Design* 57: 211–217. https://doi.org/10.1016/j.matdes.2013.12.047
13. Singh PK, Kumar A, Choudhary S, Verma RK. "A Crashbox for vehicle". Indian Complete Patent Application No. 202131010196. Filed March 2021.
14. Kohar CP, Mohammadi M, Mishra RK, Inal K. 2015. "Effects of elastic-plastic behaviour on the axial crush response of square tubes". *Thin-Walled Structures* 93: 64–87. https://doi.org/10.1016/j.tws.2015.02.023

11 Role of Notch Location and Loading Rate on Tensile Strength of Glass/ Epoxy Composites

Ruchir Shrivastava and K. K. Singh

11.1 INTRODUCTION

Composite materials have gained acceptance in transport vehicles for their cheap cost, high strength-to-weight ratio, high stiffness, and easy manufacturing techniques [1–5]. However, poor damage predictability and quasi-brittle nature pose limitations on their use. Moreover, structural discontinuity is inherent in any design, but creating a notch may cause colossal strength loss. A few articles related to the present research have been included here. Aljibori et al. [6] investigated the compression behavior of glass/epoxy laminate with and without holes. It appeared that cross-ply laminates behaved better than the other selected stacking sequence. Moreover, the specimen could bear a greater load with a smaller hole size. Kazemahvazi et al. [7] investigated the tensile strength of glass/epoxy composite with different holes and positions. It was seen that when the number of holes is high, it causes shear failure, while with a lesser number of holes, it causes global net-section failure. Pothnis et al. [8] dispersed multi-walled carbon nanotube (MWCNT) particles at the periphery of a hole in glass/epoxy composites in random, parallel, and perpendicular arrangements. It was seen that the vertical alignment of particles concerning fiber direction resulted in a 27% enhancement in tensile strength. Han et al. [9] worked with MWCNT-embedded open-hole strength of carbon fiber reinforced polymer composite (CFRP) laminate. It was seen that the inclusion of MWCNT reduced the stress concentration around the hole. It eventually betters the load-bearing ability of the sample. Wisnom et al. [10] examined the role of delamination on tensile strength with different thicknesses of ply blocks. It was concluded that higher thicknesses inverted the hole effect, but with increasing diameter, strength improved. Ozaslan et al. [11] demonstrated experimentally and numerically that stress concentration factor (SCF) varies with width/diameter (w/d) ratio in 2×2 twill weave carbon fabric. Ubaid et al. [12] demonstrated that hole pairs aligned along load direction in CFRP have better strength than holes positioned perpendicular or at an angle. The distance between the notch also affects SCF. A reduced SCF is found in longitudinally aligned holes with a decrease in hole spacing. Similar is the case with diagonally aligned holes, whereas in transversely located holes, SCF rises. Batista et al. [13] worked with glass/carbon and glass/aramid hybrid laminates. It was noticed that maximum load-bearing capacity falls considerably with increasing hole diameter. However, the modulus of elasticity increases in glass/aramid composite and reduces to some extent in glass/carbon composite. Ghezzo et al. [14] experimented with an interacting hole in the in-plane load direction. The results indicate that the laminate stress concentration factor does not influence it. Ozaslan et al. [15] worked with two interacting holes in different orientations and positions. It was noticed that the holes located in line with the loading direction depict a lesser stress concentration factor. But it is higher when they are perpendicular to the applied load. Azadi et al. [16] examined the open-hole tensile strength of carbon/epoxy laminate. The chosen loading rate

DOI: 10.1201/9781003352358-11

was 1, 2, 20, and 200 mm/min. It was seen that, with the loading rate, the tensile strength of the specimen increased, whereas the tensile strain decreased. Belgacem et al. [17] worked with different sizes of notches in glass/carbon hybrid laminates. Notably, a small contribution of carbon fiber (25%) significantly improves laminate performance. It was also noticed that the strength properties fall with an increase in notch size, especially in glass fiber reinforced polymer (GFRP) composite. Compared to the pristine sample, these dropped by 23.46, 26.45, and 45.04%. Sebaey et al. [18] worked with the flexural performance of carbon/aramid composites. It was seen that the sandwich structure depicts slower growth of damage. However, as the notch size increases, intra-ply and sandwich structure performance is the same. Ozaslan et al. [19] evaluated the impact of a hole on CFRP composite. It was seen that with a reduction in notch size, the strength of the composite increases tremendously. It was also noticed that the SCF increases with the hole diameter. The point stress criterion is found helpful for strength prediction in these cases. Jebri et al. [20] stated that in CFRP composite, the presence of a hole changes the failure pattern. It was seen that fiber failure is more dominant in these cases. Wang et al. [21] worked with different sizes and shapes of notch effects on CFRP composite. It was noticed that both parameters impact the damage pattern. It was also seen that the stacking sequence in terms of on-axis and off-axis changes the damage phenomenon from fiber fracture to damage initiation and unstable propagation. Dasari et al. [22] worked with glass/carbon hybrid composite. It was seen that few alterations in ply sequencing turn out to be beneficial in improving material elastic constant, strain at failure, and other associated properties.

In the present work, the role of the location of a pair of notches has been investigated on glass/epoxy composite. In addition, the role of the loading rate has also been investigated. The glass/epoxy composites were prepared using hand layup assisted by a compression molding machine. All the specimens were dimensioned as per ASTM D3039 [23] and had a thickness of 4 mm and fiber volume fraction of 48.13%. The hole spacing was chosen as 20, 30, and 40 mm, whereas the loading rate was 1, 10, and 100 mm/min. The digital image correlation method was employed to understand the strain field distribution in the specimen.

BOX 11.1   Nomenclature

CFRP	Carbon fiber reinforced polymer composite
DIC	Digital image correlation
GFRP	Glass fiber reinforced polymer composite
MWCNT	Multi-walled carbon nanotubes
SCF	Stress concentration factor

11.2   MATERIALS AND FABRICATION

In the present work, composite plates were cured in a compression molding machine, whereas the dry fabric layup was impregnated using the hand-layup technique (Figure 11.1). The laminate had the stacking sequence of $(0°/90°)_{4S}$. The curing pressure was maintained constant at 1.98 MPa, and it took a time of 24 hours at room temperature. The plates were left over a week, after which samples were prepared and tabbed using an aluminum strip of Araldite. The tabbed specimens were again left for seven days to enable strong bonding. The notches in the specimens were made using a bench mill drill machine (SX2.7, SIEG industrial group make) with a rotational speed of 1,000 rpm. A pair of holes is introduced along the gauge section of the specimen. The hole diameter was maintained at 2 mm, whereas the center-to-center (cc) distance of spacing was taken as 20, 30, and 40 mm (Figure 11.2). These are referred to as cc20, cc30, and cc40 afterward. The drilling parameters were kept constant for the entire experimental setup. Therefore, its impact on surface roughness and delamination has been omitted. Moreover, the surface roughness of the hole interior has been measured using a 3D optical surface profiler (Newview 900 Zygo, USA-made). The average

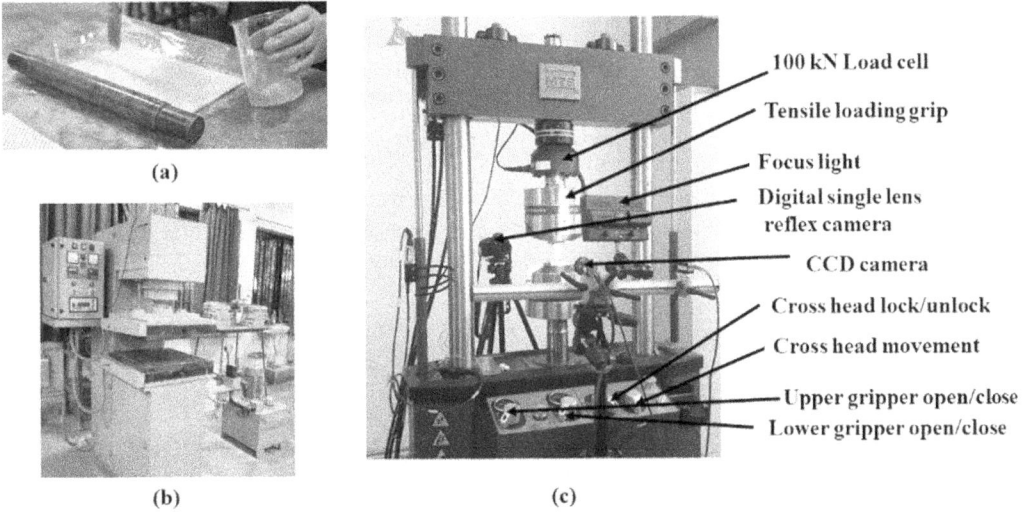

FIGURE 11.1 (a) Hand-layup technique; (b) compression molding machine; (c) experimental setup.

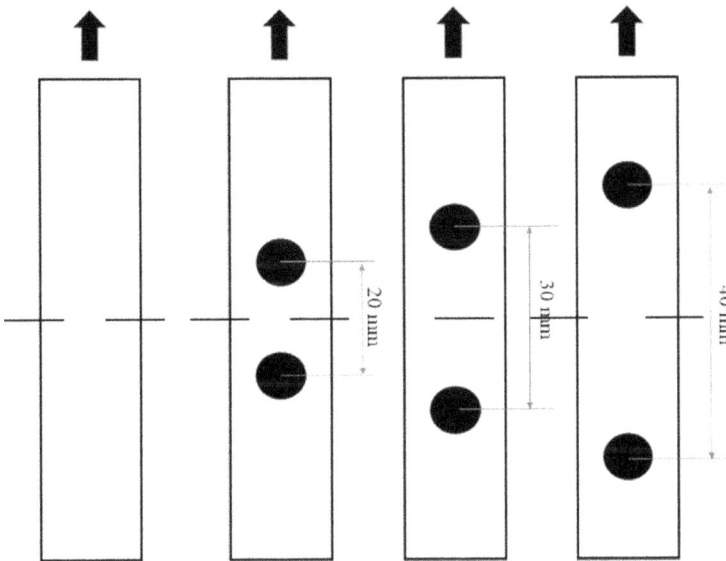

FIGURE 11.2 Notch configurations.

and maximum surface roughness were found at 36.774 and 113.992 μm, respectively. The damage morphology has been studied using a table-top microscope (Celestron, China). It has a resolution of 5 MP with a capacity to zoom by 20–200X.

11.3 EXPERIMENTATION

The tensile tests were conducted in three test specimens from each slot in a computerized universal testing machine following ASTM D3039 [23] standard. The chosen loading rate was 1, 10, and 100 mm/min. The w/d ratio was kept constant at 13 for the entire experimental work. The facility was equipped with a digital image correlation (DIC) facility to evaluate in-situ strain monitoring in x-, y-, and xy-directions. This has been evaluated by examination of the speckle pattern in the specimen. These are created by first applying white spray paint and then, upon drying, applying black spray paint. These patterns are random in nature and must be carefully made as it directly impacts DIC results. The digital image correlation was done using in-situ arrangement provided by Dantec with processing software ISTRA4.4.7.66. It consisted of a camera with a resolution of 12 megapixels. The camera was first focused appropriately to get high-quality images. Calibration of the camera position was then done using the blackboard. A well-calibrated setup was then initiated alongside the experimentation process. One picture every two seconds was captured to get accurate information on strain field distribution in the sample.

11.4 RESULTS AND DISCUSSION

11.4.1 EXPERIMENTAL TENSILE PROPERTIES

The load-displacement graph reveals that the pristine specimen, i.e., without a notch, performs comparatively better than all the notched specimens (Figures 11.3–11.5). It reflects in the ultimate tensile strength (UTS) of the specimen. At a 1 mm/min loading rate, the ultimate tensile strength of cc20, cc30, and cc40 was reduced by 13.81, 21.06, and 29.27% to its pristine specimen configuration (Figure 11.6). Similarly, at 10 and 100 mm/min, these were reduced by (6.79, 9.28, and 10.72%) and (4.07, 9.44, and 17.24%). A similar trend has been observed in Young's modulus, which was reduced by (4.8, 9.95, 16.39%), (1.13, 2.77, 3.28%), and (1.77, 4.08, 5.41%) at 1, 10, and 100 mm/min loading rates (Figure 11.7). The results demonstrate that the tensile strength reduced significantly when the holes were placed around the mid-span. The damage morphology (Figure 11.8) demonstrates fiber pull-out, yarn pull-out, fiber breakage, debonding, delamination, and matrix cracking. It was also seen that the notched specimens always fail from one of the notch locations [24]. The pristine specimens have global stretching of the gauge section, whereas the notched composites have local stretching of the gauge section. One keen observation is the more vivid matrix cracking at one surface of the specimen. As loading begins in the specimen, a resistance is developed, which restricts its deformation. But it was soon overcome by force applied, and due to various factors, such as manufacturing defects, void, etc., crack nucleation began. In woven composites, the crack propagation direction is normal to the force applied to the specimen [25]. The increase in load causes cracks propagation resulting in matrix cracking. Later, the specimen breaks depicting fiber stretching.

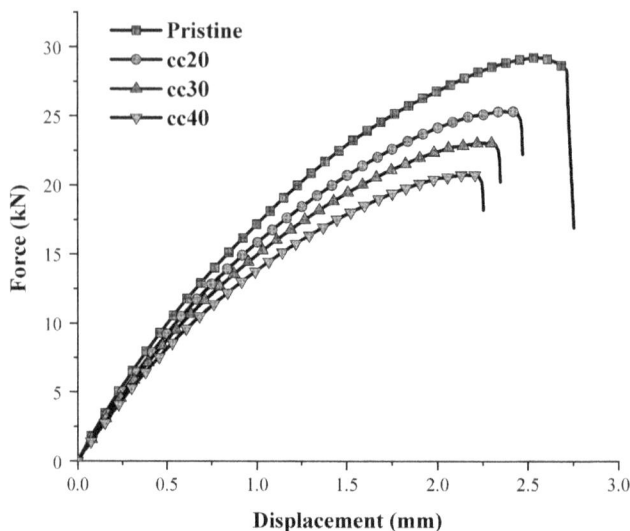

FIGURE 11.3   Load-displacement response of glass/epoxy composite at 1 mm/min loading rate.

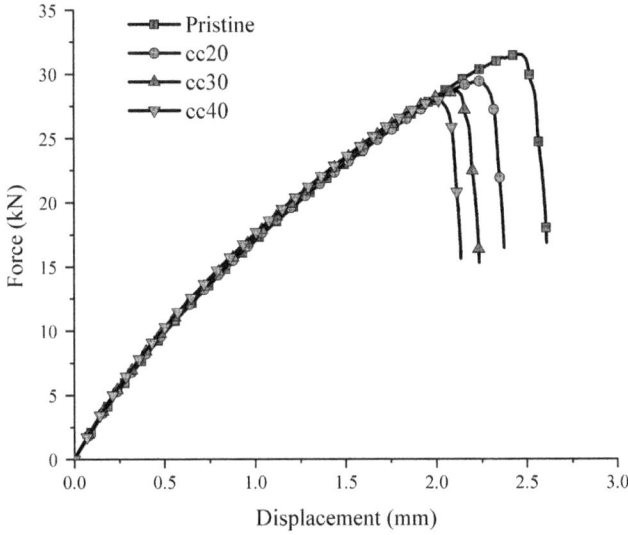

FIGURE 11.4   Load-displacement response of glass/epoxy composite at 10 mm/min loading rate.

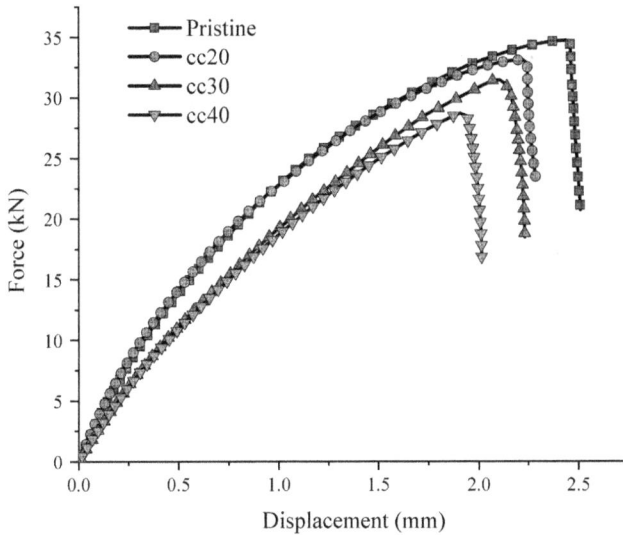

FIGURE 11.5   Load-displacement response of glass/epoxy composite at 100 mm/min loading rate.

All eight layers of glass/epoxy attempt to hold the specimen together, but extensive load extends them to different degrees. The specimen fails with ample fiber and yarn stretching. The notched specimen depicts prolonged fiber stretching and pull-out. The material has less time to respond to ever-increasing load at a high loading rate. Therefore, it is seen that data points from the graph are lesser, and also the extension appears lesser. The microscopic examination around the notch depicts that the matrix cracks enormously around it compared to the rest of the specimen. The evaluation of notched composite strength is generally done by the two most popular methods, i.e., point stress criterion and average stress criterion. It states that material fails when the stress value equals that of unnotched strength at a certain distance. This distance is known as characteristic length, which depends upon hole size, specimen width, type of material, etc. As for the present study, all these

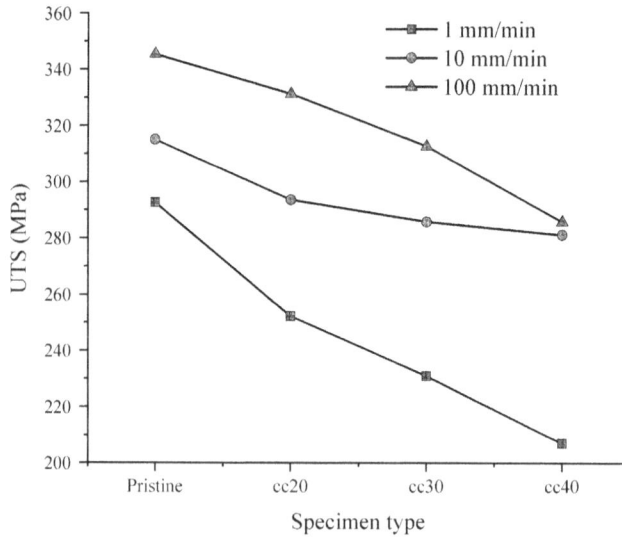

FIGURE 11.6   Ultimate tensile strength (UTS) of glass/epoxy composite.

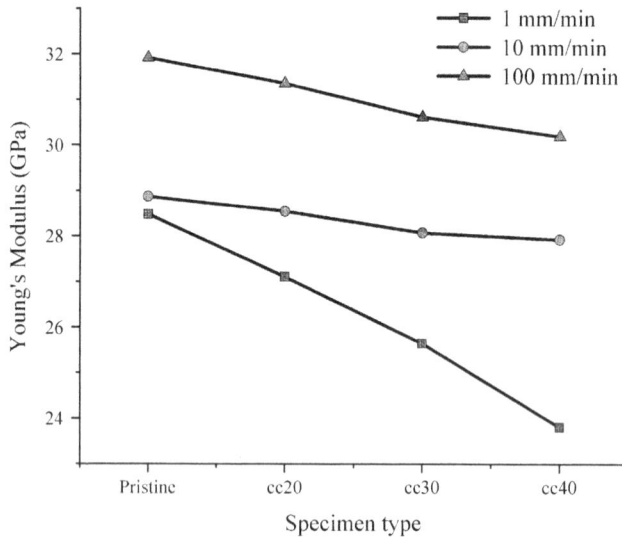

FIGURE 11.7   Young's modulus (E) of glass/epoxy composite.

parameters are the same. The characteristic length is also the same irrespective of the notch location. Therefore, in all the notched composites, the damage pattern appears similar regardless of its loading rate [26]. The results demonstrate a maximum reduction in UTS at a 1 mm/min loading rate in notched composites, which is lesser at higher loading rate scenarios. It is presumed that the load application for comparatively less time helps in this regard. The unnotched specimen depicted broom failure mode, whereas the notched composites depicted extensive yarn splitting [17]. The localized failure observed in the notched specimen was due to the stress concentration effect due to the presence of notches.

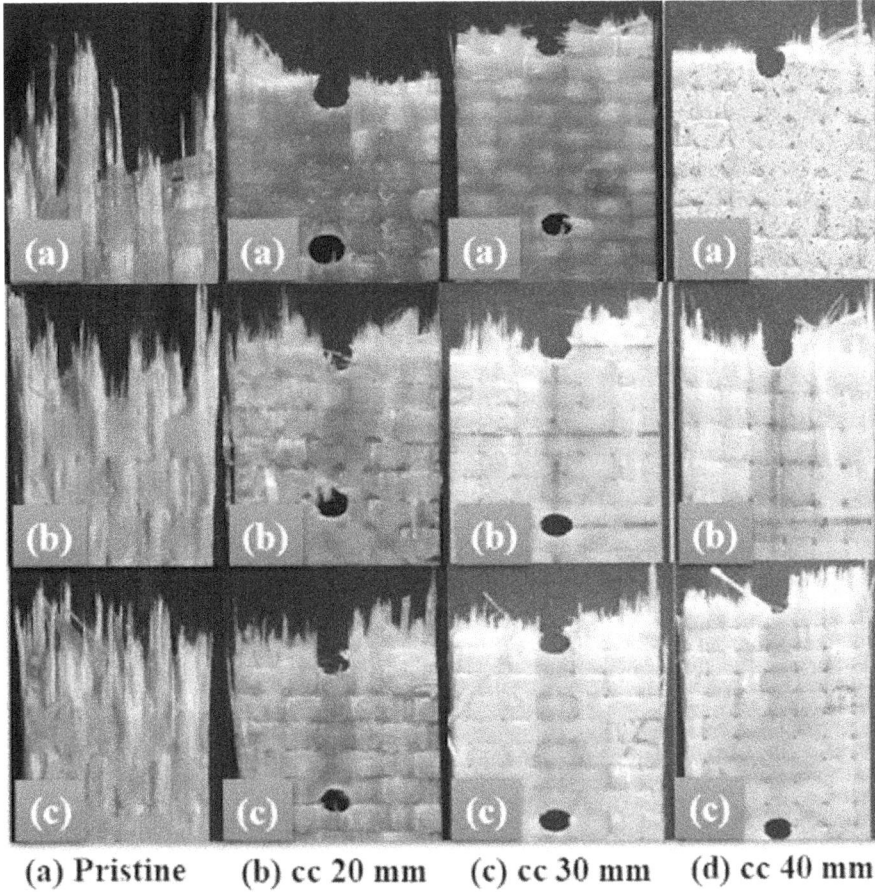

(a) Pristine   (b) cc 20 mm   (c) cc 30 mm   (d) cc 40 mm

FIGURE 11.8 Damage morphology in glass/epoxy composite: (1) 1 mm/min; (2) 10 mm/min; (3) 100 mm/min.

11.4.2 DIC STRAIN FIELDS

The engineering strain fields in the specimen have been observed in the x-, y-, and xy-direction. It was seen that with the application of load, the localized strain fields grow in size. It indicates the growing damage in the specimen, which occurs in the form of matrix cracking, fiber pulling, and breakage. The strain distribution in pristine specimens has been depicted in two stages. The images are indicative of the fact that specimen deformation takes place throughout the gauge section. It is not located on a single point [27]. Strains are lower for pristine specimens, whereas the notched composite is very high (Figures 11.9–11.12).

11.5 CONCLUSION

- The creation of a notch drastically impacts the specimen's tensile strength, which drops in the range of 4.07–29.27%. A similar trend was noticed in Young's modulus, which was reduced by 1.13–16.39%.
- Notch spacing causes changes in both ultimate tensile strength and Young's modulus, which reduces with an increase in hole spacing.

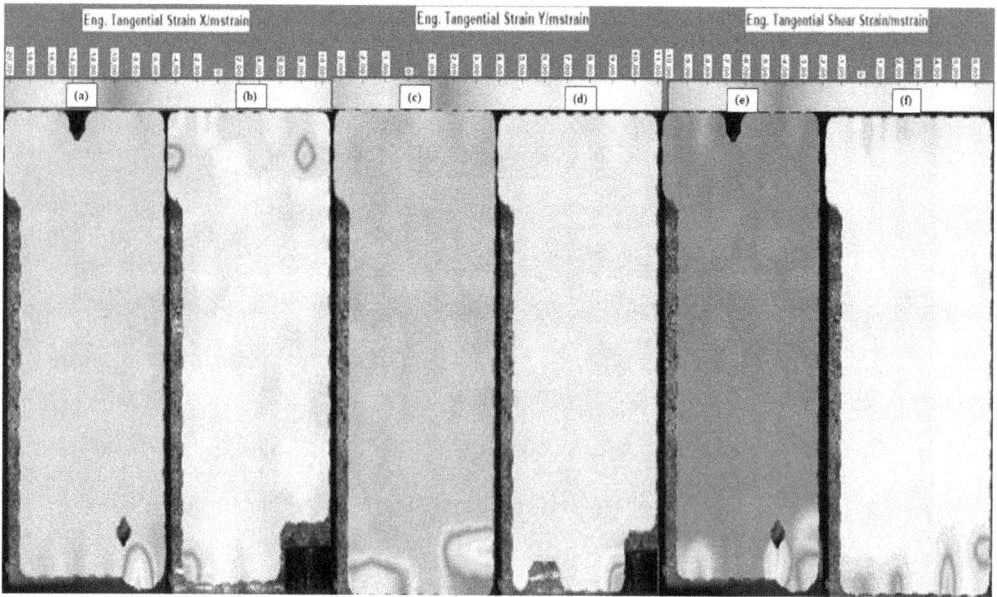

FIGURE 11.9 DIC images of pristine glass/epoxy composite at 1 mm/min loading rate: (a–b) engineering tangential strain in x-direction; (c–d) engineering tangential strain in y-direction; (e–f) engineering tangential shear strain in xy-direction.

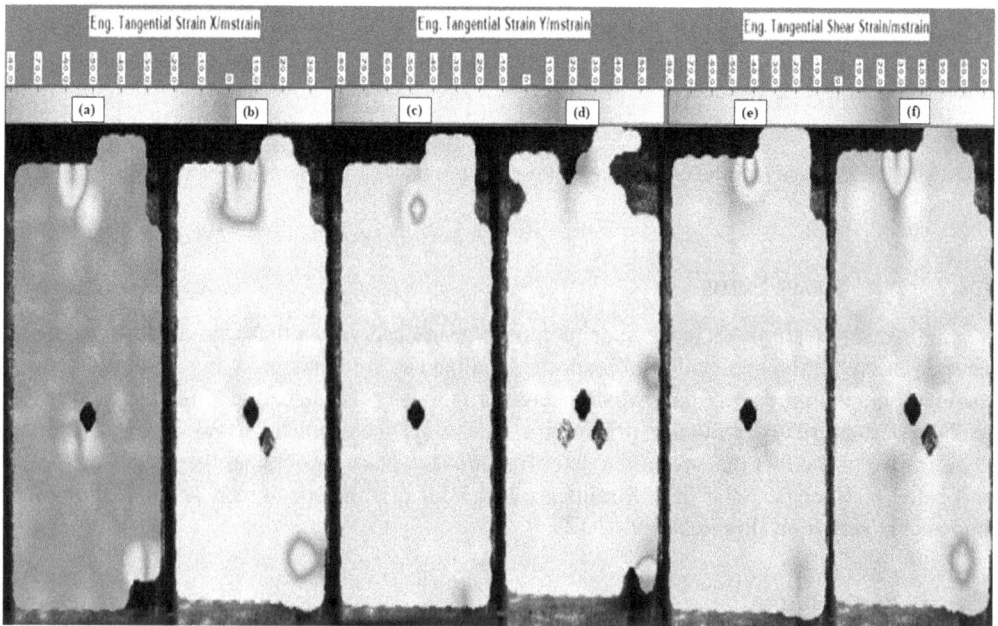

FIGURE 11.10 DIC images of cc20 glass/epoxy composite at 1 mm/min loading rate: (a–b) engineering tangential strain in x-direction; (c–d) engineering tangential strain in y-direction; (e–f) engineering tangential shear strain in xy-direction.

FIGURE 11.11 DIC images of cc30 glass/epoxy composite at 1 mm/min loading rate: (a–b) engineering tangential strain in x-direction; (c–d) engineering tangential strain in y-direction; (e–f) engineering tangential shear strain in xy-direction.

FIGURE 11.12 DIC images of cc40 glass/epoxy composite at 1 mm/min loading rate: (a–b) engineering tangential strain in x-direction; (c–d) engineering tangential strain in y-direction; (e–f) engineering tangential shear strain in xy-direction.

- A notched specimen always fails from one of the notch locations as it experiences local stretching of the specimen, whereas pristine specimens experience global extension and fail by broom failure.
- The loading rate significantly impacted tensile strength, and properties were improved. Compared to the specimen tested at a 1 mm/min loading rate, it improved in the range of 7.51–37.11%. Similarly, the modulus improved by 1.37–26.8%.

REFERENCES

1. Shrivastava R, Singh KK. Mechanical property characterization of glass/epoxy composite with varying fiber percentage and mid-plane ply orientation. *J Brazilian Soc Mech Sci Eng* 2022;44:1–16. doi:10.1007/s40430-022-03402-4.
2. Shrivastava R, Singh KK. Interlaminar fracture toughness characterization of laminated composites: A review. *Polym Rev* 2020;60:542–593. doi:10.1080/15583724.2019.1677708.
3. Modi V, Singh KK, Shrivastava R. Effect of stacking sequence on interlaminar shear strength of multi-directional GFRP laminates. *Mater Today Proc* 2020;22:2207–2214. doi:10.1016/j.matpr.2020.03.301.
4. Shrivastava R, Singh KK. Fracture toughness of symmetric and asymmetric layup GFRP laminates by experimental and numerical methods. In: Singh I, Bajpai PK, Panwar K, editors. Singapore: Springer Singapore; 2019, pp. 13–22. doi:10.1007/978-981-13-9016-6_2.
5. A. Praveen Kumar,•Tatacipta Dirgantara, P. Vamsi Krishna editors. *Advances in Lightweight Materials and Structures*. 2020 ISSN 2662-3161 ISSN 2662-317X (electronic), Springer Proceedings in Materials, ISBN 978-981-15-7826-7 ISBN 978-981-15-7827-4 (eBook), https://doi.org/10.1007/978-981-15-7827-4.
6. Aljibori HSS, Chong WP, Mahlia TMI, Chong WT, Edi P, Al-qrimli H, et al. Load-displacement behavior of glass fiber/epoxy composite plates with circular cut-outs subjected to compressive load. *Mater Des* 2010;31:466–474. doi:10.1016/j.matdes.2009.07.005.
7. Kazemahvazi S, Kiele J, Zenkert D. Tensile strength of UD-composite laminates with multiple holes. *Compos Sci Technol* 2010;70:1280–1287. doi:10.1016/j.compscitech.2010.04.005.
8. Pothnis JR, Kalyanasundaram D, Gururaja S. Enhancement of open hole tensile strength via alignment of carbon nanotubes infused in glass fiber - Epoxy - CNT multi-scale composites. *Compos Part A Appl Sci Manuf* 2021;140:106155. doi:10.1016/j.compositesa.2020.106155.
9. Han K, Zhou W, Qin R, Wang G, Ma LH. Effects of carbon nanotubes on open-hole carbon fiber reinforced polymer composites. *Mater Today Commun* 2020;24:101106. doi:10.1016/j.mtcomm.2020.101106.
10. Wisnom MR, Hallett SR. The role of delamination in strength, failure mechanism and hole size effect in open hole tensile tests on quasi-isotropic laminates. *Compos Part A Appl Sci Manuf* 2009;40:335–342. doi:10.1016/j.compositesa.2008.12.013.
11. Özaslan E, Yetgin A, Acar B. Stress concentration and strength prediction of 2×2 twill weave fabric composite with a circular hole. *J Compos Mater* 2019;53:463–474. doi:10.1177/0021998318785994.
12. Ubaid J, Kashfuddoja M, Ramji M. Strength prediction and progressive failure analysis of carbon fiber reinforced polymer laminate with multiple interacting holes involving three dimensional finite element analysis and digital image correlation. *Int J Damage Mech* 2014;23:609–635. doi:10.1177/1056789513504123.
13. ACMC B, SRL T, RS F, SHS N, EMF A. Analytical, experimental and finite element analysis of the width/diameter hole ratio effect in vinylester/carbon hybrid twill weave composites. *Compos Part C Open Access* 2020;2:100033. doi:10.1016/j.jcomc.2020.100033.
14. Ghezzo F, Giannini G, Cesari F, Caligiana G. Numerical and experimental analysis of the interaction between two notches in carbon fibre laminates. *Compos Sci Technol* 2008;68:1057–1072. doi:10.1016/j.compscitech.2007.07.023.
15. Özaslan E, Güler MA, Yetgin A, Acar B. Stress analysis and strength prediction of composite laminates with two interacting holes. *Compos Struct* 2019;221. doi:10.1016/j.compstruct.2019.04.041.
16. Azadi M, Sayar H, Ghasemi-Ghalebahman A, Jafari SM. Tensile loading rate effect on mechanical properties and failure mechanisms in open-hole carbon fiber reinforced polymer composites by acoustic emission approach. *Compos Part B Eng* 2019;158:448–458. doi:10.1016/j.compositesb.2018.09.103.
17. Belgacem L, Ouinas D, Viña Olay JA, Amado AA. Experimental investigation of notch effect and ply number on mechanical behavior of interply hybrid laminates (glass/carbon/epoxy). *Compos Part B Eng* 2018;145:189–196. doi:10.1016/j.compositesb.2018.03.026.
18. Sebaey TA, Wagih A. Flexural properties of notched carbon–aramid hybrid composite laminates. *J Compos Mater* 2019;53:4137–4148. doi:10.1177/0021998319855773.

19. Özaslan E, Acar B, Güler MA. Experimental and numerical investigation of stress concentration and strength prediction of carbon/epoxy composites. *Procedia Struct Integr* 2018;13:535–541. doi:10.1016/j.prostr.2018.12.088.

20. Jebri L, Abbassi F, Demiral M, Soula M, Ahmad F. Experimental and numerical analysis of progressive damage and failure behavior of carbon Woven-PPS. *Compos Struct* 2020;243:112234. doi:10.1016/j.compstruct.2020.112234.

21. Wang C, Yao L, He W, Cui X, Xie D, Lu S. Mechanical response and critical failure mechanism characterization of notched carbon fiber reinforced polymer laminate subjected to tensile loading. *Polym Compos* 2020;41:4221–4242. doi:10.1002/pc.25706.

22. Dasari S, Lohani S, Gangineni PK, Prusty RK. Effects of cryogenic aging on flexural behavior of advanced inter-ply hybrid fiber-reinforced polymer composites. *Trans Indian Inst Met* 2021;74:2171–2183. doi:10.1007/s12666-021-02288-5.

23. ASTM. Standard test method for tensile properties of polymer matrix composite materials. *Annu B ASTM Stand* 2014:1–13. doi:10.1520/D3039.

24. Hallett SR, Green BG, Jiang WG, Wisnom MR. An experimental and numerical investigation into the damage mechanisms in notched composites. *Compos Part A Appl Sci Manuf* 2009;40:613–624. doi:10.1016/j.compositesa.2009.02.021.

25. Muc A, Romanowicz P. Effect of notch on static and fatigue performance of multilayered composite structures under tensile loads. *Compos Struct* 2017;178:27–36. doi:10.1016/j.compstruct.2017.07.004.

26. Srivastava VK. Notched strength prediction of laminated composite under tensile loading. *Mater Sci Eng A* 2002;328:302–309. doi:10.1016/S0921-5093(01)01759-2.

27. Khechai A, Tati A, Guerira B, Guettala A, Mohite PM. Strength degradation and stress analysis of composite plates with circular, square and rectangular notches using digital image correlation. *Compos Struct* 2018;185:699–715. doi:10.1016/j.compstruct.2017.11.060.

12 Impact Analysis of Concrete Structure Using a Rate-Dependent Damage Model

K. Akshaya Gomathi and Amirtham Rajagopal

12.1 INTRODUCTION

The dynamic load causes a devastating effect on the reinforced concrete (RC) structure. There are a number of incidents where the RC structure is subjected to explosive events. The blast load is produced due to chemical, nuclear or physical events. Even a small quantity of explosives acting on the RC structure causes severe damage to the RC structure. Thus, it is important to design military, power plant, bridge and headquarter structures to be blast resistant. This can protect humans and important facilities against extremist events. The impactor or drop weight filled with explosive charge can cause impact and blast loading on the RC structure. The loading pattern of the blast load is of the distributed form and the impact load is of the concentrated form. The failure mechanism of the structure under blast and impact loading is different. Therefore, it is necessary to understand the behavior of RC structure to dynamic loading. Understanding the behavior of concrete under dynamic loading conditions allows us to design a blast and impact-mitigated concrete structure. In certain cases, there is a likelihood of RC structures undergoing combined loading cases. Hence, a detailed study is carried out to know the combined dynamic loading acting on the RC structure. Thus, the components of the structure can be improved. The dynamic load acting on the concrete structure should be considered in the design standards in order to prevent the complete collapse of the building under extreme loading conditions. The blast load acts in the form of a shock wave. The release of energy is in terms of pressure. The shock wave is of very high amplitude and acts in millisecond duration. The structure is totally engulfed by the shock wave as the distance from the blast source increases. The shock wave gets reflected from the structure and amplification of the shock wave occurs. The shock wave has both positive and negative phases. The reduction of peak overpressure occurs with increasing distance from the explosion. During the negative phase, the already weakened structure may be subjected to impact by debris. This causes additional damage to the already weakened structure causing impact and blast load to act together. The behavior of RC structure to impact load depends on the thickness of components, reinforcement ratio, velocity of impact mass and concrete properties. Therefore, a detailed parametric study needs to be carried out in order to predict the influence of these parameters under impact loading. It is well known that the higher strain rate causes the gain of strength of the concrete. There are experiments to study the material characteristics under dynamic loading. The split Hopkinson pressure bar (SHPB) is used for this purpose. It helps in finding the dynamic strength of the material. The strength of the material under dynamic load increases because of various physical mechanics. One is the strain rate sensitivity of the material and the other is the inertial effect. The dynamic to static strength of the material is related to using the dynamic increase factor (DIF). The concrete shows reduced tensile strength under static loading. But the DIF in tension shows a greater influence on concrete strength under dynamic loading.

DOI: 10.1201/9781003352358-12

BOX 12.1   Nomenclature

$\bar{\sigma}_{ij}$ Effective stress

D Damage parameter

H Heaviside function

K Bulk modulus

ε_{kk}^e Elastic volumetric strain tensor

G Shear modulus

ε_{ij}^e Elastic shear strain tensor

η Viscosity coefficient

$\dot{\varepsilon}_{ij}^p$ Plastic strain rate tensor

F Yield function

σ_m Maximum failure surface

σ_r Residual failure surface

C Fitting coefficient

$\dot{\varepsilon}^*$ Dimensionless strain rate

P Hydrostatic pressure

D_c Compression damage

D_t Tension damage

$\bar{\varepsilon}_p^t$ Effective plastic strain in tension

$\bar{\varepsilon}_p^c$ Effective plastic strain in compression

ε_{frac} Fracture strain

12.1.1   Physical Mechanism of Increase in Dynamic Strength of Concrete

There is an increase in the strength of concrete under dynamic loading because of the strain rate dependency of concrete and the inertial effect. The classification of strain rate is made in three regions. They are low, intermediate and high ranges if the strain rate is less than $10^{-1}/s$, $10^{-1} - 10^1/s$ and greater than $10^1/s$. The strain rate-dependent behavior is caused by the thermal vibration of the atom and the viscosity of the material as shown in Figure 12.1. The thermal vibration of the atom breaks the atomic bonds and the micro-cracks are formed. As the strain rate increases, the atomic vibration and more cracks are produced. In region I, a smaller number of cracks are produced. The crack propagates through the weakest path. The strength of concrete is lower in region I and there is only a slight increase in strength with increasing strain rate. More cracks are produced when the strain rate experienced by the material is in region II. There is no time for the energy in the concrete to release. Thus, the strength of the material increases to a greater extent. The cracks propagate through the aggregate. In region III, the concrete crushes and fails by forming rubble.

The moisture present in the concrete causes additional strength to the material with respect to the strain rate. The film of moisture present in the concrete exerts the return force promotional to the velocity of separation of the material. This induces an additional increase in the material strength due to the effect of the strain rate produced in the material.

There is a gain in the strength of concrete with the higher value of the rate of strain the material is undergoing [1]. This is caused due to the acceleration of particles even after the concrete crushes into rubble. The inertial effect is related to the size and shape of the concrete specimen. This is not true and has to be avoided while defining the DIF definition. This inertial effect can be avoided in the DIF definition by providing the limiting value of DIF or by giving a better definition of strain rate enhancement.

12.1.2　In-Built Model

The characteristics of concrete under dynamic loading are analyzed using the available models. The following in-built models, namely K&C Concrete Model (KCCM) [2], Continuous Surface Cap Model (CSCM) [3] and Winfrith Concrete Model (WCM) [4], are taken for the analysis. The description of the yield surface of the model is given in Figure 12.2. The KCCM considers the effect of strain rate using the DIF definition as given in the Comite Euro-International Du Beton (CEB) code. The damage law considers only the deviatoric part of stress. The limitation of the model is the automatic generation of the input parameters, which are applicable only for low hydrostatic pressure acting on the material. The DIF definition assumes that the strength of the material increases with

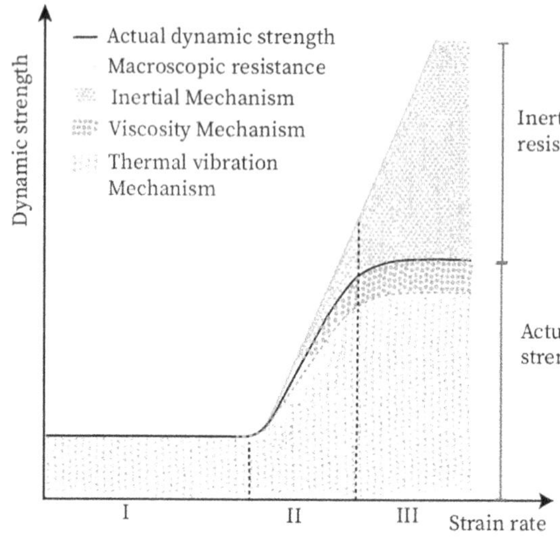

FIGURE 12.1　Mechanism of gain in strength of concrete.

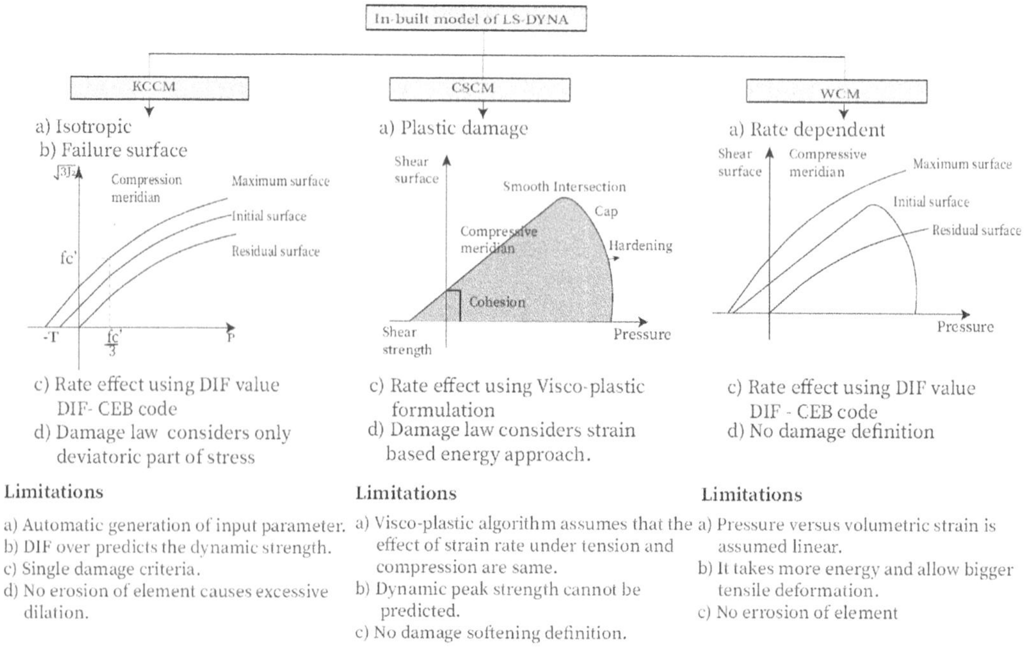

FIGURE 12.2　In-built model with its limitations.

increasing strain rate. This over-predicts the ultimate strength of the material. There is no erosion of the element; this causes excessive dilation of the material. The CSCM considers the effect of strain rate using visco-plastic formulation. The damage law is defined considering the strain-based energy approach. The limitation of the CSCM is that visco-plastic formulation assumes that the effect of strain rate under tension and compression is the same. Therefore, the dynamic ultimate strength of the material cannot be predicted accurately. There is no damage softening definition and this leads to instability in material behavior. The WCM model also considers the DIF definition to include the effect of strain rate like the KCCM model. This over-predicts the ultimate strength of the material. There is no damage definition and the equation of state (EOS) relationship is assumed to be linear. The model takes more energy and allows bigger tensile deformation of the material. There is no erosion of elements. But excessive dilation is avoided in the formulation. The fracture strain improves with strain rate; this has not been considered in the in-built model under dynamic loading.

12.2 MATERIAL MODEL

12.2.1 CONSTITUTIVE EQUATION

The constitutive equation is given using effective stress.

$$\bar{\sigma}_{ij} = \left[(1-D) + DH\left(\text{sign}\left(\sigma_{kk} \right) \right) \right] K \varepsilon_{kk}^e \delta_{ij} - \frac{2}{3}(1-D)G\varepsilon_{kk}^e \delta_{ij} + 2(1-D)G\varepsilon_{ij}^e \tag{12.1}$$

The above expressions K, G, ε_{kk}^e and ε_{ij}^e are the bulk modulus, shear modulus, elastic volumetric strain tensor and elastic shear strain tensor. H denotes the Heaviside function of hydrostatic pressure, with $H = 1$ under compressive loading $\left(\sigma_{kk} > 0 \right)$. Thus, the reversal of the stress state is taken into account. The damage parameter has a bound, $0 \le D \le 1$, where 0 and 1 represent an undamaged and fully damaged material state, respectively. If we consider the elastoplastic spring along with the viscous damper in the system, the elastoplastic spring will consider the elastoplastic behavior of concrete. The viscous damper will account for the incremental strength of the concrete. Thus, the total stress in the system will be given as

$$\sigma_{ij} = C_{ijkl}^{epd} \varepsilon_{kl}^e + \eta \dot{\varepsilon}_{ij}^p \tag{12.2}$$

where $C_{ijkl}^{epd} = \left[\left\{ (1-D) + DH\left(-\sigma_{kk} \right) \right\} K - \frac{2}{3}(1-D)G \right] \delta_{ij}\delta_{kl} + (1-D)G\left[\delta_{ik}\delta_{jl} + \delta_{il}\delta_{jk} \right]$. Thus, the following form of stress rate tensor is obtained to find the time derivative of the stress function by considering the viscous formulation and is given as

$$\dot{\sigma}_{ij} = \left(C_{ijkl}^{epd} \dot{\varepsilon}_{kl} - C_{ijkl}^{epd} \dot{\varepsilon}_{kl}^p \right) - \dot{D}C_{ijkl}\left(\varepsilon_{kl} - \varepsilon_{kl}^p \right) + \frac{\eta}{dt}\dot{\varepsilon}_{ij}^p \tag{12.3}$$

12.2.2 FAILURE SURFACE

The framework of the developed material model for the failure surface can be given in terms of hydrostatic pressure, strain rate and damage [5] as shown in Figure 12.3. The specific yield surface expression is of the form

$$Y = \left[\sigma_m (1-D) + \sigma_r D \right]\left[1 + C\ln\left(\dot{\varepsilon}^* \right) \right] \tag{12.4}$$

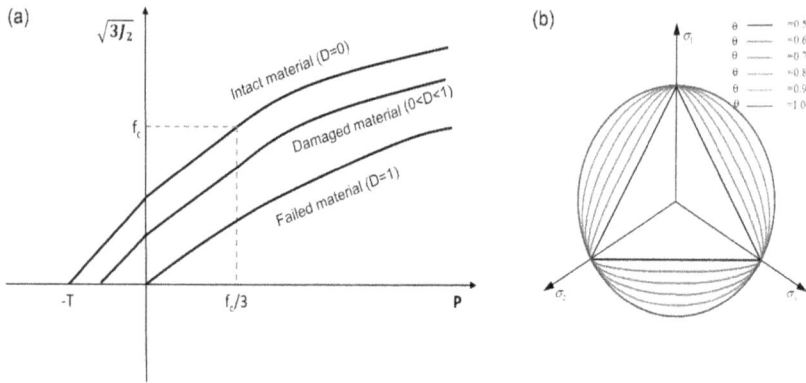

FIGURE 12.3 Failure surface in hydrostatic and deviatoric plane.

where $\dot{\varepsilon}^*$ is the dimensionless strain rate defined as $\dot{\varepsilon}^* = \left(\dfrac{\dot{\varepsilon}}{\dot{\varepsilon}_0}\right)$, and σ_m and σ_r are the strength surface at maximum and residual state. D is the damage parameter. $C = 0.007$ is the fitting coefficient.

$$\sigma_m = \begin{cases} (3T + 3P)r(\theta,e) & P \le 0 \\ \left[3T + 3\left(f_c' - 3T\right)P^*\right]r(\theta,e) & 0 < P < f_c'/3 \\ \left[f_c' + Bf_c'\left(P^* - \dfrac{1}{3}\right)^N\right]r(\theta,e) & P \ge f_c'/3 \end{cases} \tag{12.5}$$

$$\sigma_r = \begin{cases} 0 & P \le 0 \\ B_1 f_c \left(P^*\right)^{N_1} r(\theta,e) & P > 0 \end{cases} \tag{12.6}$$

$$cos(3\theta) = \frac{3\sqrt{3}}{2}\frac{J_3}{J_2^{3/2}} \tag{12.7}$$

The ratio of the compressive and current meridian (r) is given as

$$r(\theta,e) = \frac{2\left(1 - e^2\right)cos\theta + (2e - 1)\sqrt{4\left(1 - e^2\right)cos^2\theta + 5e^2 - 4e}}{4\left(1 - e^2\right)cos^2\theta + (1 - 2e)^2}$$

where e is the ratio of tensile to compressive meridian which depends on the hydrostatic pressure.

The failure surface of concrete is defined as

$$F = \sigma_{eq} - \left[\sigma_m\left(1 - D\right) + \sigma_r D\right]\left[1 + C\ln\left(\dot{\varepsilon}^*\right)\right] \tag{12.8}$$

Where $\sigma_{eq} = \sqrt{3J_2}$, J_2 is the second stress tensor for the deviatoric stress of the material.

12.2.3 DAMAGE MODELING

The compressive damage is defined as

$$D_c = \frac{\alpha \bar{\varepsilon}_p^c}{1 + \bar{\varepsilon}_p^c} \tag{12.9}$$

The damage in tension is given using the fracture strain and effective plastic strain [6]. The fracture strain is defined to know the complete damage of the material. The exponential form of tensile damage definition is given.

$$D_t = 1 - \left(1 + \left(c_1 \frac{\bar{\varepsilon}_p^t}{\varepsilon_{frac}}\right)^3\right) \exp\left(-c_2 \frac{\bar{\varepsilon}_p^t}{\varepsilon_{frac}}\right) + \frac{\bar{\varepsilon}_p^t}{\varepsilon_{frac}}\left(1 + c_1^3\right) \exp\left(-c_2\right) \tag{12.10}$$

where c_1 and c_2 are constants taken from the quasi-static test under the tensile state of loading. The values are 3 and 6.93 respectively. Under the triaxial state of loading, the value of stress is measured to be zero. The additional tensile damage ΔD_t is considered while defining the damage.

$$\Delta D_t = d_3 \times f_d \times \Delta \varepsilon_v$$

$$f_d = \begin{cases} 1 - \left|\sqrt{3J_2} / P\right| / 0.1 & 0 \leq \left|\sqrt{3J_2} / P\right| < 0.1 \\ 0 & 0.1 \leq \left|\sqrt{3J_2} / P\right| \end{cases} \tag{12.11}$$

f_d gives the damage nearer to triaxial tension. $\Delta \varepsilon_v$ is the change in the volumetric strain value. The total damage is defined as

$$D = 1 - \left(1 - D_c\right)\left(1 - D_t\right) \tag{12.12}$$

12.3 IMPACT ANALYSIS

The slab considered for the impact analysis is given in this section. The impact analysis is done on a reinforced cement concrete (RCC) slab of 1000×1000×75 mm dimension with a $8\,mm$ reinforcement bar [7]. The reinforcement is given $75\,mm$ center-to-center spacing. The concrete has a density of $2340\,kg/m^3$ and Young's modulus of $30.91\,GPa$. Poisson's ratio is taken as 0.2. The uniaxial compressive and tensile strength of concrete is taken as $29.70\,MPa$ and $2.88\,MPa$, respectively. The aggregate size of $12.50\,mm$. The steel has $105\,kg$ mass and is given a $2.5\,m$ drop height. The fixed boundary condition is provided along all the boundaries. The reinforcement has a modulus of elasticity of $198\,GPa$, ultimate strength of $501\,MPa$ and yield strength of $422\,MPa$. Table 12.1 gives the property of the material used in the RCC slab.

The damage of the RC slab under impact load is accurately predicted to that of the experiment as shown in Figure 12.4. It is observed that the developed model is able to get the localized damage on the top surface of the slab and the scabbing of the bottom surface of the slab when subjected to the impact load, whereas the WCM model cannot predict the behavior.

TABLE 12.1
Material Property under Impact Loading

Material	Property	Value
Reinforcement	Modulus of elasticity	198 GPa
	Ultimate strength	501 Mpa
	Yield strength	422 Mpa
Drop weight	Failure strain	0.15
	Density	7,850 kg/m^3
	Poisson's ratio	0.3
	Young's modulus	198 GPa
	Velocity	6.86 m/s

12.4   PARAMETRIC STUDY

The parametric study is carried out by increasing the impact velocity. It is observed that the depth of localized damage is proportional to the velocity of impact load as shown in Figure 12.5. The impact velocity is increased from 6.86 *m/s* to 13.72 *m/s*. It is observed that the slab undergoes less damage, and localized damage is formed on the top surface when it is subjected to lower impact velocity. The impact mass penetrates through the slab when it is subjected to 13.72 *m/s*. The slab should be given a larger thickness or more reinforcement ratio to reduce the damage to the slab.

The analysis is carried out by modifying the duration of impact on the structure. It is seen that as the impact duration acting in the slab increases, the localized damage of the slab increases as shown in Figure 12.6. The impact duration is changed and the damage behavior of the slab is observed. The impact duration of 3 *ms*, 6 *ms* and 20 *ms* is taken for the analysis. It is observed that the slab undergoes larger localized damage when it is subjected to 20 *ms* duration.

12.5   SHPB ANALYSIS

The split Hopkinson pressure bar (SHPB) analysis is carried out to validate the experimental results of Guo et al. [8]. SHPB works on the principle of elastic stress wave propagation in one direction. The setup has a striker bar, incident bar and transmitter bar. The specimen is sandwiched between the incident and transmitter bar. The striker bar is made to strike the incident bar at different velocities of 11.32 m/s, 15.02 m/s and 15 m/s to understand the behavior of concrete subjected to various high strain rates. The damage mechanism of the developed model is compared with the experimental results as shown in Figure 12.7. The developed model considers the viscosity behavior of the material and CSCM includes the strain rate dependency using the visco-plastic approach. Therefore, CSCM is considered among the various model available. Because of the DIF definition used in the material formulation, CSCM over-predicts the ultimate stress value. CSCM cannot find the softening nature of concrete. The developed model gets the ultimate and softening behavior of the material accurately to that of the experiment. The evolution of damage when the concrete is under various impact velocities is shown in Figure 12.7 using the developed model. It is observed that with increasing striker velocity, the strain rate experienced by the material increases. The material is subjected to strain rates of 62.03/s, 126.31/s and 163/s as those of experimental observation. The total damage of the material increases with the strain rate. At a lower strain rate, the concrete splits into two halves, and with an increasing strain rate, the material undergoes pulverized damage. The developed model can predict the behavior of the material in line with the experimental observation.

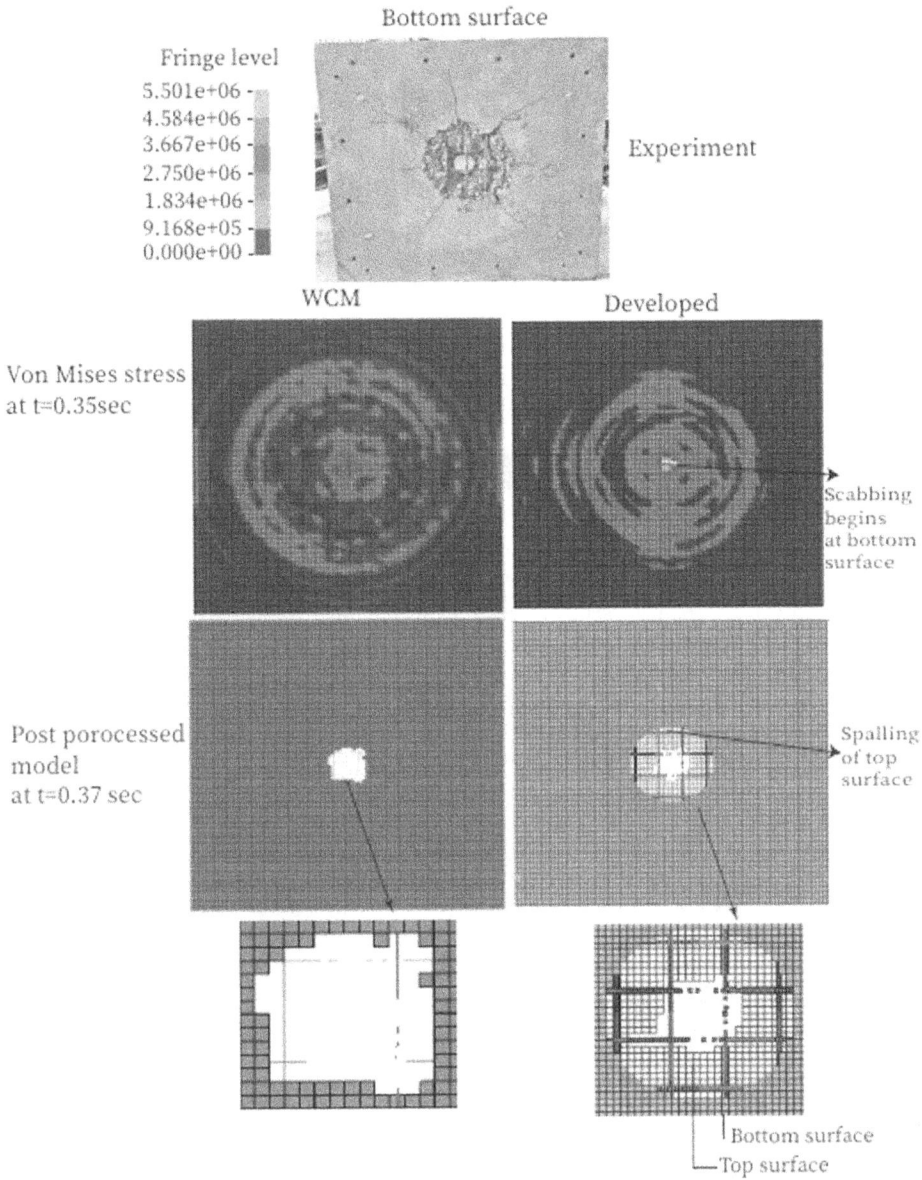

FIGURE 12.4 Failure of concrete slab under impact loading.

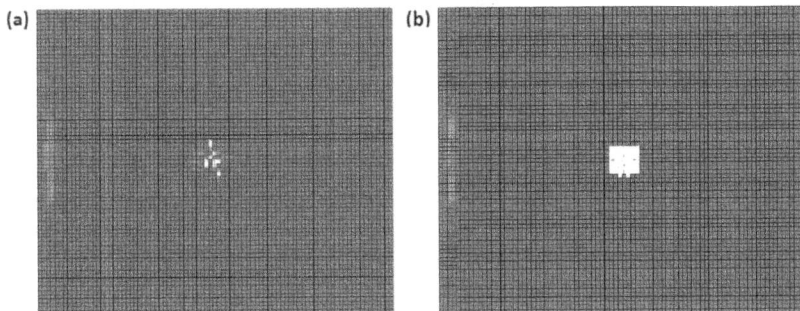

FIGURE 12.5 Slab subjected to (a) 6.86 *m/s* impact velocity and (b)13.72 *m/s* impact velocity.

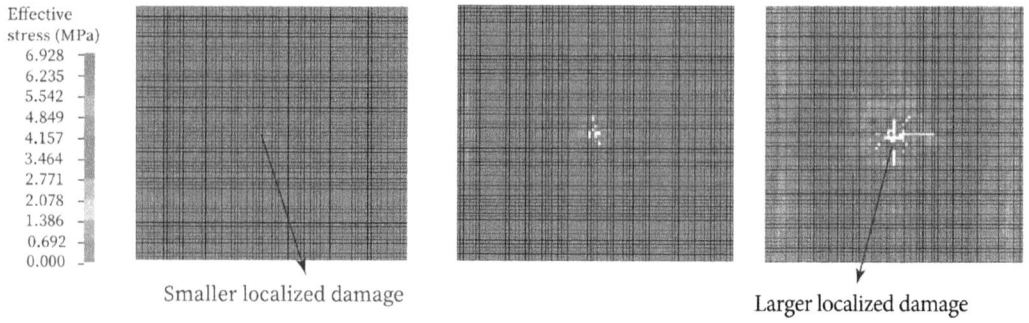

Effective
stress (MPa)
6.928
6.235
5.542
4.849
4.157
3.464
2.771
2.078
1.386
0.692
0.000

Smaller localized damage   Larger localized damage

FIGURE 12.6   Slab subjected to increased impact duration of (a) 3 *ms*, (b) 6 *ms* and (c) 20 *ms*.

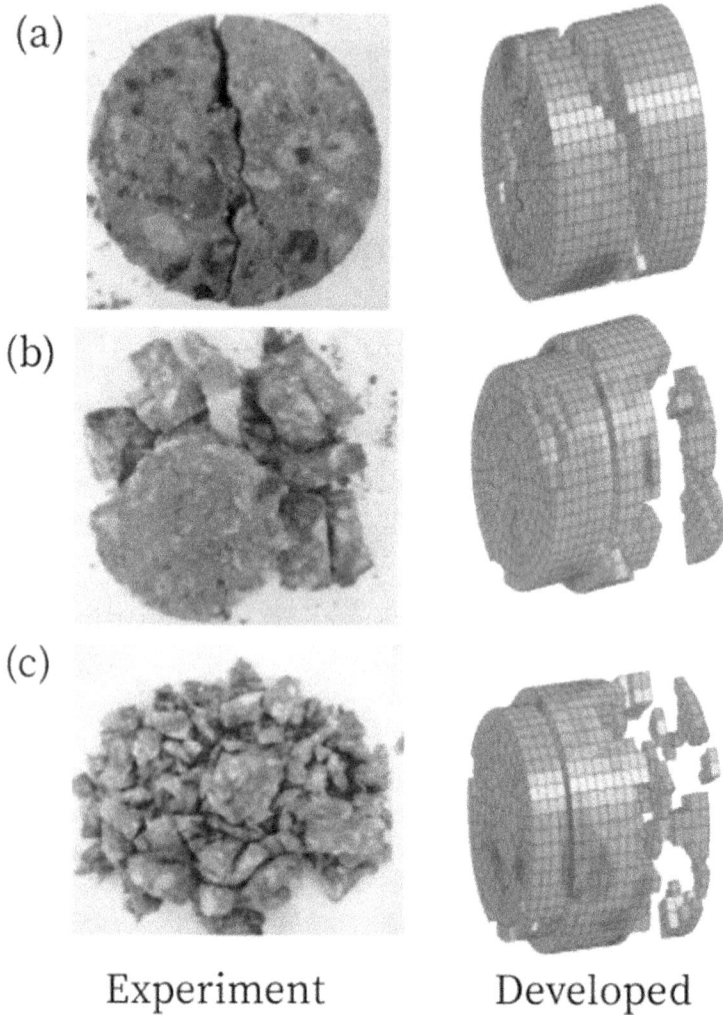

(a)

(b)

(c)

Experiment   Developed

FIGURE 12.7   Effective stress obtained using the developed model to understand the failure behavior of concrete under SHPB analysis when the striker bar is given different incident velocities of (a) 11.32 *m/s*, (b) 15.02 *m/s* and (c) 20 *m/s*.

12.6 CONCLUSION

- The plasticity-based damage model is able to predict the exact results of the experimental prediction.
- The yield surface expands with the increase in strain rate acting on the material. The maximum value is limited by giving the maximum value. This helps in avoiding the infinite increase in the strength of concrete.
- The damage definition helps in predicting the spallation and scabbing effect when subjected to impact loading.
- The maximum strength of the material increases with increasing strain rate. The developed model can predict the accurate ultimate strength of the material.
- The SHPB analysis shows that with increasing strain rate experienced by the material, the damage mechanism of the material changes. The concrete splits into two halves at a low strain rate and the material undergoes pulverized damage under a high strain rate.

REFERENCES

1. K.A. Gomathi, A. Rajagopal, K.S.S. Reddy, B. Ramakrishnan, Plasticity based material model for concrete subjected to dynamic loadings, *International Journal of Impact Engineering*, 2020;142:103581.
2. L.J. Malvar, K.B. Morrill, J.E. Crawford, Numerical modeling of concrete confined by fiber reinforced composites, *Journal of Composite Construction*, 2004;8(4):315–22.
3. LSTC. LS-DYNA Theory manual, Vol.18, 2002.
4. R.M. Brannon, S. Leelavanichkul, *Survey of Four Damage Models for Concrete*. California: Sandia National Laboratories; 2009.
5. K.A. Gomathi, A. Rajagopal, Dynamic performance of reinforced concrete slabs under impact and blast loading using plasticity based approach, *International Journal of Structural Stability and Dynamics*, 2020;20(14):2043015.
6. K.A. Gomathi, A. Rajagopal, S.S. Prakash, Predicting the failure mechanism of RC slabs under combined blast and impact loading, *Theoretical and Applied Fracture Mechanics*, 2022;119:103357.
7. H. Sadraie, A. Khaloo, H. Soltani, Dynamic performance of concrete slabs reinforced with steel and GFRP bars under impact loading, *Engineering Structures*, 2019;191(81):62.
8. J. Guo, Q. Chen, W. Chen, J. Chen, Tests and numerical studies on strain-rate effect on compressive strength of recycled aggregate concrete, *Journal of Material in Civil Engineering*, 2019;31(11):0401928113.

13 A Novel Pulse-Shaping Technique to Forecast the Behavior of Brittle Material Using Split Hopkinson Pressure Bar

Shivani Verma, M. D. Goel and N. N. Sirdesai

13.1 INTRODUCTION

Currently, the exponential growth of dynamic activities (e.g., blast and impact) has made the analysis of construction materials [1–3] the primary focus of research. The split Hopkinson pressure bar (SHPB) is commonly used to analyze the behavior of different materials under high-impact loading [4–6]. Under dynamic loading, brittle materials (such as concrete and rock) behave differently than ductile materials (such as steel) [7–9]. Such brittle materials must be studied under dynamic loading conditions, and safety precautions must be taken accordingly.

In SHPB, the size of the subject specimen is particularly important for brittle materials. If the specimen is too small, it will not produce representative results, whereas if it is too large, it will not reach equilibrium and will fail prematurely [10]. Eventually, both instances would yield unreliable outcomes. This can be remedied with the aid of pulse-shaping techniques, which are employed in conventional SHPBs to modify the generated pulse. Limited research has been conducted on modifying the pulse shape for brittle materials. The majority of studies were limited to the use of pulse shapers [11–16], and few of them considered striker variation as a means of pulse shape correction [17–21]. Additionally, the striker modifications that are discussed in the literature either apply to all materials or are only applicable to ductile materials [22, 23]. Therefore, the literature lacks an adequate analysis of brittle material under dynamic loading with pulse shaping.

This study aims to determine the optimal striker shape for producing a desired pulse for brittle materials under dynamic compression loading conditions. This study focuses primarily on the numerical solution to the aforementioned issue. Once the numerical study yields the desired results, this method could also be applied to experimental analysis. The results of the novel striker shape developed in this study were also compared to those of other available striker shapes. It has been established that the striker shape described in this study produces positive outcomes and complies with the essential criteria of the equilibrium condition for brittle material under dynamic loading circumstances.

13.2 SPLIT HOPKINSON PRESSURE BAR (SHPB)

SHPB is a tool for characterizing materials that deform at high strain rates (10^2–10^4 s^{-1}). The typical SHPB test setup is depicted schematically in Figure 13.1(a), where the striker bar impacts the incident bar to generate a one-dimensional compressive stress wave. The stress wave travels along

DOI: 10.1201/9781003352358-13

the incident bar in the direction of the specimen positioned between the pressure bars. At the interface between the incident bar and the specimen, a portion of the tensile stress wave is reflected. Simultaneously, the remainder travels through the specimen and reaches the specimen-transmitter bar interface. The wave of compressive stress leaves the specimen, propagates along the transmitter bar, and is measured by a gauge on the transmitter bar.

The type of wave propagation analysis is dependent on the material type and its dimensions. Small-diameter SHPB can be used to analyze both ferrous and fibrous materials. The diameter of SHPB typically ranged between 10 and 25 mm. However, special material characterizations may necessitate a SHPB with a larger diameter. In addition, the relatively large specimen size for the large-diameter SHPB experiment limits the strain rate. The SHPB with 50–100 mm bars is generally utilized for these materials. The large-diameter bars exacerbate stress wave dispersion [24]. The following equations come from the literature and can be used to determine the stresses and strains on small and large-diameter specimens:

a. *One-wave analysis method for small-diameter split Hopkinson pressure bar*

$$\sigma_s(t) = \frac{A_B E_B}{A_{s0}} \varepsilon_T(t) \tag{13.1}$$

$$\varepsilon_s(t) = \int \dot{\varepsilon} dt = \frac{2c_b}{H_{s0}} \int \varepsilon_R(t) dt \tag{13.2}$$

For brittle materials, such as rock and concrete-like materials, a large-diameter SHPB with a high impact velocity capability is necessary to achieve desired strain rate. The larger specimens are required because these are representative of the real materials.

b. *Three-wave analysis method for large-diameter split Hopkinson pressure bar*

$$\dot{\varepsilon}_s(t) = \frac{c_b}{H_{s0}} \left(-\varepsilon_I + \varepsilon_R + \varepsilon_T \right) \tag{13.3}$$

$$\varepsilon_s(t) = \int \dot{\varepsilon} dt = \frac{c_b}{H_{s0}} \int \left[-\varepsilon_I(t) + \varepsilon_R(t) + \varepsilon_T(t) \right] dt \tag{13.4}$$

$$\sigma_s(t) = \frac{A_B E_B}{2 A_{s0}} \left[-\varepsilon_I(t) + \varepsilon_R(t) + \varepsilon_T(t) \right] \tag{13.5}$$

Here, A_B represents the cross-section area of the bars, E_B represents the elastic modulus of the bars, A_{s0} shows the original area of the specimen face in contact with the bar, c_b represents the wave velocity of steel bars, H_{s0} represents the original length of the specimen, σ_s is stress in the specimen, ε_s is strain in the specimen, ε_s is strain rate in the specimen, ε_I is incident strain, ε_R is reflected strain and ε_T is transmitted strain.

13.3   FINITE ELEMENT (FE) MODELING AND MODEL GEOMETRY

This study compares the behavior of rock material under the influence of different striker shapes. LS-DYNA® is used to simulate the Split Hopkinson Pressure Bar (SHPB) at full scale [25]. As shown in Figure 13.1, the SHPB configuration consists of three components: the striker, the incident bar and the transmission bar. The length of the incident and transmission bars is 3 m, while the length of the striker varies based on its shape. The incident and transmission bar element size are considered as ten units.

(a)

(b)

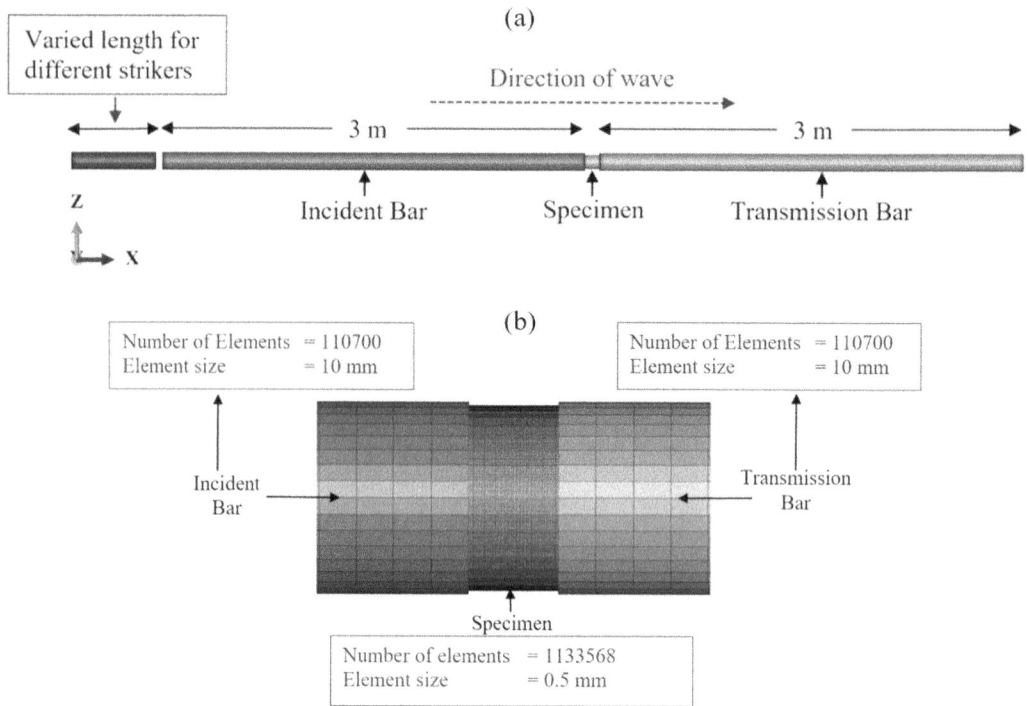

FIGURE 13.1   (a) Setup for split Hopkinson pressure bar. (b) Meshing details of SHPB and the specimen.

To maintain a constant volume for comparative analysis, the striker length differs. Considering the experimental scenario of the SHPB, each component is modeled as a solid cylinder. This study develops a novel striker shape to analyze the behavior of brittle materials. Figure 13.2 depicts the geometry and meshing details of this striker. For the purpose of optimizing the results, three variations of the special-shaped striker were considered. Three standard shapes, namely cylindrical, conical, and spindle, were also considered in order to compare the results with those of the proposed special-shaped striker. Figure 13.3 depicts the geometric and meshing details of these strikers. The element size remains five units for all striker shapes. For the optimization of results, the variation in the end diameter of the striker's front face is also considered. The SHPB is constrained axially in two directions and rotationally in all directions in order to achieve an experiment-like one-dimensional wavefront. The specimen is free from all constraints comparable to those of the experimental test. The load is applied by providing velocity to the striker.

The contact between the bar and specimen is AUTOMATIC_SURFACE_TO_SURFACE. In addition, a parametric study is conducted based on the variation in striker velocity and striker face diameter.

13.4   MATERIAL MODEL AND PROPERTIES

13.4.1   Johnson-Holmquist (JH-2) Model

This model considers the material's softening characteristics, pressure-dependent strength, fracture generation strength, evaluation of damage and strain rate effects. This model is suitable for simulating ceramics, glass and other brittle materials. It consists of three fundamental parameters: the strength model, the damage model and the EOS (Equation of State) model.

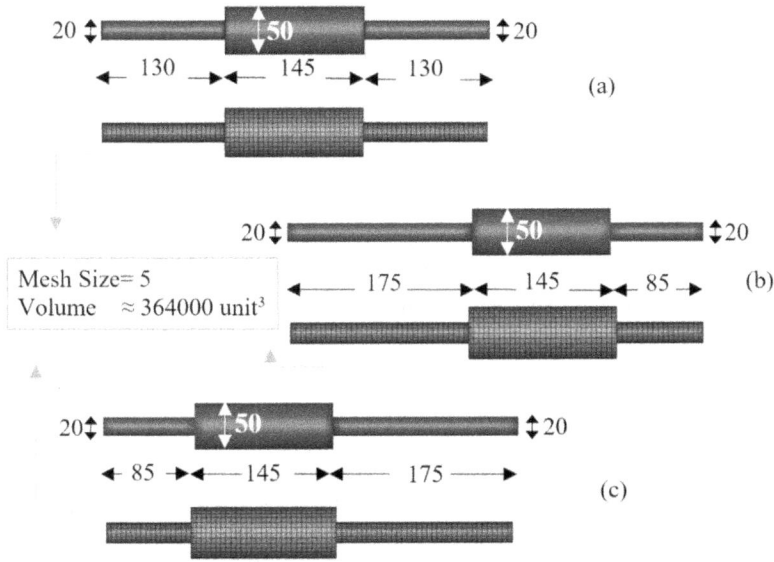

FIGURE 13.2   Geometric and meshing details of special-shaped strikers: (a) type A1, (b) type A2 and (c) type A3 (note: all dimensions are in mm).

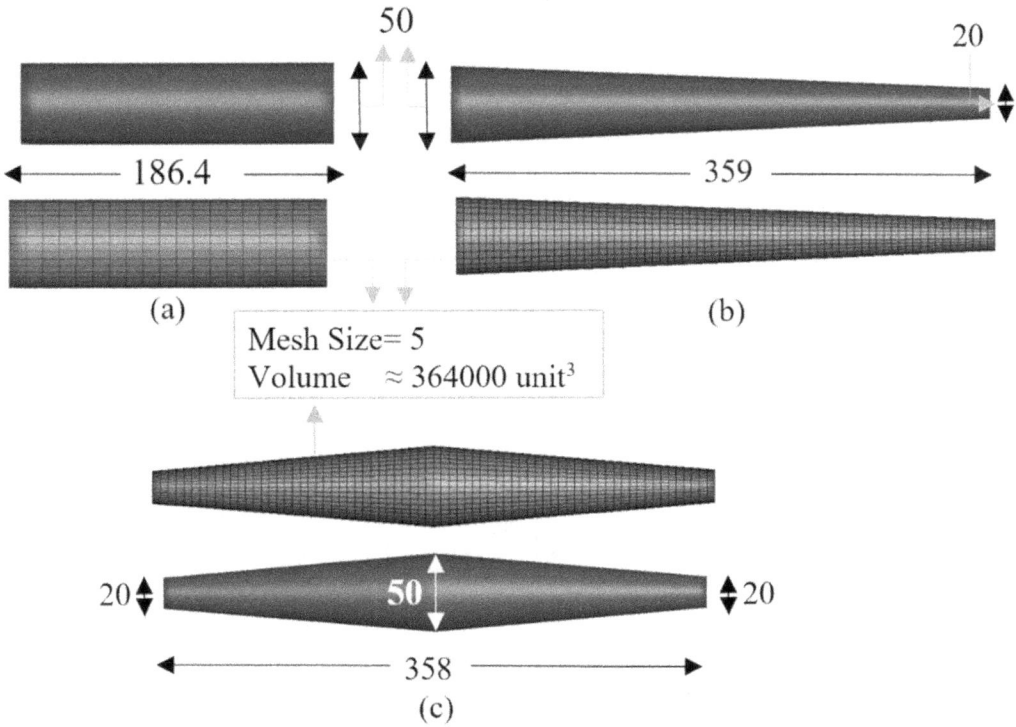

FIGURE 13.3   Geometric and meshing details of strikers: (a) cylinder, (b) conical and (c) spindle shape (note: all dimensions are in mm).

13.4.1.1 Strength Model

The JH-2 model is based on the relation between normalized values of equivalent stress and pressure [26]. The normalized intact strength is thus represented by (see Figure 13.4),

$$\sigma_i^* = A\left(P^* + T^*\right)^n \left(1 + C \ln \dot{\varepsilon}^*\right) \tag{13.6}$$

Here, $\sigma_i^* = \sigma_i / HEL$ is normalized equivalent strength, σ_i is actual intact strength and HEL is the Hugoniot elastic limit of material. $P^* = P / P_{HEL}$ is normalized pressure, P is actual pressure and P_{HEL} is the pressure at Hugoniot elastic limit. $T^* = T / P_{HEL}$ is the normalized static tensile limit (HTL) that a material can resist. $\dot{\varepsilon}^* = \dot{\varepsilon} / \varepsilon_0$ is normalized strain rate, $\dot{\varepsilon}$ is actual equivalent strain rate and $\varepsilon_0 = 1.0$ s^{-1} is reference strain rate, and A, n and C are material parameters.

The normalized fracture strength of the material is given as,

$$\sigma_f^* = B\left(P^*\right)^m \left(1 + C \ln \dot{\varepsilon}^*\right) \tag{13.7}$$

Here, B and m are residual strength constants. In JH-2, an upper limit (σ^*_{fmax}) for the yield surface of the material is considered, giving the flexibility to control the location of the fracture strength. The yield surface of the damaged material is given as,

$$\sigma_D^* = \sigma_i^* - D\left(\sigma_i^* - \sigma_f^*\right) \tag{13.8}$$

Here, D is the damage parameter, and its values range between 0 and 1.

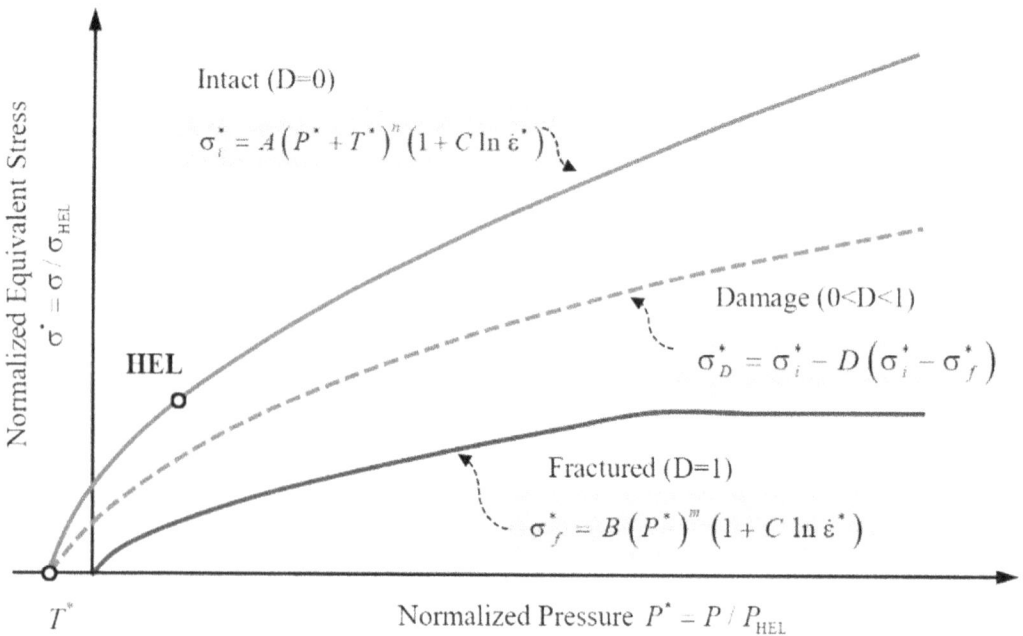

FIGURE 13.4 Depiction of intact, damaged and fractured surfaces of material in JH-2 model. Source: www.sciencedirect.com/science/article/abs/pii/S0734743X19310917.

13.4.1.2 Damage Model

Figure 13.5(a) depicts the nonlinear increase in material damage explained by the JH-2 model. Plastic deformation changes the state of the material from intact to fractured, and the amount of plastic strain required to change the state of a material is dependent on the pressure. This equivalent fracture plastic strain, ε_f^p, is given as,

$$\varepsilon_f^p = D_1 \left(P^* + T^* \right)^{D_2} \tag{13.9}$$

Here, D_1 and D_2 are the damage constants.

13.4.1.3 Equation of State (EOS) Model

As shown in Figure 13.5(b), the polynomial EOS in the JH-2 model represents the relationship between hydrostatic pressure and volumetric strain.

The plot between P and μ consists of elastic and plastic stages of the material. The hydrostatic pressure before the fracture starts $(D = 0)$ is given as,

$$P = K_1\mu + K_2\mu^2 + K_3\mu^3 \tag{13.10}$$

Here, K_1, K_2 and K_3 are material constants (K_1 is the bulk modulus of the material), and $\mu = \rho / \rho_0 - 1$, wherein ρ is the current density and ρ_0 is the initial density.

13.5 VALIDATION OF FE MODEL

Analysis of the outcomes reported by Peng [27] has corroborated the efficacy of the results from the current investigation. The incident and transmitter bars are 2 m and 1.5 m in length, respectively, and the material properties and dimensions of the numerical simulation are maintained to be identical to those of the experimental setup [27]. The incidence bar, transmitter bar, and sample all have a diameter of 50 mm. The length-to-diameter (L/D) ratio of the specimen in the experimental study is 0.5, and the same L/D ratio has been maintained in the numerical analysis in order to compare the results. For the specimen, MAT_JOHNSON_HOLMQUIST_CERAMICS (110) is used, whereas MAT_PLASTIC_KINEMATIC (003) has been utilized for the steel bars. The contact between the bars and the specimen is defined using AUTOMATIC_SURFACE_TO_SURFACE.

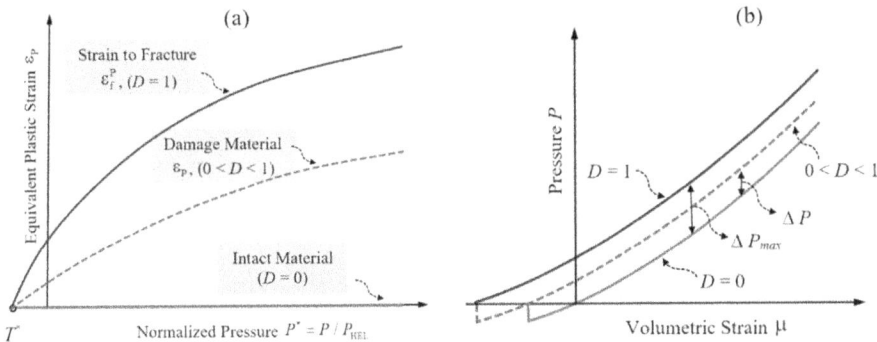

FIGURE 13.5 (a) Damage evaluation of material (damage model) and (b) relation between pressure and volumetric strain in JH2 model (EOS). Source: www.sciencedirect.com/science/article/abs/pii/S0734743X19310918.

By comparing the fracture evolution phases of the experimental and numerical specimens, as shown in Figure 13.6, the effectiveness of the numerical model was examined.

The numerical analysis's findings have been found to be in strong agreement with those of Peng's [27] experimental findings. Peak stresses in the experimental test and the current numerical analysis were both 165.25 MPa and 171.9 MPa, respectively. Considering genuine experimental and ideal numerical settings as a reason, it can be said that the results are in good agreement with those of Peng [27] with a discrepancy of only 4%. When considering the reliability of the FE results, this error rate is well within the acceptable range. Both the experimental and the numerical data indicate that the brittle specimen will fail via axial splitting. The material model and its parameters are kept the same as those of the experimental study [27] in order to increase the certainty of the results achieved in the current investigation.

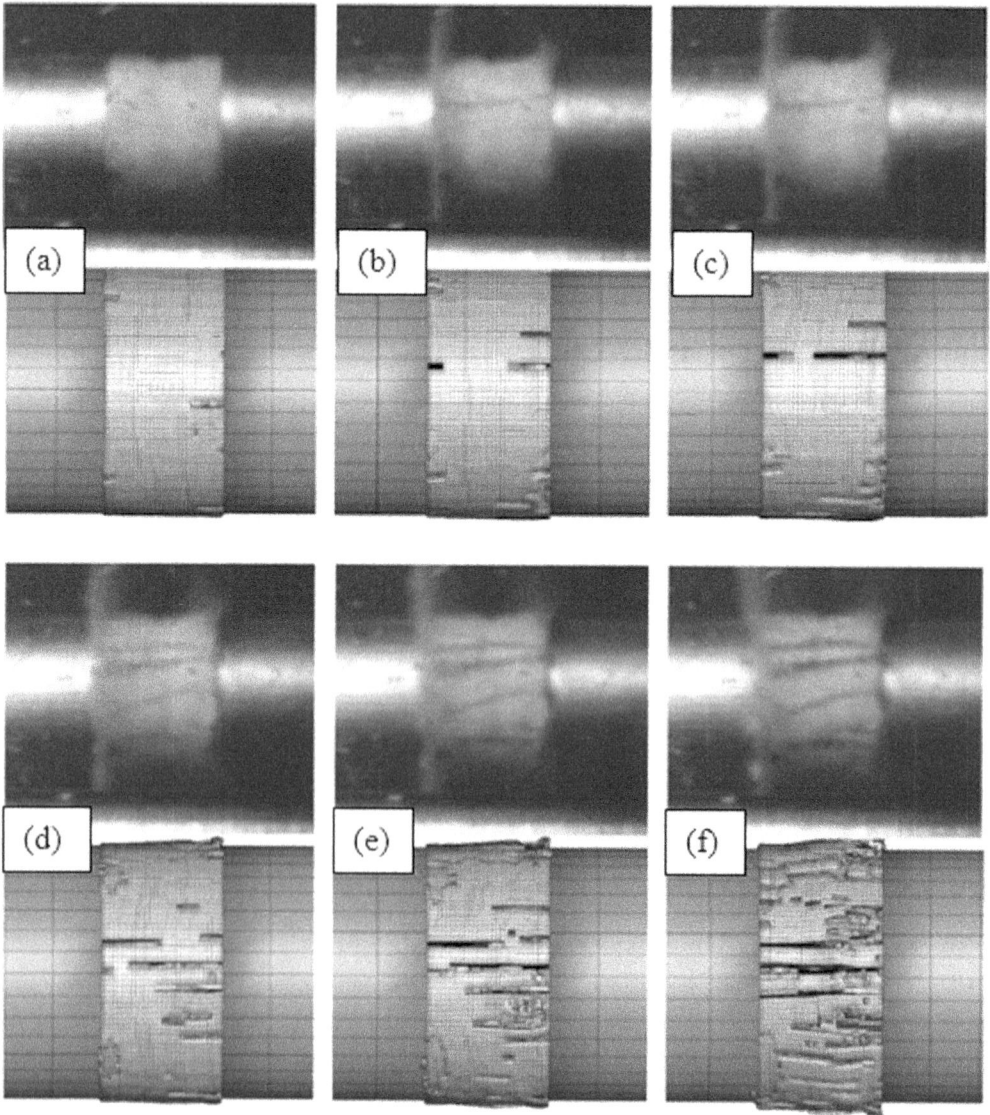

FIGURE 13.6 Fracture evaluation in the specimen for the experimental and the present numerical results considering the length-diameter ratio of the specimen as 0.5. Source: www.hindawi.com/journals/amse/2017/2048591/.

13.6 RESULTS AND DISCUSSION

Under high strain rates, the brittle material behaves very differently than the ductile material. When considering brittle materials, the size and shape of the specimens vary significantly. In general, specimens with a large diameter are useful for brittle materials under high strain rate loading conditions. In case of large-diameter specimens, however, the fundamental assumptions of the SHPB theory do not hold. To circumvent this constraint, the incident pulse must be modified so that the brittle material achieves stress equilibrium prior to failure. The subsequent section discusses the behavior of rock specimens under the influence of a novel striker shape.

13.6.1 EFFECT OF PROPOSED SHAPES OF STRIKER (STEPPED A1, A2 AND A3 STRIKERS)

For the purpose of comparison, the volume and velocity of these strikers were held constant. Figure 13.7 clearly demonstrates that the A1 and A2 strikers provide a more realistic output-history profile than the A3 striker, given the same input parameters. Under the given input parameters, the A2 striker is able to generate a profile that is clearer and smoother. For additional analysis, the A2 striker was compared to standard striker shapes used for the analysis of brittle materials. For the above comparative analysis, the striker velocity is 10 m/s, and variations in striker velocity are also discussed in the sections that follow.

13.6.2 COMPARISON OF PROPOSED STRIKER WITH THE CONVENTIONAL STRIKER

This research aims to determine the optimal striker shape for brittle materials such as concrete and rocks. Figure 13.8 depicts the output parameters of the four distinct strikers, including the cone, double cone, cylinder and stepped A2. One of the three stepped striker shapes in the previous section produced more realistic results than the other two. Due to the novelty of this striker's shape, the

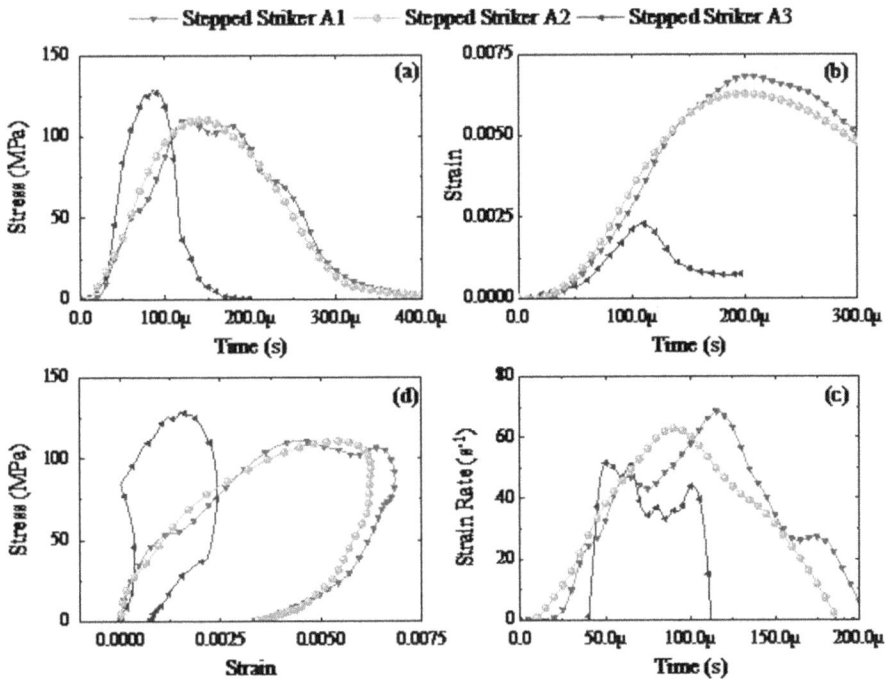

FIGURE 13.7 Effect of various special-shaped striker on specimen's (a) stress-time history plot, (b) strain-time history plot, (c) stress-strain plot and (d) strain-rate-time history plot.

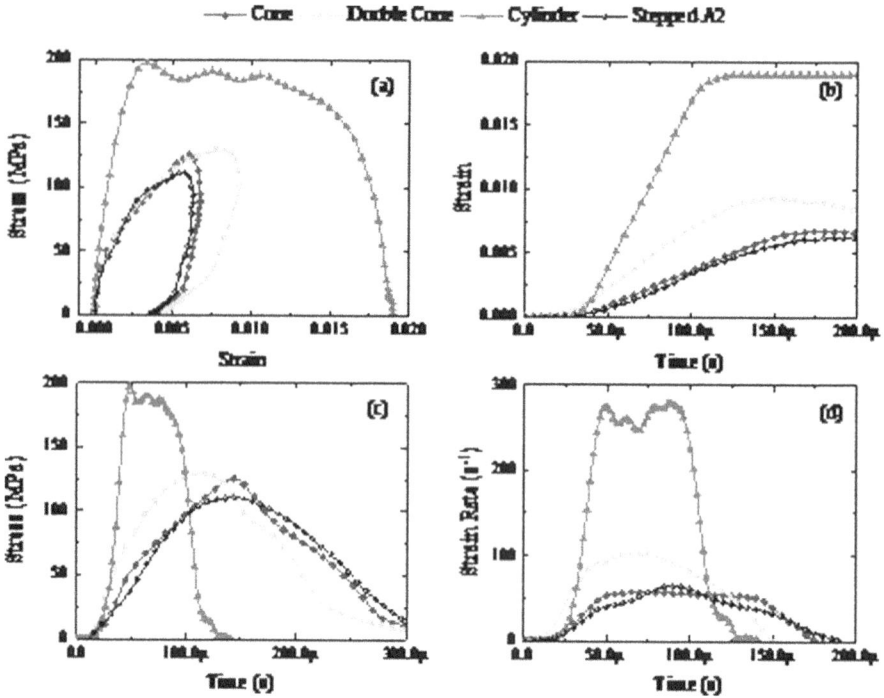

FIGURE 13.8 Comparison of conventional and novel striker on specimen's (a) stress-time history plot, (b) strain-time history plot, (c) stress-strain plot and (d) strain-rate-time history plot.

authors compared it to standard striker shapes to evaluate its performance and dependability under high strain rates. For a better understanding of how the shape of these strikers affects the specimen's behavior, the volume and velocity of the strikers are held constant. A few conclusions can be drawn from Figure 13.8. The cylinder-shaped striker provides the least reliable results for brittle materials because the rock specimen does not have sufficient time to reach equilibrium and therefore fails prematurely. The double cone striker overestimates the results for the given input parameters, but the overall trend is comparable to that of the cone and stepped A2 striker. The cone and stepped A2 strikers produce comparable results, but the stepped A2 produces fewer oscillations than the cone striker. For a better understanding of the results, see Figure 13.9, which depicts the specimen's damage profile as a result of various strikers. The figure clearly demonstrates that with the cylinder striker, the specimen is completely crushed, whereas the other three strikers produce a more favorable damage profile. The specimen damage caused by the stepped A2 striker is the lowest compared to other striker shapes. This implies that the specimen will have sufficient time to reach equilibrium before failing at its maximum strength.

13.6.3 EFFECT OF VARYING STRIKER VELOCITY

For the parametric study, the striker's velocity was varied to determine the specimen's behavior. In this study, four velocities of 10 m/s, 20 m/s, 30 m/s and 40 m/s are considered. As it was observed in the previous section that the stepped A2 striker produces more realistic results than the other two striker shapes, only the stepped A2 striker was considered for this parametric study.

13.6.3.1 Effect on Stress and Strain Rate of Specimen

Figure 13.10(a) illustrates the stress-time history of the brittle specimen under the influence of four different velocities using a stepped A2 striker. It is observed that as the striker's velocity increases,

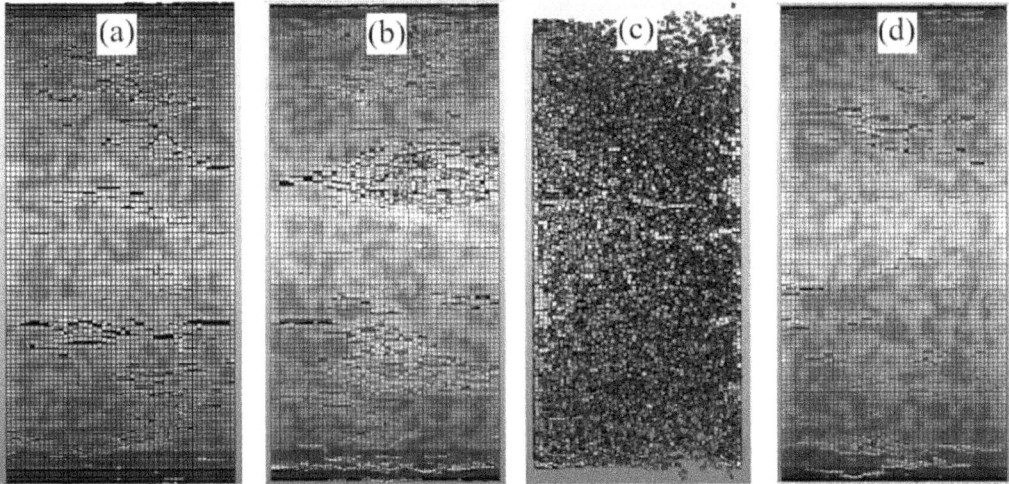

FIGURE 13.9 Damage profile of the rock specimen at t = 600 μs for (a) cone striker, (b) double cone striker, (c) cylinder striker and (d) stepped A2 striker.

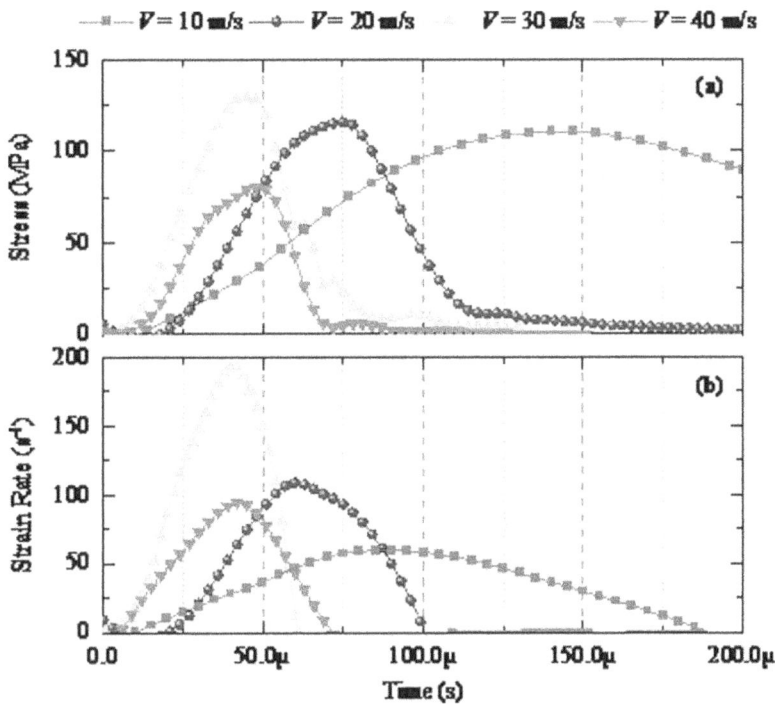

FIGURE 13.10 (a) Effective stress-time history and (b) strain-rate-time history of the specimen using stepped A2 striker.

the specimen's stress also increases, but this increase is capped at 30 m/s. Beyond 30 m/s, the specimen will experience the reverse phenomenon, so it can be concluded that 30 m/s is the threshold velocity of the striker for the given set of input parameters under this condition.

In addition, as the velocity increases, so does the stress slope, and the specimen fails rapidly. In contrast, for lower velocity, the area under the curve is greater and the slope is flatter, giving the specimen sufficient time to fail. Figure 13.10(b) depicts the strain-rate-time history of the brittle specimen under the influence of four distinct velocities with a stepped A2 striker shape. It is evident from the graph that the strain rate increases as striker velocity increases. The velocity threshold for an increase in strain rate is 30 m/s, and the strain rate falls as velocity increases further. This variation in the specimen's strain rate corresponds with the developed stresses in the specimen.

13.6.3.2   Stress-Strain Relation of Specimen

Figure 13.11 depicts the specimen's stress-strain response under the influence of different striker velocities. It can be seen from the graphs that as the striker's velocity increases, the slope of the stress-strain curve also increases, indicating an increase in the rate of change of the specimen's stress with respect to its strain. The stress-strain relationship of the specimen exhibits a similar pattern to the stress and strain rate behavior of specimens at various velocities. For the specified input parameter, the threshold value of the striker velocity is 30 m/s. The high strain rate effect on the specimen is not observed beyond this threshold.

The strain rate has a more significant impact on the dynamic compressive strength of the subject specimen. The greater the strain rate, the greater the specimen's dynamic compressive strength. For a better understanding of the results, see Figure 13.12, which depicts the damage profile of a specimen at different times using the stepped A2 striker studied in this study. It is found that as the velocity of the striker increases, the damage to the specimen gets more severe (Figure 13.12). Damage to the specimen also changes from axial splitting to a completely pulverized state (see Figure 13.12).

13.7   SUMMARY AND CONCLUSIONS

This study compares the suggested novel striker with traditional striker forms documented in the literature. The authors tested the innovative striker's reliability by altering its velocity. The numerical analysis yields these conclusions:

1. A novel stepped A2 striker form is proposed in this study as a reliable and effective method of simulating the behavior of brittle materials such as concrete and rocks under high strain rate loading conditions.
2. The proposed innovative striker shape (stepped A2) creates a stress-strain, stress-time and strain-rate-time profile that is generally smooth. The standard striker shape's curves oscillate violently and do not produce realistic results. Therefore, the usual rectangular loading

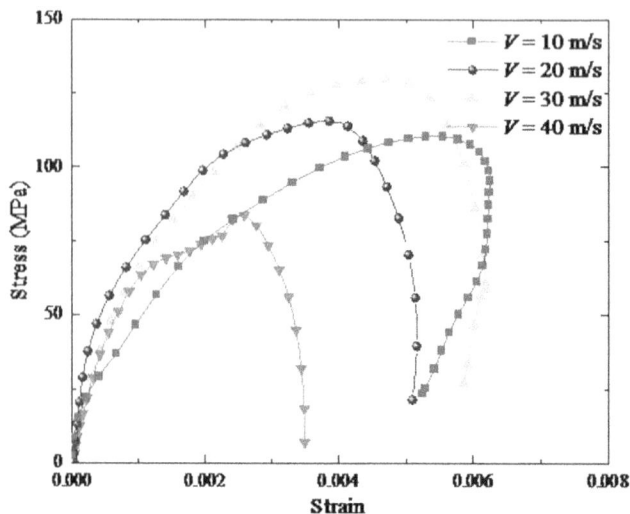

FIGURE 13.11   Stress-strain history profile of the specimen using stepped A2 striker.

$t = 600$ µs	$t = 800$ µs	$t = 1000$ µs	$t = 1200$ µs

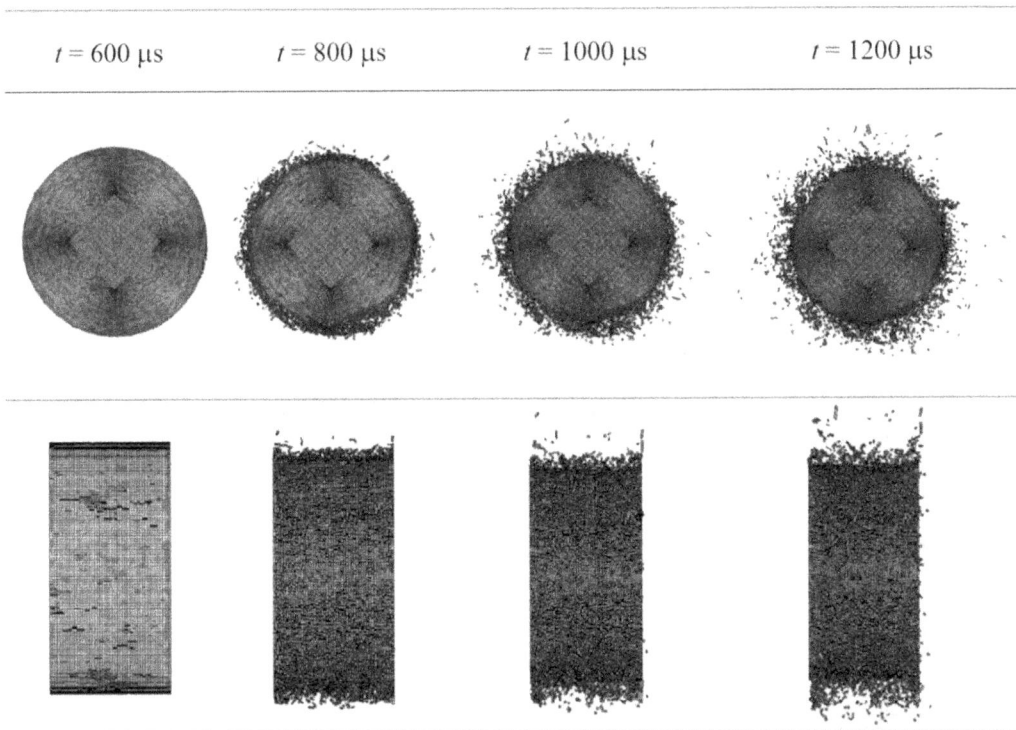

FIGURE 13.12 Damage evaluation of specimen using stepped A2 striker under various time intervals for striker velocity $V = 10$ m/s.

    waveform is inappropriate for dynamic tests on rock, concrete and other brittle materials utilizing a large-diameter split Hopkinson pressure bar (SHPB).

3. A parametric investigation into varying striker velocity (10 m/s, 20 m/s, 30 m/s and 40 m/s) is also conducted to compare the outcomes of the novel striker (stepped A2) to those of other traditional strikers (cylinder, double cone and cone). There are fewer oscillations even at high velocities with the innovative striker compared to the other strikers.

The authors suggest a technique for assessing brittle materials subjected to high strain rates. This technology is a superior alternative to those currently utilized for the pulse shaping of brittle materials.

REFERENCES

1. Goel MD. Blast: Characteristics, Loading and Computation-An Overview. *Adv. Struct. Eng*, 2015; 1: 417–434.
2. Goel MD and Matsagar VA. Blast-Resistant Design of Structures. *Pract. Period. Struct. Des. Constr.*, 2014; 19(2).
3. Goel MD, Matsagar VA and Gupta AK. An Abridged Review of Blast Wave Parameters. *Def. Sci. J.*, 2012; 62(5): 300–306.
4. Gama BA, Lopatnikov SL and Gillespie Jr JW. Hopkinson Bar Experimental Technique: A Critical Review. *Appl. Mech. Rev.*, 2004; 57(4): 223–250.
5. Xia K and Yao W. Dynamic Rock Tests using Split Hopkinson (Kolsky) Bar System–A Review. *J. Rock Mech. Geotech. Eng.*, 2015; 7(1): 27–59.
6. Peng K, Gao K, Liu J et al. Experimental and Numerical Evaluation of Rock Dynamic Test with Split-Hopkinson Pressure Bar. *Adv. Mater. Sci. Eng.*, 2017, Article ID 2048591.

7. Wang QZ, Zhang S and Xie HP. Rock Dynamic Fracture Toughness Tested with Holed-Cracked Flattened Brazilian Discs Diametrically Impacted by SHPB and its Size Effect. *Exp. Mech.*, 2010; 50(7): 877–885.
8. Huang S, Xia K and Zheng H. Observation of Microscopic Damage Accumulation in Brittle Solids Subjected to Dynamic Compressive Loading. *Rev. Sci. Instrum.*, 2013; 84(9), PMID: 24089837.
9. Chai J, Liu Y, OuYang Y et al. Application of Digital Image Correlation Technique for the Damage Characteristic of Rock-like Specimens under Uniaxial Compression. *Adv. Civ. Eng.*, 2020, Article ID 8857495.
10. Shu DW, Luo CQ and Lu GX. Numerical Simulations of the Influence of Striker Bar Length on SHPB Measurements. *Int. J. Mod. Phys. B*, 2008; B (22): 5813–5818.
11. Frew DJ, Forrestal MJ and Chen W. Pulse Shaping Techniques for Testing Brittle Materials with a Split Hopkinson Pressure Bar. *Exp. Mech.*, 2002; 42(1): 93–106.
12. Vecchio KS and Jiang F. Improved Pulse Shaping to Achieve Constant Strain Rate and Stress Equilibrium in Split-Hopkinson Pressure Bar Testing. *Metall. Mater. Trans. A*, 2007; 38(11): 2655–2665.
13. Naghdabadi R, Ashrafi MJ and Arghavani J. Experimental and Numerical Investigation of Pulse-Shaped Split Hopkinson Pressure Bar Test. *Mater. Sci. Eng. A.*, 2012; 539: 285–293.
14. Baranowski P, Janiszewski J, Malachowski J. Study on Computational Methods Applied to Modelling of Pulse Shaper in Split-Hopkinson Bar. *Arch. Mech.*, 2014; 66(6): 429–452.
15. Panowicz R, Janiszewski J and Kochanowski K. Influence of Pulse Shaper Geometry on Wave Pulses in SHPB Experiments. *J. Theor. Appl. Mech.*, 2018; 56.
16. Pang S, Tao W, Liang Y et al. A Modified Method of Pulse-Shaper Technique Applied in SHPB. *Compos. B. Eng.*, 2019; 165: 215–221.
17. Cloete TJ, Van Der Westhuizen A, Kok S et al. A Tapered Striker Pulse Shaping Technique for Uniform Strain Rate Dynamic Compression of Bovine Bone. *EDP Sci.*, 2009; (1): 901–907.
18. Li X, Zou Y and Zhou Z. Numerical Simulation of the Rock SHPB Test with a Special Shape Striker based on the Discrete Element Method. *Rock Mech. Rock Eng.*, 2014; 47(5): 1693–1709.
19. Forrestal MJ and Warren TL. A Conical Striker Bar to Obtain Constant True Strain Rate for Kolsky Bar Experiments. *J. Dyn. Behav. Mater.*, 2021; 7(1): 161–164.
20. Baranowski P, Malachowski J, Gieleta R et al. Numerical Study for Determination of Pulse Shaping Design Variables in SHPB Apparatus. *Bulletin of the Polish Academy of Sciences. Tech. Sci.*, 2013; 61(2): 459–466.
21. Gerlach R, Sathianathan SK, Siviour C et al. A Novel Method for Pulse Shaping of Split Hopkinson Tensile Bar Signals. *Int. J. Impact Eng.*, 2011; 38(12): 976–980.
22. Gupta MK. Numerical Simulation of AA7075 under High Strain Rate with Different Shape of Striker of Split Hopkinson Pressure Bar. *Mater. Today Commun.*, 2021, 26, Article ID 102178.
23. Dvořák R, Koudelka P and Fíla T. Numerical Modelling of Wave Shapes during SHPB Measurement. *Acta Polytech. CTU Proc.*, 2019; 25: 25–31.
24. Gama BA, Lopatnikov SL and Gillespie Jr JW. Hopkinson Bar Experimental Technique: A Critical Review. *Appl. Mech. Rev.*, 2004; 57(4): 223–250.
25. Hallquist JO. LS-DYNA® Keyword User's Manual Volume I. LSTC, Version, 971, 2009.
26. Baranowski P, Kucewicz M, Gieleta R et al. Fracture and Fragmentation of Dolomite Rock using the JH-2 Constitutive Model: Parameter Determination, Experiments and Simulations. *Int. J. Impact Eng.*, 2020; 140, Article ID 103543.
27. Peng K, Gao K, Liu J et al. Experimental and Numerical Evaluation of Rock Dynamic Test with Split-Hopkinson Pressure Bar. *Adv. Mater. Sci. Eng.*, 2017, Article ID 2048591.

14 Experimental Investigation of Dynamic Behaviour of Ceramic Material and the Effectiveness of Pulse Shapers

V. B. Brahmadathan and C. Lakshmana Rao

14.1 INTRODUCTION

Ceramic materials like boron carbide, silicon carbide, and alumina undergo deformation at a high strain rate when used in impact loading and ballistic applications such as body armour and vehicle and building protection [1–3]. Boron carbide, silicon carbide and alumina are used as strike-face materials in body armour systems. Crack generation, crack propagation and coalescence of various cracks cause the failure of ceramic material when the bullet hits the strike face material. The dynamic behaviour of such materials depends on multiple parameters like strain rate, confinement pressure, nature of loading, constituents used in the manufacture, the microstructure, etc. The strength and other properties of these materials have significant variations even if they are manufactured or processed similarly. This is due to the deviations in the distribution, size and shape of cracks and other defects in the material; thus, experimental studies on ceramic material are always important. Several researchers tried to characterise the dynamic behaviour of various ceramic materials. Nanoindentation measurements are performed to study the effect of the boron/carbon ratio on the microstructure and mechanical properties [4]. Planar impact experiments are used to study the re-shock/release behaviour [5] and also to find various constitutive model constants for boron carbide [6] and silicon carbide [7]. Alumina ceramic is tested using a sphere impact test to study its failure mechanism [8]. Another widely used experimental technique for dynamic testing is split Hopkinson pressure bar (SHPB) experiments. Conventionally produced alumina and 3D printed alumina [9] were dynamically tested, and failure mechanisms were studied using SHPB [10, 11]. Pulse shapers are also required for the testing of brittle materials like ceramics for achieving constant strain rate and dynamic equilibrium in SHPB experiments [12]. The strain rate achieved is very high in these studies and the way of expressing the effectiveness of the pulse shaper also needs some improvements.

In the present study, characterisation of the dynamic behaviour of alumina (Al_2O_3) is done using split Hopkinson pressure bar experiments and a method for evaluating the effectiveness of pulse shaper is also discussed.

14.2 HIGH STRAIN RATE EXPERIMENTS

14.2.1 Split Hopkinson Pressure Bar Test

Split Hopkinson pressure bar (SHPB) can be used for high strain rate testing of ceramics in the range of 10^2 to 10^4 s^{-1} strain rate [13]. The experimental setup mainly consists of three bars; striker,

DOI: 10.1201/9781003352358-14

incident and transmitter bars with lengths of 400 mm, 2000 mm, and 2,000 mm, respectively. The specimen is fixed between the incident bar and transmitter bar, as shown in Figure 14.1.

The striker bar hits the incident bar with a velocity v_{st}, which is controlled by the pressure inside the gas gun. This impact will generate a stress wave, and it will propagate through the incident bar. Based on the impedance mismatch between the bar and specimen, some waves will reflect, and the remaining wave will transmit. Strain gauges are placed at the centre of the incident and transmitter bars to measure the strain in the bars in the form of voltage. The strain gauges are connected to the amplifier unit, and the amplified signal is viewed through an oscilloscope. The signals are recorded in the data acquisition unit.

Alignment checking is very important in testing. This can be done by conducting a test without the specimen. For properly aligned bars, the incident wave and transmitted wave will match; theoretically, there should not be any reflected wave, but in a practical case, acceptable alignment between the bar can be considered if the area under the reflected wave is less than 5% of the area under the incident wave. The SHPB works based on one-dimensional wave propagation theory. The equations used for calculating input stress, output stress, strain rate and strain are:

$$Output\ stress, \sigma_o = \varepsilon_t E_b \frac{A_b}{A_s} \tag{14.1}$$

$$Input\ stress, \sigma_i = \left(\varepsilon_i + \varepsilon_r\right) E_b \frac{A_b}{A_s} \tag{14.2}$$

$$Strain\ rate, \dot{\varepsilon} = \frac{\left(\varepsilon_i - \varepsilon_r - \varepsilon_t\right)}{L_s} C_o \tag{14.3}$$

$$Strain, \varepsilon = \int \dot{\varepsilon}\, dt \tag{14.4}$$

Where,

ε_i = incident strain signal
ε_r = reflected strain signal
ε_t = transmitted strain signal
E_b = elastic modulus of the bar
L_s = length of the specimen
C_o = wave speed of the bar.
A_b and A_s are areas of the cross-section of the bar and the specimen.

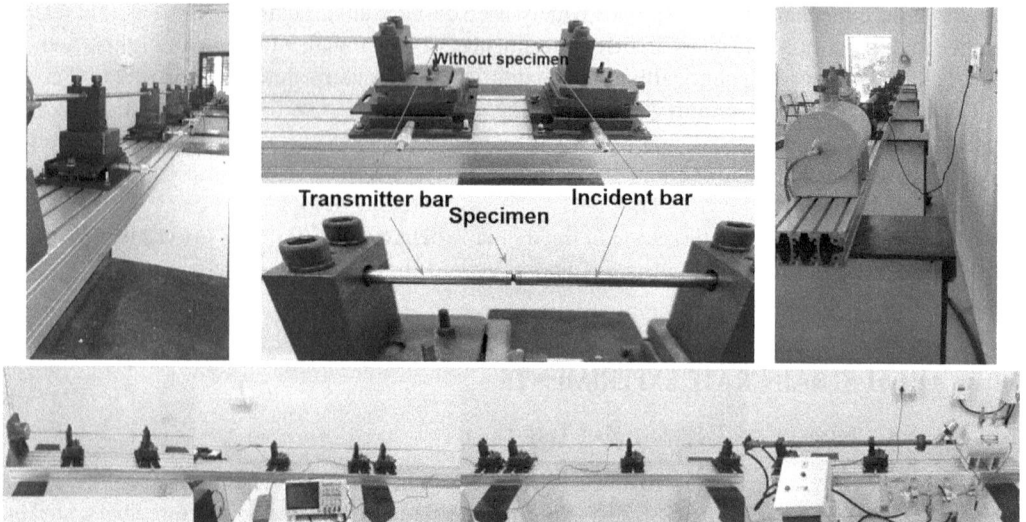

FIGURE 14.1 Split Hopkinson pressure bar (SHPB) setup.

14.2.2   REQUIREMENTS OF PULSE SHAPING TECHNIQUE

The conventional SHPB is modified to test brittle materials like ceramic by using pulse shapers (Figure 14.2). For brittle materials, to achieve a constant strain rate and dynamic equilibrium, ramp pulses are recommended because the ceramic material fails within a short duration of time and requires a minimum of four reflections to achieve dynamic stress equilibrium in the specimen [14].

$$\textit{Time required for achieving stress equilibrium, } t = n\frac{L_s}{C_s} \qquad (14.5)$$

Where,
   n = number of reflections required for equilibrium
   L_s = length of the specimen
   C_s = wave speed in the specimen

   Ramp pulses have more rise time than rectangular pulses developed in the conventional SHPB (Figure 14.2).
   Rectangular-shaped pulses can be converted to the ramp pulse by using a pulse shaper to achieve a constant strain rate and dynamic equilibrium [15].
   In this study, three conditions are tested: 1) without any pulse shaper, 2) a small amount of grease and 3.) with a pulse shaper (Figure 14.3). Copper material is used for pulse shapers with a square cross-section (8 mm × 8 mm) and thickness of 0.5 mm.
   A typical set of pulses without pulse shapers, with a small amount of grease at the incident bar and with pulse shapers at the same gas gun pressure of 1.5 bar are shown in Figures 14.4, 14.5 and 14.6. In each case, the rise time for the incident pulse is different, and pulses are smooth when pulse shapers are used.

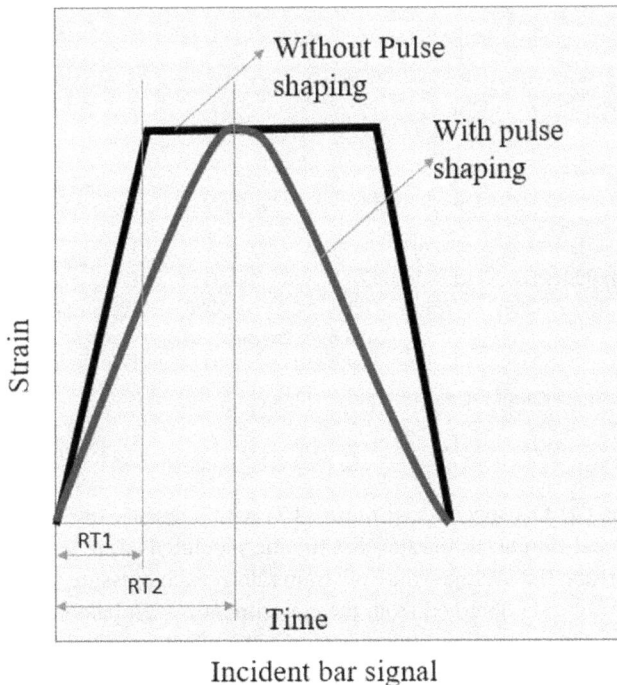

Incident bar signal

FIGURE 14.2   Conventional incident signal and pulse-shaped signal.

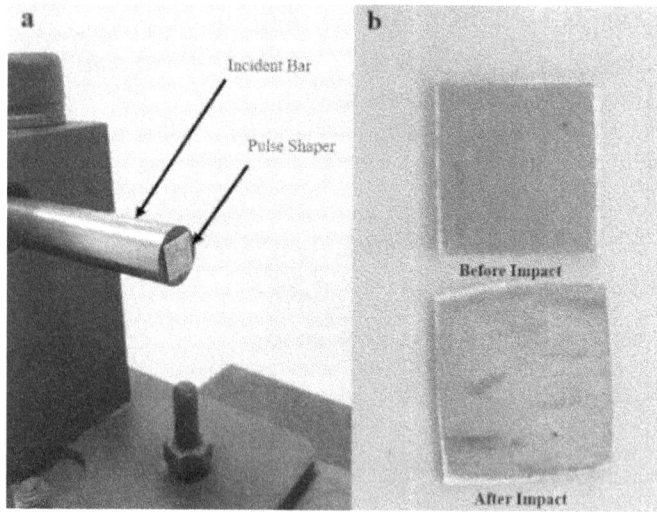

FIGURE 14.3   (a) Copper pulse shaper placed at incident bar end; (b) copper pulse shaper before and after impact.

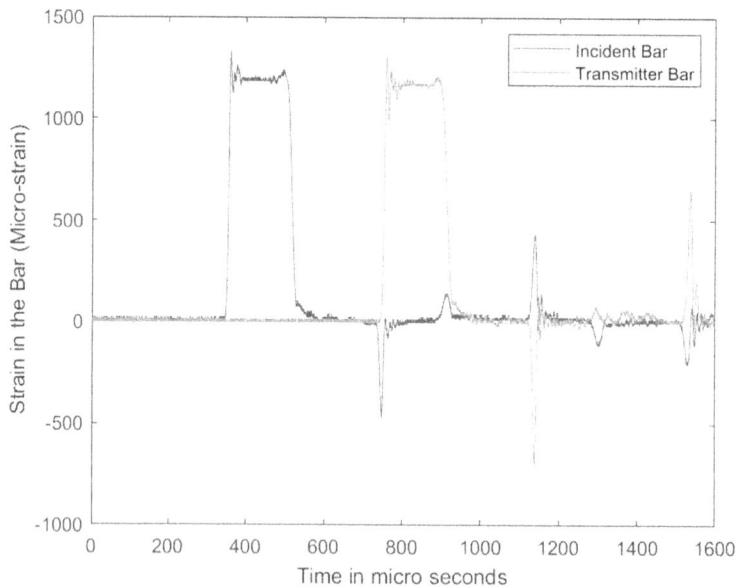

FIGURE 14.4   Bar strains without the use of a pulse shaper.

14.2.3 MATERIALS

A cylindrical alumina (Al_2O_3) specimen (Figure 14.7) with a density of 3,890 kg/m^3 and purity of 99.9% is tested at three different strain rates. Six to nine specimens were tested at gas gun pressures of 0.5, 1.5 and 2.5 bar, and the average value of strain rate at each pressure was 101.96 s^{-1}, 136.96 s^{-1} and 243.78 s^{-1}. The strain rate obtained from the experiment is well below the transition strain rate (1,591.47 s^{-1}), which is calculated by equation (14.6) [13]. The transition strain rate is the strain rate at which the strain rate dependency on strength changes drastically. Alumina shows negligible strain rate sensitivity below the transition strain rate; therefore, strain rate dependency on the strength of alumina is minimal in this study [16].

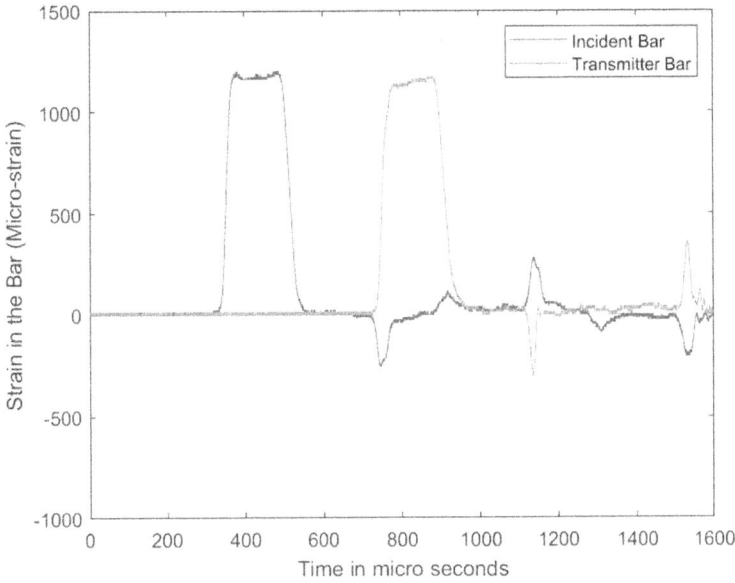

FIGURE 14.5   Bar strains without the use of pulse shaper, but a small amount of grease at the incident bar.

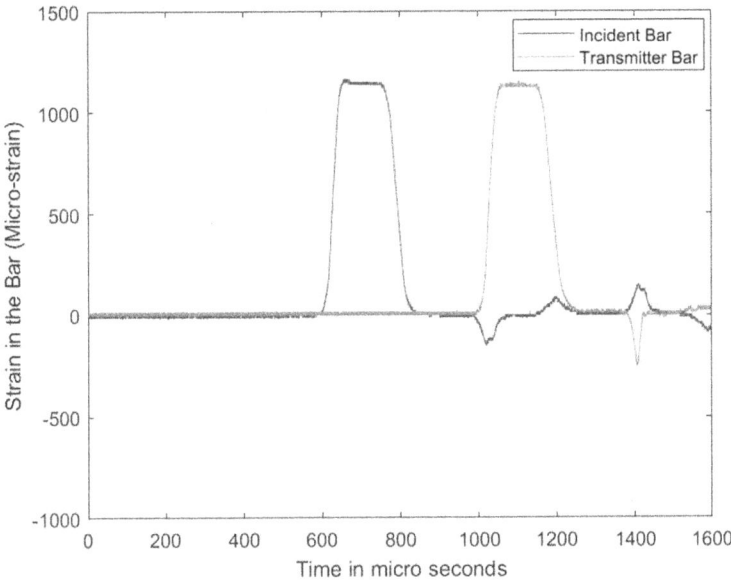

FIGURE 14.6   Bar strains with the use of a pulse shaper.

Transition strain rate:

$$\dot{\varepsilon}_t = \frac{\sigma_0}{E} \frac{C_*}{L_s}$$ (14.6)

Where, σ_0 = quasi-static strength
   C_* = characteristic speed
   E = elastic modulus
   L_s = specimen length

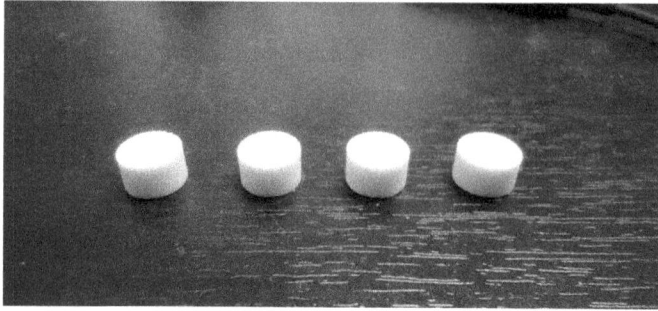

FIGURE 14.7   Alumina specimens.

The average diameter and length of test specimens are 5 mm and 8.8 mm, respectively, with an L/D ratio of 0.6.

14.3   RESULTS AND DISCUSSIONS

Three different strain rates, 101.96 s^{-1}, 136.96 s^{-1} and 243.78 s^{-1}, are achieved by setting a gas gun pressure of 0.5 bar, 1.5 bar and 2.5 bar; corresponding average striker velocities are 6.52 m/s, 11.6 m/s and 15.03 m/s.

14.3.1   EFFECTIVENESS OF PULSE SHAPER

The signal obtained from SHPB with and without a pulse shaper with a gas gun pressure of 1.5 bar is shown in Figures 14.4 and 14.6. For achieving a constant strain rate and dynamic equilibrium, more rise time is required. The rise time of the wave without a pulse shaper is 13.6 µs, and with a pulse shaper is 69.28 µs.

The pulse shaper changes the incident pulse into a ramp-shaped pulse to increase the rise time. The effectiveness of the pulse shaper is evaluated based on its capacity to achieve a constant strain rate and dynamic equilibrium. For this, the peak strain rate to the average strain rate ratio is used as a parameter to evaluate the effectiveness of the pulse shaper. A bit of grease is also used at the incident bar to check the changes in the incident wave.

Figure 14.8 shows the strain rate vs time for the test with and without the pulse shaper and only grease at the incident bar end. The variation in peak strain rate to average strain rate ratio decreases with a pulse shaper. Therefore, the pulse shaper effectively reduces the strain rate variation with respect to the average value.

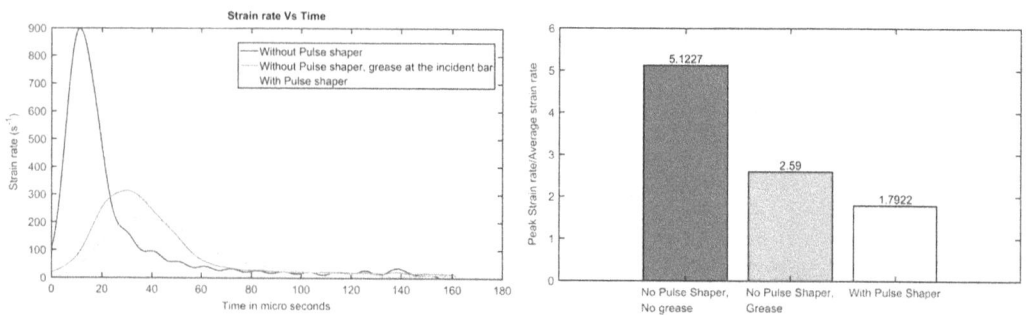

FIGURE 14.8   (a) Strain rate vs time; b) peak strain rate to average strain rate ratio.

The input force and output force on the specimen without a pulse shaper are not matching at the initial stages of the loading, and curves are not smooth, but dynamic equilibrium is achieved in the case of the test with a pulse shaper (Figure 14.9).

The quality of the stress-strain curve also depends on the incident, reflected and transmitted waves. The expected linear elastic curve is obtained using a pulse shaper (Figure 14.10).

14.3.2 STRESS-STRAIN RESPONSE

Stress-strain responses of the alumina are shown in Figure 14.11. The quasi-static strength for the alumina is 3.5 GPa, and since the specimen is not even loaded up to its quasi-static strength, the obtained stress-strain curves are insufficient to calculate the peak strength values at each strain rate. The pulse shapers are effective in achieving the constant strain rate condition up to a certain level and dynamic equilibrium, but more loading time is required to estimate the peak stress of the material.

Figure 14.12a shows a specimen which is not fractured after the test; the gas gun pressure for this test was 0.5 bar (average strain rate = 101.96 s^{-1}). The other two specimens (Figure 14.12b and c) are the post-test specimens corresponding to the gas gun pressure of 1.5 bar (average strain rate = 136.96 s^{-1}) and 2.5 bar (average strain rate = 243.78 s^{-1}).

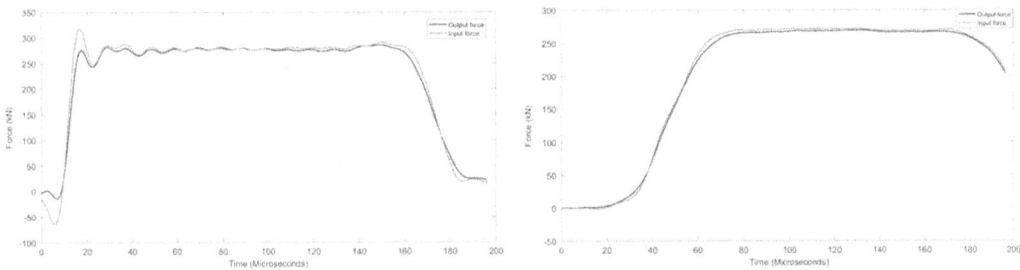

FIGURE 14.9   (a) Force vs time, without pulse shaper; (b) force vs time, with pulse shaper.

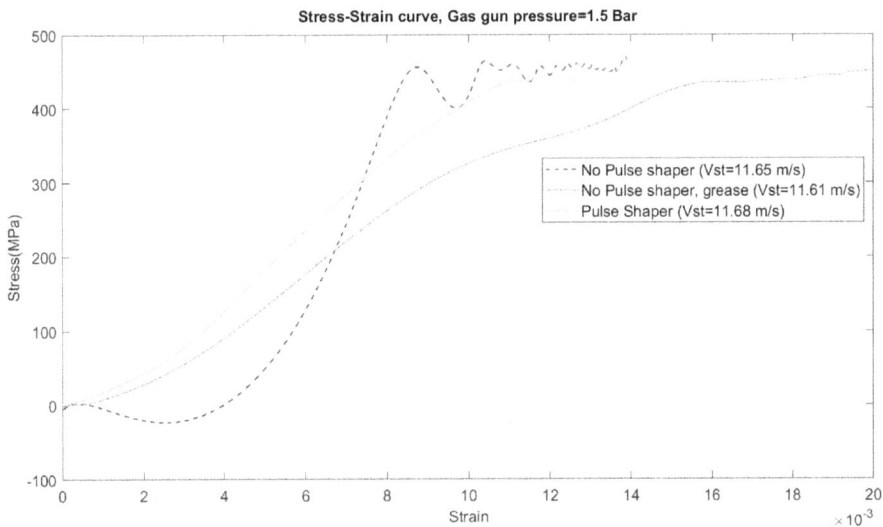

FIGURE 14.10   Comparison of stress-strain curve without pulse shaper, with grease and with pulse shaper.

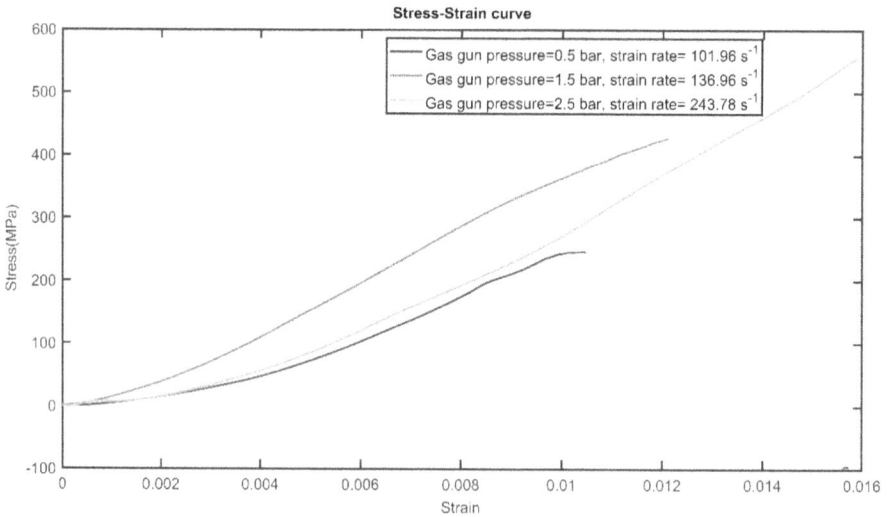

FIGURE 14.11   Average stress-strain curve for three different strain rates.

FIGURE 14.12   Post-test specimens.

The main failure mechanism is the axial splitting of the specimens. Even if the gas gun pressure is kept constant, the number of fragments created after the impact is different. The reason can be the differences in the crack density and orientation of the crack inside the specimen.

14.4   CONCLUSION

The dynamic behaviour of alumina was studied using a split Hopkinson pressure bar under three different strain rates well below the transition strain rate. Copper is used as the pulse shaper material. The effectiveness of the pulse shaper is evaluated based on its ability to achieve a constant strain rate and dynamic equilibrium.

- The peak strain rate to average strain rate ratio is used as a parameter for evaluating constant strain rate conditions.
- For achieving a constant strain rate condition, the peak strain rate to average strain rate ratio should be close to 1.
- With a pulse shaper, the peak strain rate to average strain rate ratio is reduced from 5.12 to 1.72; therefore, pulse shapers are effective.
- Input stress and output stress match well in the experiments with the pulse shaper.

- A linear stress-strain curve is obtained in the case of experiments with the pulse shaper.
- Stress-strain curves are plotted against three different strain rates. But more study is required on the pulse shaper's effectiveness to estimate the stress-strain response of the material since the loading couldn't continue up to the specimen's peak strength.
- The specimens failed by axial splitting.

REFERENCES

1. I. G. Crouch, "Body armour – New materials, new systems," *Defence Technology*, vol. 15, no. 3, pp. 241–253, 2019, doi: 10.1016/j.dt.2019.02.002.
2. B. Tepeduzu and R. Karakuzu, "Ballistic performance of ceramic/composite structures," *Ceram Int*, vol. 45, no. 2, pp. 1651–1660, Feb. 2019, doi: 10.1016/j.ceramint.2018.10.042.
3. M. S. Boldin, N. N. Berendeev, N. V. Melekhin, A. A. Popov, A. V. Nokhrin, and V. N. Chuvildeev, "Review of ballistic performance of alumina: Comparison of alumina with silicon carbide and boron carbide," *Ceramics International*, vol. 47, no. 18. Elsevier Ltd, pp. 25201–25213, Sep. 15, 2021. doi: 10.1016/j.ceramint.2021.06.066.
4. C. Cheng, K. M. Reddy, A. Hirata, T. Fujita, and M. Chen, "Structure and mechanical properties of boron-rich boron carbides," *J Eur Ceram Soc*, vol. 37, no. 15, pp. 4514–4523, 2017, doi: 10.1016/j.jeurceramsoc.2017.06.017.
5. T. J. Vogler, W. D. Reinhart, and L. C. Chhabildas, "Dynamic behavior of boron carbide," *J Appl Phys*, vol. 95, no. 8, pp. 4173–4183, 2004, doi: 10.1063/1.1686902.
6. T. J. Holmquist and G. R. Johnson, "Characterization and evaluation of boron carbide for plate-impact conditions," *J Appl Phys*, vol. 100, no. 9, 2006, doi: 10.1063/1.2362979.
7. T. J. Holmquist and G. R. Johnson, "Characterization and evaluation of silicon carbide for high-velocity impact," *J Appl Phys*, vol. 97, no. 9, 2005, doi: 10.1063/1.1881798.
8. E. C. Simons, J. Weerheijm, G. Toussaint, and L. J. Sluys, "An experimental and numerical investigation of sphere impact on alumina ceramic," *Int J Impact Eng*, vol. 145, no. June, p. 103670, 2020, doi: 10.1016/j.ijimpeng.2020.103670.
9. M. DeVries *et al.*, "Quasi-static and dynamic response of 3D-printed alumina," *J Eur Ceram Soc*, vol. 38, no. 9, pp. 3305–3316, Aug. 2018, doi: 10.1016/j.jeurceramsoc.2018.03.006.
10. Z. Wang and P. Li, "Dynamic failure and fracture mechanism in alumina ceramics: Experimental observations and finite element modelling," *Ceram Int*, vol. 41, no. 10, pp. 12763–12772, Apr. 2015, doi: 10.1016/j.ceramint.2015.06.110.
11. S. Acharya, S. Bysakh, V. Parameswaran, and A. Kumar Mukhopadhyay, "Deformation and failure of alumina under high strain rate compressive loading," *Ceram Int*, vol. 41, no. 5, pp. 6793–6801, 2015, doi: 10.1016/j.ceramint.2015.01.126.
12. Y. B. Lu and Q. M. Li, "Appraisal of pulse-shaping technique in split hopkinson pressure bar tests for brittle materials", *International Journal of Protective Structures*. 2010;1(3): 363–390. doi:10.1260/2041-4196.1.3.363.
13. G. Ravichandran and G. Subhash, "A micromechanical model for high strain rate behavior of ceramics," *Int J Solids Struct*, vol. 32, no. 17–18, pp. 2627–2646, 1995, doi: 10.1016/0020-7683(94)00286-6.
14. G. Ravichandran and G. Subhash, "Critical appraisal of limiting strain rates for compression testing of ceramics in a split Hopkinson pressure bar," *Journal of the American Ceramic Society*, vol. 77, no. 1, pp. 263–267, 1994, doi: 10.1111/j.1151-2916.1994.tb06987.x.
15. D. J. Frew, M. J. Forrestal, and W. Chen, "Pulse shaping techniques for testing brittle materials with a split Hopkinson pressure bar." *Experimental Mechanics* 42, 93–106 (2002). https://doi.org/10.1007/BF02411056
16. J. Lankford, "Mechanisms responsible for strain-rate-dependent compressive strength in ceramic materials," *Journal of the American Ceramic Society*, vol. 64, no. 2, pp. C-33-C-34, 1981, doi: 10.1111/j.1151-2916.1981.tb09570.x.

15 Optimizing Pulse Shaper Dimensions for Testing Rocks in Split Hopkinson Pressure Bar

Venkatesh M. Deshpande and Tanusree Chakraborty

15.1 INTRODUCTION

The split Hopkinson pressure bar (SHPB) is widely used to measure the mechanical properties of materials subjected to high strain rate loading, usually in the range of 10^2–10^4/s (Hopkinson 1872). For a valid and reliable SHPB test, the specimen should be in a uniaxial stress equilibrium and deform at a constant strain rate (Xia and Yao 2015). The attainment of these conditions depends on the shape of the incident pulse, which defines the loading rate. Conventionally, the shape of the incident pulse in SHPB tests is trapezoidal or rectangular. The pulse has a short rise time and large amplitude Pochhamer-Chree oscillations resulting in wave dispersion (Parry et al. 1995). Brittle materials like rock require a slowly rising or half-sine wave-type pulse. It facilitates dynamic force equilibrium and constant strain rate loading in the rock specimen (Frew et al. 2002; Li et al. 2000). To this end, various techniques collectively known as pulse shaping techniques have been employed to modify the shape of the incident pulse.

So far, pulse shaping has been performed using three methods which are i) changing the shape of the striker bar, ii) using a dummy specimen or preloading bar or both in between the striker and input bars, and iii) using a suitable 'tip' material on the incident bar face at the impact end. Christensen et al. (1972) and Li et al. (2000) used a truncated conical striker bar to produce a ramping pulse for dynamic tests on rocks. Li et al. (2004) and Wang et al. (2018) tested granite specimens and created a shaped pulse using a spindle-shaped striker bar. Ellwood et al. (1982) shaped incident pulse by adding a preloading bar and a dummy specimen before the incident bar. Parry et al. (1995) eliminated the use of a dummy specimen by using a preloading bar of lower strength than that of input bars. However, both tested ductile materials only. Employing a shaped striker bar, a dummy specimen, or an extra bar increases the operational or manufacturing cost. A relatively economical and less cumbersome pulse shaping method is the use of a 'tip' material, also known as pulse shaper (PS). A PS is placed on the incident bar face near the striking end using a lubricant. The material can be paper, metals like copper, brass, and aluminum, rubber, or the same material as the testing specimen (Frantz et al. 1984). A part of the energy due to the impact of the striker bar is used in the plastic deformation of the PS. It increases the rise time and filters out the high-frequency oscillations of the incident wave. This gives the specimen enough time to achieve stress uniformity and deform at a constant strain rate loading before experiencing failure.

The PS material and size must be decided based on the testing material, strain rate to be achieved, striker bar length (L), and its velocity (V_s) (Follansbee and Frantz 1983; Naghdabadi et al. 2012). Many researchers have investigated the behavior of pulse shapers by varying their dimensions and highlighting their role in obtaining a desirable incident pulse. Frantz et al. (1984) first attempted to

DOI: 10.1201/9781003352358-15

use PS by using a brass disc to test steel specimens in SHPB and showed that by using a suitable material, the rise time of the incident pulse could be prolonged. Nemat-Nasser et al. (1991) used a metal cushion and provided a complete analysis of the plastic deformation of the PS by using a power-law relationship. Togami et al. (1996) employed plexiglass, Hsiao et al. (1998) employed rubber, and Song et al. (2007b) employed polyurethane foam as pulse shapers to test metals, composite, and polymeric isocyanate (PMDI) foam specimens, respectively. A dual PS of copper and steel was used by Frew et al. (2005) and Song et al. (2007a) to test high-strength elastic-plastic material like steel. Forrestal et al. (2002) used C11000 annealed copper discs as pulse shapers. Ramírez and Rubio-Gonzalez (2006) performed numerical simulations to test the efficacy of different materials, viz. aluminum, copper, and iron, as pulse shapers. Optimum dimensions to reduce dispersion were suggested.

Naghdabadi et al. (2012) investigated the effects of a copper disc on the shape of the incident pulse experimentally and numerically. It was found that the rise time and the duration of the incident pulse increase by increasing the PS thickness or by decreasing its diameter. A relatively small-diameter PS was recommended for testing brittle materials. However, the dynamic tests were limited to copper and cast-iron specimens, and brittle materials like rock were not tested. Song et al. (2016) dynamically loaded ice specimens in SHPB and used copper pulse shapers. Contrary to Naghdabadi et al. (2012), Song et al. (2016) found that the rise time and the duration of the incident pulse increase by increasing the PS diameter. Based on the experimental and numerical data from the tests on concrete specimens, Shemirani et al. (2016) provided guidelines for choosing appropriate PS. Chen et al. (2016) chose three different dimensions of copper PS, two of them disc type and one annular type. It was concluded that a smaller copper PS could extend the rise time and was better than a large and annular copper PS.

Many research works on dynamic tests of rocks in SHPB have employed pulse shapers (Chakraborty et al. 2016; Dai et al. 2013; Frew et al. 2001; Malik et al. 2017; Mishra et al. 2019; Shi et al. 2018; Wang et al. 2018; Wu et al. 2020; Xia et al. 2008; Zwiessler et al. 2017). However, none of them provide a concrete explanation as to why the given size of the PS is used except that it helps achieve stress equilibrium and constant strain rate loading in the specimen. The effects of PS diameter, thickness, and striker bar length and velocity on the incident pulse shape and rise time are not studied. All the tests are performed on bar diameters less than 76 mm except for Mishra et al. (2019). A large diameter of SHPB, such as 76 mm, will allow a bigger rock specimen to be used and hence provide a better representation of its dynamic properties. The objective of the present study is to be able to choose appropriate dimensions of pulse shaper for testing rock specimens in SHPB. A large-diameter (76 mm) SHPB is employed to perform dynamic tests on Kota sandstone. The diameter of PS is varied, and its effects on the rise time and the duration of the incident pulse are studied. The role of the PS dimensions in achieving dynamic force equilibrium is appraised.

15.2 MATERIALS AND METHODOLOGY

15.2.1 PULSE SHAPING STUDY PLAN

Copper is used as a PS in the current study. The pulse shapers are circular with 15, 20, and 25 mm diameters and a thickness of 1.6 mm.

15.2.2 DYNAMIC TEST METHOD

Dynamic tests are performed in the SHPB setup shown in Figure 15.1. All the bars are made of C350 maraging steel. The diameter of the bars is 76 mm, while the length of the incident and transmission bars is 3.048 m each. For all the tests, a striker bar of length 609.6 mm is launched at a gas gun pressure of 0.24 MPa (35 psi). The samples have an aspect ratio of 0.5 to optimize friction and inertia effects (ISRM 2015). The stress, strain, and strain rate in the specimen are calculated based

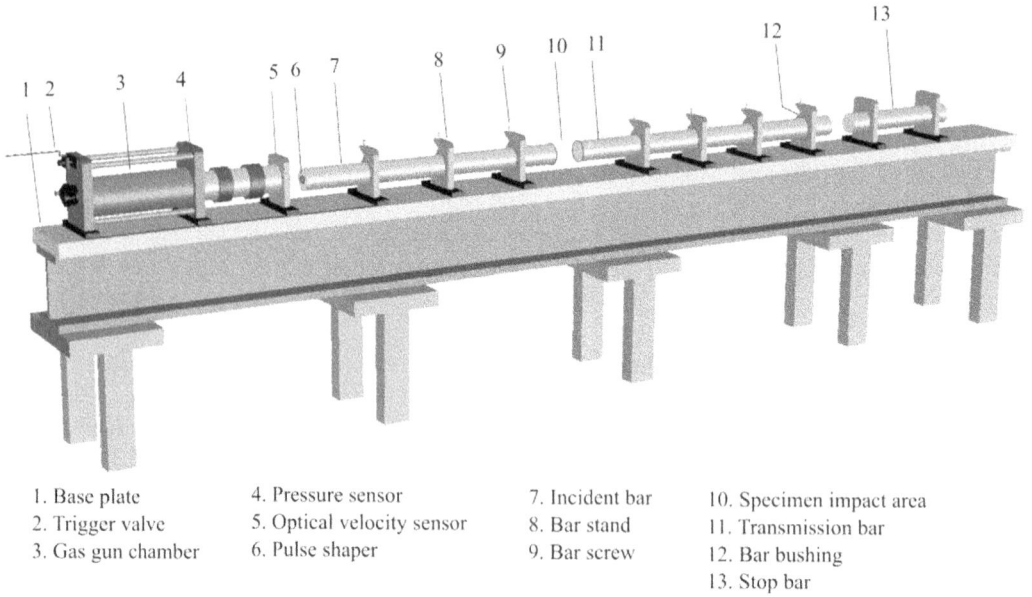

1. Base plate
2. Trigger valve
3. Gas gun chamber
4. Pressure sensor
5. Optical velocity sensor
6. Pulse shaper
7. Incident bar
8. Bar stand
9. Bar screw
10. Specimen impact area
11. Transmission bar
12. Bar bushing
13. Stop bar

FIGURE 15.1 Split Hopkinson pressure bar (SHPB) setup used for dynamic testing. Source: Deshpande et al. (2022).

on the one-dimensional stress wave propagation theory (Kolsky 1964) and the assumption that force equilibrium is achieved between both the faces of the specimen during dynamic loading, i.e. $\varepsilon_i + \varepsilon_r = \varepsilon_t$, using the following equations.

$$\varepsilon(t) = -\frac{2C}{l} \int_0^t \varepsilon_R dt \tag{15.1}$$

$$\dot{\varepsilon} = \frac{d\varepsilon(t)}{dt} = -\frac{2C}{l} \varepsilon_R \tag{15.2}$$

$$\sigma = Y \frac{A}{a} \varepsilon_T \tag{15.3}$$

Here specimen response quantities are ε = strain, $\dot{\varepsilon}$ = strain rate, σ = stress, a = cross-sectional area of the specimen, l = specimen length. Bar properties are C = wave speed in the bar, ε_R = reflected strain, ε_T = transmitted strain, Y = Young's modulus, A = cross-sectional area. The bar strains ($\varepsilon_I, \varepsilon_R, \varepsilon_T$) are obtained by converting the measured voltage signals using the formula given below (REL Inc. 2014). The voltage signals are captured at a high resolution of 20 million per second.

$$\text{Strain} = \text{Voltage reading} \times \frac{120\Omega}{2V \times 2 \times (120\Omega + 59940\Omega)} \tag{15.4}$$

The equilibrium in the specimen is checked by calculating the front face force (or front force) F_f and back face force (or back force) F_b at the incident bar-specimen (IB-S) and specimen-transmission bar interface (S-TB), respectively. The front force at the IB-S interface is calculated using incident and reflected strains in the incident bar. The back force at the S-TB interface is calculated using transmitted strain in the transmission bar, as shown in the equations below.

$$F_f = AY(\varepsilon_I - \varepsilon_R) \tag{15.5}$$

$$F_b = AY\varepsilon_T \tag{15.6}$$

15.3 RESULTS AND DISCUSSION

15.3.1 DYNAMIC FORCE EQUILIBRIUM

Dynamic force equilibrium achieved for specimens using PS of various diameters is shown in Figure 15.2. Out of the three diameters used, the best force equilibrium is achieved for $d = 15$ mm. In contrast, for $d = 25$ mm, the front force oscillates throughout the experiment and does not match the back force. It is, therefore, concluded that the pulse shaper with $d = 15$ mm is appropriate for testing rocks in SHPB. The effects of PS diameter are explained in the next section.

15.3.2 PULSE SHAPER DIAMETER EFFECTS

The effect of PS diameter on the incident pulse is illustrated in Figure 15.3. The incident pulse duration is close to 300 μs without a pulse shaper, whereas it increases to 580 μs approximately when pulse shapers of $d = 15$, 20, or 25 mm are used. When no PS is employed, the rise time of the incident pulse is short, ~37 μs. Ravichandran and Subhash (1994) stated that it takes around three to four reverberations inside the sample to reach force equilibrium. Considering this fact, the rise time of the pulse without a pulse shaper is insufficient. The rise time increases to 290 μs when a PS of $d = 15$ mm is employed. As the diameter further increases to 20 and 25 mm, there is no significant improvement in the rise time, and it remains approximately constant. However, it is observed that the rising portion of the pulses generated using $d = 20$ and 25 mm is not smooth and has several kinks/oscillations. The kinks are more in number when $d = 25$ mm. It is anticipated that these kinks do not allow the rock specimens to achieve force equilibrium. The kinks can be attributed to inertia-induced stresses in the pulse shapers (Heard et al. 2014). More investigation is needed in this regard to form a firm conclusion. It must be noted that the primary aim of the current study is to study the effects of PS dimensions. Hence dynamic stress-strain curves of the rock specimens are not presented here.

15.4 CONCLUSION

In the present work, the effects of PS diameter are investigated experimentally. An attempt is made to optimize the PS dimension for a large-size SHPB. It is found that without a PS, brittle materials like rocks cannot be tested under dynamic loading in SHPB. To achieve dynamic force equilibrium, a PS is essential. The PS must be designed based on the testing conditions. When the diameter of

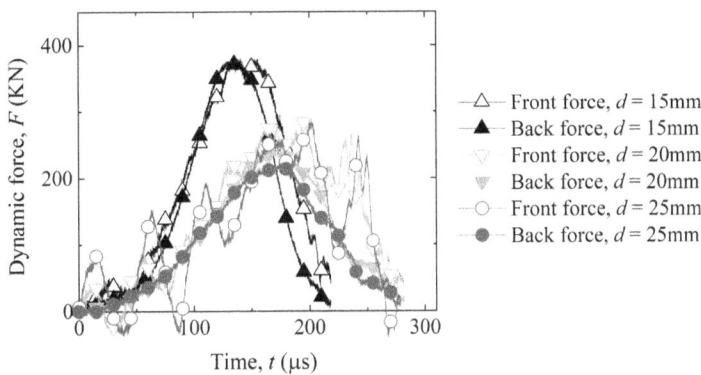

FIGURE 15.2 Dynamic force equilibrium achieved for pulse shaper of various diameters – variation of front and back forces with time.

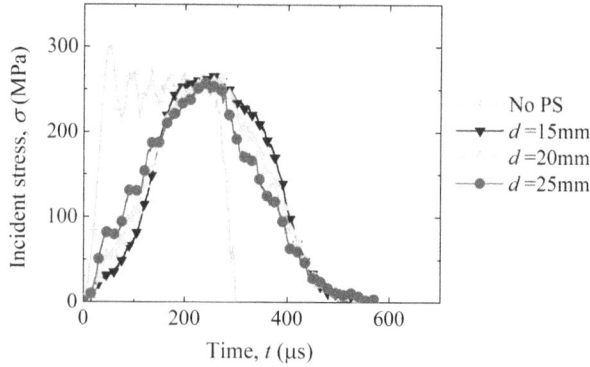

FIGURE 15.3 Incident pulse generated using pulse shaper of diameters 15, 20, and 25 mm.

PS equals 15 mm, dynamic force equilibrium is achieved. Further increase in the diameter does not improve the rise time. Thus, for the given test conditions, the optimum diameter and thickness of the PS are 15 mm and 1.6 mm. It must be noted that tests were conducted with a single PS thickness. Future investigation will include studying the effects of thickness on the incident pulse and performing a numerical investigation.

REFERENCES

Chakraborty, T., Mishra, S., Loukus, J., Halonen, B., and Bekkala, B. (2016). Characterization of three Himalayan rocks using a split Hopkinson pressure bar. *International Journal of Rock Mechanics and Mining Sciences*, 85, 112–118. https://doi.org/10.1016/j.ijrmms.2016.03.005

Chen, X., Ge, L., Zhou, J., and Wu, S. (2016). Experimental study on split Hopkinson pressure bar pulse-shaping techniques for concrete. *Journal of Materials in Civil Engineering*, 28(5), 4015196. https://doi.org/10.1061/(asce)mt.1943-5533.0001494

Christensen, R. J., Swanson, S. R., and Brown, W. S. (1972). Split-hopkinson-bar tests on rock under confining pressure. *Experimental Mechanics*, 12(11), 508–513. https://doi.org/10.1007/bf02320747

Dai, F., Xia, K., Zuo, J. P., Zhang, R., and Xu, N. W. (2013). Static and dynamic flexural strength anisotropy of Barre granite. *Rock Mechanics and Rock Engineering*, 46(6), 1589–1602. https://doi.org/10.1007/s00603-013-0390-y

Deshpande, V. M., Chakraborty, P., Chakraborty, T., and Tiwari, V. (2022). Application of copper as a pulse shaper in SHPB tests on brittle materials- experimental study, constitutive parameters identification, and numerical simulations. *Mechanics of Materials*, 171, 104336. https://doi.org/10.1016/j.mechmat.2022.104336

Ellwood, S., Griffiths, L. J., and Parry, D. J. (1982). Materials testing at high constant strain rates. *Journal of Physics E: Scientific Instruments*, 15(3), 280. http://stacks.iop.org/0022-3735/15/i=3/a=009

Follansbee, P. S., and Frantz, C. (1983). Wave propagation in the split Hopkinson pressure bar. *Journal of Engineering Materials and Technology*, 105(1), 61. https://doi.org/10.1115/1.3225620

Forrestal, M. J., Frew, D. J., and Chen, W. (2002). The effect of sabot mass on the striker bar for split Hopkinson pressure bar experiments. *Experimental Mechanics*, 42(2), 129–131. https://doi.org/10.1007/bf02410873

Frantz, C., Follansbee, P. S., and Wright, W. J. (1984). *New experimental techniques with the split Hopkinson pressure bar*. Office of Scientific and Technical Information (OSTI). https://doi.org/10.2172/6854600

Frew, D. J., Forrestal, M. J., and Chen, W. (2001). A split Hopkinson pressure bar technique to determine compressive stress-strain data for rock materials. *Experimental Mechanics*, 41(1), 40–46. https://doi.org/10.1007/bf02323102

Frew, D. J., Forrestal, M. J., and Chen, W. (2002). Pulse shaping techniques for testing brittle materials with a split Hopkinson pressure bar. *Experimental Mechanics*, 42(1), 93–106. https://doi.org/10.1007/BF02411056

Frew, D. J., Forrestal, M. J., and Chen, W. (2005). Pulse shaping techniques for testing elastic-plastic materials with a split Hopkinson pressure bar. *Experimental Mechanics*, 45(2), 186–195. https://doi.org/10.1007/bf02428192

Heard, W. F., Martin, B. E., Nie, X., Slawson, T., and Basu, P. K. (2014). Annular pulse shaping technique for large-diameter Kolsky bar experiments on concrete. *Experimental Mechanics, 54*(8), 1343–1354. https://doi.org/10.1007/s11340-014-9899-6

Hopkinson, J. (1872). On the rupture of iron wire by a blow. *Proceedings of Manchester Literary Philosophical Society, 11*, 40–45.

Hsiao, H. M., Daniel, I. M., and Cordes, R. D. (1998). Dynamic compressive behavior of thick composite materials. *Experimental Mechanics, 38*(3), 172–180. https://doi.org/10.1007/bf02325740

ISRM. (2015). *The ISRM suggested methods for rock characterization, testing and monitoring: 2007–2014* (R. Ulusay (ed.)). https://link.springer.com/book/10.1007%2F978-3-319-07713-0

Kolsky, H. (1964). Stress waves in solids. *Journal of Sound and Vibration, 1*(1), 88–110. https://doi.org/10.1016/0022-460x(64)90008-2

Li, X. B., Lok, T. S., and Zhao, J. (2004). Dynamic characteristics of granite subjected to intermediate loading rate. *Rock Mechanics and Rock Engineering, 38*(1), 21–39. https://doi.org/10.1007/s00603-004-0030-7

Li, X. B., Lok, T. S., Zhao, J., and Zhao, P. J. (2000). Oscillation elimination in the Hopkinson bar apparatus and resultant complete dynamic stress-strain curves for rocks. *International Journal of Rock Mechanics and Mining Sciences, 37*(7), 1055–1060. https://doi.org/10.1016/S1365-1609(00)00037-X

Malik, A., Chakraborty, T., Rao, K. S., Kumar, D., Chandel, P., and Sharma, P. (2017). Dynamic response of Deccan trap basalt under Hopkinson bar test. *Procedia Engineering, 173*, 647–654. https://doi.org/10.1016/j.proeng.2016.12.124

Mishra, S., Chakraborty, T., Basu, D., and Lam, N. (2019). Characterization of sandstone for application in blast analysis of tunnel. *Geotechnical Testing Journal, 43*(2), 20180270. https://doi.org/10.1520/gtj20180270

Naghdabadi, R., Ashrafi, M. J., and Arghavani, J. (2012). Experimental and numerical investigation of pulse-shaped split Hopkinson pressure bar test. *Materials Science and Engineering: A, 539*, 285–293. https://doi.org/10.1016/j.msea.2012.01.095

Nemat-Nasser, S., Isaacs, J. B., and Starrett, J. E. (1991). Hopkinson techniques for dynamic recovery experiments. *Proceedings of the Royal Society A: Mathematical, Physical and Engineering Sciences.* https://doi.org/10.1098/rspa.1991.0150

Parry, D. J., Walker, A. G., and Dixon, P. R. (1995). Hopkinson bar pulse smoothing. *Measurement Science and Technology, 6*(5), 443. http://stacks.iop.org/0957-0233/6/i=5/a=001

Ramírez, H., and Rubio-Gonzalez, C. (2006). Finite-element simulation of wave propagation and dispersion in Hopkinson bar test. *Materials and Design, 27*(1), 36–44. https://doi.org/10.1016/j.matdes.2004.08.021

Ravichandran, G., and Subhash, G. (1994). Critical appraisal of limiting strain rates for compression testing of ceramics in a split Hopkinson pressure bar. *Journal of the American Ceramic Society, 77*(1). https://doi.org/10.1111/j.1151-2916.1994.tb06987.x

REL Inc. (2014). *SHPB Instruction Manual.* https://www.relinc.net/split-hopkinson-bar-kolsky-bars/reference-material/

Shemirani, A. B., Naghdabadi, R., and Ashrafi, M. J. (2016). Experimental and numerical study on choosing proper pulse shapers for testing concrete specimens by split Hopkinson pressure bar apparatus. *Construction and Building Materials, 125*, 326–336. https://doi.org/10.1016/j.conbuildmat.2016.08.045

Shi, X., Liu, D., Yao, W., Shi, Y., Tang, T., Wang, B., and Han, W. (2018). Investigation of the anisotropy of black shale in dynamic tensile strength. *Arabian Journal of Geosciences, 11*(2), 42. https://doi.org/10.1007/s12517-018-3384-y

Song, B., Chen, W., Antoun, B. R., and Frew, D. J. (2007a). Determination of early flow stress for ductile specimens at high strain rates by using a SHPB. *Experimental Mechanics, 47*(5), 671–679. https://doi.org/10.1007/s11340-007-9048-6

Song, B., Syn, C. J., Grupido, C. L., Chen, W., and Lu, W.-Y. (2007b). A long split Hopkinson pressure bar (LSHPB) for intermediate-rate characterization of soft materials. *Experimental Mechanics, 48*(6), 809–815. https://doi.org/10.1007/s11340-007-9095-z

Song, Z., Wang, Z., Kim, H., and Ma, H. (2016). Pulse shaper and dynamic compressive property investigation on ice using a large-sized modified split Hopkinson pressure bar. *Latin American Journal of Solids and Structures, 13*(3), 391–406. https://doi.org/10.1590/1679-78252458

Togami, T. C., Baker, W. E., and Forrestal, M. J. (1996). A split Hopkinson bar technique to evaluate the performance of accelerometers. *Journal of Applied Mechanics, 63*(2), 353–356. https://doi.org/10.1115/1.2788872

Wang, Z. L., Li, H. R., Wang, J. G., and Shi, H. (2018). Experimental study on mechanical and energy properties of granite under dynamic Triaxial condition. *Geotechnical Testing Journal, 41*(6), 20170237. https://doi.org/10.1520/gtj20170237

Wu, R., Li, H., Li, X., Xia, X., and Liu, L. (2020). Experimental study and numerical simulation of the dynamic behavior of transversely isotropic phyllite. *International Journal of Geomechanics*, *20*(8), 4020105. https://doi.org/10.1061/(asce)gm.1943-5622.0001737

Xia, K., Nasseri, M. H. B., Mohanty, B., Lu, F., Chen, R., and Luo, S. N. (2008). Effects of microstructures on dynamic compression of Barre granite. *International Journal of Rock Mechanics and Mining Sciences*, *45*(6), 879–887. https://doi.org/10.1016/j.ijrmms.2007.09.013

Xia, K., and Yao, W. (2015). Dynamic rock tests using split Hopkinson (Kolsky) bar system – A review. *Journal of Rock Mechanics and Geotechnical Engineering*, *7*(1), 27–59. https://doi.org/10.1016/j.jrmge.2014.07.008

Zwiessler, R., Kenkmann, T., Poelchau, M. H., Nau, S., and Hess, S. (2017). On the use of a split Hopkinson pressure bar in structural geology: High strain rate deformation of Seeberger sandstone and Carrara marble under uniaxial compression. *Journal of Structural Geology*, *97*, 225–236. https://doi.org/10.1016/j.jsg.2017.03.007

16 Dynamic Strength Enhancement of Concrete in Split Hopkinson Pressure Bar Test

Kavita Ganorkar, M. D. Goel and Tanusree Chakraborty

16.1 INTRODUCTION

In the present scenario, extreme loading conditions such as impact, projectile penetration, and blast events have been so frequent that there is an urgent need to understand the behavior of the construction materials such as concrete under such circumstances. Some of these scenarios are (i) tunnel lining subjected to vehicular accident or blast, (ii) any possible projectile attack on a nuclear power plant, (iii) airport concrete runways undergoing impact loading due to aircraft landing, and (iv) other military and important structures subjected to impact or blast load.

The concrete material behaves differently under dynamic loading conditions than static loading. Hence concrete material properties at high strain rates are essential to understand their complex stress state for the analysis and design of structure against impact or blast load. Split Hopkinson pressure bar (SHPB) is a generally used test method to characterize brittle material's dynamic properties. The mechanical properties of different structural materials such as concrete, rocks, metals, and rubber are experimentally investigated within the strain rate range of order 10^2/s to 10^4/s using the SHPB setup. The fundamental principles behind this experimental analysis are one-dimensional stress wave propagation and stress equilibrium during the SHPB experiment. SHPB was first developed by Hopkinson (1914) and later modified by Herbert Kolsky (1949), and it was then further extended with recent advances by Chen and Song (2011).

Numerous researchers have used split Hopkinson pressure bar techniques to investigate the dynamic compressive strength of concrete (Bischoff and Perry, 1991; Grote et al., 2001; Li and Meng, 2003; Xu and Wen, 2013; Ganorkar et al., 2021a). When using the SHPB system, the hypothesis of stress homogenization should be met. As a result, large-diameter SHPB test systems must be used for dynamic testing of materials containing large-size aggregate particles, such as concrete (Huang and Xiao, 2020; Wang et al., 2021; Lv et al., 2022). Many researchers have investigated the dynamic properties of concrete using such large diameters.

The objective of the present study is to find out the strain rate effect of 28 days of compressive strength of 30 MPa concrete using 76 mm diameter SHPB. The current research looks at the dynamic behavior of concrete specimens under strain rates ranging from 100 to 382/s.

16.2 PRINCIPLES OF THE SHPB SYSTEM

The SHPB system comprises three bars: a striker bar, an incident bar, and a transmission bar of 76 mm diameter, as shown in Figure 16.1. The striker bar is propelled through the gas gun and impacts the free end of the incident bar. It induces a longitudinal compressive stress wave in the incident bar. Upon reaching the bar specimen interface, a part of the incident wave is reflected as the reflected wave. The

DOI: 10.1201/9781003352358-16

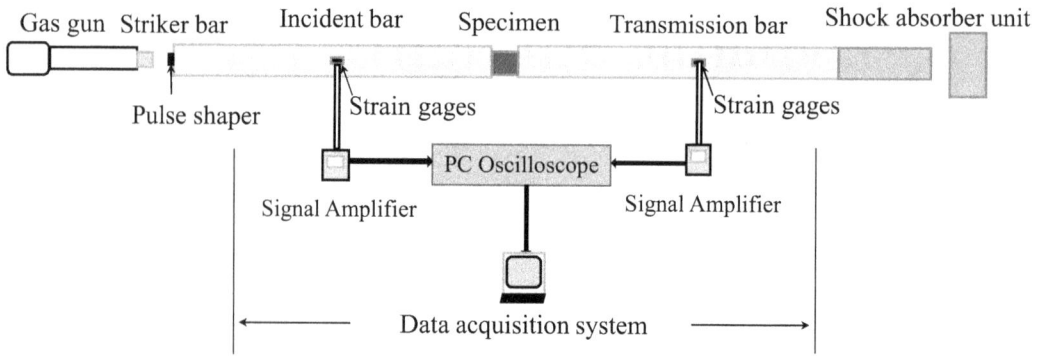

FIGURE 16.1   Schematic diagram of split Hopkinson pressure bar (SHPB).

remainder passes through the specimen to the transmitted bar as the transmitted wave. Strain gauges are attached to the center of the incident and transmission bars to record the stress wave pulse.

Based on one-dimensional loading conditions and assuming that the specimen gets deformed uniformly, the strain, stress, and strain rate in the specimen can be determined from the following equations,

$$\varepsilon(t) = \frac{-2c_{bar}}{L} \int_0^t \varepsilon_r dt \qquad (16.1)$$

$$\sigma = \frac{A_{bar} E_{bar} \varepsilon_t}{A_{sp}} \qquad (16.2)$$

$$\dot{\varepsilon}(t) = \frac{-2c_{bar}\varepsilon_r}{L} \qquad (16.3)$$

where A_{bar} and A_{sp} are the cross-sectional areas of the bar and specimen, and ε_r and ε_t represent reflected and transmitted strains, respectively. The E_{bar} is the elastic Young's modulus of the bar and $\dot{\varepsilon}$ is the loading strain rate of the tested specimen of length L.

16.3   EXPERIMENTAL DETAILS

The 28 days of uniaxial compressive strength of 30 MPa concrete is used in the present experimental work. The literature review shows that specimens with a slenderness ratio of 0.5–1 are suitable for SHPB testing of brittle materials (Lv et al., 2022; Huang and Xiao, 2020; Ganorkar et al., 2021b). Additionally, Fu et al. (2018) suggested that the specimen's transverse inertia and end friction confinement effect (between specimen and bars) influence on the test results could be minimized for a 0.5 slenderness ratio of the specimen. As a result, specimens measuring 54 mm in diameter and 108 mm in length are prepared for the current investigation. After that, they are cut into 54 mm diameter, 0.5 slenderness ratio pieces, and finally, the front and back faces of the specimens are ground to make them parallel. In order to reduce the end friction confinement effect, a small layer of grease is applied on both ends of the specimen before testing. The specimen is positioned between tungsten carbide (WC) platens before being sandwiched between the incident and transmission bars. Platens are crucial in lowering the concentration of stress and preventing indentation on the surface where the specimen and incident bar face, as well as specimen and transmission bar surfaces, come into contact. For the current study, platens with a 76 mm diameter were chosen so that their impedance would equal that of the SHPB bar material.

Figure 16.2 shows the typical waveform of incident reflected and transmitted pulses. The pulse shaper of 20 mm diameter and 1.5 mm thickness is employed on the impact end of the incident bar (Ganorkar et al., 2021b). The dimension of the pulse shaper is chosen such that it will increase the rise time and total duration of the incident pulse. This ensures the accuracy of the test results by providing sufficient time for the specimen to achieve stress equilibrium (Ravichandran and Subhash, 1994). The rising time of the incident pulse is up to 0.2–0.3 ms, and the total incident pulse duration is measured in the range of 0.45–0.54 ms, which gives the specimen enough time to reach stress equilibrium. Constant strain rate is another essential characteristic for the validity of the test results. It is characterized by the strain plateau in the reflected wave over the full testing duration. These conditions can be approximated by the appropriate selection of pulse shaper. Figure 16.2 shows the constant strain rate achieved during the testing.

The validity of an SHPB test is justified by the achievement of force equilibrium at the two ends of the specimen, as shown in Figure 16.3. All the experiments show a perfect force equilibrium demonstrating the test results' validity. In the current experimental work, the strain rate is defined as the mean strain rate from the beginning of the elastic response of the stress-strain curve to the maximum stress achieved by the specimen (Zwiessler et al., 2017).

16.4   RESULTS AND DISCUSSION

The concrete samples are tested in the SHPB device for dynamic compressive strength from 100/s to 382/s. An important index to describe the dynamic properties of the concrete is the stress-strain curve. Figure 16.4 presents the stress-strain curves for different strain rate values, and the peak stress is noted for each strain rate. It is seen from Figure 16.4 that the strength of the concrete increases with the increase in loading rate. It increases from 57 to 84 MPa. The maximum strain achieved by the sample is nearly the same up to the 214/s strain rate. But beyond the 214/s strain

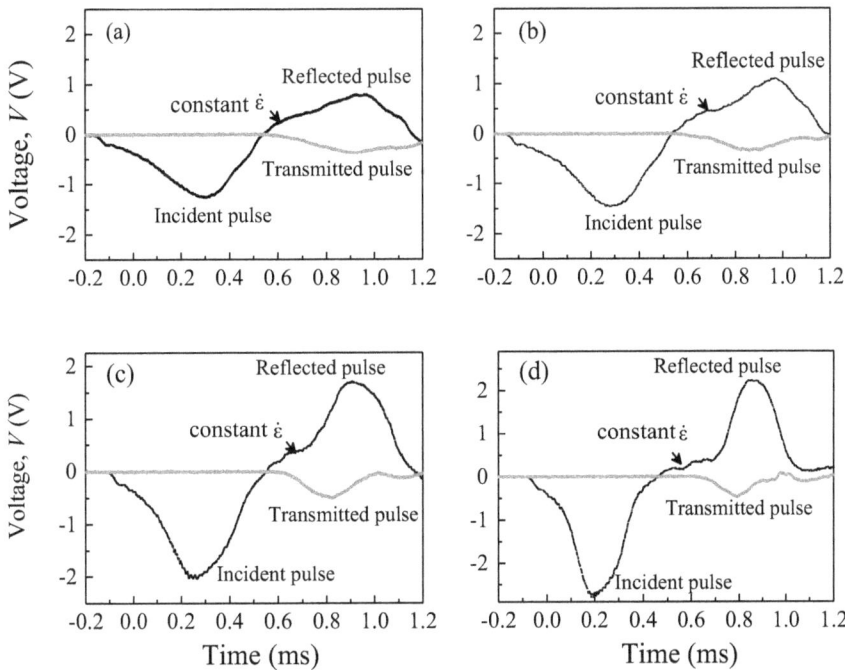

FIGURE 16.2   Incident, reflected, and transmitted waveforms from the SHPB testing.

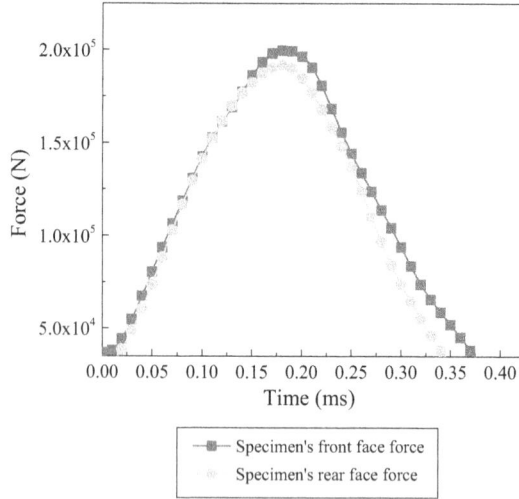

FIGURE 16.3 Dynamic force equilibrium achieved for the specimen tested using SHPB.

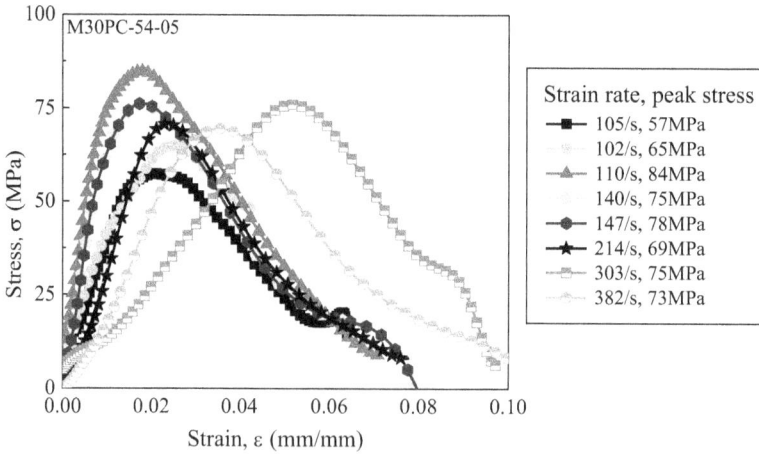

FIGURE 16.4 Stress-strain curve for M30 grade concrete.

rate, it is observed that maximum strain increase with the increase in strain rate. It indicates that the mode of failure of concrete changes from brittle to ductile at a high strain rate. Also, until the strain rate reaches 150/s, strength increases with the loading rate, but after that, the strength remains nearly constant.

16.4.1 DYNAMIC INCREASE FACTOR AND PROPOSED CORRELATION EQUATION

The dynamic increase factor (*DIF*) for concrete has been determined by comparing the dynamic peak stress with the static strength of the concrete. The DIF values are plotted in Figure 16.5, corresponding to the natural logarithm of the strain rate. The figure shows that the dynamic strength of concrete is 1.9 to 2.8 times that of the static strength for the strain rates varying from 100/s to 382/s. A correlation equation for the M30 grade of concrete has been developed for the calculated DIF with the natural logarithmic of the strain rate. The correlation equation is given by

$$DIF = 0.27\log\dot{\varepsilon} + 1 \quad for \quad 100/s \le \dot{\varepsilon} \le 382/s \left(R^2 = 0.986\right) \tag{16.4}$$

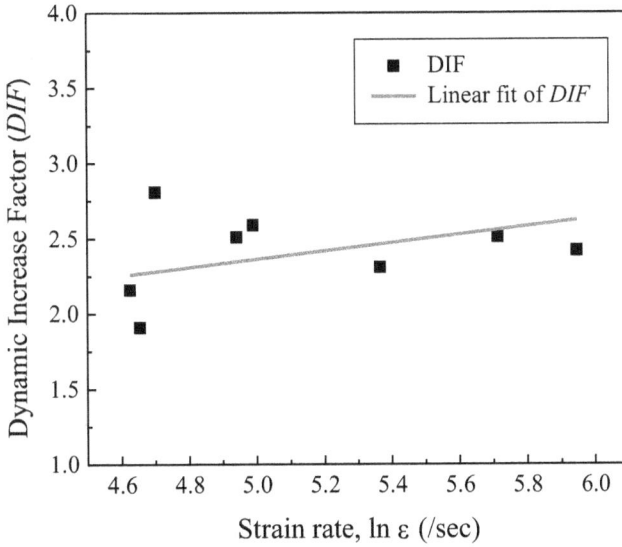

FIGURE 16.5 Dynamic increase factor comparison for M30 grade concrete.

This proposed correlation equation will be applicable to the strain rate range considered in the current work.

16.4.2 CRITICAL STRAIN AND ULTIMATE STRAIN

The deformation behavior of the specimen is identified from the critical strain achieved by the sample. It is the axial strain corresponding to the maximum stress of the stress-strain curve. Moreover, the ultimate strain is the maximum strain achieved by the sample. Figure 16.6 shows the critical and ultimate strain corresponding to the strain rate. The critical and ultimate strain of the specimen increase with the strain rate growth. It shows how the concrete material will behave depending on the strain rate. The sample's critical strain is discovered to be between 1.7 and 5.2% for the stated strain rate, while the ultimate strain is reported to be between 4.7 and 16%.

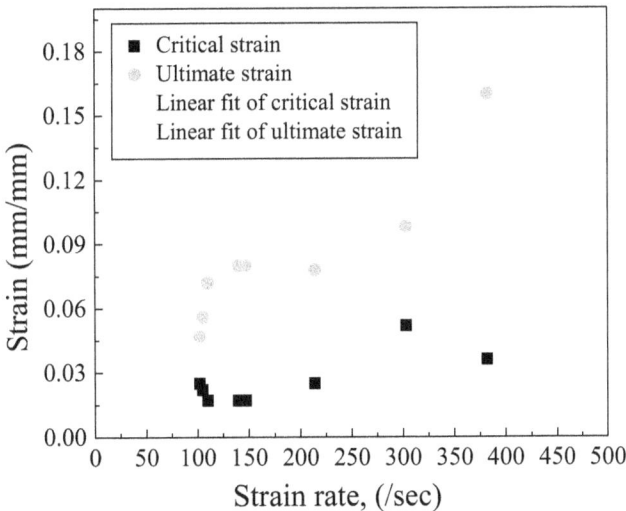

FIGURE 16.6 Critical and ultimate strain corresponding to the strain rate.

16.4.3 ENERGY ABSORPTION CAPACITY

The ability of the concrete material to withstand fracture damage under impact loading is characterized by its energy absorption capacity. The capacity of the specimen to absorb energy serves as a measure of the specimen's toughness. According to Huang and Xiao (2020), it is the kinetic energy that the concrete specimen absorbs throughout the fragmentation process. The present study calculates it as the area enclosed by the stress-strain curve. The energy absorption capacity of the specimen is found to be in the range of 2–4 MJ/m³ for the 100–382/s strain rate range, as depicted in Figure 16.7. The maximum energy absorption capacity of the specimen is noted to be 4.1 MJ/m³, which corresponds to the 303/s strain rate of the specimen. However, for the 110–160/s strain rate, the achieved energy absorption capacity of the specimen is nearly the same, which is noted as 3.2–3.6 MJ/m³.

16.5 CONCLUSION

Characterization of concrete has been performed in the present work for a strain rate range varying from 100/s to 382/s through a uniaxial compressive 76 mm diameter SHPB test. The dynamic stress-strain response of the concrete specimens is studied.

*   The stress-strain curves conclude that the mode of failure of the concrete specimen is a ductile type of failure at a high strain rate and brittle failure at a low strain rate.
*   The dynamic increase factor varies from 1.9 to 2.8.
*   A correlation equation for the M30 grade of concrete has been developed for the calculated DIF with the natural logarithmic of the strain rate.
*   Energy absorption capacity, critical strain, and ultimate strain are the strain-rate-dependent properties of the concrete.
*   Energy absorption capacity varies in the range of 2–4 MJ/m³ for the 100/s to 382/s strain rates.

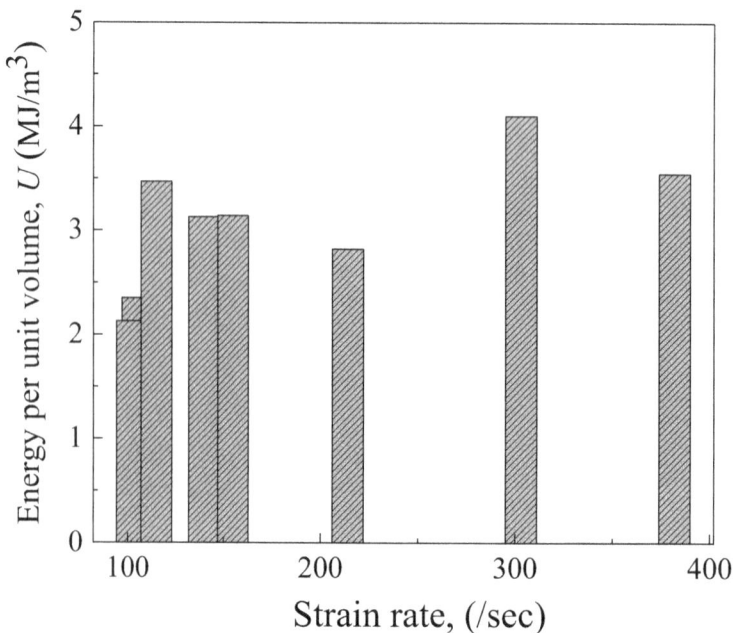

FIGURE 16.7 Energy absorption capacity.

REFERENCES

Bischoff, P. H. and Perry, S. H. (1991). Compressive behavior of concrete at high strain rates. *Materials and Structures*, 24(6), 425–450.

Chen, W. and Song, B. (2011). *Split Hopkinson (Kolsky) bar- Design, testing and applications.* Springer, New York.

Fu, Q., Niu, D., Li, D., Wang, Y., Zhang, J. and Huang, D. (2018). Impact characterization and modelling of basalt-polypropylene fibre reinforced concrete containing mineral admixtures. *Cement and Concrete Composites*, 93, 246–259.

Ganorkar, K., Arora, K., Gaur, L., Goel, M. and Chakraborty, T. (2021a). High strain rate characterization of concrete using split Hopkinson pressure bar. *Indian Concrete Journal*, 95(11), 28–35.

Ganorkar, K., Goel, M. D. and Chakraborty, T. (2021b). Specimen size effect and dynamic increase factor for basalt fiber reinforced concrete using split Hopkinson pressure bar. *Journal of Materials in Civil Engineering*. ASCE. https://doi.org/10.1061/(ASCE)MT.1943-5533.0003992.

Grote, D. L., Park, S. W. and Zhou, M. (2001). Dynamic behavior of concrete at high strain rates and pressures: I. experimental characterization. *International Journal of Impact Engineering*, 25(9), 869–886.

Huang, B. and Xiao, Y. (2020). Compressive impact tests of lightweight concrete with 155-mm-diameter spilt Hopkinson pressure bar. *Cement and Concrete Composites*, 114(May), 103816. https://doi.org/10.1016/j.cemconcomp.2020.103816

Li, Q. M. and Meng, H. (2003). About the dynamic strength enhancement of concrete-like materials in a split Hopkinson pressure bar test. *International Journal of Solids and Structures*, 2(40), 343–360.

Lv, N., Wang, H., Rong, K., Chen, Z. and Zong, Q. (2022). The numerical simulation of large diameter split Hopkinson pressure bar and Hopkinson bundle bar of concrete based on a mesoscopic model. *Construction and Building Materials*, 315, 125728, 1-17.

Ravichandran, G. and Subhash, G. (1994). Critical appraisal of limiting strain rates for compression testing of ceramics in a split Hopkinson pressure bar. *Journal of the American Ceramic Society*, 77(1), 263–267.

Wang, J., Li, W., Xu, L., Du, Z., Gao, G. and Alves, M. (2021). Experimental study on pulse shaping techniques of large diameter SHPB apparatus for concrete. *Latin American Journal of Solids and Structures*, 18(1), 1–15. https://doi.org/10.1590/1679-78256268

Xu, H. and Wen, H. M. (2013). Semi-empirical equations for the dynamic strength of concrete-like materials. *International Journal of Impact Engineering*, 60, 76–81.

Zwiessler, R., Kenkmann, T., Poelchau, M. H., Nau, S. and Hess, S. (2017). On the use of a split Hopkinson pressure bar in structural geology: High strain rate deformation of Seeberger sandstone and Carrara marble under uniaxial compression. *Journal of Structural Geology*, 97, 225–236.

17 Static and Dynamic Strength and Failure in Fiber-Reinforced Ultra-High-Performance Concrete

Shanideo N. Jadhav, Prakash Nanthagopalan and Krishna Jonnalagaddaa

17.1 INTRODUCTION

Ultra-high-performance concrete (UHPC) is an extensively used material in both civil and military structures. Its composition has been evolving in order to assure a great level of structural reliability against various extrinsic factors. Concrete structures [1] such as nuclear power plants, dams, bridges, and bunkers can be exposed to extreme and intensive loadings such as earthquakes, blast waves, explosions, or projectile impacts. The UHPC is a quasi-brittle material that exhibits brittle failure with relatively low tensile strength and strain capacity. The typical compressive strength of UHPC is at least 120 MPa. Its tensile strength is typically 8–15% of compressive strength [2]. Fiber reinforcement improves its tensile properties and ductility by arresting and bridging formed cracks, enhancing its toughness and energy absorption. Often, the Brazilian disc test, an indirect test method of tensile strength, is used for brittle materials. On the other hand, the center-cracked circular disc (CCCD) specimens can be used to investigate brittle fracture behavior [3] to estimate fracture toughness as well as understand crack propagation, crack branching, and process zone formation and size. In this work, a UHPC composition with fiber reinforcement is investigated for its mechanical behavior under tension and compression along with measurement of fracture toughness. A split Hopkinson pressure bar (SHPB) setup is used on the same specimen geometry to extract dynamic tensile strength. In-situ experiments with 3D-DIC were performed to measure stress-strain behavior, tensile strength, fracture toughness, and the initiation and evolution of cracks. The outline of this chapter is as follows: section 17.2 elaborates on the material composition, casting technique, specimen preparation, and different tests performed. In section 17.3, the experimental results are presented and discussed. Finally, in section 17.4, the summary and conclusions from the results are reported.

NOMENCLATURE

BD	Brazilian Disc
SHPB	Split Hopkinson pressure Bar
CCCD	Centrally Cracked Circular Disc
DIC	Digital Image Correlation
DIF	Dynamic Increase Factor
K_{IC}	Fracture toughness

DOI: 10.1201/9781003352358-17

17.2 MATERIAL AND EXPERIMENTS

17.2.1 MATERIAL AND COMPOSITION

In this study, an Ordinary Portland Cement (OPC) 53 grade conforming to IS 269 [4] was used with fine sand having a maximum particle size of 2.36 mm, ultrafine slag, polycarboxylate ether-based superplasticizer, and micro-steel fibers of 6 mm in length and 220 μm in diameter as primary reinforcement in the preparation of UHPC. Table 17.1 shows the quantity of the ingredients used to cast the specimens.

17.2.2 SPECIMEN PREPARATION AND CURING

Cement and ultrafine slag were mixed together in a twin-shaft mixer for two minutes. Then the mixture of water and superplasticizer was added and mixed for two minutes. Once the added material is mixed and starts appearing like a paste, the fine aggregate sand is added and mixed for three minutes. Finally, the steel fibers are added and mixed for another three minutes. After mixing, the mixture of UHPC is cast into the molds and stored at $27 \pm 2°C$, RH > 90%, for 24 hours. Afterward, the specimens are demolded and cured in water. The Brazilian disc specimens of 48 mm in diameter and 19 mm in thickness are prepared using waterjet cutting.

17.2.3 TENSILE EXPERIMENTS

A Brazilian disc was employed to perform a split tensile test with a diameter of 48 mm and 19 mm in thickness. It is placed between the flat platens of a servo-hydraulic UTM. A load is applied diametrically based on ASTM C496 [8] to the specimen leading to maximum principal stress at the center and tensile failure by splitting. The maximum tensile stress of the specimen can be calculated based on the assumption that failure occurs at the point of maximum tensile stress using,

$$\tilde{A} = \frac{2P}{\pi DL} \tag{17.1}$$

where P is the force on the specimen at failure; D and L are the diameter and height of the concrete specimen, respectively.

The split tensile experiments are performed under quasi-static loading using UTM. In addition, the deformation of the circular face of the sample in the quasi-static experiments is captured using optical imaging. Whole field deformation was extracted using digital image correlation (DIC).

17.2.4 COMPRESSION TEST

Uniaxial compression experiments are performed on two different geometries of cube specimens with dimensions 12 X 12X 19 mm^3 and 15X 10X 19 mm^3, respectively. A UTM is employed to

TABLE 17.1
Composition of UHPC (kg/m^3)

UHPC	Kg/m^3
Cement	800
Ultrafine slag	200
Sand	1,300
Fiber	100
Superplasticizer	20
Water	200

perform the test by compressing the specimen between two platens. Again, in-situ optical imaging and DIC are performed to extract deformation-time history.

17.2.5 FRACTURE TEST

The CCCD specimen geometry is taken for fracture test with 48 mm diameter and 19 mm height with a center notch (a/R = 0.3) of size 15 mm length and 0.8 mm width. The specimen is placed between two flat platens with the notch aligned along the compression axis. The peak load required for the initiation of crack propagation is recorded and the mode I stress intensity factor of the specimen is calculated.

17.2.6 HIGH STRAIN RATE TENSILE EXPERIMENTS

The split Hopkinson pressure bar (SHPB) setup is used to characterize the mechanical behavior of the material under high strain rate loading. It consists of two bars and a projectile made of high-strength material. The specimen is sandwiched between the bars and a projectile is impacted at one end of the input bar using a gas gun. Upon impact, a stress wave is generated in the incident bar and travels toward the specimen. After loading the specimen, a part of the wave is transmitted to the transmission bar, and another part is reflected in the incident bar, depending on the impedance mismatch at the specimen-bar interface. In this study, a SHPB bar setup with 20 mm pressure bars was used for testing Brazilian disc specimens. Based on the one-dimensional wave propagation analysis, the stress (σ), stress rate ($\dot{\sigma}$), and strain rate ($\dot{\varepsilon}$) in the Brazilian disc can be calculated as follows:

$$p(t) = A_b E_b \varepsilon_t(t); \ \sigma_t(t) = \frac{2P(t)}{\Pi DL}; \dot{\sigma}_t(t) = \frac{d\sigma(t)}{dt}; \dot{\varepsilon}_t(t) = \frac{\dot{\sigma}_t(t)}{E_s} \tag{17.2}$$

where Ab and E_b are the cross-sectional area and elastic modulus of the pressure bar, respectively, t is the time, and E_s is the elastic modulus of the specimen. The stress rate ($\dot{\sigma}$) can be also calculated from the slope of the stress history, which is often used in the case of Brazilian disc samples, since the strain is not uniform.

17.2.7 DIGITAL IMAGE CORRELATION (DIC)

Digital image correlation is a non-contact technique used to measure the deformation of a specimen subjected to external loading. An optical setup consisting of a camera and lens mounted on a tripod in front of the specimen with a speckle pattern and capturing the images throughout the test. These images are processed and give precise results using an analysis tool.

17.3 RESULTS AND DISCUSSION

17.3.1 QUASI-STATIC LOADING RESPONSE

In Figure 17.1, the load-displacement response of the split tensile test is shown along with the optical images taken at specific time intervals. The load-displacement curve markers show that by the time the peak load is achieved, there is already a diametral crack. Although this is a surface crack seen in the image, the gap between the fractured surfaces indicates that the crack is across the thickness of the sample. Beyond initial peak stress, the load is sustained by the sample with some variation, perhaps due to the crushing of the sample in the vicinity of the contacts with the platens and fiber bridging. Upon further compression, there is another peak, representing the highest load taken by the sample, with significant contact region damage seen by the multiple cracks near the lower platen. Even beyond this latter peak, the specimen did not crush and pulverize due to the presence

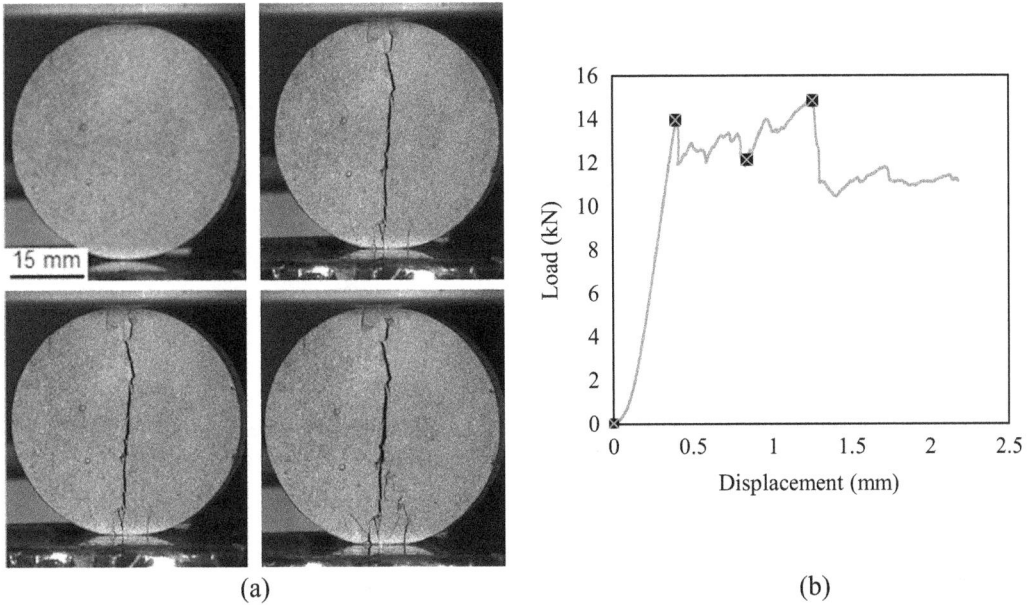

FIGURE 17.1   (a) Diametral failure of the specimen with increasing load under quai-static loading and (b) load-displacement curve showing the response and indicating the load corresponding to the pictures in (a).

of steel fibers, which held the two parts of the sample together. The loading was stopped at around 2 mm of total diametral displacement to further study the damage associated with the fibers. For the specimen shown in Figure 17.1, the first peak resulted in a split tensile stress of 10.56 MPa. However, this cannot be considered as the tensile strength of the material owing to the already existing diametral crack as discussed above. So, the actual tensile stress in these fiber-reinforced steel samples should be evaluated at the occurrence of the first crack at the center of the disc. This is the reason the current study employed in-situ imaging during loading.

Further, split tensile tests under quasi-static loading were conducted on UHPC specimens cured for different time periods, viz., seven and 14 days. Figure 17.2(a) shows their load-displacement response, where it can be seen that the 14-day curing resulted in higher peak strength of 9 ± 1 MPa compared to the seven-day curing tensile strength of 7 ± 2 MPa. On the other hand, the UHPC composition used in this study seems to achieve most of its strength within seven days given the insignificant increase after 14 days of curing. The tensile strength in both cases is in the range reported in the literature [5]. The effect of the addition of fibers to the composition sustained the capacity to load beyond the peak and did not cause pulverization of the sample leading to a sudden drop in load to zero.

Representative results from uniaxial compression experiments are presented in Figure 17.2(b). There are two curves corresponding to the different cross-section areas of the sample (see legend) and the same length of 19 mm along the loading axis, to elucidate the size effect seen on the mechanical response. Although the stiffness is not much different (due to nominal strain computed from cross-head displacement), the peak strength does suffer in the case of smaller samples. The peak compressive strength for larger and smaller cross-section samples was 120 MPa and 80 MPa, respectively. Further work, including in-situ imaging and strain measurement, is required to understand the size effect in compression.

From the in-situ fracture toughness experiments conducted using CCCD specimen geometry, using optical imaging, the load-time response is shown in Figure 17.2(c). The response curve, just like in the case of the split tensile test, is linear until the peak load. There is a small unloading event occurring in this linear region, and this is when the crack from the notch is seen along the loading

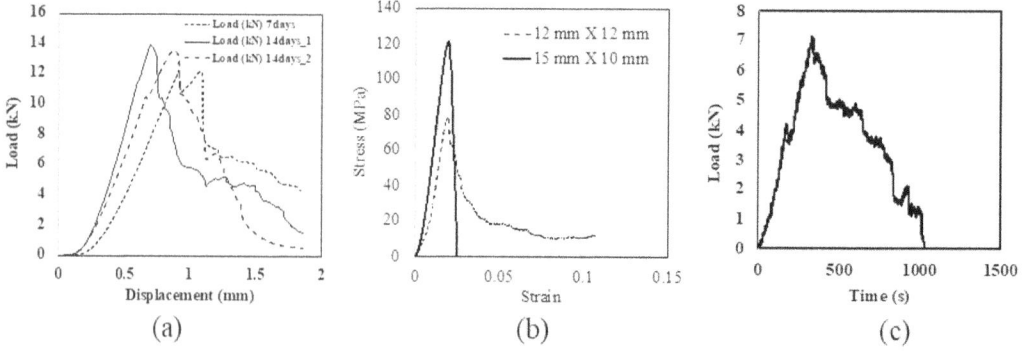

FIGURE 17.2 (a) Load-displacement response under quasi-static split tensile loading as a function of curing time; (b) stress-strain response under quasi-static loading and uniaxial compression for specimens with different cross-section areas; and (c) load-time curve for CCCD sample performed to compute the fracture toughness.

axis, spanning the entire diametral axis (see Figure 17.3 for the time-resolved images of the CCCD specimen). To calculate the mode I fracture toughness, the following equation is used:

$$K_{IC} = \frac{P}{t\sqrt{R}} f(a/W) \tag{17.3}$$

where P is the fracture load, and R and t are the radius and thickness of the CCCD specimen. The finite geometry shape factor $f(a/W)$ was found to be 0.14 [7].

Using this intermediate load of $P \sim 4.37$ kN, the mode-I stress intensity factor was calculated and the value was 0.208 MPa $\sqrt{m}$. Beyond this intermediate peak, the specimen sustained further load, reaching a peak, which is twice the load, when the crack appeared on the surface. Again, the in-situ imaging helped in identifying the correct load to be used in the calculation of fracture toughness. The computed value in this study for steel fiber-reinforced UHPC is similar to the values reported in the literature, thus affirming the load chosen based on surface observation. Note that the root radius of the notch employed in this study is smaller than the fiber, but is much larger than the constituents of the aggregate as well as porosity, which is expected to be similar to the value reported by Gurusideswar et al. [5]. Further, the post-peak strength is similar to the split tensile test and the overall area under the curves is small due to the smaller ligament where only the steel fibers are holding the two ligaments of the specimen. The final drop in the load corresponds to the damage in the contact region as seen in the split tensile experiment.

In order to further understand the reason behind this crack propagation below the peak load in the fracture experiments, the images captured during the fracture toughness test on samples with

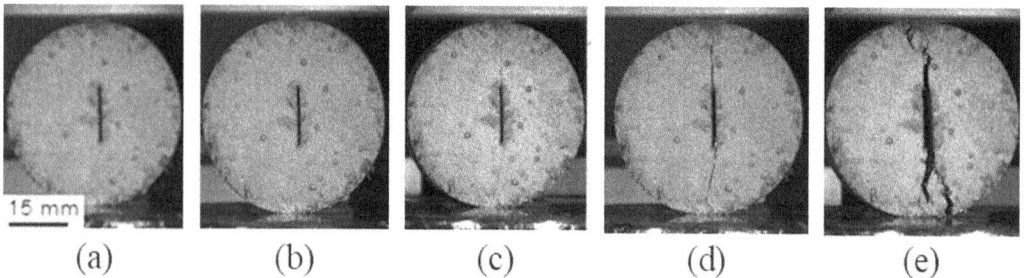

FIGURE 17.3 Images showing the initiation and propagation of a crack from the notch tip in a CCCD fracture specimen loaded under diametral compression.

speckle patterns were processed using digital image correlation. The DIC contours for the horizontal normal strain component, ε_{xx}, are shown in Figure 17.4, at different time intervals. It is surprising to see that from the beginning of loading, the strain is higher near the contact region and is tensile in nature. The strain contours in Figure 17.4(b) and (c) suggest that the crack may have nucleated near the contact region and propagated to the notch. This is contrary to our expectation for a fractured specimen. But, some caution is required as this observation is purely based on the surface strain on one side of the disc. Two reasons can be attributed to this behavior seen in the fracture experiment. One, the notch root radius is too large for the crack to initiate there and propagate. The second more interesting and plausible reason is that the stress and strain near the contact region with the platens are higher due to the specimen edge that is not flattened out to eliminate stress concentration. The latter explanation is potentially acceptable based on the DIC calculations performed on the split tensile experiments also. Figure 17.4(d) to (f) shows the evolution of the ε_{xx}, which is seen to be higher near the contact region and not at the center of the sample. With an increase in load, the tip of the maximum strain contour moves from the contact region to the center of the disc. So, although the apparent strength and toughness reported in this work seem to agree with those in other work, the origin of the failure is not where it is supposed to happen. It is not clear whether this phenomenon is due to the presence of fibers, which arrest the crack and bridge the cracked surfaces.

17.3.2 DYNAMIC SPLIT TENSILE RESPONSE

Figure 17.5(a) shows the specimen mounted in an SHPB setup. The axis of the bar is aligned with a diametral line drawn on the sample surface. This line is useful in post-mortem analysis to establish the diametral loading and failure of the sample. Based on the predetermined pulse width and shape,

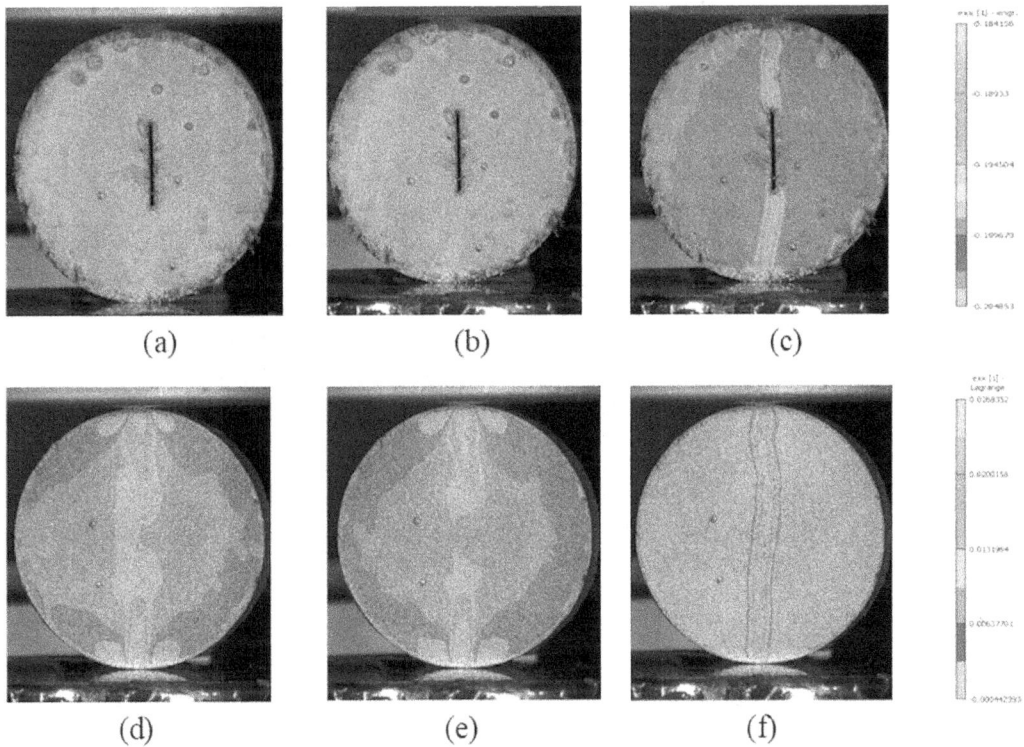

FIGURE 17.4 (a), (b), (c) DIC contours for the horizontal normal strain component (ε_{xx}) from the CCCD specimen used for fracture toughness experiments; (d), (e), (f) strain contours from the split tensile experiment.

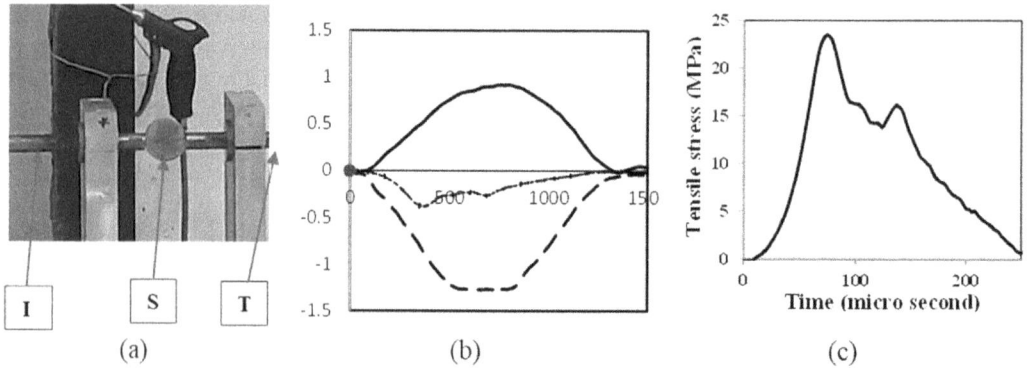

FIGURE 17.5 (a) A Brazilian disc specimen mounted on SHPB setup; (b) the strain gauge signals obtained from the incident and transmitted bars; and (c) stress history at the center of the disc sample, calculated from the boundary force applied. (I, S, and T stand for incident bar, specimen, and transmit bar.)

the sample is loaded and the signal thus recorded is shown in Figure 17.5(b). The incident pulse has a long rise time to ensure that the specimen reaches the peak load slowly and the maximum tensile stress is at the center of the sample after stress homogenization. Further, the reflected signal is representative of the net velocity of the loading along the sample diameter. Due to the sample geometry, the stress and strain in the sample are not uniform and therefore, only the stress at the center of the disc from the applied load at the boundary is calculated, and its time history is plotted in Figure 17.5(c). Based on the strain histories in the SHPB bars, a tensile strength of 23.69 MPa is achieved under dynamic loading. This value is much higher than that calculated under quasi-static loading. The calculated dynamic increase factor (DIF) for this UHPC composition was 2.42, which is slightly lower than the values reported in the literature [3, 5], which can be attributed to the absence of costly ultrafine constituents in this study.

17.3.3 STRAIN LOCALIZATION AND EVOLUTION OF UHPC SPECIMEN UNDER DYNAMIC LOADING

The high-speed camera was programmed at 140,000 fps with an image resolution of 128×104 pixels to understand the strain localization and evolution during dynamic loading. A camera stored all digital images at high speed and later these digital images are processed using software (Vic-2D) to determine the strain field in the specimen.

A specimen of size 48 mm × 19 mm is patterned using a fine-tip marker and placed between the bars (incident and transmit) shown in Figure 17.6(a). A striker bar is initiated and a stress wave propagates through the specimen and bars, which is captured by the strain gauges mounted on both bars. Further computing these strain histories, a tensile stress vs time plot is generated for the tested specimen shown in Figure 17.6(b).

The test is operated with the striker velocity of 11.62 m/s, and it is difficult to visualize the evolution of the crack in the specimen. Here, the images from the high-speed camera make the visualization possible. These digital images are then exported to the software (Vic-2D) and post-processed using DIC parameters. Figure 17.7(a), (b), and (c) shows DIC contours for the vertical normal strain component (ε_{yy}) of the UHPC specimen and also helps in understanding the propagation of the crack.

17.4 CONCLUSION

In this study, a fiber-reinforced UHPC composition without costly ultrafine aggregates was cast, cut into various sample geometries, and tested under tension and compression with in-situ imaging and digital image correlation calculation. In addition, CCCD fracture geometry-based toughness

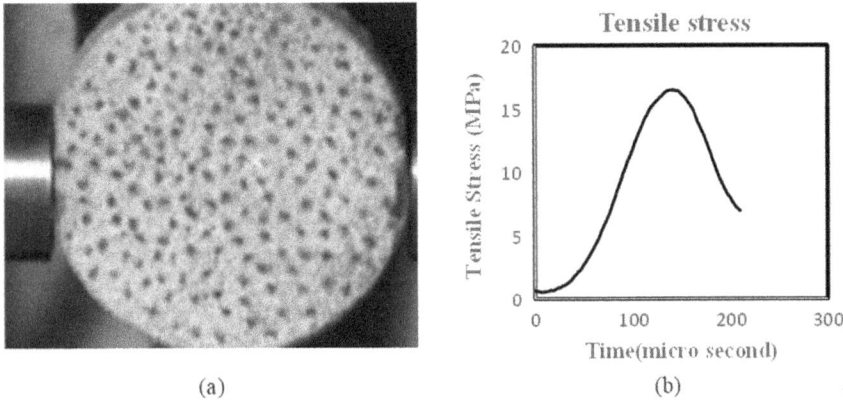

(a)

(b)

FIGURE 17.6   (a) Speckle-patterned Brazilian disc mounted between two bars on SHPB setup; (b) tensile stress vs time plot calculated at the center of the disc.

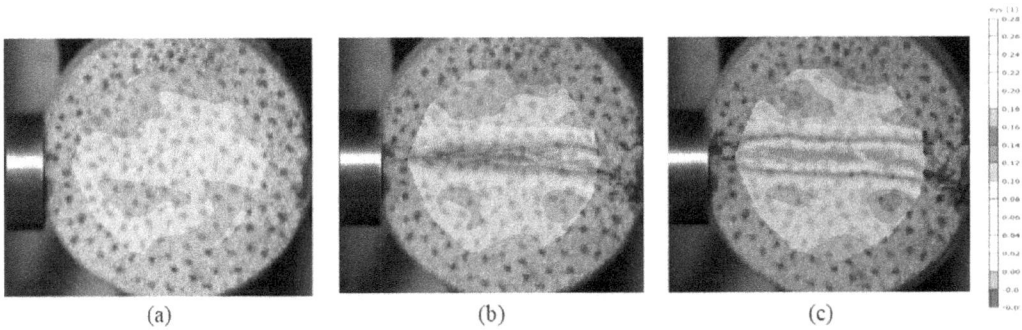

(a)   (b)   (c)

FIGURE 17.7   (a), (b), (c) DIC contours for the vertical normal strain component (ε_{yy}) of a UHPC specimen tested at a loading rate of 44.6GPa/s.

experiments were also conducted to evaluate critical stress intensity factors. From the results of the study, the tensile and compressive strength were 9.778 MPa and 120 MPa, respectively. The curing time period chosen in this work did not result in drastic changes in strength beyond seven days. The fracture toughness at the initiation of the mode I crack propagation calculated was 0.208 MPa m$^{0.5}$. These values obtained are comparable to the literature results [6,7] from similar or large sample sizes. A notable observation from this study with in-situ imaging was that both the tensile failure stress and fracture toughness corresponding to the initial crack do not correspond to the peak load applied at the boundary. Further, the addition of steel fibers has a significant effect not only on the post-peak stress and toughness but also on the peak strength as seen from the load history vs strain contours. Further, the dynamic split tensile experiments showed a significant increase in strength, indicating the strain rate sensitivity of the UHPC material. The dynamic increase factor obtained for this UHPC composition with fibers was 2.42. This value is slightly lower compared to compositions including pozzolanic materials, which leads us to believe that fibers play a far more important role in the strengthening of UHPC.

REFERENCES

1. Shah SP, Swartz SE, Ouyang C. *Fracture mechanics of concrete*. New York: John Wiley & Sons, Inc, 1995.
2. Wu Z. et al. "Investigation of mechanical properties and shrinkage of ultra-high-performance concrete: Influence of steel fiber content and shape," *Composites Part B: Engineering*, vol. 174, Elsevier Ltd, Oct 2019. https://doi.org/10.1016/j.compositesb.2019.107021
3. Hou C, Wang Z, Liang W, Li J, Wang Z. Determination of fracture parameters in center cracked circular discs. China, 2016. http://dx.doi.org/10.1016/j.tafmec.2016.04.006
4. IS 269, Ordinary Portland Cement – Specification, Bureau of Indian Standards.
5. Gurusideswar S, Shukla A, Jonnalagadda KN, Nanthagopalan P. Tensile strength and failure of ultra-high performance concrete (UHPC). India, 2020. https://doi.org/10.1016/j.conbuildmat.2020.119642
6. Wang Q-Z, Xing L. Determination of fracture toughness KIC by using the flattened Brazilian Disc for Rocks China, 1999. https://doi.org/10.1016/S1365-1609(03)00093-5
7. Akbardoost MR, Ayatollahi J. Size and geometry effects on rock fracture toughness: Mode I fracture, 2014. https://doi.org/10.1007/s00603-013-0430-7
8. C496-96, ASTM. Standard Test Method for Splitting Tensile Strength of Cylindrical Concrete Specimens.

18 Dynamic Characterization of Goat Tibia

Ankit Malik, Nishant K. Sahu, Sudipto Mukherjee,
Anoop Chawla and Naresh V. Datla

18.1 INTRODUCTION AND MOTIVATION

18.1.1 NEED FOR DYNAMIC CHARACTERIZATION OF BONE

Bones are subjected to high strain rate loading conditions in traumatic events such as blast injury scenarios and automotive crashes. Tibia [1, 2] and femur [3] are the most frequently injured long bones during a vehicle crash, especially in car-to-pedestrian crashes. For instance, 84% and 37% of the injuries in the tibia and femur, respectively, involve the front bumper as the impacting surface [4]. Gupta emphasized that the most common bony injuries during trauma were the tibia/fibula (41.39%) and femur (41.47%) [5].

During traumatic events, deformation of the bone tissues occurs beyond their failure limits due to high strain rate loading, resulting in altered physiological function or anatomical lesions. Earlier studies focused on the characterization of small machined specimens (cuboid or cylindrical specimens) and provided essential insights into the strain rate effects on the mechanical behavior of bones [6–11]. Mather conducted the drop-weight test on the human femur, concluded that the energy required for a bone fracture depends on the loading rate, and emphasized the need for dynamic testing of bones [12].

18.1.2 METHODS FOR DYNAMIC CHARACTERIZATION OF BONES

Although token experiments provided essential insights into the mechanical behavior of bones, they characterized localized material properties of bone rather than whole bone properties. Machining-induced artifacts also affect such experiments; therefore, whole bone testing is considered the most relevant representation of bone response and failure [13]. Therefore, whole bone specimens are used in this study to understand the dynamic response of the bone under impact. Many studies on whole bone used linear beam theory to estimate the material properties of bone [14]. Lenthe pointed out that beam theory underestimates the tissue modulus by 29% [15]. Linear beam theory completely neglects the yielding and damage of the bone, which are its well-documented behaviors [16–18]. Kourtis estimated correction factors to predict bone material properties using the Euler-Bernoulli beam theory [19]. Zhang compared the simulation results with the experimental result and characterized the material properties of bone using the inverse finite element (FE) method [20]. They concluded that the inverse FE method is a more appropriate method to characterize bone mechanical behavior [20].

Figure 18.1 shows the flowchart of the inverse characterization methodology. Dynamic properties are estimated by optimizing the material parameters of the FE model so that the computational force-time response can match the experimental force-time response. Therefore, it is essential to understand the forces during the dynamic impact scenario. This knowledge will help to minimize the possibility of injury and to design better preventive measures. In this work, only the force-time

DOI: 10.1201/9781003352358-18

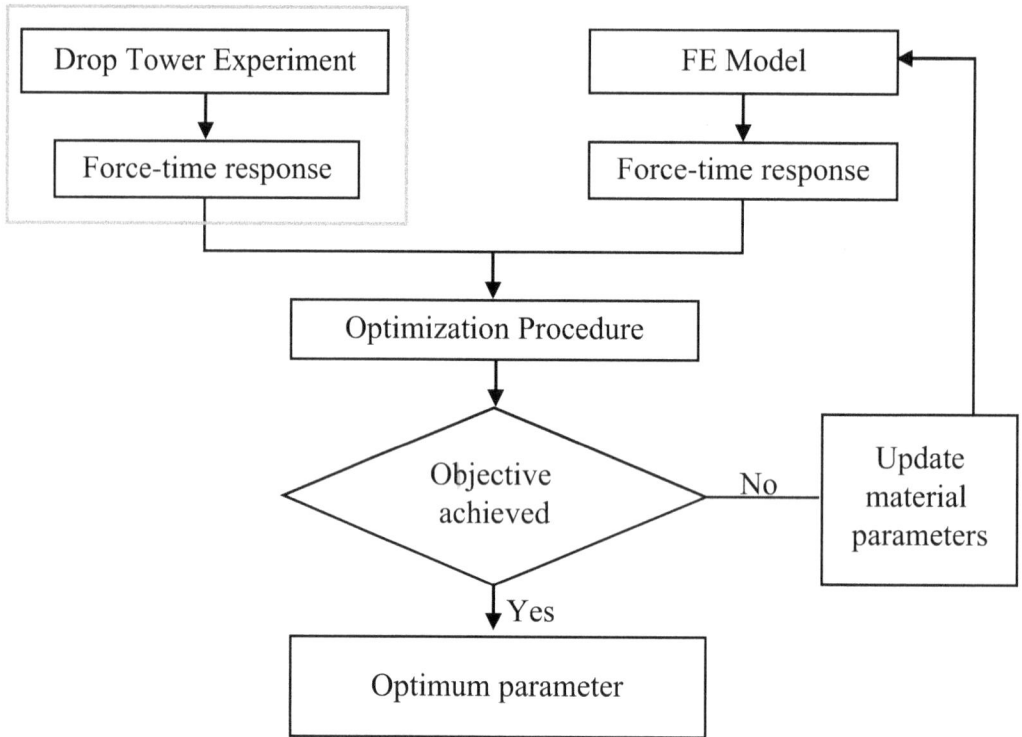

FIGURE 18.1   Inverse characterization methodology.

response (highlighted in Figure 18.1) of goat tibia bones was studied in three-point bending using a drop tower setup with impact speeds up to 3.25 m/s. The results of this work will be used as the training set for the FE model of the bone. The FE model and experimental results will be used to characterize the properties of the bones using the inverse characterization method.

18.1.3   GOAT TIBIA BONE

Due to their inherent inter-specimen variability, large numbers of cadaveric human bone specimens are often needed to conduct meaningful studies [21, 22]. The ethical constraints associated with using cadaveric human bone specimens are the storage, usage, procurement, and disposal of human specimens. Goat tibia bones are used due to their low cost, easy availability, ease of accommodation, and not being subject to extensive ethical procedures.

The goat tibia (Figure 18.2) is the prominent bone of the lower leg, and after the femur bone, it is the second-largest bone in the body. It is located at the front side of the lower leg and consists

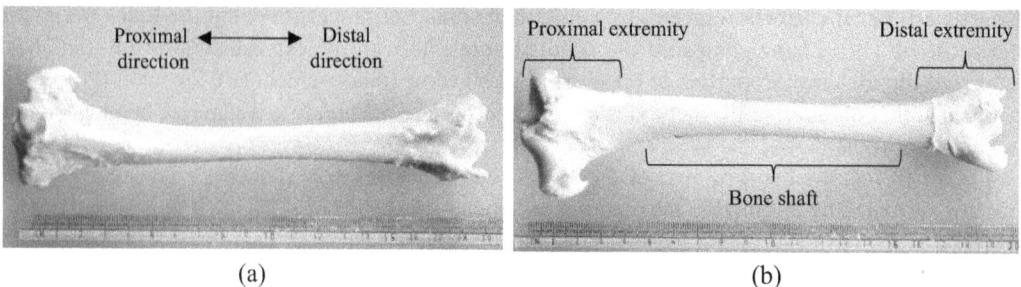

FIGURE 18.2   (a) Anterior view and (b) posterior view of the left goat tibia bone.

of a shaft and two extremities. Its proximal extremity is larger than the distal. The medial side of the distal end, i.e., the side facing towards the body's midline, projects downwards beyond the rest of the bone and is called the medial malleolus. When the medial malleolus is facing medially and observed from the proximal side, the anterior part of the bone shaft is most prominent and is crest-like and sinuously curved, terminating below the anterior border of the medial malleolus.

18.2 METHODOLOGY

18.2.1 Specimen Preparation

Before testing the goat, the bone specimen was wrapped in plastic wraps and stored in a freezer at −20°C. Before testing, the bone was thawed in warm water for 1 h so that soft tissues could be easily removed. All soft tissues (muscle, fat, and tendons) were removed from the bone specimen with particular attention to the bone ends, which were to be potted later. Specific attention was also given to avoid damaging bone and creating artificial stress concentrators. Specimens were regularly sprayed with 0.9% W/V saline water solution to prevent dehydration.

To ensure the fixed orientation of bone specimens during the experiment, they were potted at their proximal and distal ends. A special fixture was used to facilitate the potting procedure. Before the potting process, the bone was wrapped in a cotton gauze bandage and sprayed with 0.9% W/V saline water solution to prevent dehydration due to heat generation from the setting of bone cement. Figure 18.3 shows the steps for specimen preparation. The bone specimen was first clamped in the fixture in pre-defined orientation and alignment. Bone specimens were fixed in hollow aluminum fixtures using quick-setting bone cement (polymethyl methacrylate, PMMA). The bone specimen was clamped in the fixture for 30 min to allow the bone cement to dry. Simply supported boundary conditions were achieved by resting the potting mounts on the rollers mounted on top of two side supports.

18.2.2 Drop Tower Experimental Setup

The three-point bending test is widely used to characterize the mechanical behavior of materials [23]. Figure 18.4 shows a detailed three-dimensional model of the drop tower setup, which is widely used to conduct a three-point bending test (Figure 18.5) on bones [12, 24–27]. A heavy crosshead is dropped from a variable height, falling along two guiding rods. A piezoelectric load cell fixed between the crosshead and the impactor records the load during the impact. Specimens are positioned in the fixed orientation by "potting" their ends as per specific experimental protocol, and the impactor approaches the specimen from the top. After the failure of the specimen, two shock absorbers arrest the fall of the crosshead. The load-time history is recorded using data acquisition systems at 20 kHz sampling rates.

A heavy crosshead of 30.5 kg is dropped from variable heights along two guiding channels. Table 18.1 gives the details of the experiment and goat tibia bone specimens studied in these experiments. Nine tests were performed on the goat tibia at impact velocities of 1.0, 2.0, and 3.25 m/s in the posterior to anterior impact direction at the mid-shaft of the bone. The average span length was 130 mm, and the average weight of bone specimens was 73 g.

18.2.3 Data Acquisition and Processing

The impact load was recorded using a piezoelectric load cell fixed between the crosshead and the impactor. Two piezoelectric load cells were also fixed at the two supports. The load-time history is recorded using the data acquisition system at a sampling rate of 20 kHz. High-frequency components of load-time data were filtered using a Butterworth low-pass filter with a cut-off frequency of 2.5 kHz. The displacement-time history of the test was captured using a high-speed digital camera

(a) Goat Tibia specimen

(b) After soft tissues are removed

(c) Tibia wrapped with wet cotton gauze bandage before potting

Aluminum fixtures

Bone cement

(d) Tibia potted with cement in fixture

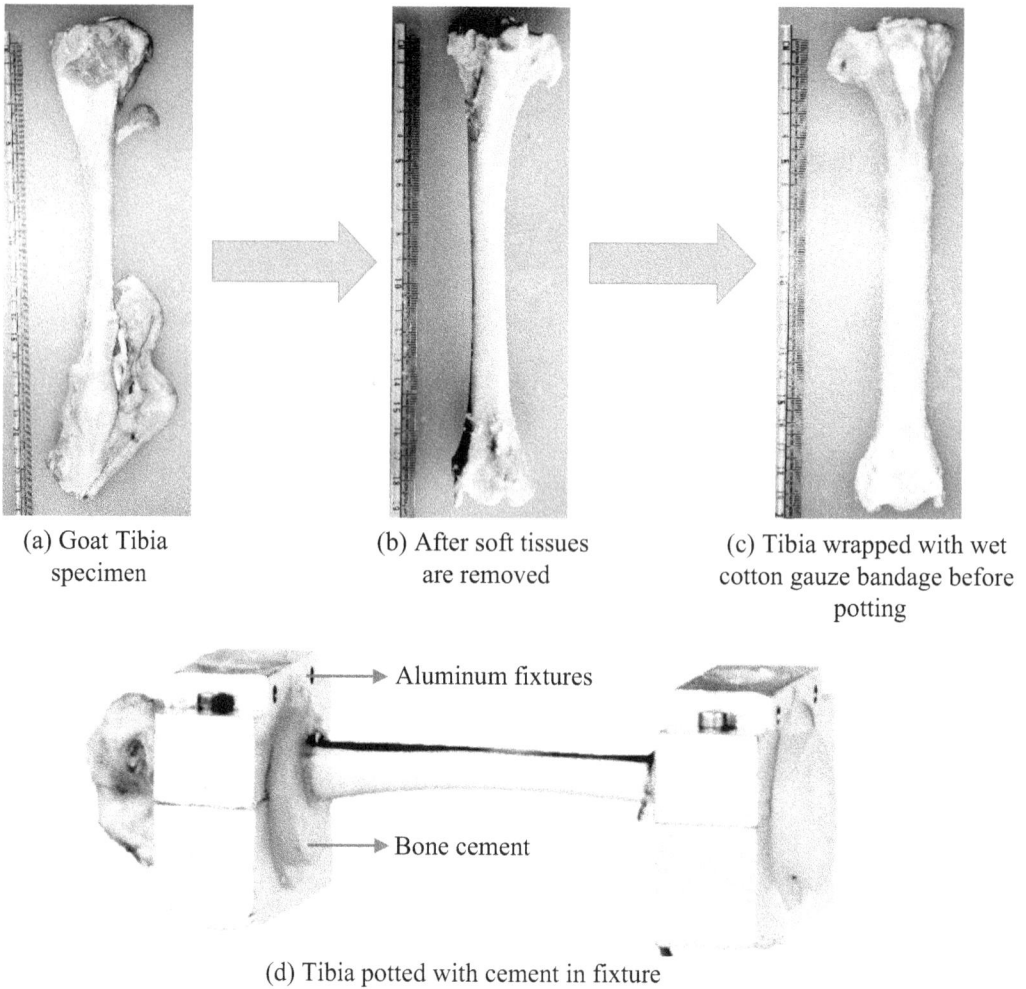

FIGURE 18.3 Steps to prepare the whole bone specimen for impact test.

at a sampling rate of 20,000 frames/s. Figure 18.5 shows the experimental setup of the three-point bending test setup with goat tibia bone. Thin copper tapes were pasted on the bone and bottom side of the impactor to help trigger the data acquisition system and high-speed digital camera. The circuit is triggered when the copper tapes contact each other during the experiment.

18.3 EXPERIMENTAL RESULTS AND DISCUSSION

Figures 18.6, 18.7, and 18.8 show variations of impact force with time and impactor displacement for average impact velocities of 1.17, 2.04, and 3.24 m/s, respectively. All the plots are plotted from the trigger time to the visible fracture initiation time. The fracture initiation time was visually identified from the high-speed videos. A non-linear response in force-time plots was shown by all the goat bones tested in this study, which the non-linearity of the bone tissue material may influence.

Table 18.2 summarizes the observations from all the experiments. The maximum load before failure initiation does not increase with higher impact velocities except for GT 1, GT 4, and GT 9, which have similar physical characteristics. It shows that factors other than impact velocities, like cortical thickness, bone mid-shaft diameter, and bone mass distribution across the bending axis, play an essential role in determining maximum load before failure initiation [28, 29]. For 1.17 m/s

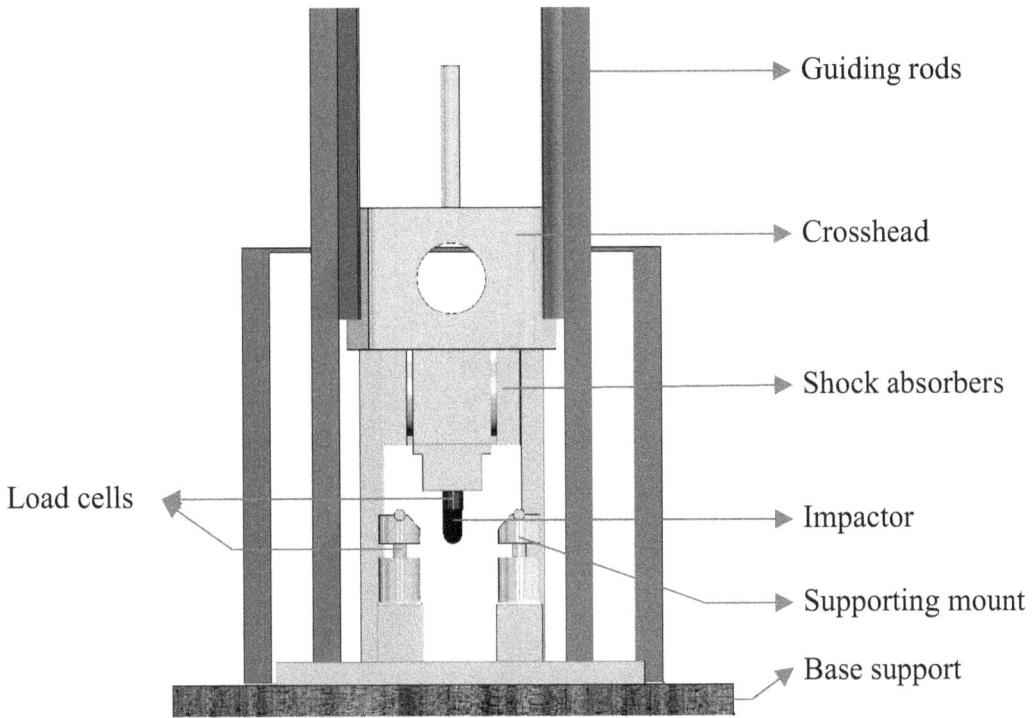

FIGURE 18.4 Detailed model of the drop tower setup.

average impact velocity, the average time to fracture was 4.48 ms. Increasing the impact velocities

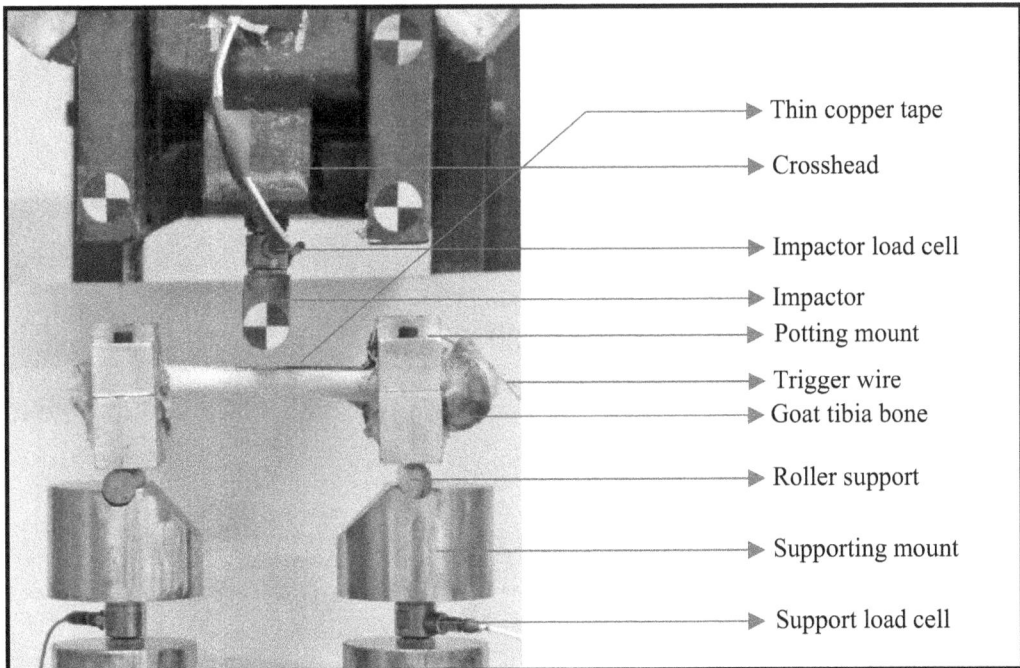

FIGURE 18.5 Three-point bend test setup along with the goat tibia bone.

TABLE 18.1

Bone Length and Weight and the Impact Velocities Used in the Impact Tests

Specimen ID	Bone length (mm)	Bone weight (g)	Impact velocity (m/s)
GT 1	180	77	1.0
GT 2	180	67	
GT 3	180	67	
GT 4	180	77	2.0
GT 5	180	67	
GT 6	180	63	
GT 7	210	83	3.25
GT 8	210	82	
GT 9	180	77	

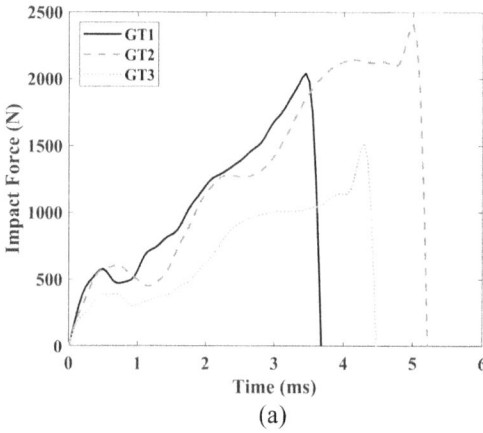

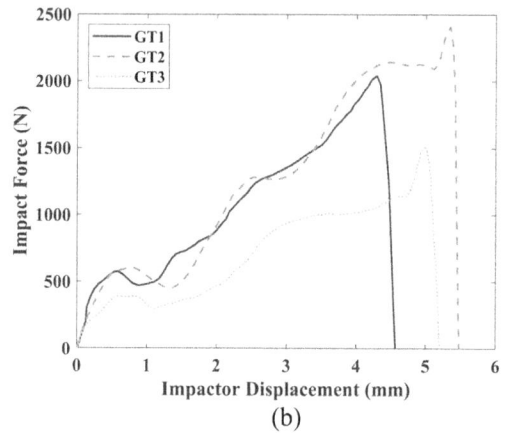

FIGURE 18.6 (a) Impact force vs. time and (b) impact force vs. impactor displacement data for 1.17 m/s average impact velocity.

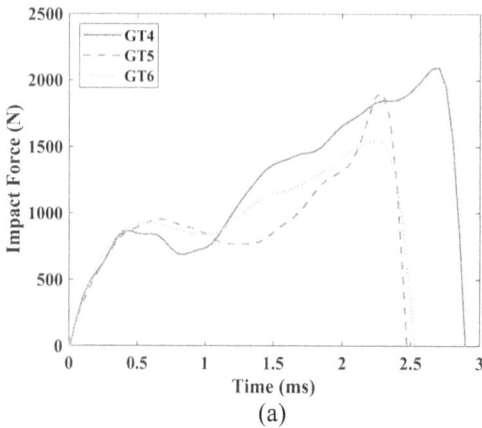

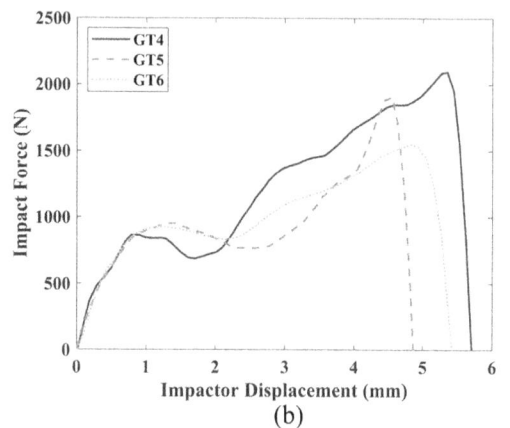

FIGURE 18.7 (a) Impact force vs. time and (b) impact force vs. impactor displacement data for 2.04 m/s average impact velocity.

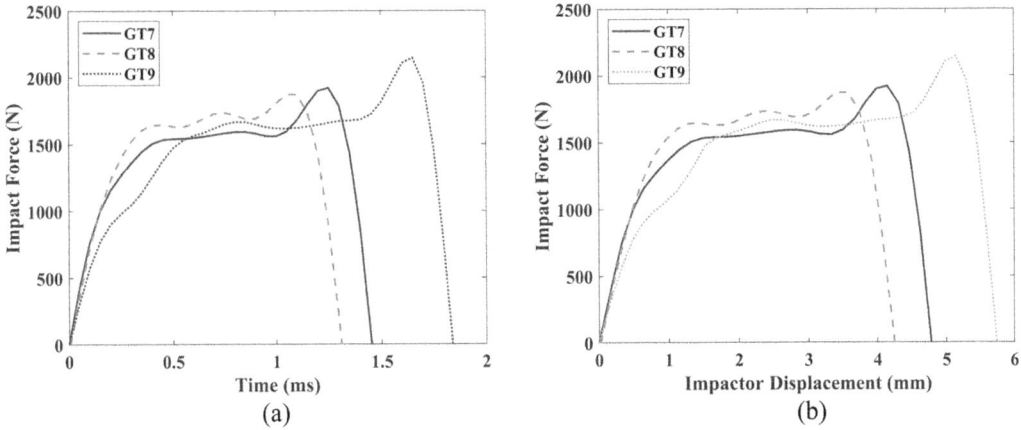

FIGURE 18.8 (a) Impact force vs. time and (b) impact force vs. impactor displacement data for 3.24 m/s average impact velocity.

to 2.04 m/s and 3.24 m/s decreased the average time to fracture to 2.65 ms and 1.58 ms, respectively. For 1.17 m/s average impact velocity, the average energy absorption capacity was 5.13 J. Increasing the impact velocities to 2.04 m/s and 3.24 m/s increased the average energy absorption capacity to 5.7 J and 7.0 J, respectively. The deflection of bone at the point of contact before visible fracture initiation for all cases is $5.17\,mm \pm 0.44$.

Figure 18.9 compares the average impact force vs. time and average impact force vs. average impactor displacement for all average impact velocities. It can be observed in Figure 18.9(a) that with increasing impact velocity, the initial slope (just after the impact) of the average impact force vs. time curve is also increasing. The initial slope of the average impact force vs. time curve was highest at 6.5 kN/ms for the average impact velocity of 3.24 m/s. The lowest slope of 1.7 kN/ms was recorded at the average impact velocity of 1.17 m/s. A similar increase in the initial slope (just after the impact) of the average impact force vs. the average impactor displacement curve is also observed with increasing impact velocity. The highest stiffness (2.2 kN/mm) was recorded at the average impact velocity of 3.24 m/s, while the lowest (1.6 kN/mm) stiffness was observed at the

TABLE 18.2
Experimental Impact Velocity, Time Before Fracture, Maximum Load Just Before Fracture, Deflection of Bone Before Fracture, and Energy Absorption Capacity for All Specimens

Specimen ID	Experimental impact velocity (m/s)	Time before fracture (ms)	Maximum load just before fracture (N)	Deflection of bone before fracture (mm)	Energy absorption capacity (J)
GT 1	1.23	3.7	2,041	4.6	4.8
GT 2	1.12	5.25	2,409	5.5	6.9
GT 3	1.17	4.5	1,521	5.2	3.7
GT 4	2.01	2.9	2,095	5.7	6.9
GT 5	1.98	2.5	1,907	4.9	4.7
GT 6	2.14	2.55	1,549	5.5	5.5
GT 7	3.34	1.55	1,921	4.9	6.7
GT 8	3.25	1.35	1,872	4.4	6.1
GT 9	3.13	1.85	2,145	5.7	8.2

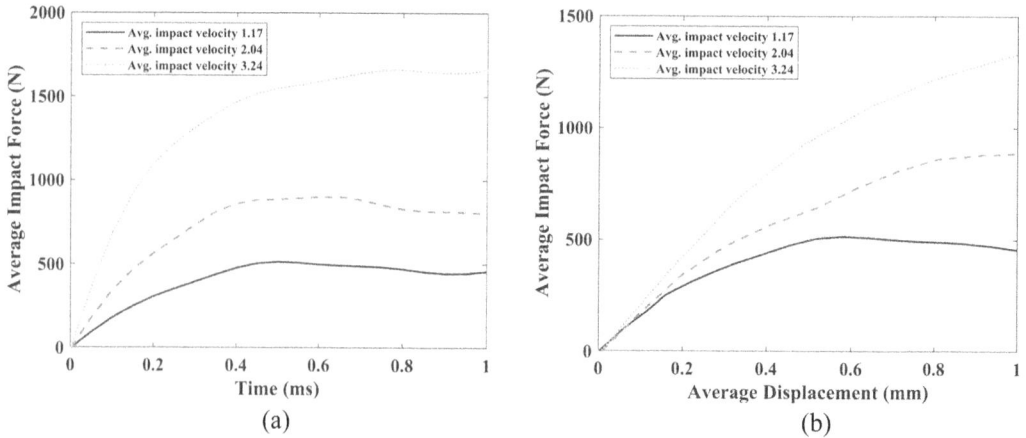

FIGURE 18.9 Comparison of (a) average impact force vs. time and (b) average impact force vs. average impactor displacement for impact velocities of 1.17, 2.04 and 3.24 m/s.

average impact velocity of 1.17 m/s. This increase in the loading rate and stiffness suggests an increase in bone stiffness with increased impact velocity and indicates the strain rate dependency of bone stiffness.

18.4 CONCLUSION

This work has discussed the dynamic behavior of the goat tibia bone subjected to a three-point bending test. Tests were performed on nine specimens at three different impact velocities. A non-linear response in force-time plots was shown by all the goat bones tested in this study, which may be influenced by the non-linearity of the bone tissue material. The fact that the maximum load before failure initiation does not always increase with impact velocities suggests that additional parameters may also play a role in determining the maximum load before failure initiation. Time to visible failure initiation decreased with increased impact velocity, while the average energy absorption capacity increased with increased impact velocities. In all cases, the deflection of bone at the point of contact before visible fracture initiation remains the same with minimal variation. Goat tibia bones exhibited strain rate dependency in stiffness.

ACKNOWLEDGMENTS

The authors acknowledge the financial support received from Defense Research and Development Organization (DRDO) Industry Academia Centre of Excellence (DFTM/03/3203/M/01/JATC).

REFERENCES

1. Chichom-Mefire A, Palle-Ngunde J, Fokam PG, Mokom-Awa A, Njock R, Ngowe-Ngowe M. Injury patterns in road traffic victims comparing road user categories: Analysis of 811 consecutive cases in the emergency department of a level I institution in a low-income country. *Int J Surg Open*. 2018;10:30–36. doi:10.1016/j.ijso.2017.11.005
2. Burns ST, Gugala Z, Jimenez CJ, Mileski WJ, Lindsey RW. Epidemiology and patterns of musculoskeletal motorcycle injuries in the USA. *F1000Research*. 2015;4(May). doi:10.12688/f1000research.4995.1
3. Aloudah AA, Almesned FA, Alkanan AA, Alharbi T. Pattern of fractures among road traffic accident victims requiring hospitalization: Single-institution experience in Saudi Arabia. *Cureus*. 2020;12(1):1–8. doi:10.7759/cureus.6550
4. Ashton SJ, Pedder JB, MacKay GM. Pedestrian injuries and the car exterior. *SAE Tech Pap*. 1977;86:357–374. doi:10.4271/770092

5. Gupta A, Mishra S, Uikey S, Maravi D. A study – Incidence and pattern of musculoskeletal injuries among patients attending the emergency of tertiary health care center in Central India. *J Orthop Dis Traumatol*. 2019;2(1):11. doi:10.4103/2665-9352.263462

6. Crowninshield RD, Pope MH. The response of compact bone in tension at various strain rates. *Ann Biomed Eng*. 1974;2(2):217–225. doi:10.1007/BF02368492

7. McElhaney JH. Dynamic response of bone and muscle tissue. *J Appl Physiol*. 1966;21(4):1231–1236. doi:10.1152/jappl.1966.21.4.1231

8. Hansen U, Zioupos P, Simpson R, Currey JD, Hynd D. The effect of strain rate on the mechanical properties of human cortical bone. *J Biomech Eng*. 2008;130(1):1–8. doi:10.1115/1.2838032

9. Wood JL. Dynamic response of human cranial bone. *J Biomech*. 1971;4(1). doi:10.1016/0021-9290(71)90010-8

10. Wright TM, Hayes WC. Tensile testing of bone over a wide range of strain rates: Effects of strain rate, microstructure and density. *Med Biol Eng*. 1976;14(6):671–680. doi:10.1007/BF02477046

11. Adharapurapu RR, Jiang F, Vecchio KS. Dynamic fracture of bovine bone. *Mater Sci Eng C*. 2006;26(8):1325–1332. doi:10.1016/j.msec.2005.08.008

12. Mather BS. Observations on the effects of static and impact loading on the human femur. *J Biomech*. 1968;1(4):331–335. doi:10.1016/0021-9290(68)90027-4

13. Nalla RK, Kruzic JJ, Kinney JH, Ritchie RO. Mechanistic aspects of fracture and R-curve behavior in human cortical bone. *Biomaterials* 2005; 26(2), 217–231. doi.org/10.1016/j.biomaterials.2004.02.017

14. Kress TA, Snider JN, Porta DJ, Fuller PM, Wasserman JF, Tucker GV. Human femur response to impact loading. *Int Res Counc Biomech Inj*. 1993;21:93–104.

15. van Lenthe GH, Voide R, Boyd SK, Müller R. Tissue modulus calculated from beam theory is biased by bone size and geometry: Implications for the use of three-point bending tests to determine bone tissue modulus. *Bone*. 2008;43(4):717–723. doi:10.1016/j.bone.2008.06.008

16. Currey JD. What determines the bending strength of compact bone? *J Exp Biol*. 1999;202(18):2495–2503. doi:10.1242/jeb.202.18.2495

17. Wright TW, Ramesh KT, Molinari A. Status of statistical modeling for damage from nucleation and growth of voids. *AIP Conf Proc*. 2006;845 I(2006):690–693. doi:10.1063/1.2263416

18. Wright TW, Ramesh KT. Dynamic void nucleation and growth in solids: A self-consistent statistical theory. *J Mech Phys Solids*. 2008;56(2):336–359. doi:10.1016/j.jmps.2007.05.012

19. Kourtis LC, Carter DR, Beaupre GS. Improving the estimate of the effective elastic modulus derived from three-point bending tests of long bones. *Ann Biomed Eng*. 2014;42(8):1773–1780. doi:10.1007/s10439-014-1027-3

20. Zhang G, Xu S, Yang J, Guan F, Cao L, Mao H. Combining specimen-specific finite-element models and optimization in cortical-bone material characterization improves prediction accuracy in three-point bending tests. *J Biomech*. 2018;76:103–111. doi:10.1016/j.jbiomech.2018.05.042

21. Hobatho MC, Rho J and Ashman RB. Atlas of mechanical properties of human cortical and cancellous bone. *J Biomech*. 1992;25:669.

22. Cowin SC. *Bone Mechanics Handbook*. CRC Press; 2001.

23. Lobo H, Lorenzo J. High speed stress-strain material properties as inputs for the simulation of impact situations. IBEC Proceedings, Stuttgart, Germany (1997).

24. Kerrigan JR, Bhalla KS, Madeley NJ, Funk JR, Bose D, Crandall JR. Experiments for establishing pedestrian-impact lower limb injury criteria. *SAE Technical Paper*. 2003:2003-01-0895. doi:10.4271/2003-01-0895

25. Funk JR, Kerrigan JR, Crandall JR. Dynamic bending tolerance and elastic-plastic material properties of the human femur. *Annu Proc Assoc Adv Automot Med*. 2004;48:215–233.

26. Rahmoun J, Naceur H, Morvan H, Drazetic P, Fontaine C, Mazeran PE. Experimental characterization and micromechanical modeling of the elastic response of the human humerus under bending impact. *Mater Sci Eng C*. 2020;117:111276. doi:10.1016/j.msec.2020.111276

27. Schubert A, Erlinger N, Leo C, Iraeus J, John J, Klug C. Development of a 50th percentile female femur model. *Conf Proc Int Res Counc Biomech Inj IRCOBI*. 2021;2021-September:308–332.

28. Gibson LJ. The mechanical behaviour of cancellous bone. *J Biomech*. 1985;18(5):317–328. doi:10.1016/0021-9290(85)90287-8

29. Hart NH, Nimphius S, Rantalainen T, Ireland A, Siafarikas A, Newton RU. Mechanical basis of bone strength: Influence of bone material, bone structure and muscle action. *J Musculoskelet Neuronal Interact*. 2017;17(3):114–139.

19 FE Study on the Effect of Angle of Impact on the Performance of Aluminium Plate against an Ogive Projectile

Vimal Kumar and Aman Kumar

19.1 INTRODUCTION

The speed of innovations and advances in ballistic technology has increased the threat of missiles and projectile attacks on defence vehicles and structures. However, research is in progress for the design of defence systems, shelters and equipment like bulletproof jackets, helmets and vehicles. In most cases, the projectile strikes the target at some angle. The angle of impact, defined as the angle subtended by the velocity vector with the plane of the target, is an important parameter that affects the ballistic resistance of the target. Aluminium alloys, due to their superior mechanical properties like strength, stiffness, durability and lightweightness, are being used in defence systems. Backman and Goldsmith (1978) investigated the impact at different angles for conical and blunt-nosed projectiles (diameter 12.7 mm) with impact velocities ranging from 20 to 1,025 m/s. Tests were performed to evaluate the change in velocity and rotation caused by projectiles impacting target plates at varied obliquity angles. The results confirmed that the obliquity at the exit increased rapidly with increasing impact obliquity for 3.175 mm thick aluminium plates, but it changed linearly for 6.35 mm thick plates of all materials studied. Iqbal et al. (2010) studied the effects of projectile shapes and sizes on 12 mm thick steel targets. In the study, a 20 mm diameter conical-nosed projectile and a 19 mm diameter ogive-nosed projectile were used. The impact velocities of the projectile varied from 229 and 600 m/s. Under normal impact, a circular hole which was almost equal to the diameter of the projectile formed with a circular bulge formation on the back side of the plate (Iqbal et al. 2010). At oblique impacts, the hole and bulge were noticed to have an elliptical shape. The aluminium target failed by the formation of the petal. As the obliquity increased, the top two petals shrank and the bottom two were found to have grown. Iqbal et al. (2017) conducted a numerical investigation to determine the ballistic properties and energy absorption capability of 1 mm thick aluminium 1100-H12 plates. Results showed that projectiles with a flat nose deviated away from the point of contact in the opposite direction from those with an ogive nose. The ballistic limit of a double-nosed bullet is higher than that of an ogive or flat projectile. The ballistic limit of oblique projectiles grew larger with increasing obliquity, while that of other projectiles shrank. In Khaire et al. (2021), the effect of projectile eccentricity and obliquity impact on the ballistic performance of the aluminium shell was investigated against the ogive nose projectile. The result showed that the ballistic performance of the target was influenced by both eccentricity and obliquity. The target was reported to have failed in ductile hole enlargement with the formation of petals; however, the number of petals formed was dependent upon eccentricity and obliquity. Jones and Paik (2012) and Rusinek et al. (2008) explored the energy dissipated by the projectile and concluded that when the velocity was close to the ballistic limit, the target absorbed the maximum kinetic energy of the projectile. The increase in the span

DOI: 10.1201/9781003352358-19

diameter of the target resulted in an increase in the ballistic limit due to more energy absorption and increased global deformation. The penetration trajectory of an ogive-nosed projectile penetrating a semi-infinite aluminum alloy target was studied by Wei et al. (2021). Experiments with two kinds of ogive-nosed projectiles obliquely penetrating targets were conducted with impact velocities ranging from 700 to 1,100 m/s where oblique angles were from 0 to 40°. The results showed that when the oblique angle increased, the deflection angle of the projectile increased rapidly and the stability of the penetration trajectory become poorer. The present study is carried out to determine the ballistic performance of 7075-T6 aluminium alloy subjected to oblique impact by ogive-nosed projectiles with calibre radius head (CRH) 1 and 3 at an impact velocity of 800–1,500 m/s. The ballistic limit, residual velocity, energy dissipation and failure pattern were studied and discussed in subsequent sections.

19.2 CONSTITUTIVE MODELLING

The materials used for the target plate and projectile were AA7075-T6 and steel of grade AISI 4340, respectively. The empirical Johnson-Cook (JC) model (Johnson 1983) was used to model impact and penetration phenomena because of the model's simplicity and suitability. The Johnson-Cook model is a function of von Mises tensile flow stress, following strain hardening, strain rate hardening and thermal softening (Johnson 1983; Sundaram et al. 2020). The model has the advantage of expressing the softening effect of heating. In addition, the Johnson-Cook model can be easily applied to aluminium and steel alloys. In addition, the parameters can be easily obtained by a few experiments and are simple to implement (Johnson and Cook 1985). The equivalent von Mises stress ($\tilde{A}$) in this model can be expressed by equation (19.1) shown below:

$$\bar{\sigma} = \left[A + B\left(\bar{\varepsilon}^{pl}\right)^n \right]\left[1 + C\ln\frac{\dot{\bar{\varepsilon}}^{pl}}{\dot{\varepsilon}_0} \right]\left[1 - \hat{T} \right] \tag{19.1}$$

Where $\bar{\varepsilon}^{pl}$ is the equivalent plastic strain; A, B, n, C and m are the specific material parameters; $\frac{\dot{\bar{\varepsilon}}^{pl}}{\dot{\varepsilon}_0}$ is the dimensionless plastic strain rate, where $\dot{\bar{\varepsilon}}^{pl}$ is the equivalent plastic strain rate; and $\dot{\varepsilon}_0$ is the user-defined reference strain rate. The homologous temperature $\hat{T}$ is the non-dimensional temperature defined in equation (19.2):

$$\hat{T} = \frac{\left(T - T_0\right)}{\left(T_{melt} - T_0\right)}, \quad T_0 \leq T \leq T_{melt} \tag{19.2}$$

where T is the absolute temperature, T_0 is the room temperature and T_{melt} is the melting temperature. The fracture model proposed by Johnson and Cook (1985) takes into account the effect of stress triaxiality, strain rate and temperature on the equivalent fracture strain; see equation (19.3).

$$\bar{\varepsilon}_f^{pl} = \left[D_1 + D_2 \exp\left(D_3 \frac{\sigma_m}{\bar{\sigma}}\right) \right]\left[1 + D_4 \ln\left(\frac{\dot{\bar{\varepsilon}}^{pl}}{\dot{\varepsilon}_0}\right) \right]\left[1 + D_5 \hat{T} \right] \tag{19.3}$$

where D_1, D_2, D_3, D_4 and D_5 are material damage parameters, $\frac{\sigma_m}{\bar{\sigma}}$ is the stress triaxiality ratio and σ_m is the mean stress.

19.3   FINITE ELEMENT MODELLING

The finite element simulations were performed in ABAQUS/Explicit to determine the ballistic performance of the aluminium alloy 7075-T6 plate. The aluminium plate and projectile were modelled as three-dimensional deformable bodies. The target plate having the size of 100 mm × 100 mm × 6 mm was impacted under oblique impact at angles of 75°, 60° and 45° by steel bullets; see Figure 19.1. The ogive-shaped projectiles with CRH values 1 and 3 and a constant diameter of 7.62 mm were considered. The fixed boundary condition was assigned to all four edges of the target plate. The contact between the projectile and the target was modelled as surface-to-surface contact. The target region of the plate was meshed with 0.15–1.0 mm solid elements. The meshing of the plate and the projectile is shown in Figure 19.2(a) and (b), respectively.

19.4   RESULTS AND DISCUSSION

19.4.1   BALLISTIC LIMIT

It was found that the nose shape of the projectile and the angle of impact had a noticeable influence on the ballistic resistance of the aluminium plate. As shown in Figure 19.3, the velocity at the ballistic limit for the CRH3 projectile was decreased from 10 to 6% as compared to CRH1 projectiles within a difference of 50–100 m/s at different impact angles, i.e., from 45° to 60°. It was affirmed that the residual velocity for the CRH1 projectile was higher than that of the CRH3 projectile for all impact angles. The ballistic limits of the CRH1 and CRH3 projectiles at 90° impact angle were 750 and 850 m/s respectively. It was affirmed from this observation that the ballistic limit of the CRH3 projectile at 75° and 90° impact angles was found to be the same. In general, the ballistic limit from 75° to 90° did not vary so much.

Figure 19.4 shows the initial impact velocity versus residual velocity data points and fitting curves obtained from the simulations for CRH1 and CRH3 projectiles. In general, it was observed that, at the highest impact velocity, the velocity drop was found to be 10–20% and 13–21% with a decrease in angle of impact, i.e., from 75° to 45° for CRH1 and CRH3 projectiles respectively. On the other hand, at the ballistic limit, the velocity drop was found to be 55–85% and 86–70% with a decrease in angle of impact, i.e., from 75° to 45° for CRH1 and CRH3 projectiles respectively. At higher impact velocities, the difference between the residual velocity was relatively smaller which increased with a reduction in impact velocity. It was found that the residual velocity was always highest in the case of a 90° impact angle and it was found to be almost the same up to 75°.

19.4.2   ENERGY ABSORPTION

The energy absorption in the target is one of the important parameters that implicate its ballistic resistance when subjected to a high-velocity impact. A projectile carries a large amount of kinetic energy that dissipates during the impact process by the deformation of the target as well as the projectile. The process of energy absorption also led to the development of different deformation modes under ballistic impact. Figure 19.5 shows the energy dissipated under impact against various impact energies. It was observed that the energy dissipation was higher for the CRH3 projectile as compared to the CRH1 projectile. Also, it was observed that the energy absorption increased with the increasing obliquity of the projectile. The least energy absorption of kinetic energy occurred at 75°. That means the energy losses get minimised due to less contact force in case of smaller obliquity. The pattern of energy absorption was not found to be linear with angle variation; as can be seen in Figure 19.5, there was a non-linear increase in energy absorption that occurred below a 60° impact angle. The energy absorption also has an inverse relation with the impact energy of the projectile. In general, the impact energy absorption was found to be decreasing with an increase in the kinetic energy of the projectile.

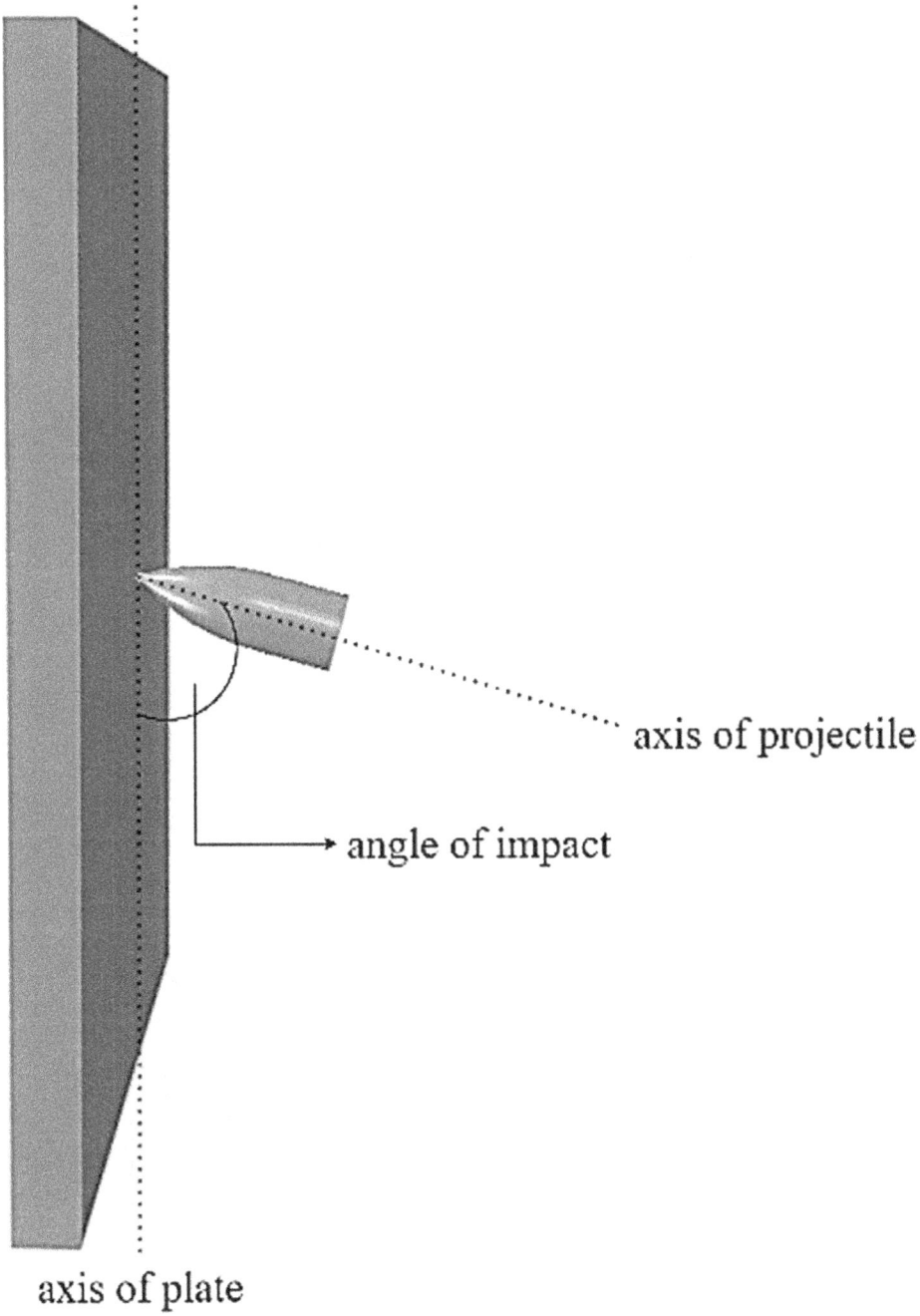

axis of projectile

angle of impact

axis of plate

FIGURE 19.1   Geometry showing the angle of impact.

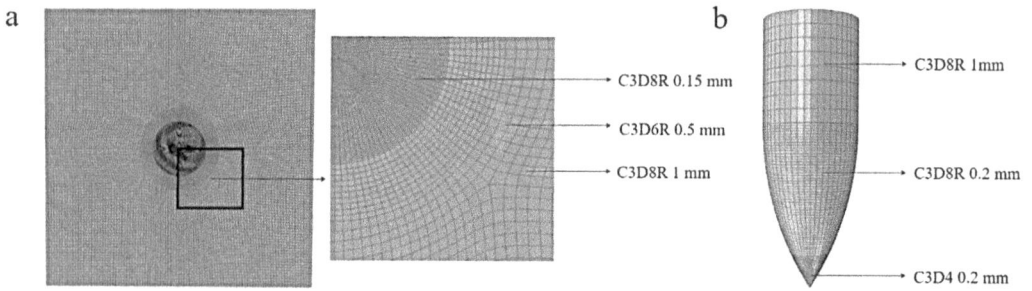

FIGURE 19.2 Meshing in the finite element model: (a) target plate and (b) projectile.

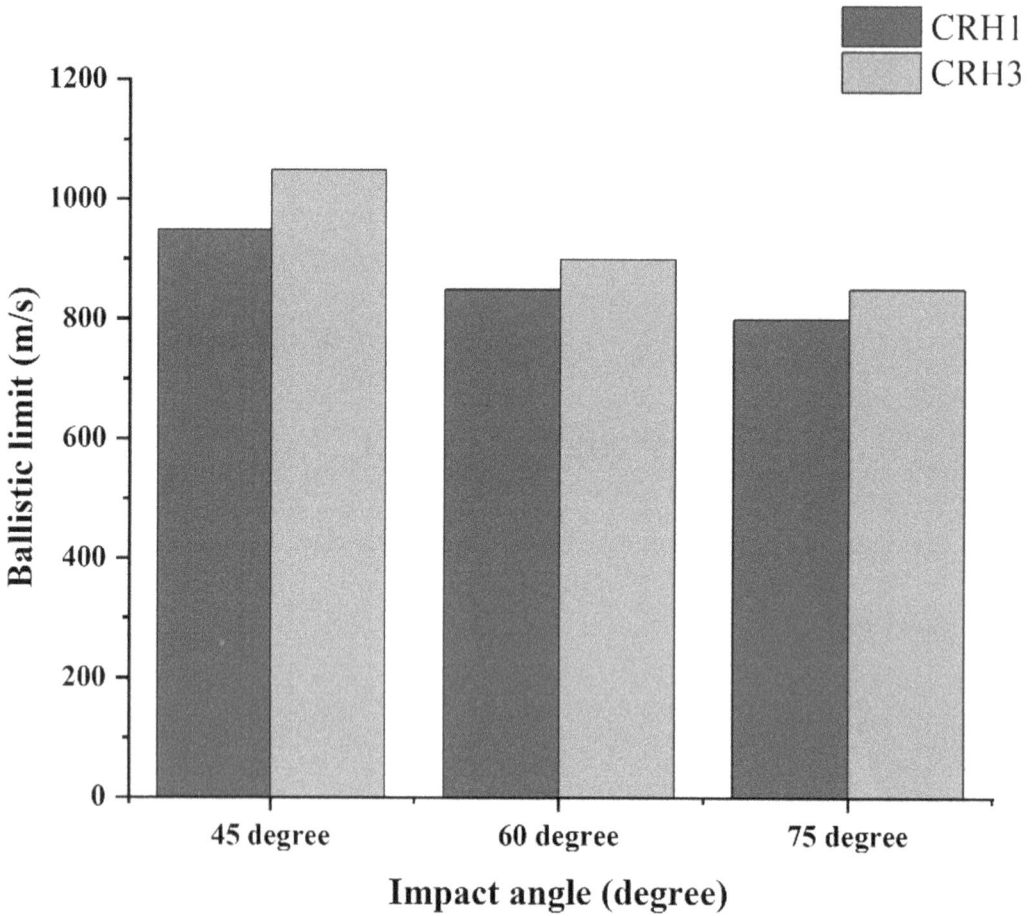

FIGURE 19.3 Ballistic limits for various impact angles of CRH1 and CRH3 projectiles.

19.4.3 FAILURE PATTERN

Figure 19.6 shows the failure patterns of the aluminium plate which indicates that failure occurs with the formation of petals irrespective of the obliquity. Under the oblique impact, the projectile deviates from its central axis during perforation. While entering the target, it deviates away from the normal plate and while leaving the target it deviates towards the normal plate, and as a consequence,

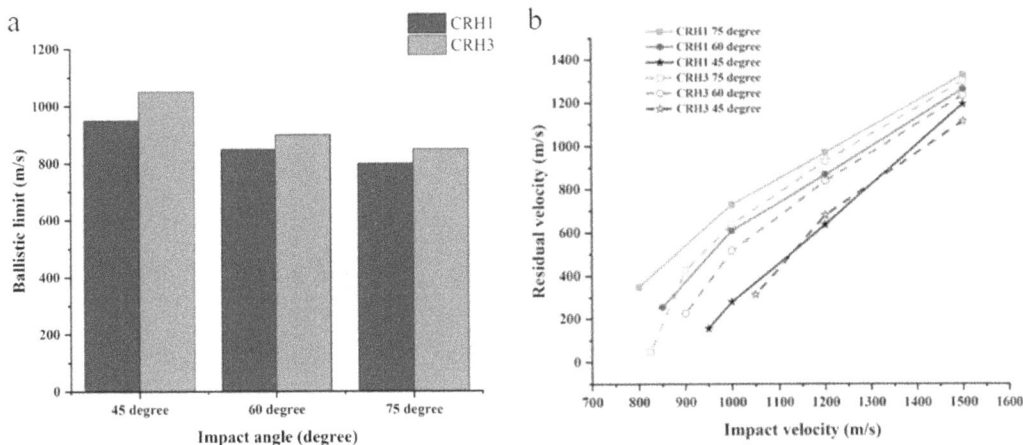

FIGURE 19.4 Impact versus residual velocity curves for CRH1 and CRH3 projectiles.

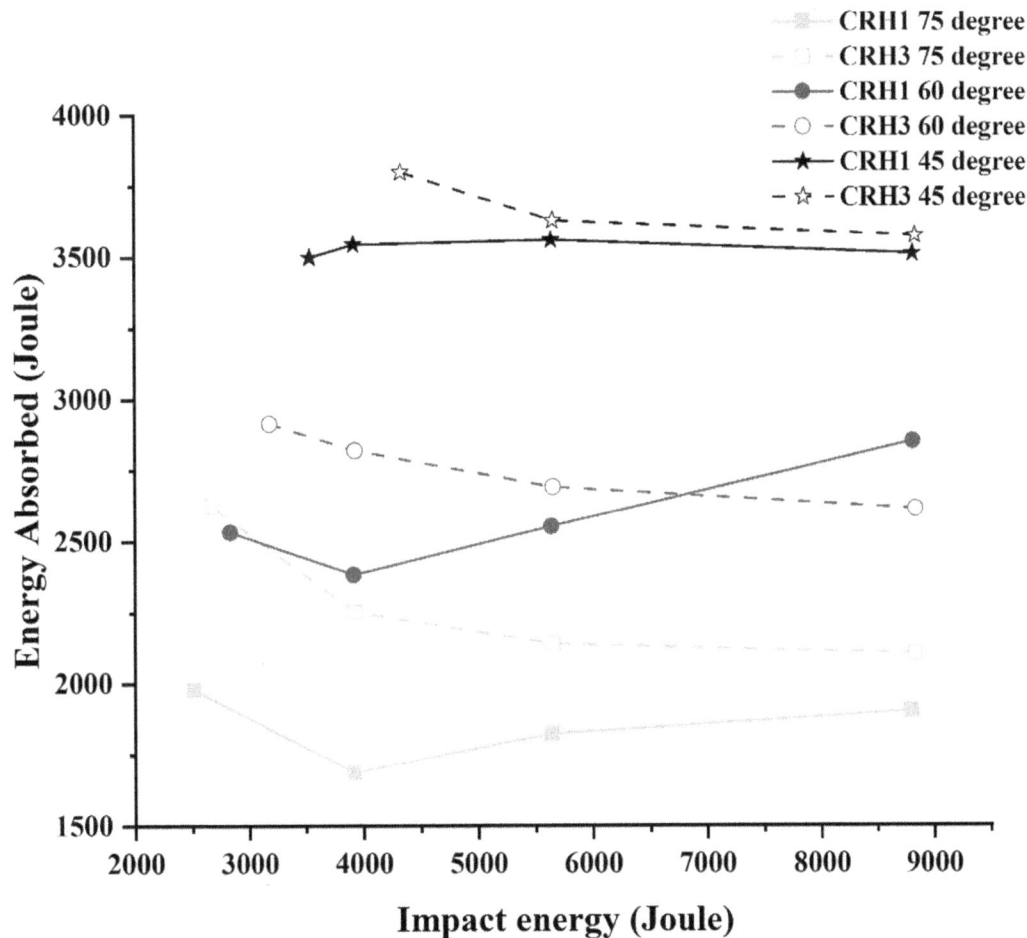

FIGURE 19.5 Energy absorption by the target for CRH1 and CRH3 projectiles.

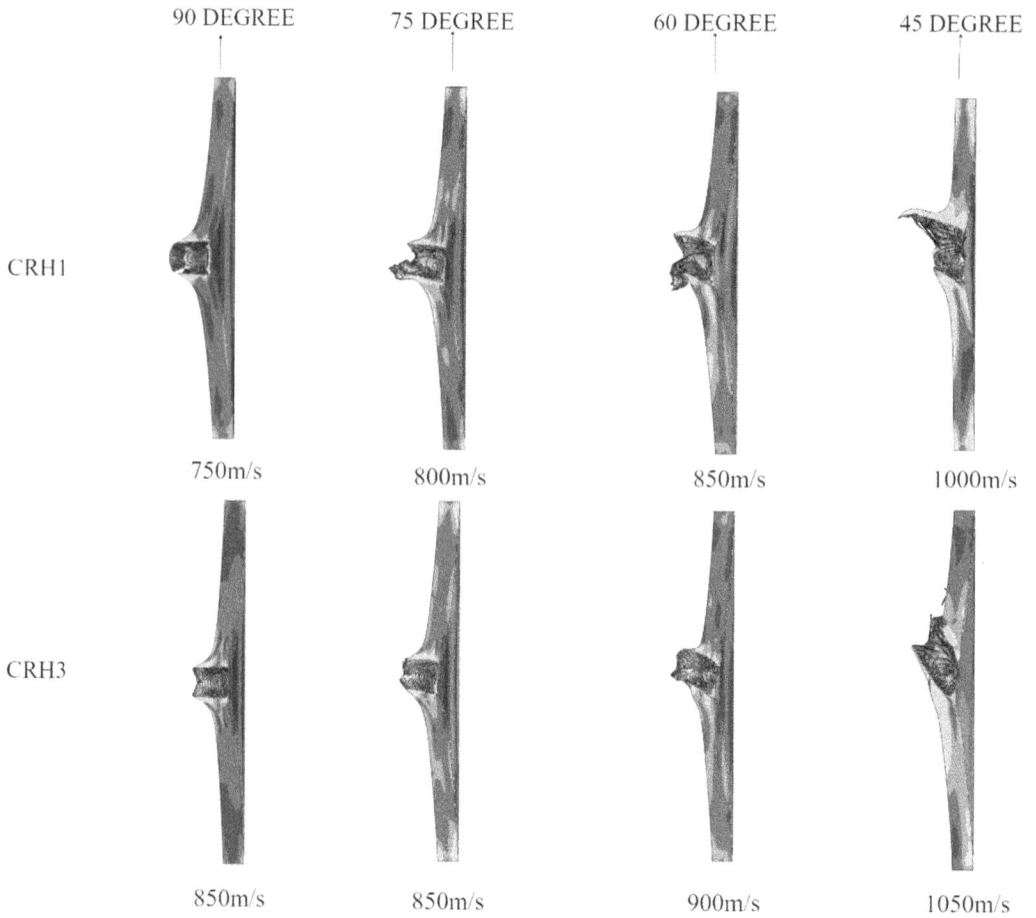

FIGURE 19.6   Failure pattern of plates at the ballistic limit for impact angles 90°, 75°, 60° and 45°.

the final angle of obliquity at the exit becomes lesser than that of the initial obliquity. Similar findings were noticed in the available study by Iqbal et al. (2010). It can be seen that more damage was caused by the CRH1 projectile as compared to the CRH3 projectile.

19.4.4   DAMAGE PROFILE

Figure 19.7 indicates the damage profile of the plates impacted due to different CRH projectiles at the exit of the projectile. The outward bulge on the back side of the plate was observed and the size of this bulge was found to be increasing as the angle of obliquity impact was increased. In general, the shape of the perforation hole was changed from circular to ovular or elliptical configuration with the increase in impact obliquity. The size of the perforation also increased from 75° to 45° impact. The CRH1 projectile caused more damage to the plate.

19.5   CONCLUSION

The numerical study was performed to determine the ballistic performance of the aluminium plate subjected to oblique impact by ogive-nosed projectiles with an impact velocity of 700–1,500 m/s. The effects of projectile nose shape, as well as the angle of impact and velocity, were studied. The plates were impacted by an AISI 4340 steel projectile with CRH values 1 and 3 at different impact

45 DEGREE   75 DEGREE

CRH1

1000m/s   800m/s

CRH3

1050m/s   850m/s

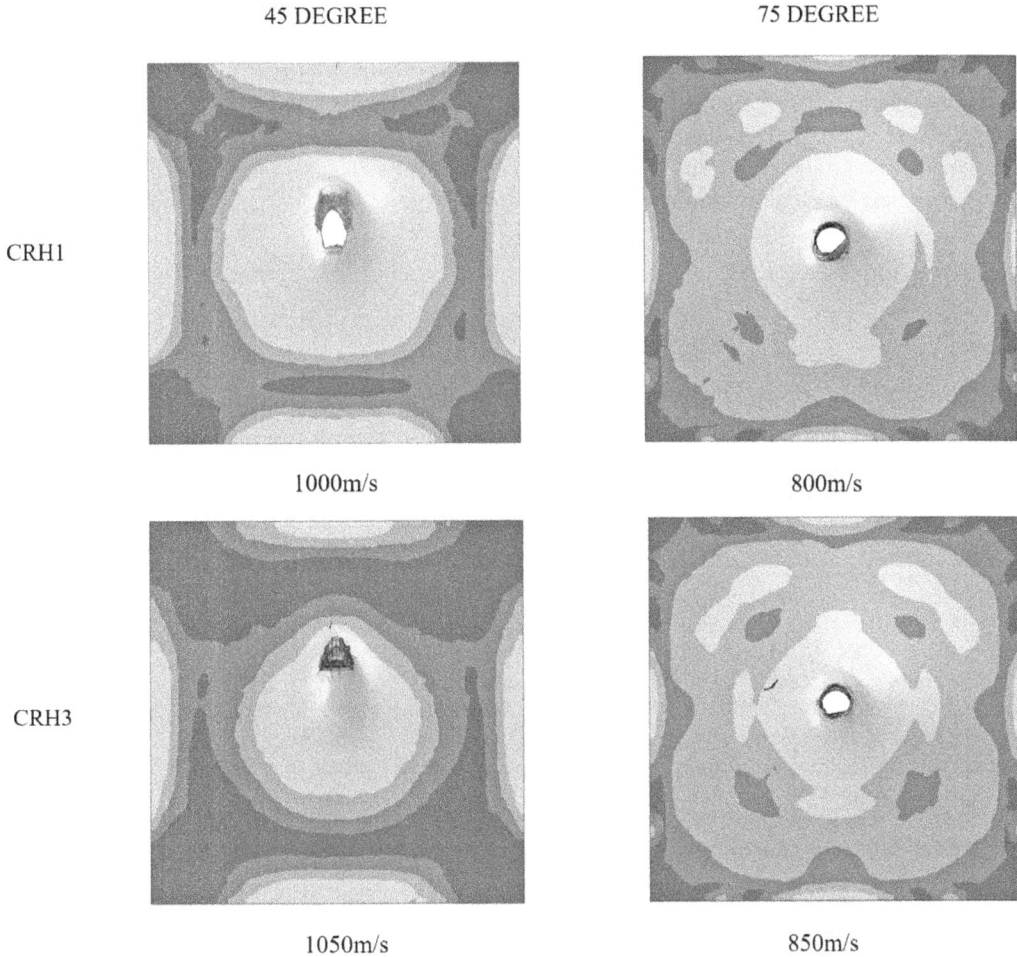

FIGURE 19.7 Damage profile of plates at the ballistic limit for impact angles 90°, 75°, 60° and 45°.

angles, i.e., 75°, 60° and 45°. The comparisons were also made with the 90° (normal) impact. The ballistic limit, residual velocity, impact energy and damage profile were discussed. The following conclusions are made based on this study:

1. The velocity at the ballistic limit for CRH3 projectiles was decreased by up to 10% as compared to CRH1 projectiles. The reduction was up to 100 m/s. The ballistic limit of the CRH1 projectile was increased by 10–15% with an increase in the angle of impact from 45° to 75°. At 75°, the lowest ballistic limit, i.e., 800 m/s for CRH1 and 850 m/s for CRH3 projectiles, was observed.

2. The velocity drop of the projectile is found to be relatively faster with the reduction in impact velocity. The exit velocity of the projectile has increased with an increase in impact velocity.

3. The target plate absorbed more energy for the CRH3 projectile compared to the CRH1 projectile. Also, energy absorption increased with increasing obliquity. The pattern of energy absorption was shown to be non-linear with angle variation as a non-linear increase in energy absorption by the target plate occurred at an impact angle of less than 60°.

4. The energy absorbed by the target plate was found to have an inverse relationship with the projectile's impact energy. In general, the energy absorption was shown to decrease as the projectile's kinetic energy increased.

5. The damage caused by the CRH1 projectile was noticed to be higher as compared to the CRH3 projectile. The shape of the perforation hole changed from circular to ovular/elliptical configuration with the increase in impact obliquity. The size of the perforation also increased with a reduction in impact angle from 75° to 45°.

ACKNOWLEDGEMENTS

The authors are highly grateful to Dr Abhishek Rajput, IIT Indore, and Dr Venkatesan J., CSIR-SERC Chennai, for helping to carry out simulations.

REFERENCES

Backman, M.E. and Goldsmith, W., 1978. The mechanics of penetration of projectiles into targets. *International Journal of Engineering Science*, 16 (1), 1–99.

Iqbal, M.A., Gupta, G., and Gupta, N.K., 2010. 3D numerical simulations of ductile targets subjected to oblique impact by sharp nosed projectiles. *International Journal of Solids and Structures*, 47 (2), 224–237.

Iqbal, M.A., Senthil, K., Madhu, V., and Gupta, N.K., 2017. Oblique impact on single, layered and spaced mild steel targets by 7.62 AP projectiles. *International Journal of Impact Engineering*, 110, 26–38.

Johnson, G.R., 1983. A constitutive model and data for materials subjected to large strains, high strain rates, and high temperatures. *Proc. 7th Inf. Sympo. Ballistics*, 541–547.

Johnson, G.R. and Cook, W.H., 1985. Fracture characteristics of three metals subjected to various strains, strain rates, temperatures and pressures. *Engineering Fracture Mechanics*, 21 (1), 31–48.

Jones, N. and Paik, J.K., 2012. Impact perforation of aluminium alloy plates. *International Journal of Impact Engineering*, 48, 46–53.

Khaire, N., Tiwari, G., and Iqbal, M.A., 2021. Effect of eccentricity and obliquity on the ballistic performance and energy dissipation of hemispherical shell subjected to ogive nosed. *Thin-Walled Structures*, 161, 107447.

Rusinek, A., Rodríguez-Martínez, J.A., Arias, A., Klepaczko, J.R., and López-Puente, J., 2008. Influence of conical projectile diameter on perpendicular impact of thin steel plate. *Engineering Fracture Mechanics*, 75 (10), 2946–2967.

Sundaram, S.K., Bharath, A.G., and Aravind, B., 2020. Influence of target dynamics and number of impacts on ballistic performance of 6061-T6 and 7075-T6 aluminum alloy targets. *Mechanics Based Design of Structures and Machines*, 50(3), 993–1011

Wei, H., Zhang, X., Liu, C., Xiong, W., Chen, H., and Tan, M., 2021. Oblique penetration of ogive-nosed projectile into aluminum alloy targets. *International Journal of Impact Engineering*, 148, 103745.

20 Ballistic Performance of Oblique Impact on Monolithic and Layered in-Contact Metallic Targets against Ogival-Shaped Projectiles

Manoj Kumar and Nidhi Kumari

20.1 INTRODUCTION

Many engineering applications use normal and oblique impacts, such as vehicle crashworthiness, the design of lightweight body armor, and domestic usage. Many parameters affect the failure mode and ballistic resistance of thin metallic targets, including target and projectile material, shape of projectiles, incidence velocity of the projectile, and impact velocity. In place of a monolithic or single-plate design, multilayer configurations containing several parallel plates are proposed. For multilayered thin plates, the plates' thickness, sequence, and numbering affect the failure modes of the target, which vary the ballistic resistance between different target configurations. Several studies on the ballistic behavior of multilayered target plates have been conducted, but compared to monolithic target plates, their ranges were limited. Furthermore, the research on multilayered target plates is still in progress. Awerbuch et al. [1] investigated the impact of projectiles on a metallic target plate in an experimental and analytical study. The experiments were conducted using a 0.22-in caliber lead bullet impacted on pure aluminum and aluminum alloy with a thickness of 2.0 to 6.0 mm target plate. The authors' earlier model for normal perforation was adjusted to include the influence of impact angle and the experimental finding showed satisfactory agreement. The goal of this research was to see how the velocity drop changed with an obliquity for a projectile and target with certain mechanical and physical parameters. Goldsmith and Finnegan [2] showed experimentally the normal and oblique impact in aluminum and steel plates. Various angles of obliquity were studied while cylindrical and cone-shaped projectiles were used to impact target plates. The target damage, which consisted of dishes, petal plug, and band separated from the crater, was observed, and it was found that the petal formation pattern completely changed with obliquity in target plates.

Corran et al. [3] investigated the penetration of different thicknesses of steel and aluminum plates. They looked for two projectiles, first blunt and then cylindroconical projectiles, and found that the plate's ballistic resistance varies depending on the projectile's mass and nose configuration. According to Radin and Goldsmith [4], layered aluminum and polycarbonate targets offer less impact resistance than monolithic target plates. According to Madhu et al. [5], layered steel targets have higher ballistic resistance that is comparable to monolithic targets of equal thickness. Senthil and Iqbal [6] concluded that monolithic aluminum target plates show enhanced ballistic performance compared to double-layer in-contact target plates struck by ogive nosed projectiles.

Zhou and Stronge [7] studied experimentally and numerical analysis on monolithic and multilayer thin metallic targets hit by hemispherical and blunt projectiles at various angles of obliquity.

DOI: 10.1201/9781003352358-20

They discovered that layered targets absorb more energy as compared to monolithic targets for blunt projectiles, but hemispheric targets behave differently.

Several studies were done on oblique impact; Gupta and Madhu [8, 9] and Iqbal et al. [10, 11] analyzed the influence of obliquity on metallic targets and found that obliquity had a considerable impact on target ballistic resistance.

The numerical analyses are carried out on Q235 steel over an ogival-nosed projectile with normal and oblique impact in the proposed investigation. The impact velocity of the projectile changes between 134.93 and 500.00 m/s. All of the simulations are solved using an ABAQUS/Explicit solver. This research determines the target plates' residual velocity and energy absorption during impact.

20.2   METHODOLOGY

A three-dimensional finite element model of a 2 mm thick monolithic and layered target of Q235 steel subjected to an ogival projectile with a diameter of 12.62 mm and 34.5 g mass is modeled using the ABAQUS/CAE module. The layered target has been configured in a double layer, and each layer's thickness is 1 mm. The projectile is modeled as a rigid body because of negligible deformation of the projectile upon impact whereas the target plate is modeled as a deformable body. Figure 20.1 depicts a typical 3D FE model of a target and a projectile.

The 'encastré' boundary condition keeps the target plate at the peripheral fixed. The surface-to-surface contact or node-to-surface contact is introduced between the target plate and projectile and is modeled utilizing a kinematic contact algorithm which is defined in ABAQUS/CAE2020 [13] in which the target plate the contact region is defined as a node-based slave surface and the projectile's outer surface is assigned as the master surface.

The plate element type is C3D8R. To save computational time, the target is partitioned into inner and outside contact zones. However, because significant plastic deformation and cracks occur in this area, the target's impact region meshes elaborately, as seen in Figure 20.1. The target plate is

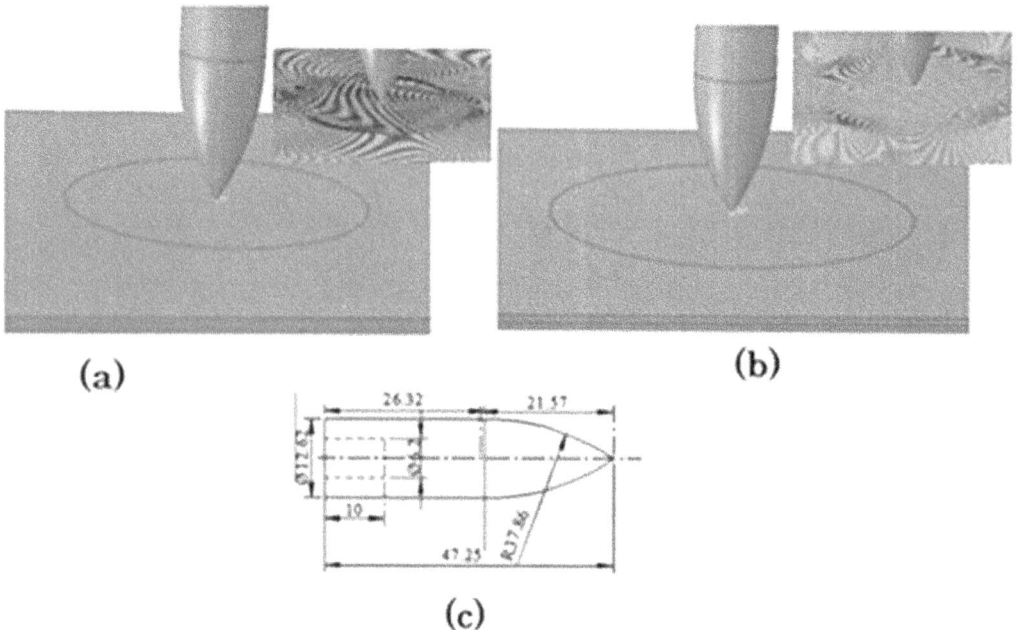

FIGURE 20.1   FE model for target: (a) monolithic, (b) layered, (c) dimensions of ogival-nosed projectile.
Source: [12]; https://drive.google.com/file/d/18EtIn883XNttcn6yBnmbWPijsxDozovK/view?usp=share_link.

meshed up of uniform elements with a one-to-one aspect ratio and a size of $0.3 \times 0.3 \times 0.3\ mm^3$. As illustrated in Figure 20.1, in the plate's impact region, mesh size is selected by performing a mesh convergence analysis. The simulations are performed using 100, 150, 200, 250, and 300 elements in the target, corresponding to three elements in the impact region at an impact velocity of 126.22 m/s.

Figure 20.2 shows that after 250 elements in mesh, residual velocity is nearly constant. In conclusion, the target impact zone's number of elements is considered to be 250, and the matching element $0.3 \times 0.3 \times 0.3\ mm^3$ is chosen in this analysis.

20.2.1 CONSTITUTIVE MATERIAL MODELING

The Johnson-Cook constitutive material model is utilized to analyze the characteristic of a material of the Q235 steel target [14] and is included with ABAQUS. The Johnson-Cook model's equivalent von Mises stress is defined as follows:

$$\sigma_{eq} = \left[A + B\left(\varepsilon_{eq}\right)^n \right]\left[1 + C\ln\dot{\varepsilon}_{eq}^* \right]\left[1 - \hat{T}^m \right] \tag{20.1}$$

Here A, B, n, C, and m are defined as material characteristics identified through mechanical testing, and the dimensionless strain rate is denoted by the symbol $\dot{\varepsilon}_{equivalent}^* = \dfrac{\dot{\varepsilon}_{equivalent}}{\dot{\varepsilon}_a}$. Here $\dot{\varepsilon}_a$ is the user-defined strain rate and $\hat{T}$ is introduced as a nondimensional temperature characterized as:

$$\hat{T} = \left(T - T_a\right)/\left(T_{melting} - T_a\right) \qquad\qquad T_a \le T \le T_{melting} \tag{20.2}$$

Where T is room temperature, $T_{melting}$ denotes melting point temperature, and T_a denotes ambient temperature.

Johnson and Cook's fracture criterion model accounts for the effects of strain rate, strain triaxiality, and temperature on equivalent fracture strain. At the point when the damage parameter reaches unity, the material is damaged.

According to the fracture criterion:

$$\omega = \sum\left(\frac{\Delta\varepsilon_{equivalent}}{\varepsilon_f}\right) \tag{20.3}$$

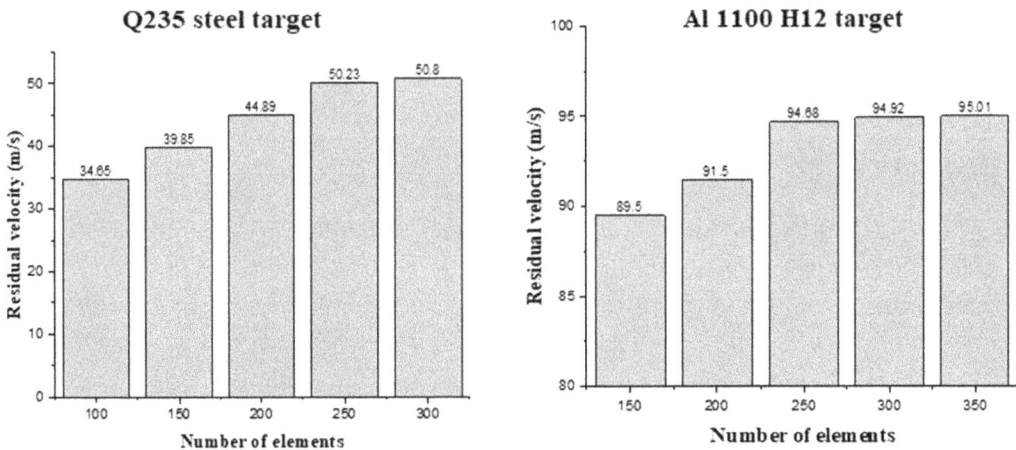

FIGURE 20.2   Mesh convergence test at 126.22 m/s incidence velocity.

Where $\Delta\varepsilon_{\text{equivalent}}$ is the proportional increase in equivalent plastic strain ε_f, which is known as the strain caused by failure.

The fracture strain is determined as follows:

$$\overline{\varepsilon}_f = \left[D_1 + D_2 \exp\left(D_3 \frac{\sigma_a}{\overline{\sigma}} \right) \right]\left[1 + D_4 \ln\left(\dot{\varepsilon}^*_{\text{eq}} \right) \right]\left[1 + D_5 \hat{T} \right] \tag{20.4}$$

Where D_1, D_2, D_3, D_4, D_5 are the material parameters, and $\dfrac{\sigma_a}{\overline{\sigma}}$ is stress triaxiality where σ_a is mean stress and $\overline{\sigma}$ is equivalent stress.

When a material fails, the stress-strain relationship fails to effectively reflect the material's true behavior, resulting in severe mesh size dependence based upon strain localization, which lowers energy loss as mesh refinement increases.

When a material fails, the stress-strain function earlier characterizes the property of a material, and strain localization creates a significant mesh size dependence, leading energy loss to decrease as the mesh size is refined. By initiating a stress-displacement response after failure start, Hillerborg's failure energy criterion is employed to reduce mesh dependence [15]. This even takes into account the overall effects of many degradation processes occurring simultaneously on the same material. As a result, as a damage evolution criterion, the fracture energy idea is integrated with the Johnson-Cook damage model in this study. Table 20.1 illustrates the material attributes utilized to define the target material behavior.

20.3   RESULTS AND DISCUSSION

Using an ogival-nosed projectile with a normal and oblique incidence angle, the ballistic resistance and energy absorption capacity of a 2 mm target with single and layer plates are examined. ABAQUS/CAE2020 is used to model the target and projectile. The impact velocities of the

TABLE 20.1
Material Parameters of Q235 Steel Target Plate

Elasticity modulus	$E(N/m^2)$	200×10^6
Density	$\rho\left(kg/m^3 \right)$	7,800
Poisson's ratio		0.3
Yield stress constant	$A(N/m^2)$	229×10^6
Strain hardening constant	$B(N/m^2)$	439×10^6
	n	0.503
Viscous effect	C	0.1
Reference strain rate	$\dot{\varepsilon}_0\,(s^{-1})$	1.1×10^{-1}
Thermal softening constant	m	0.55
Transition temperature	$\theta_{\text{Transition}}\,(K)$	293
Melting temperature	$\theta_{\text{Melting}}\,(K)$	1,795
Fracture strain constant	D_1	0.3
	D_2	0.9
	D_3	−2.8
	D_4	0
	D_5	0.0

Source: [16].

projectile range from 134.93 m/s to 500 m/s. The outcomes of simulation as shown in Figure 20.3(b) for the monolithic target at 126.22 m/s impact velocity at normal to target are confirmed by the experimental results of the monolithic target given in Deng et al. [12]. As depicted in Figure 20.3 as a result, the simulation results in Figure 20.3(c) align very near to the experimental result. The number of petals formed in the numerical result is similar to the experimental result which is shown in Figure 20.3(a).

Table 20.2 presents the target plate's ballistic limit at different impact angles. From the table, it is observed that with the increasing impact angle, the target plates' ballistic limit increases. For monolithic plates, the ballistic limit is found to be 2.1%, 5.2%, and 9.6% as compared to the normal incidence angle. For the layered target when the angle of impact is varied from normal to 15°, 30°, and 45°, the ballistic resistance increases by 1.6%, 7.7%, and 15% respectively.

The von Mises stress distribution for the monolithic and layered target at 500 m/s impact velocity is shown in Figure 20.4.

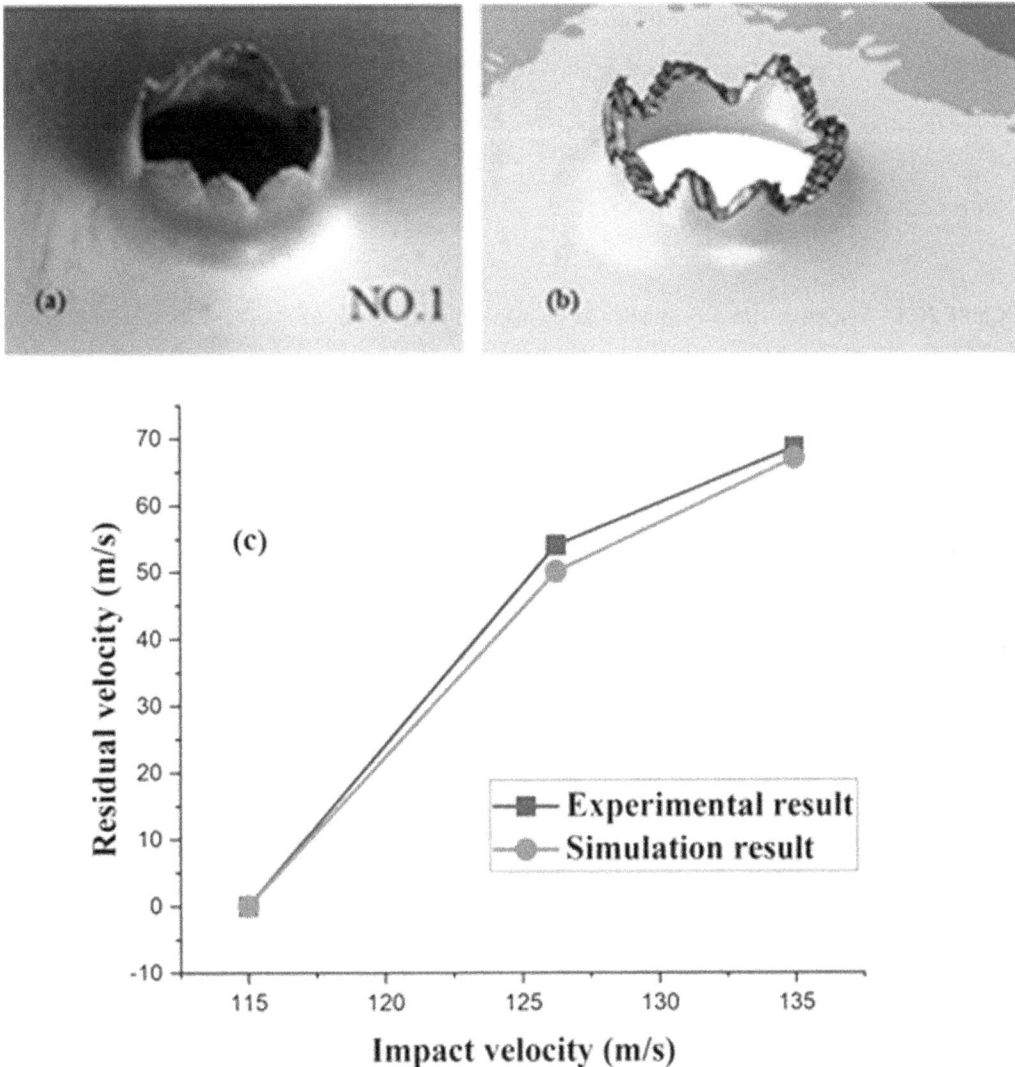

FIGURE 20.3 Deformed profile of target: (a) experimental, (b) numerical result at 126.22 m/s impact velocity, (c) graphical comparison between experimental and numerical results. Source: [12]; https://drive.google.com/file/d/18EtIn883XNttcn6yBnmbWPijsxDozovK/view?usp=share_link.

TABLE 20.2

Ballistic Limit Values at Various Obliquities

	0°	15°	30°	45°
Monolithic	115 m/s	117.5 m/s	121.2 m/s	127.1 m/s
Layered	100 m/s	101.6 m/s	108.3 m/s	115 m/s

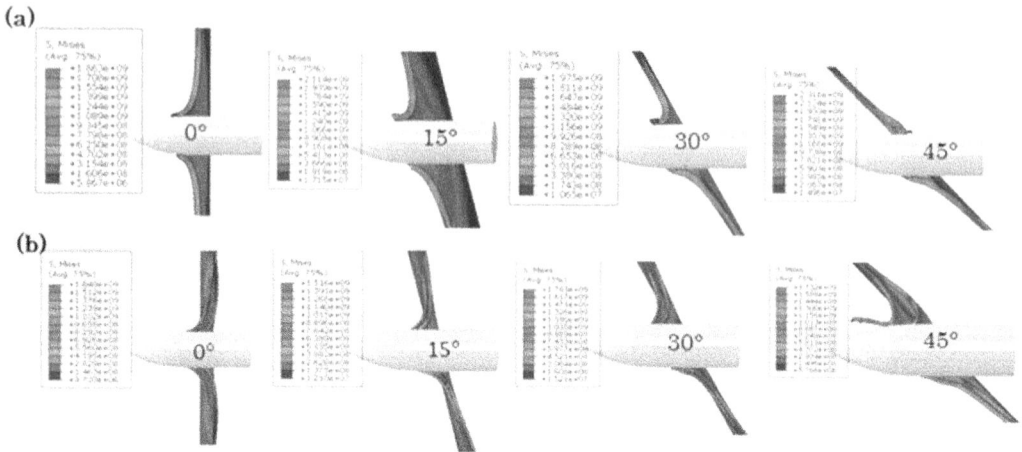

FIGURE 20.4   Von Mises stress distribution for (a) monolithic and (b) layered target plates at 500m/s impact velocity.

Figure 20.5 shows the typical deformation profile of single and layered targets. It can be observed that the projectile is a failure because of enlarging ductile and producing petals. The obliquity changes the failure mode of the target plate. For normal impact, a circular plug is produced, although for oblique impact elliptical plugs are formed. Whenever the target plate is struck at an angle of 0 degrees, four petals form. The size of the lower petals decreases when the impact angle is changed from 0° to 45°, resulting in the production of two long lips and an elliptical hole because of obliquity.

Table 20.3 shows that the velocity drop is lower at a higher velocity, although the velocity drop is higher at a lower incidence velocity. The reason behind this is that at the higher velocity, the

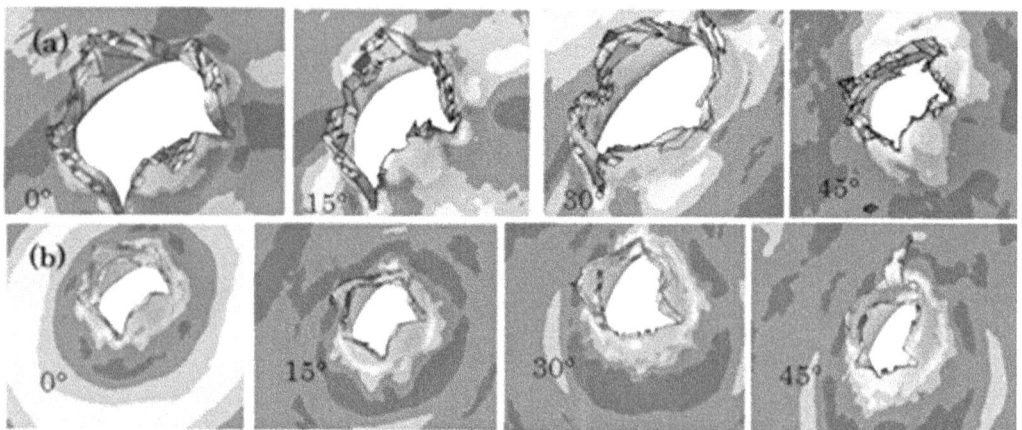

FIGURE 20.5   Deformation profile for (a) monolithic and (b) layered target plate.

TABLE 20.3
Residual Velocity of Monolithic and Layered Targets at Various Obliquities

Impact velocity (m/s)	Experimental velocity (m/s)	Present study							
		0°		15°		30°		45°	
		Monolithic (m/s)	Layered (m/s)	Monolithic (m/s)	Layered (m/s)	Monolithic (m/s)	Layered (m/s)	Monolithic (m/s)	Layered (m/s)
500		481.5	484	479.5	483	474.6	479.1	471	473.85
400		379.2	383.8	378.3	382.5	372.5	381.7	365.1	370.8
300		275.1	280	273.5	278.8	267.4	275.6	254.1	266.8
200		162.7	171	160.3	168.5	151.1	165.1	132.4	150
134.93	68.65*	67	86.5	64	82.7	43.8	74.5	0	46

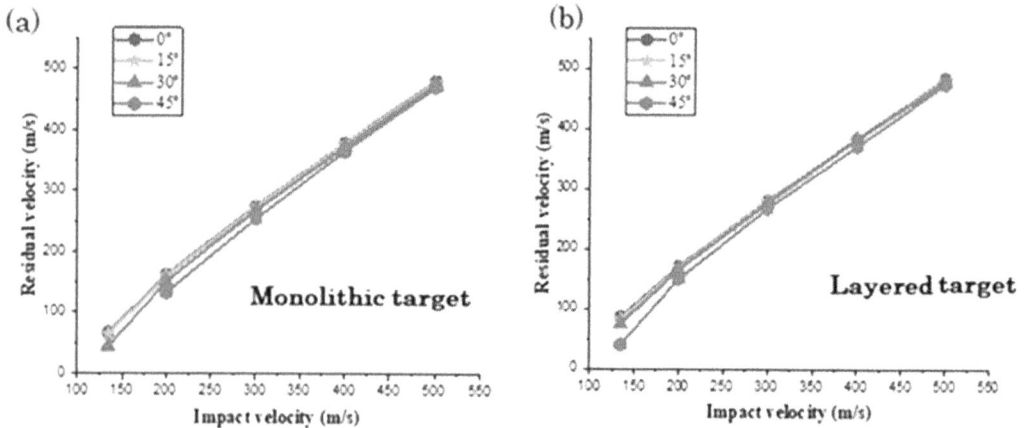

FIGURE 20.6 Correlation of the residual velocity at various impact angles: (a) monolithic and (b) multilayered targets.

contact time between target and projectile is much less as compared to lower incidence velocity during impact. Therefore in the case of lower incidence velocity, the target gets more time to deform in a particular failure mode which is the reason it provides better ballistic resistance. However, it is also found that when the incidence angle increases, the drop in impact velocity increases. The experimental residual velocity for 134.93 m/s at 0° on a monolithic target [17] is 68.65 m/s and the simulation's residual velocity is 67 m/s, which is very near to the experimental result.

Single-layer and double-layer targets' residual velocities are shown in Figure 20.6 at different oblique angles. With increasing obliquity, both single-layer and double-layer targets perform better.

The energy dissipation that occurs during projectile penetration is difficult to measure due to the complex stress wave transfer that occurs during impact [17]. Using the conservation of energy, the total energy received by targets can be calculated. The initial kinetic energy for each projectile transformed into work is plotted against its initial impact velocity.

Figure 20.7 illustrates that the energy absorbed is increased with increasing impact velocity for both target plates. This behavior is primarily seen during impact due to a higher impact velocity, which increases a larger contact surface between the target plate and projectile. It is also found that at 45°, the monolithic target is not fully penetrated by the projectile, so total kinetic energy is absorbed in the plastic deformation of the target.

20.4 CONCLUSIONS

This study examines the ballistic resistance and energy absorption of Q235 steel target plates with a thickness of 2 mm single-layer and double-layer targets with the same thickness. The target plate is subjected to an ogival projectile at various angles of 0°, 15°, 30°, and 45°. The residual velocities due to preformation obtained by numerical simulation in the ABAQUS/Explicit solver are validated with experimental results. Based on numerical research of monolithic and layered targets with increasing obliquity, both targets' ballistic resistance increases significantly.

Some of the conclusions reached as a result of the study are:

- Depending on obliquity, monolithic and layered targets have significantly greater ballistic resistance.
- The drop in residual velocity is lower at a higher impact velocity.
- The failure pattern for single- and double-layer target plates differs as obliquity increases.

(a) Monolithic target

(b) Layered target

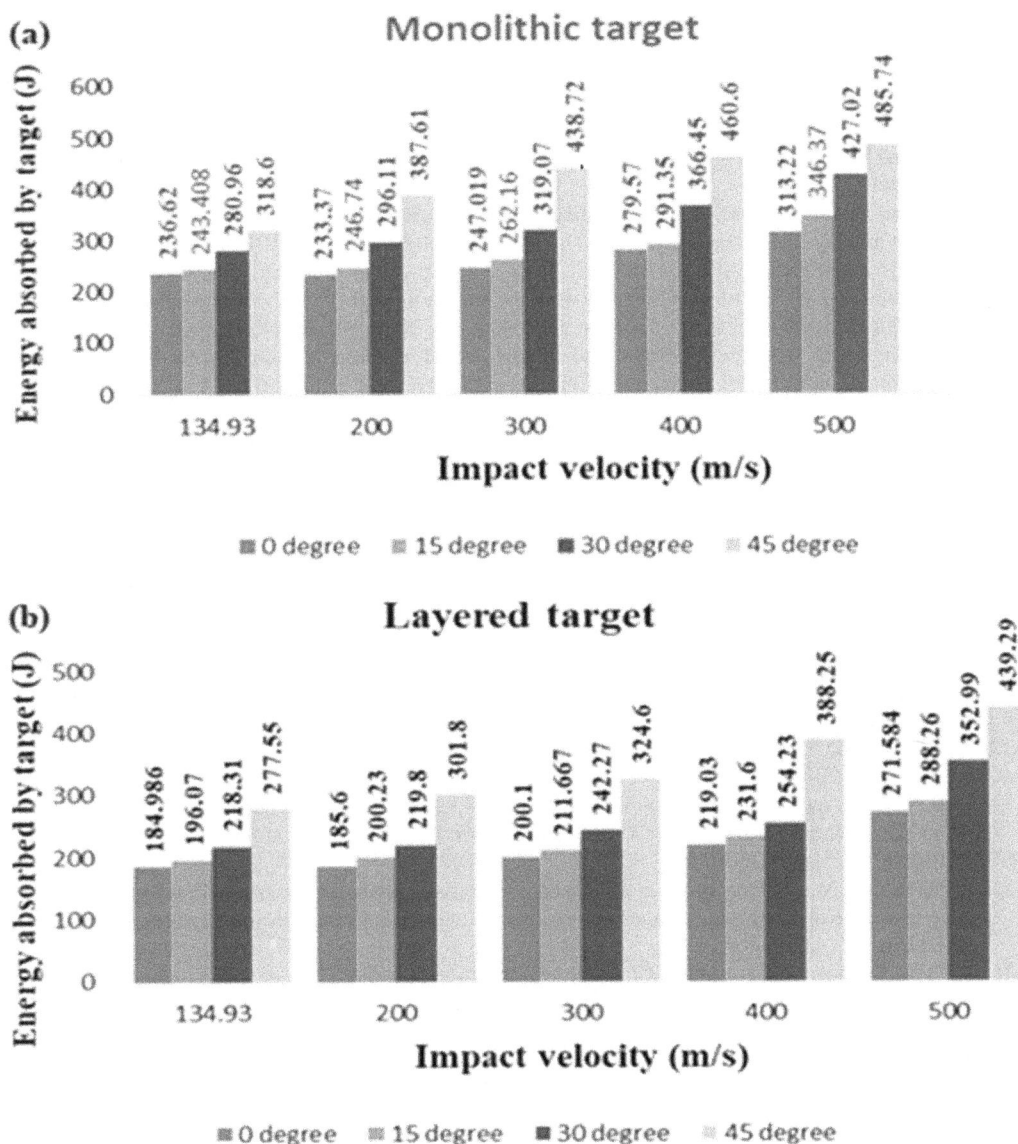

FIGURE 20.7 Energy absorbed by (a) monolithic and (b) multilayered targets at various impact angles.

- Compared to a layered target of the same thickness, the monolithic target offers better ballistic resistance.
- The layered as well as single-layer targets absorb more energy with an increased impact angle, i.e., in oblique impact, more energy is absorbed by both the targets.

REFERENCES

1. J. Awerbuch and S. R. Bodner, "An investigation of oblique perforation of metallic plates by projectiles," *Exp. Mech.*, vol. 17, no. 4, pp. 147–153, 1977, doi: 10.1007/bf02324213.
2. W. Goldsmith and S. A. Finnegan, "Normal and oblique impact of cylindro-conical and cylindrical projectiles on metallic plates," *Int. J. Impact Eng.*, vol. 4, no. 2, pp. 83–105, 1986, doi: 10.1016/0734-743X(86)90010-2.

3. R. S. J. Corran, P. J. Shadbolt, and C. Ruiz, "Impact loading of plates – An experimental investigation," *Int. J. Impact Eng.*, vol. 1, no. 1, pp. 3–22, 1983, doi: 10.1016/0734-743X(83)90010-6.

4. J. Radin and W. Goldsmith, "Normal projectile penetration and perforation of layered targets," *Int. J. Impact Eng.*, vol. 7, no. 2, pp. 229–259, 1988, doi: 10.1016/0734-743X(88)90028-0.

5. V. Madhu, T. Balakrishna Bhat, and N. K. Gupta, "Normal and oblique impacts of hard projectile on single and layered plates – An experimental study," *Def. Sci. J.*, vol. 53, no. 2, pp. 147–156, 2003, doi: 10.14429/dsj.53.2139.

6. K. Senthil and M. A. Iqbal, "Effect of projectile diameter on ballistic resistance and failure mechanism of single and layered aluminum plates," *Theor. Appl. Fract. Mech.*, vol. 67–68, pp. 53–64, 2013, doi: 10.1016/j.tafmec.2013.12.010.

7. D. W. Zhou and W. J. Stronge, "Ballistic limit for oblique impact of thin sandwich panels and spaced plates," *Int. J. Impact Eng.*, vol. 35, no. 11, pp. 1339–1354, 2008, doi: 10.1016/j.ijimpeng.2007.08.004.

8. N. K. Gupta and V. Madhu, "Normal and oblique impact of a kinetic energy projectile on mild steel plates," *Int. J. Impact Eng.*, vol. 12, no. 3, pp. 333–343, 1992, doi: 10.1016/0734-743X(92)90101-X.

9. N. K. Gupta and V. Madhu, "An experimental study of normal and oblique impact of hard-core projectile on single and layered plates," *Int. J. Impact Eng.*, vol. 19, no. 5–6, pp. 395–414, 1997, doi: 10.1016/s0734-743x(97)00001-8.

10. M. A. Iqbal, G. Gupta, and N. K. Gupta, "3D numerical simulations of ductile targets subjected to oblique impact by sharp nosed projectiles," *Int. J. Solids Struct.*, vol. 47, no. 2, pp. 224–237, 2010, doi: 10.1016/j.ijsolstr.2009.09.032.

11. M. A. Iqbal, S. H. Khan, R. Ansari, and N. K. Gupta, "Experimental and numerical studies of double-nosed projectile impact on aluminum plates," *Int. J. Impact Eng.*, vol. 54, pp. 232–245, 2013, doi: 10.1016/j.ijimpeng.2012.11.007.

12. Y. Deng, W. Zhang, and Z. Cao, "Experimental investigation on the ballistic resistance of monolithic and multi-layered plates against ogival-nosed rigid projectiles impact," *Mater. Des.*, vol. 44, pp. 228–239, 2013, doi: 10.1016/j.matdes.2012.06.048.

13. S. Michael, SIMULIA, ABAQUS, and Standard Version. "6.9 Analysis User's Manual." Pawtucket, Rhode Island, 2009.

14. G. R. Johnson and W. H. Cook, "Fracture characteristics of three metals subjected to various strains, strain rates, temperatures and pressures," *Eng. Fract. Mech.*, vol. 21, no. 1, pp. 31–48, 1985, doi: 10.1016/0013-7944(85)90052-9.

15. A. Hillerborg, M. Modeer, and P. E. Petersson, "Analysis of crack formation and crack growth in concrete by means of fracture mechanics and finite elements," *Am. Concr. Institute, ACI Spec. Publ.*, vol. SP-249, pp. 225–237, 2008.

16. S. Chen, B. Pang, W. Cao, and R. Chi, "Energy absorption characteristic of thin monolithic q235 steel plates under oblique penetrating of ogive-nosed projectiles," *J. Phys. Conf. Ser.*, vol. 1855, no. 1, 2021, doi: 10.1088/1742-6596/1855/1/012014.

17. T. Børvik, M. Langseth, O. S. Hopperstad, and K. A. Malo, "Ballistic penetration of steel plates," *Int. J. Impact Eng.*, vol. 22, no. 9. 1999, doi: 10.1016/S0734-743X(99)00011-1.

21 Parametric Study on Ballistic Impact Response of Ceramic-Composite Armour

Sagar Ghatke, Sunil Nimje, Ashish Mohan,
Ramdas Chennamsetti and Rajendra Gupta

21.1 INTRODUCTION

As it is known, weapons are an important aspect of the defence system of any country. A weapon is something used to contend with another thing, a means to gain an advantage. It is a device used for the purpose of target damage. As a weapon is used by one defence force on another, the other force is also similar in inflicting damage back. To protect our combat vehicles, warhead machinery, artillery systems and most importantly the soldiers and officers of armed forces in war from the damage inflicted by the weapons, the system of armour is used.

In a battlefield scenario, there are three types of activities performed, namely marching, manoeuvring and supporting. Marching is done with the help of infantry and its purpose is to capture locations and points during the drill. Supporting activity includes the use of artillery. It can use guided or unguided operations and sources. Manoeuvring includes the use of armour and is the main activity that defines the strength of any army. Its main function is to judge the potential for surprise and outsmart the enemy. Highly trained soldiers and officers are deployed for these functions. During manoeuvring operations, there is a great risk of the army getting attacked, and it is the job of various armours to protect the combat vehicles, warhead machinery, artillery systems and most importantly the soldiers and officers.

Hence armour comes into the picture. Armour is a protective covering that is used to prevent damage from being imposed on an object, an individual or a vehicle by direct contact with weapons or projectiles, usually during battle, or from the damage induced by a potentially dangerous environment or action. Areal densities for various types of armours made of different material configurations are shown in Table 21.1.

A finite element (FE) model with a well-defined material model, as established by K. Krishnan et al. [2], is essential for understanding the numerous nuances of projectile–armour interaction and devising efficient lightweight solutions.

To create a thin, light and cost-effective armour package, an ad hoc design optimization is carried out. Fawaz et al. [3] used the LS-Dyna 3D code to perform a numerical analysis of ceramic-composite armours. The impact of a conical projectile form on the alumina-13 carbon/epoxy combination is simulated. For this type of study, several material models are employed for different configurations and impact circumstances. Based on LS-Dyna code, a new FE simulation of the ballistic perforation of the ceramic/composite targets, which are impacted by cylindrical tungsten projectiles, has been presented by Feli et at. [4].

In this work, a three-dimensional quarter-symmetric ceramic-composite armour is modelled using LS-DYNA FE code. The parametric study is carried out to find out sensitive parameters of the armour during projectile impact.

DOI: 10.1201/9781003352358-21

TABLE 21.1
Different Types of Armour and Their Areal Density

Armour type	Areal density (kg/m²)
Ceramic-composite	98
Ceramic-aluminium	195
Titanium	220
Aluminium	267
Steel	283

Source: [1].

21.2 SIMULATION MODEL

In the present work, the Lagrangian approach with the erosion contact algorithm in the LS-DYNA code has been used to solve a three-dimensional quarter-symmetric FE model. Further, in simulation model Johnson-Holmquist, the Chang-Chang theory, the Johnson-Cook model and the Mie-Gruneisen equation of state (EOS) are used for ceramic tiles, composite layers and projectile, respectively. The CONTACT_ERODING_SURFACE_TO_SURFACE contact algorithm is used between projectile and target [5]. CONTACT_TIEBREAK_SURFACE_TO_SURFACE is used in between composite layers [5]. TSSFAC is the scale factor for the computed time step default of 0.9, and in this chapter, it is taken as 0.5 since this simulation is a high-speed impact. In this case, SOFT = 2 is used to activate the segment-based penalty algorithm [5].

The boundary conditions are shown in Figure 21.1(a) where the outer edges of the armour are fixed in all directions as displacement in $u_x = u_y = u_z = 0$ and all rotations in $\theta_x = \theta_y = \theta_z = 0$.

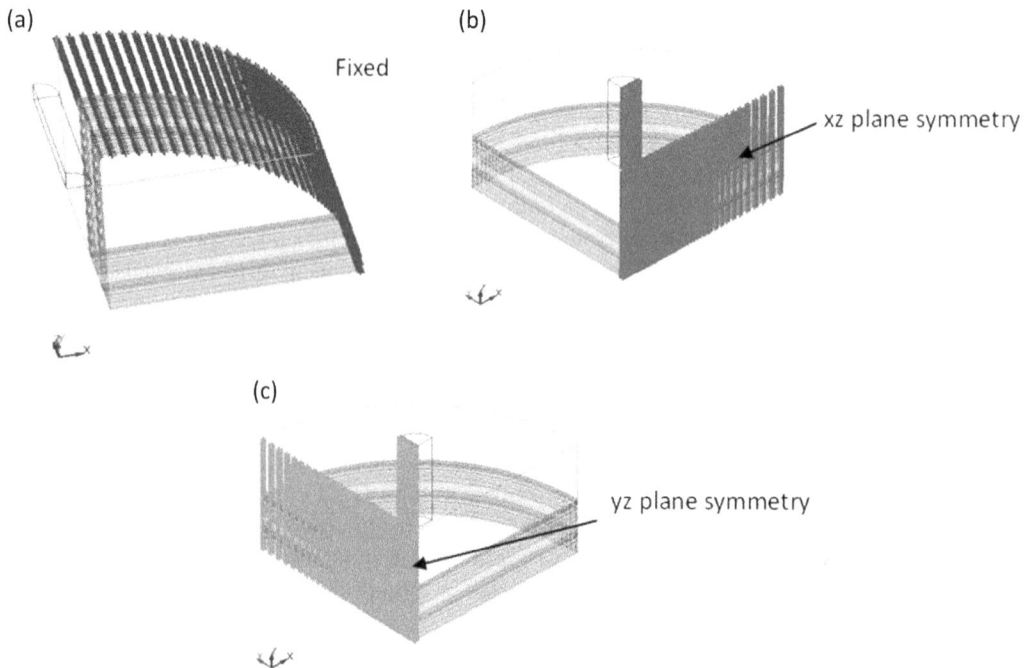

FIGURE 21.1 (a) Fixed boundary conditions, (b) symmetry of xz-plane and (c) symmetry of yz-plane of the quarter model.

To reduce the simulation time, the quarter-symmetric model is used in this study. For the symmetry of the xz-plane, the displacement in $u_y = 0$ and rotations such as $\theta_x = \theta_z = 0$ are shown in Figure 21.1(b), and for the symmetry of the yz-plane, the displacement in $u_x = 0$ and rotations viz. $\theta_y = \theta_z = 0$ are illustrated in Figure 21.1(c). The logic behind the symmetric condition is that the elements should not move out of that plane. Figure 21.2 shows a meshed three-dimensional quarter-symmetric model for validation. The mesh size of the projectile is $0.5 \times 0.5 \times 0.5$ mm and the total elements are 5,160. For the ceramic layer, the mesh size is $0.5 \times 0.5 \times 0.5$ mm at the impact zone and the total elements are 80,120. In the case of the composite, $0.5 \times 0.5 \times 0.5$ mm mesh size is used with 100,150 elements.

The model used in the current study consists of a projectile made of two parts: one is the core and the other is the jacket. The armour consists first of a composite cover then a second ceramic layer and a later composite backing material, as shown in Figure 21.4.

Similar boundary conditions as discussed in an earlier section are used in this model also, as illustrated in Figure 21.1. The initial velocity of the projectile is 695 m/s. The mesh size of the composite cover, composite backing and ceramic layer is $0.25 \times 0.25 \times 0.65$ mm and consists of 196,794 elements. For the projectile core and projectile, it is $0.175 \times 0.175 \times 0.28$, and the total elements are 18,056.

The material of the projectile core and projectile jacket are of steel and copper-coated steel and the Johnson-Cook [6] material model is used for both the projectile core and projectile jacket. Their material properties are shown in Table 21.1. The first layer of armour is a ceramic layer made of silicon carbide and for this layer, the Johnson-Holmquist material model is used in the simulation, as shown in Table 21.1. The second layer of the armour is a composite backing layer made of E-glass/epoxy and its properties are mentioned in Table 21.2. For the simulation, Chang-Chang criteria are used in LS-Dyna for this layer.

21.3 SIMULATION RESULTS AND DISCUSSION

In this chapter, the three-dimensional quarter-symmetrical model of the ceramic-composite armour is simulated using the LS-Dyna tool. The meshed model of ceramic-composite amour as shown in Figure 21.2 is modelled as per Feli et al. [4] for validation. In this simulation, the residual velocity of this projectile is plotted with four different initial velocities of the projectile. The validation FE model shows close agreement of residual velocity of the projectile with Feli et al. [4] and Chocron [7] at different initial velocities as illustrated in Figure 21.3.

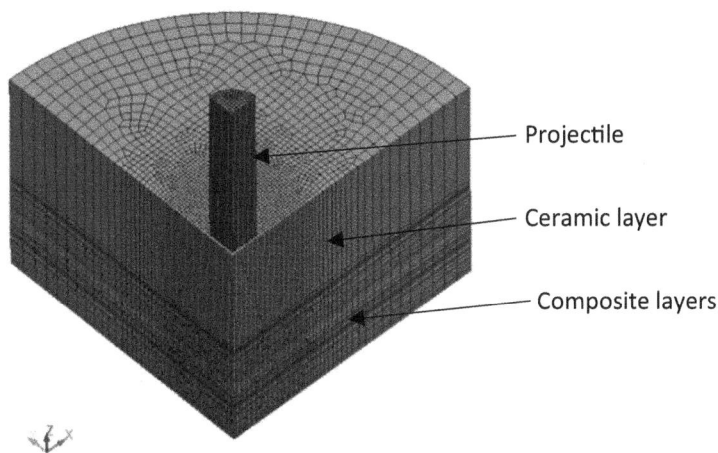

FIGURE 21.2   The meshed three-dimensional quarter-symmetric model for validation.

A new quarter-symmetric model of a ceramic-backed composite with composite cover plate is modelled and shown in Figure 21.4. Similar boundary conditions as discussed in an earlier section (Figure 21.1) are used. In this model, the projectile consists of two components of core and jacket. For the current study, the initial velocity of the projectile is 695 m/s.

A simulation of the ballistic impact of the projectile on the target at different time intervals is shown in Figure 21.5. The energy plots are shown in Figure 21.6 and Figure 21.7. The maximum tensile pressure strength of ceramic is varied in Table 21.3, which shows the projectile is perforating the armour. As the maximum tensile pressure increases, the residual velocity also keeps

TABLE 21.2

Johnson-Cook Material Property for Projectile

Material property	Projectile core	Projectile jacket
Density	7.83e–6 kg/mm³	7.83e–6 kg/mm³
G	76.3 GPa	76.3 GPa
A	0.2344 GPa	0.4482 GPa
B	0.4138 GPa	0.3034 GPa
C	0.003	0.0033
N	0.25	0.15
M	1.03	0.03
T_m	1,800 K	1,800 K
T_r	293	293
EPSO	0.001	0.001
C_p	477 (J/kg K)	477 (J/kg K)
D_1	5.625	2.25
D_2	0.3	0.0005
D_3	–7.2	–3.6
D_4	–0.0123	–0.0123
D_5	0	0

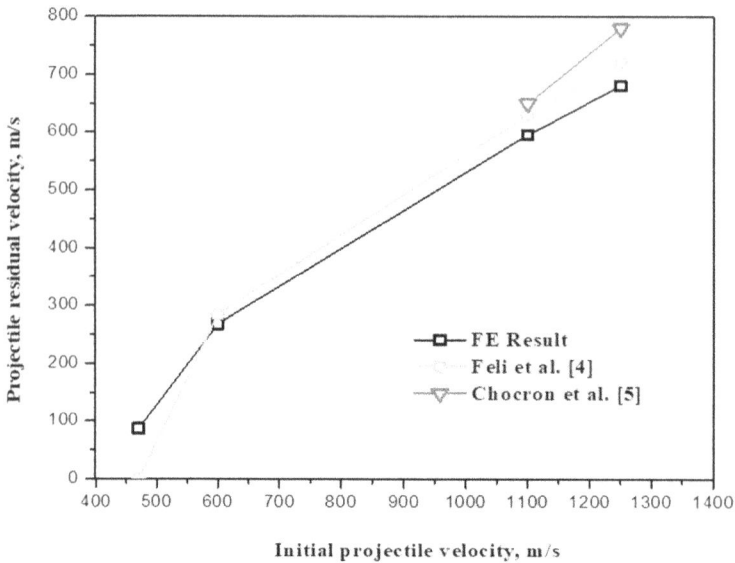

FIGURE 21.3   Validation graph of ceramic-composite armour.

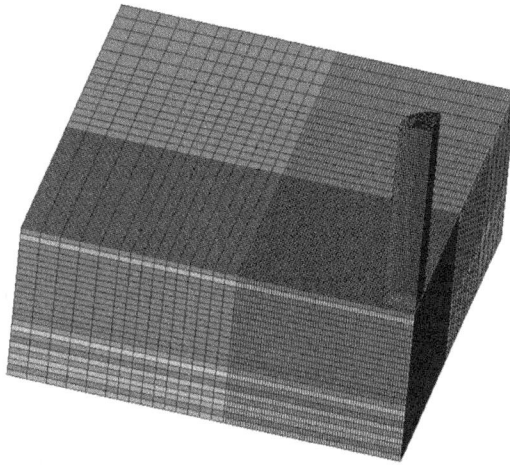

FIGURE 21.4 Meshing scheme for quarter-symmetric ceramic-composite armour considered in the present work.

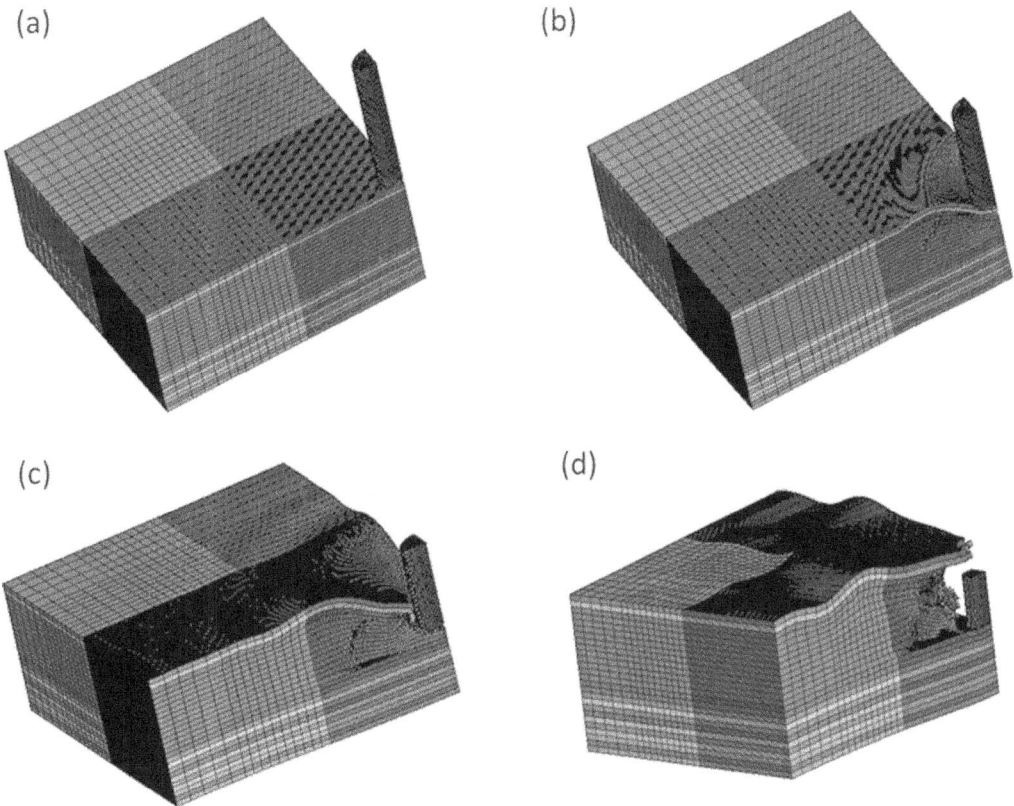

FIGURE 21.5 Ballistic impact simulation at different time intervals.

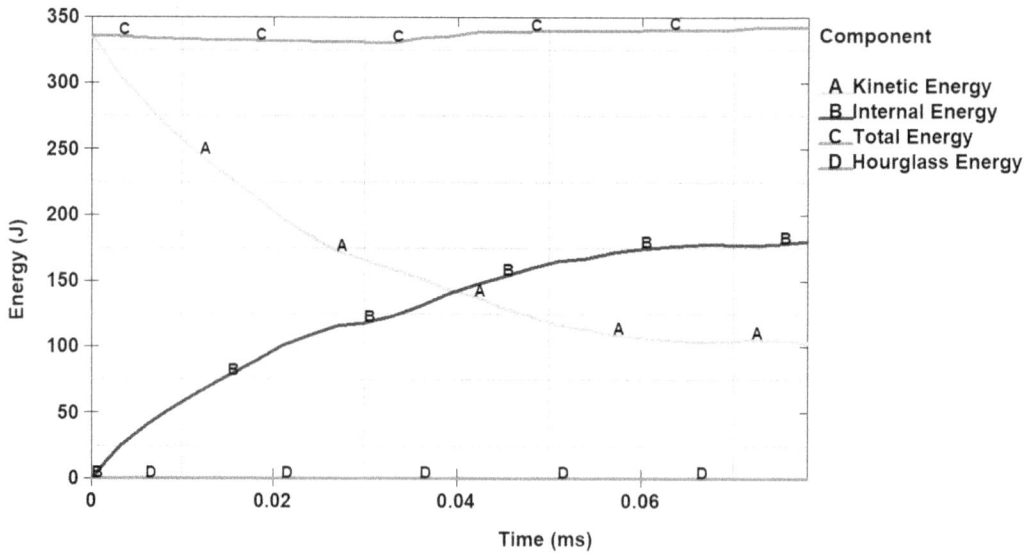

FIGURE 21.6 Total energy plot of the model.

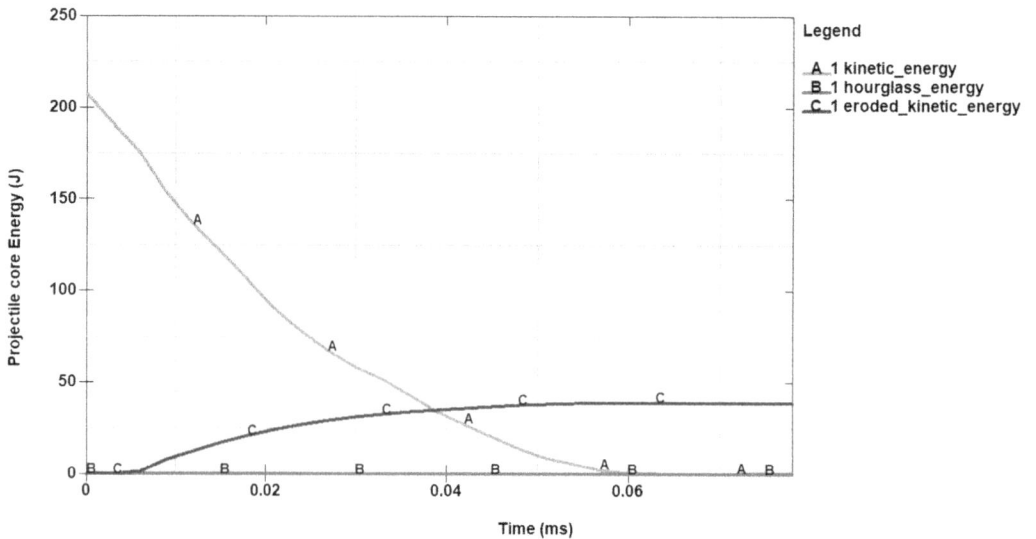

FIGURE 21.7 Energy plot of the projectile core.

on increasing. The increasing bulk modulus of the ceramic decreases the residual velocity of the bullet as shown in Table 21.4. The failure strain values of the ceramic have a significant impact on the results. With reference to Table 21.5, the increase in failure strain of the ceramic up to 0.3 indicates that it is able to stop the projectile at 13.8 mm depth of penetration and keeps decreasing with the increase of failure strain values. It is observed from Table 21.6 that the intact strength parameter of ceramic also has a significant impact. An increase in P1 value decreases the strength of the ceramic.

TABLE 21.3

Johnson-Holmquist JH1 Model for ceramic and Composite Material Properties

Ceramic	Value	Composite	Value
Density	3.91e–6 kg/mm^3	Density	1.9e–6 kg/mm^3
G	152 GPa	E_A	20.7 GPa
P_1	1 GPa	E_B	20.7 GPa
S_1	2.599 GPa	E_C	7.96 GPa
P_2	1 GPa	PR_{BA}	0.12
S_2	3.52	PR_{CA}	0.145
C	0	PR_{CB}	0.25
EPSO	0.001	G_{AB}	4.15 GPa
T	–0.379 GPa	G_{BC}	4.15 GPa
Alpha	1	G_{CA}	4.15 GPa
SFMAX	1.3 GPa	Shear strength AB	64 GPa
BETA	1	Transverse shear strength S_{YZ}	40 GPa
DP_1	1 GPa	Transverse shear strength S_{ZX}	40 GPa
EPFMAX	1	Normal tensile strength SN	244 GPa
K_1	238.14 GPa	Longitudinal tensile X_T	400 GPa
FS	0.99	Transverse tensile Y_T	400 GPa
		Transverse compressive Y_C	200 GPa

TABLE 21.4

Maximum Tensile Pressure Strength of Ceramic with Initial Projectile Velocity 695 m/s

Maximum tensile pressure strength of ceramic (GPa)	Residual velocity of projectile (m/s)
0.2650	476
0.3032	506
0.3411	502
0.4169	505
0.4548	517
0.4927	533

TABLE 21.5

Bulk Modulus Ceramic Layer with Initial Projectile Velocity 695 m/s

Bulk modulus K_1 (GPa)	Residual velocity of projectile (m/s)
238.14	507
240.57	500
245.43	519
247.86	506
250.29	463

TABLE 21.6
Ceramic Failure Strain with Initial Projectile Velocity 695 m/s

Ceramic failure strain	Residual velocity or DOP of projectile
0.1	406 m/s
0.2	277 m/s
0.3	13.8 mm
0.4	13 mm
0.5	9.3 mm

TABLE 21.7
Intact Strength of Ceramic with Initial Projectile Velocity 695 m/s

Pressure point 1 for intact material P_1 (GPa)	Residual velocity or DOP of projectile
0.8	9.3 mm
0.9	9.3 mm
1.0	9.2 mm
1.1	9.1 mm
1.2	608 m/s
1.3	610 m/s

21.4   CONCLUSION

In the present research, the three-dimensional quarter-symmetric model of ceramic-composite armour is validated by Feli et al. [2]. The conoidal failure of ceramic material is also observed in this study. A new armour model with composite-cover ceramic-backed composite is modelled. In this study, the effect of material properties variation on armour response is studied by altering the parameters of ceramic materials. It is found that failure strain values play a significant role in this impact simulation. It is observed that intact strength parameter P1 of the Johnson-Holmquist JH1 material model should be low to stop the projectile perforation. With increasing bulk modulus of ceramic, the residual velocity is found to be decreased. Maximum tensile pressure strength is found to be inversely proportional to the residual velocity of the projectile.

REFERENCES

1. K. Akella, Multilayered ceramic-composites for armour applications. In: Mahajan Y.R., Johnson R. (eds) *Handbook of Advanced Ceramics and Composites*. Springer, Cham (2020).
2. K. Krishnan, S. Sockalingam, S. Bansal, S.D. Rajan, Numerical simulation of ceramic composite armor subjected to ballistic impact, *Composites: Part B* 41 (2010) 583–593.
3. Z. Fawaz, W. Zheng, K. Behdinan, Numerical simulation of normal and oblique ballistic impact on ceramic composite armour, *Composite Structures* 63 (2004) 387–395.
4. S. Feli, M.R. Asgari, Finite element simulation of ceramic/composite armour under ballistic impact, *Composites: Part B* 42 (2011) 771–780.
5. LS-DYNA Manual R13.0 Vol I (2021).
6. LS-DYNA Manual R13.0 Vol II (2021).
7. I.S. Chocron-Benloulo, J. Rodriguez, V. Sauchez-Galvez, A simple analytical model to simulate textile fabric ballistic behavior, *Text Res J* 67(7) (1997) 520–528.

22 A Numerical Investigation on UHMWPE Composite Panel Subjected to Ballistic Impact with Lead Core Projectiles

Joseph Solomon and Puneet Mahajan

22.1 INTRODUCTION

Body armor is used to protect the human body against ballistic threats. Armor systems typically have a hard armor panel (HAP) and a soft armor panel (SAP). The need for a lightweight and flexible armor system has led to the use of composites in body armor systems. Recently, a HAP of ultra-high molecular weight polyethylene (UHMWPE) composite supported by a SAP of the same material in the form of fabric and foam has been developed [9] to deal with threat-level three as per BIS (Bureau of Indian Standards) for multiple impacts. Data from ballistic tests was taken from [9].

A 7.62 × 51 mm NATO ball weighing 9.5 g is a commonly used ammunition and consists of a soft core of alloyed lead with brass casing. UHMWPE fiber-reinforced composite is a suitable material for ballistic protection due to its high strength, stiffness, and low density. Taylor and Carr (Taylor & Carr, 2001) performed a post-failure (Taylor & Carr, 2001) analysis of ballistic-impacted UHMWPE composite at two thicknesses, 7 mm and 25 mm, impacted by a 1.1 g steel ball and various 7.62 mm armor-piercing (AP) NATO ammunition. Chocron et al. (Chocron, Anderson Jr, Grosch, & Popelar, 2001) investigated the effects of a NATO 7.62 APM2 on thin aluminum panels. Simple constitutive equations were utilized for both target and projectile materials and the numerical findings did not always match experimental results. Investigations of the influence of the mechanical properties of the lead core and brass jacket of a NATO 7.62 mm ball bullet in numerical simulations of ballistic impacts by Giglio et al. (Giglio, Gilioli, Manes, Peroni, & Scapin, 2012) and Manes et al. (Manes, Lumassi, & Giudici, 2013) also suggest that the behavior of such type of projectile is very dangerous for thin aluminum structures (after impact the soft core tends to become mushroom-shaped).

Back face deformation (BFD) is the imprint formed by the bullet being trapped on the backside of the plate, preventing it from departing and penetrating the body. In this study, BFD was measured by vernier caliper on the clay which was kept behind the armor panel. The simulation software used in the research work is Ansys Autodyn and the material model has been validated in the software. The material model could simulate experimental results of high-velocity impacts on UHMWPE panels available in open literature very well. The result of the simulations is in good agreement with experiments.

22.2 PANEL DESCRIPTION

UHMWPE is a thermoplastic polymer made up of long polyethylene molecular chains. Substantial strength may be obtained by using a gel spinning technique, which results in highly ordered and crystalline molecular structures aligned in the spinning direction. The gel spinning procedure starts

DOI: 10.1201/9781003352358-22

with a high-temperature solution of UHMWPE in a solvent. The solution is then spun into a liquid filament, which is subsequently quenched in water to create gel fibers. These fibers are pulled in hot air at high strain rates of the order of $1s^{-1}$, resulting in smooth circular cross-sections with a molecular orientation of higher than 95% and crystallinity of up to 85%. These fibers are made up of smaller macro-fibrils with a diameter of 0.5 mm to 2 mm, which are mainly (Nguyen, 2015) composed of micro-fibrils with a diameter of 20 nm. For body armor application, fibers can be woven into fabrics to produce a soft and flexible material for SAP or coated in a matrix and aligned to make uni-directional plies, which are then stacked and pressed under temperature and pressure to form rigid laminates for HAP. Figure 22.1 depicts the production process from UHMWPE to fiber-reinforced composite laminates and the armor panel schematically.

The panel configuration which successfully passed the physical ballistic test had dimensions of 250 mm × 300 mm and was slightly curved. HAP is composed by compressing around 90 layers of UHMWPE ply [0/90/0/90]. The thickness of HAP is 12.5 mm. SAP is composed by stitching 3 mm of UHMWPE woven fabric with 27 mm foam. The velocity of the NATO ball projectiles was around 825 ± 5 m/s. The HAP first takes the trauma caused by the projectile and the bulge formed behind the HAP is endured by the SAP. BFDs are obtained on clay which is kept behind the SAP [9].

22.3 NUMERICAL SIMULATION

A numerical simulation technique based on the finite element method is used for both the projectile and target. Since 90 layers of actual UHMWPE ply would increase the simulation time significantly, five layers each of 2.5 mm of UHMWPE for HAP were modeled to reduce simulation time. The geometric configuration of the panel and a cross-section of the projectile are shown in Figure 22.2.

Normal and shear sub-laminate interface strengths used were 5.35 MPa and 7.85 MPa respectively. Each layer is modeled with solid brick elements of 1 mm in size. SAPs were modeled with three layers of 1 mm thick UHMWPE with solid brick elements of 1 mm element size. The foam is 26.5 mm thick and modeled with solid brick elements of 2 mm element size. The clay is 50 mm thick and modeled with solid brick elements of 2 mm element size. The projectile is modeled with solid tetrahedral elements of 1 mm element size. Frictional contact with dynamic and static friction coefficients of 0.15 and 0.3 respectively are used between layers of SAP. The contact between foam and clay was kept frictionless. The rear face of the clay was kept fixed.

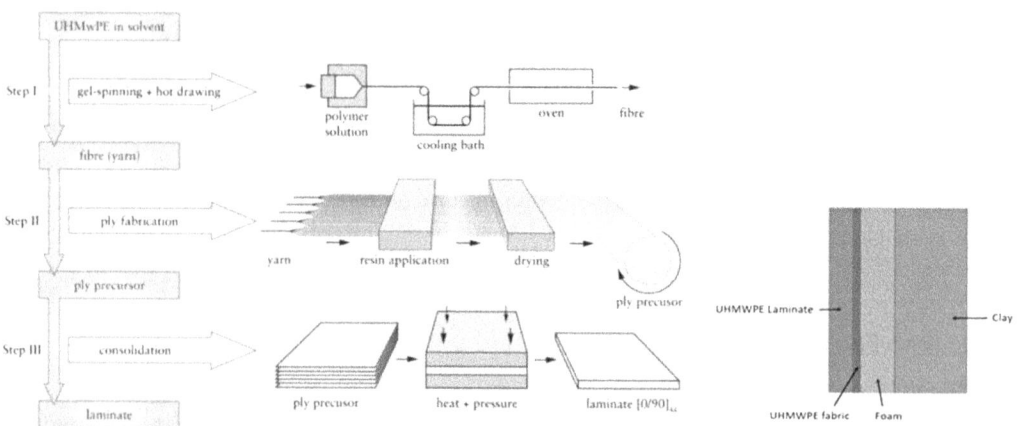

FIGURE 22.1   (a) Process involved in fabricating UHMWPE fiber to laminate; (b) panel description.

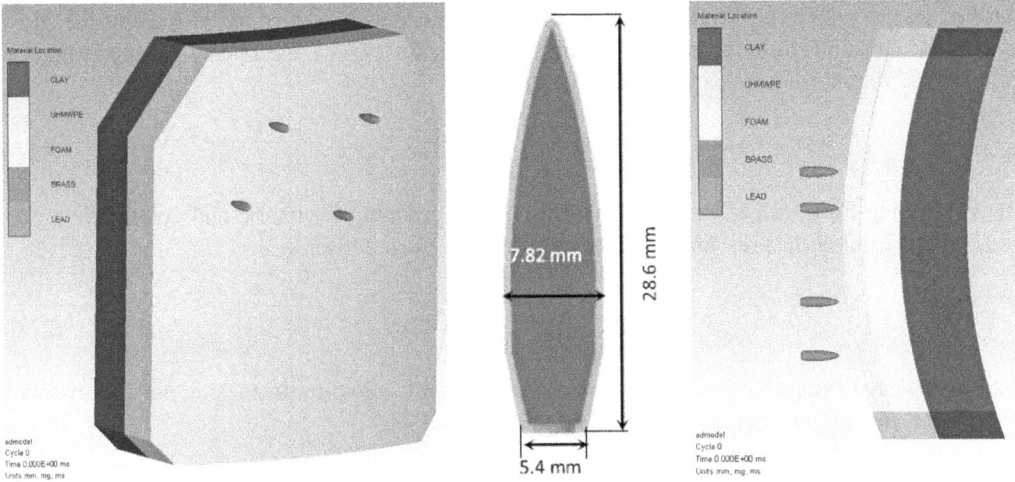

FIGURE 22.2   (a) Geometric configuration of the panel; (b) cross-section of the bullet; (c) top view of the panel.

22.3.1   MATERIAL MODEL OF UHMWPE

The non-linear orthotropic material model implemented in Ansys Autodyn is used to model the ballistic impact response of the UHMWPE composite. The incremental stress-strain relations for composites can be expressed as

$$[\sigma]^{n+1} = [\sigma]^{n} + [C][\dot{\epsilon}]\Delta t \tag{22.1}$$

The constitutive stress-strain relations for an orthotropic material are given by

$$\begin{bmatrix} \sigma_{11} \\ \sigma_{22} \\ \sigma_{33} \\ \sigma_{23} \\ \sigma_{31} \\ \sigma_{12} \end{bmatrix} = \begin{bmatrix} C_{11} & C_{12} & C_{13} & 0 & 0 & 0 \\ C_{21} & C_{22} & C_{23} & 0 & 0 & 0 \\ C_{31} & C_{32} & C_{33} & 0 & 0 & 0 \\ 0 & 0 & 0 & C_{44} & 0 & 0 \\ 0 & 0 & 0 & 0 & C_{55} & 0 \\ 0 & 0 & 0 & 0 & 0 & C_{66} \end{bmatrix} \begin{bmatrix} \epsilon_{11} \\ \epsilon_{22} \\ \epsilon_{33} \\ \epsilon_{23} \\ \epsilon_{31} \\ \epsilon_{12} \end{bmatrix} \tag{22.2}$$

The incremental linear elastic constitutive relations for an orthotropic material can be expressed as

$$\begin{bmatrix} \Delta\sigma_{11} \\ \Delta\sigma_{22} \\ \Delta\sigma_{33} \\ \Delta\sigma_{23} \\ \Delta\sigma_{31} \\ \Delta\sigma_{12} \end{bmatrix} = \begin{bmatrix} C_{11} & C_{12} & C_{13} & 0 & 0 & 0 \\ C_{21} & C_{22} & C_{23} & 0 & 0 & 0 \\ C_{31} & C_{32} & C_{33} & 0 & 0 & 0 \\ 0 & 0 & 0 & C_{44} & 0 & 0 \\ 0 & 0 & 0 & 0 & C_{55} & 0 \\ 0 & 0 & 0 & 0 & 0 & C_{66} \end{bmatrix} \begin{bmatrix} \Delta\epsilon_{11} \\ \Delta\epsilon_{22} \\ \Delta\epsilon_{33} \\ \Delta\epsilon_{23} \\ \Delta\epsilon_{31} \\ \Delta\epsilon_{12} \end{bmatrix} \tag{22.3}$$

The above linear relations do not include the non-linear shock effects. The shock effects are taken into consideration by splitting the volumetric response of the material from its shear response as

$$\Delta\epsilon_{ij} = \Delta\epsilon_{ij}^{d} + \Delta\epsilon_{ave} \tag{22.4}$$

Further, the average direct strain increment $\Delta\varepsilon_{ave}$ is defined as one-third of the trace of the strain tensor, and for small strain increments, the trace of the strain tensor is assumed to be approximately equal to the volumetric strain increment. Therefore

$$\Delta\,\varepsilon_{ave} = \left(\Delta\,\varepsilon_{vol}\right)/3. \tag{22.5}$$

By substituting Eqns. (22.4) and (22.5) in Eqn. (22.3), grouping deviatoric and volumetric terms, and further defining the pressure as a third of stress increment tensor as

$$\Delta P = -\frac{1}{3}\left(\Delta\sigma_{11} + \Delta\sigma_{22} + \Delta\sigma_{33}\right) \tag{22.6}$$

and further modifying the volumetric strain contribution to pressure to include non-linear shock effects, the final incremental pressure can be written as

$$\Delta P = \Delta P_{EOS}\left(\varepsilon_{vol}, e\right) - \frac{1}{3}\left[C_{11} + C_{21} + C_{31}\right]\Delta\,\varepsilon_{11}^{d}$$

$$- \frac{1}{3}\left[C_{12} + C_{22} + C_{32}\right]\Delta\,\varepsilon_{22}^{d} \tag{22.7}$$

$$- \frac{1}{3}\left[C_{13} + C_{23} + C_{33}\right]\Delta\,\varepsilon_{33}^{d}$$

The material model uses the shock equation of state which is defined by

$$U_{s} = C_{0} + SU_{p} \tag{22.8}$$

where U_s is the shock velocity, U_p is the particle velocity, S is the slope of the U_s–U_p relationship, and C_0 is the bulk acoustic sound speed, which is derived from effective bulk modulus K' and density ρ as

$$C_0 = \sqrt{\frac{K'}{\text{Á}}} \tag{22.9}$$

The anisotropic hardening behavior of the orthotropic material was modeled using the nine-parameter yield function available in Autodyn. This function is derived from the anisotropic yield criteria of Tsai-Hill; however, it relaxes the constant pressure assumption. The quadratic function is given as

$$f\left(\sigma_{ij}\right) = a_{11}\sigma_{11}^{2} + a_{22}\sigma_{22}^{2} + a_{33}\sigma_{33}^{2} + 2a_{12}\sigma_{11}\sigma_{22} + 2a_{23}\sigma_{22}\sigma_{33} + 2a_{13}\sigma_{11}\sigma_{33} + 2a_{44}\sigma_{23}^{2} + 2a_{55}\sigma_{31}^{2} + 2a_{66}\sigma_{12}^{2} = k \tag{22.10}$$

The nine material constants a_{ij} represent the degree of anisotropy in the material's behavior. Failure in the material model is based on combined stress-strain criteria given as

$$\left(\frac{\sigma_{ii}}{S_{ii}\left(1 - D_{ii}\right)}\right)^{2} + \left(\frac{\sigma_{ij}}{S_{ij}\left(1 - D_{ij}\right)}\right)^{2} + \left(\frac{\sigma_{ki}}{S_{ki}\left(1 - D_{ki}\right)}\right)^{2} \geq 1$$

$$\text{for } i, j, k = 1, 2, 3 \tag{22.11}$$

S_{ii} is the failure strength of the material in respective directions. Damage D_{ii} follows a linear relationship with stress and strain given as

$$D_{ii} = \frac{L\sigma_f \epsilon_{cr}}{2G_{f,ii}}$$ (22.12)

L is the characteristic length, ϵ_{cr} is the crack strain (which is the strain above failure initiation strain), and $G_{(f,ii)}$ is the fracture energy in the direction of damage. The material model parameters are taken from (Nguyen, 2015) and are given in Table 22.1.

TABLE 22.1
Non-Linear Orthotropic Material Model Parameters for UHMWPE Composite

Equation of state: orthotropic	Value	Strength: orthotropic yield	Value
Reference density ρ	980 kg/m³	Plasticity constant 11 a_{11}	0.016
Young's modulus 11 E_{11}	3.62 GPa	Plasticity constant 22 a_{22}	0.0006
Young's modulus 22 E_{22}	51.1 GPa	Plasticity constant 33 a_{33}	0.0006
Young's modulus 33 E_{33}	51.1 GPa	Plasticity constant 12 a_{12}	0.0
Poisson's ratio ν_{12}	0.013	Plasticity constant 13 a_{13}	0.0
Poisson's ratio ν_{23}	0.0	Plasticity constant 23 a_{23}	0.0
Poisson's ratio ν_{31}	0.5	Plasticity constant 44 a_{44}	1.0
Shear modulus G_{12}	2.0 GPa	Plasticity constant 55 a_{55}	1.7
Shear modulus G_{23}	0.192 GPa	Plasticity constant 66 a_{66}	1.7
Shear modulus G_{31}	2.0 GPa	Effective stress #1	1,480 kPa
Volumetric response: shock		Effective stress #2	7,000 kPa
Gruneisen coefficient	1.64	Effective stress #3	27 MPa
Parameter C1 = C_0	3,570 m/s	Effective stress #4	40 MPa
Parameter S1 = S	1.3	Effective stress #5	50 MPa
Reference temperature	293 K	Effective stress #6	60 MPa
Specific heat c_v	1,850 J/kgK	Effective stress #7	80 MPa
Failure: orthotropic softening		Effective stress #8	98 MPa
Tensile failure stress 11 S_{11}	(disabled)	Effective stress #9	200 MPa
Tensile failure stress 22 S_{22}	1.15 GPa	Effective stress #10	1.0 GPa
Tensile failure stress 33 S_{33}	1.15 GPa	Eff plastic strain #1	0.0
Max shear stress 12 S_{12}	575 MPa	Eff plastic strain #2	0.01
Max shear stress 23 S_{23}	120 MPa	Eff plastic strain #3	0.1
Max shear stress 31 S_{31}	575 MPa	Eff plastic strain #4	0.15
Fracture energy 11 G_{11c}	790 J/m²	Eff plastic strain #5	0.175
Fracture energy 22 G_{22c}	30 J/m²	Eff plastic strain #6	0.19
Fracture energy 33 G_{33c}	30 J/m²	Eff plastic strain #7	0.2
Fracture energy 12 G_{12c}	1,460 J/m²	Eff plastic strain #8	0.205
Fracture energy 23 G_{23c}	1,460 J/m²	Eff plastic strain #9	0.21
Fracture energy 31 G_{31c}	1,460 J/m²	Eff plastic strain #10	0.215
Damage coupling coefficient C	0		
Bonds: sub-laminate interface			
Normal strength S_N	5.35 MPa		
Shear strength S_S	7.85 MPa		

22.3.2 Johnson-Cook Model for Brass Casing

This model represents the strength behavior of materials, typically metals, subjected to large strains, high strain rates, and high temperatures. Such behavior might arise in problems of intense impulsive loading due to high-velocity impact. With this model, the yield stress varies depending on the strain, strain rate, and temperature. The model defines the yield stress Y as

$$Y = \left[A + B\left(\varepsilon_{eff}^p\right)^N \right]\left(1 + C \ln\left(\frac{\dot{\varepsilon}_{eff}^p}{\dot{\varepsilon}_0} \right) \right)\left[1 - \left(\frac{T - T_R}{T_M - T_R} \right)^M \right] \tag{22.13}$$

where ε_{eff}^p is the effective plastic strain, T_M is the melting temperature, T_R is the reference temperature when determining A, B, C, M, and N, $\dot{\varepsilon}_0$ is the reference strain rate, and A, B, C, N, and M are material constants. The cumulative damage is given by

$$D = \Sigma\left(\Delta\mu_{eff}^p / \varepsilon^F \right)\varepsilon_{eff}^p / \varepsilon^F \tag{22.14}$$

with
$$\varepsilon^F = \left(D_1 + D_2\exp\left(D_3\frac{P}{\sigma_{eff}} \right) \right)\left(1 + D_4\ln\left(\frac{\dot{\varepsilon}_{eff}^p}{\dot{\varepsilon}_0} \right) \right)\left(1 - D_5\left(\frac{T - T_R}{T_M - T_R} \right)^M \right) \tag{22.15}$$

where P is the pressure, σ_{eff} is the von Mises stress, and D_1, D_2, D_3, D_4, D_5 are failure parameters.

22.3.3 Steinberg-Guinan Model for Lead Core

The Steinberg-Guinan constitutive model describes shear modulus and yield strength based on the following equations:

$$G = \left[G_0\left(1 + \left(\frac{G_p'}{G_0} \right)\frac{P}{\eta^{1/3}} + \left(\frac{G_T'}{G_0} \right)(T - 300) \right) \right] \tag{22.16}$$

$$Y = Y_0\left(1 + \beta\left(\varepsilon + \varepsilon_i \right) \right)^n\left[1 + \left(\frac{Y_p'}{Y_0} \right)\frac{P}{\eta^{1/3}} + \left(\frac{G_T'}{G_0} \right)(T - 300) \right] \tag{22.17}$$

where G is shear modulus, T is temperature, η is compression or the initial specific volume divided by the specific volume, Y is yield strength, β is the work hardening parameter, ε is strain, ε_i is the initial equivalent plastic strain, and P is pressure. The subscript 0 refers to the reference state where T = 300 K, P = 0, and ε = 0. G_p' or Y_p' refer to dG/dP or dY/dP, respectively, and G_T' or Y_T' refer to dG/dT or dY/dT, respectively. Material properties for brass have been taken from (Choudhary, et al., 2020). The engineering material library provided by Ansys 19.2 has been used for lead properties. Table 22.2 and Table 22.3 represent the material properties of brass and lead respectively.

22.3.4 Hyperelastic Model for Foam

Following are several forms of strain energy potential (Ψ) provided for the simulation of hyperelastic materials. The strain energy function for the three-parameter Mooney-Rivlin model [12] is,

TABLE 22.2
Material Model Parameters for Brass

Properties	Value
E(MPa)	115,000
ϑ	0.31
ρ (kg/m^3)	8,520
A(MPa)	206
B(MPa)	505
n	0.42
c	0.01
$\dot{\mu}_0(\mathrm{s}^{-1})$	5e–4
m	1.68

TABLE 22.3
Material Model Parameters for Lead

Properties	Value
ρ (kg/m^3)	11,340
Gruneisen coefficient	2.74
Parameter C1 (m/s)	2.006e3
Parameter S1	1.42
Reference temperature (K)	295
Specific heat (J/KgK)	124
Strength	Steinberg-Guinan
Shear modulus (MPa)	8,600
Yield stress (Mpa)	8
Max yield stress (Mpa)	100
Hardening constant	110
Hardening exponent	0.52
dG/dP	1
dG/dT (Mpa/K)	–9.976
dY/dP	9.304e–4
Melting temperature (K)	760

$$\Psi = C_{10}(\overline{I_1} - 3) + C_{01}(\overline{I_2} - 3) + C_{11}(\overline{I_1} - 3)(\overline{I_2} - 3) + \frac{1}{d}(J - 1)^2 \qquad (22.18)$$

where C_{10}, C_{01}, and C_{11} are material constants and d is the material incompressibility parameter. Material properties are taken from (Ignatova & Sapozhnikov, 2015)] as given in Table 22.4.

22.3.5 ELASTOPLASTIC MODEL FOR CLAY

This model uses the original von Mises premise that the yield stress has a constant value. Consequently, the von Mises cylinder has a fixed radius. States lying inside the cylinder are elastic. States on the surface of the cylinder are plastic. This model does not generally take into account the effect of strain hardening, strain rate sensitivity, or thermal softening. The material model

TABLE 22.4
Material Model Parameters for Foam

Properties	Value
ρ (kg/m^3)	100
C_{10} (kPa)	5.927000e+04
C_{01} (kPa)	−4.474000e+04
$C_{11}\left(\text{kPa}\right)$	8.990000e+03
d (/kPa)	1.000000e–06

parameters are taken from (Carton, Roebroeks, Broos, Halls, & Zheng, 2014) and are given in Table 22.5.

22.4   RESULTS AND DISCUSSION

Four shots were fired at the target. The bullet impact locations on the armor have been decided such that each impact location is at least 51 mm from any edge of the armor and the distance between subsequent impacts is 51 mm or more as per BIS. It was observed that though it takes almost 0.10 milliseconds for the velocity of the bullet to reduce to zero, a subsequent impact after 0.12 milliseconds of the previous impact does not affect the results. The BFS (Back Face Signature) plots of the last layer of HAP for four impacts with time gaps of 0.12 milliseconds are shown in Figure 22.3. It may be noted from the BFS obtained on the last layer of HAP that the previous shot does not increase the subsequent BFD at that location. Therefore, the time gap between impacts was kept at 0.12 milliseconds for all multiple impacts. The BFD caused by the first bullet does not cause any deformation at the second, third, and fourth impact locations.

The first three layers of HAP were reported to be completely perforated in the numerical model. There is perforation in the fourth shot in the fourth layer. The last, i.e., the fifth layer of HAP does not suffer any perforation. The HAP layers one to five after all four shots are shown in Figure 22.4. Maximum BFD obtained on clay was measured experimentally after the fourth shot. BFS values obtained on clay from the simulation were compared with the experimental results as shown in Table 22.6. The simulation results have shown a good correlation with experimental results.

As discussed earlier, HAP was modeled as five layers each of 2.5 mm thickness in simulation whereas the armor panel used in experiments was manufactured by combining more than 90 layers of UHMWPE plies. Therefore, it was difficult to replicate the delamination observed in experiments. However, the cut section views confirm that the phenomena of delamination are captured well in the simulation. For all the shots there is complete delamination of the first three layers of HAP. Delamination between the fourth and fifth layers is marginal. Figure 22.5 shows the delamination in layers of HAP.

TABLE 22.5
Material Model Parameters for Clay

Properties	Value
ρ (kg/m^3)	1,539
Bulk modulus (MPa)	274
Shear modulus (MPa)	2.2
Yield stress (MPa)	0.06

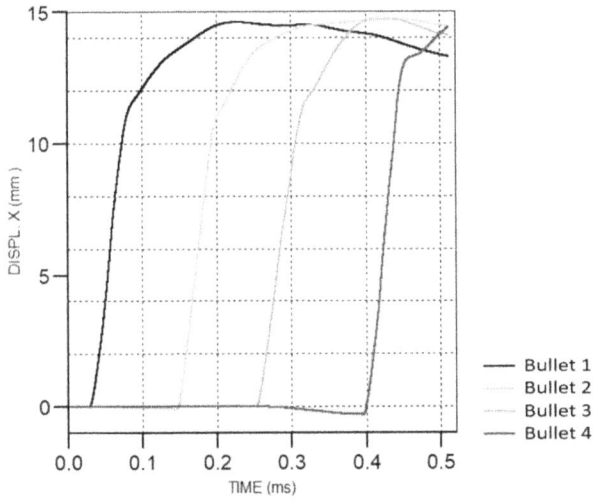

FIGURE 22.3 Back face deformation on the last layer of HAP.

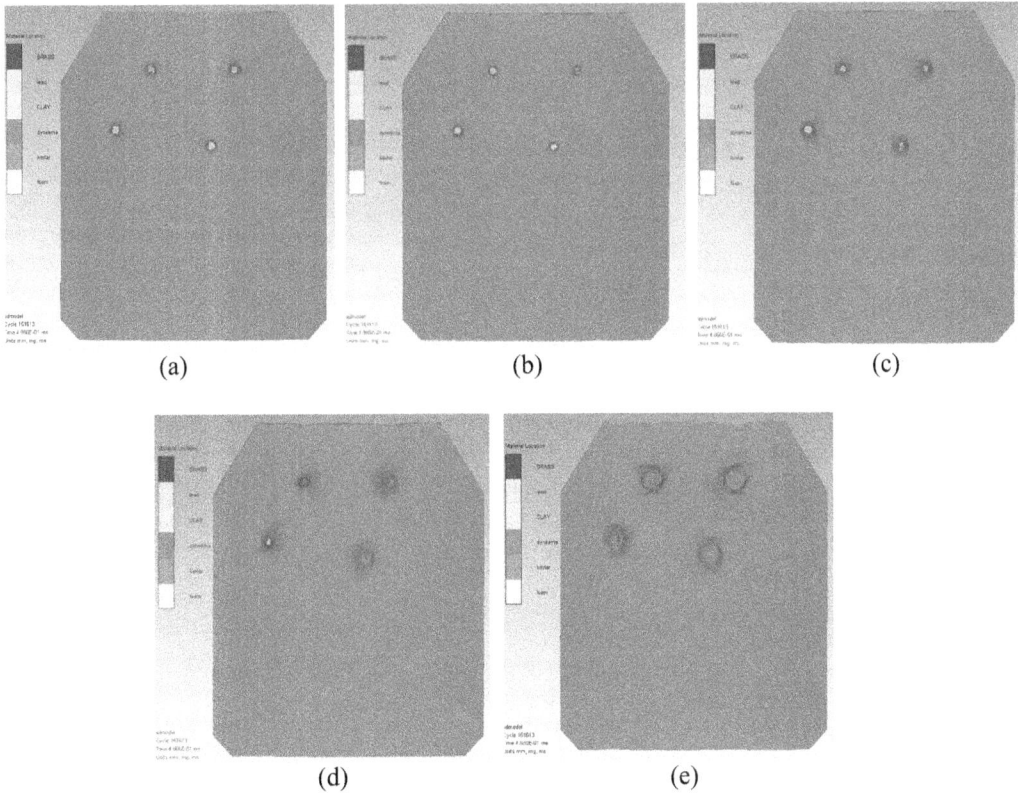

FIGURE 22.4   (a) First layer of HAP; (b) second layer of HAP; (c) third layer of HAP; (d) fourth layer of HAP; (e) fifth layer of HAP.

TABLE 22.6
Comparison of BFD Obtained Experimentally with Simulation

Shot no.	Velocity of projectile (m/s)	BFD (mm) obtained experimentally on clay	BFD (mm) obtained by simulation on clay
1.	822.3	17.5	12.8
2.	824.9	11.7	13.3
3.	823.1	7.2	14.8
4.	823.8	12.4	13.2

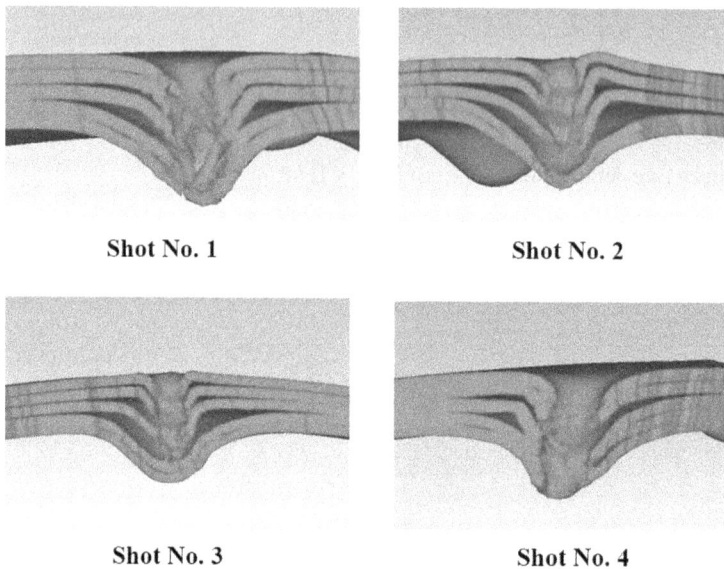

Shot No. 1   Shot No. 2

Shot No. 3   Shot No. 4

FIGURE 22.5 Delamination in the layers of HAP.

22.5 CONCLUSION

The BFS obtained experimentally were matched with the simulation. The various aspects such as delamination were studied only using simulation because it was difficult to study delamination with experiments. The simulation has given the experience of the correct mesh size, erosion criteria, and correct material model.

REFERENCES

S. Taylor and D. Carr, (1999), Post failure analysis of 0/90 degrees ultra-high molecular weight poly ethylene composite after ballistic testing, *Journal of Microscopy*, 196(2):249–256.

S. Chocron, C. E. Anderson, D. J. Grosch, and C. H. Popelar, (2001), Impact of the 7.62-mm APM2 projectile against the edge of a metallic target, *International Journal of Impact Engineering*, 25:423–437.

M. Giglio, A. Gilioli, A. Manes, L. Peroni, and M. Scapin, (2012), Investigation about the influence of the mechanical properties of lead core and brass jacket of a NATO 7.62mm ball bullet in numerical simulations of ballistic impacts, EDP Sciences.

A. Manes, D. Lumassi, L. Giudici, and M. Giglio, (2013), An experimental–numerical investigation on aluminium tubes Subjected to ballistic impact with softcore 7.62 ball projectiles, *Thin-Walled Structures*, 73: 68–80.

L. H. Nguyen, (2015), The ballistic performance of thick ultra high molecular weight polyethylene composite.

S. Choudharya, P. K. Singh, S. Khare, K. Kumar, P. Mahajan, and R. K. Verma, (2020), Ballistic impact behaviour of newly developed armour grade steel: An experimental and numerical study, *International Journal of Impact Engineering*, 140(2020):103557.

E. Carton, G. Roebroeks, H. Broos, V. Halls, and J. Zheng, (2014), Characterization of dynamic properties of ballistic clay.

A. V. Ignatova and S. B. Sapozhnikov, (2015), Two-scale modeling of the mechanical behavior of a composite foam, *Mechanics of Composite Materials*.

N. Bhatnagar, (2021), Indian Institute of Technology Delhi, (pvt. communication).

23 High-Velocity Impact Response of Titanium/ Composite Laminates
An Analytical Modeling

Ankush P. Sharma and R. Velmurugan

23.1 INTRODUCTION

An assembly of layers of thin metal alloy sheets interchangeably bonded to fiber-reinforced polymer composite is termed a fiber metal laminate (FML). Postponing and preventing crack growth, high damage tolerance to fatigue crack growth, low density, and blunt notch strength are exceptional FML properties, which make them considered in industrial structures (Sinmazçelik et al. 2011). The low- and high-velocity impact properties of GLARE (glass fiber-reinforced aluminum laminate), ARALL (aramid fiber-reinforced aluminum laminate), and CARALL (carbon fiber-reinforced aluminum laminate) have been principally investigated (Krishnakumar 1994). Aluminum as a principal structural material cannot be employed at high temperatures, viz. 180° C, on account of its poorer creep resistance than titanium alloy (Burianek and Spearing 2002). A later alloy is to be substituted for the following generation of FMLs. The impact and corrosion resistances of composites are to be further maximized and protected individually (Jakubczak, Bieniás, and Drózdziel 2020). Hybrid titanium composite laminate can replace conventional FMLs by combining enhanced mechanical and thermal properties with increased damage tolerance.

Ramadhan et al. (2013) have demonstrated that the predicted projectiles' residual velocity at penetration and energy absorptions of FMLs follow experiments. The optimum impact-resistant structure is the plate with back aluminum. Abdullah and Cantwell (2012) have examined approximately the same specific high-velocity perforation energy of laminates with growing ply number. Plastic membrane stretching, aluminum layer fracture, delamination, and fiber fracture are the damage mechanisms of the perforated plates. Further, the Reid-Wen model successfully predicts the perforation resistance of the FMLs. Chen et al. (2013) have found that the rear and front aluminum surfaces of GLAREs show linear and nearly highly curved initial cracking and circular- and elliptical-shaped plugging for projectile initial velocity larger than the ballistic limit velocity under normal and oblique ballistic impact, respectively. Also, the latter velocity is higher at normal than oblique impact. Li et al. (2016) have shown that for BARALL (basalt fiber-reinforced aluminum laminate), GLARE, and ARALL, the global deformation increases and decreases for velocity before and at penetration, separately. The localized deformation region displays similar aluminum ply tearing, fiber fracture, matrix cracks, debonding, and delamination for all three laminates. Xu et al. (2018) have scrutinized that a better penetration resistance to flat-nosed projectiles is shown by CFRP and CARALLs laminated orthogonally than to hemispherical- and sharp-nosed projectiles. Sharma et al. (2021a) have reported that the extent of opening and spreading of delamination is significantly influenced by distribution of titanium layers within the FMLs. Comparable low-velocity impact behavior of FMLs is observed (Sharma and Velmurugan 2022). Also, Sharma and Velmurugan

DOI: 10.1201/9781003352358-23

(2021a) have reported that the titanium-based FML with more metallic layers exhibits the progressive failure of composites, dissipating more membrane energy than bending energy before fracture, succeeded by FMLs with less and outside metallic layers under low-velocity impact perforation.

It is apparent from the above investigations that the high-velocity impact response of titanium-based FMLs with a constant thickness of total metal layers has not been predicted analytically particularly ahead of the first composite failure. Consequently, the present study emphasizes this direction. Four distinct FML layups entailing titanium alloy Ti-6Al-4V sheets of different thicknesses and unidirectional glass fiber-reinforced epoxy (GFRP) layers are considered. The predicted out-of-plane deformation history of aluminum-based FMLs is associated with experiments and titanium-based FMLs, assessing the validity of the projected model.

23.2   PROBLEM FORMULATION

A thin, circular, and clamped titanium (Ti)/GFRP plate having radius α and thickness h is examined. A hemispherical projectile of tip radius R, mass M_o, and preliminary kinetic energy E_k is employed to impact the plate center. A large aspect ratio α/h is supposed to acquire small shear deformation and confined indentation. A polar coordinate system (r, θ, z) is considered at the plate center. The boundary conditions satisfied by the plate lateral displacement w are $w = 0, \dfrac{\partial w}{\partial r} = 0 \ (r = \alpha)$. The total energy absorption before the first composite failure $E_{abs\,tot}$ is the addition of membrane and bending deformation energies of titanium and composite layers E_{def} and delamination energy of composites E_{del}, $E_{abs\,tot} = E_{def} + E_{del}$.

23.3   THEORETICAL FORMULATION

The high-velocity impact response of the circular FML plate is predicted in three stages using the spring-mass system (Tsamasphyros and Bikakis 2011; Hoo Fatt et al. 2003). The plate is initially deformed by the projectile until composites exhibit delamination. The resultant differential equation of motion is (Tsamasphyros and Bikakis 2009):

$$\left(M_0 + m_e\right)\ddot{w}_0 + P_0 + \left(K_b + K_{m1}\right)w_0 + K_{m2}w_0^3 = 0 \tag{23.1}$$

wherein the load through-loading $P_L\left(w_0\right)$ is:

$$P_L\left(w_0\right) = P_0 + \left(K_b + K_{m1}\right)w_0 + K_{m2}w_0^3 \tag{23.2}$$

The effective mass of the plate m_e is quantified (Tsamasphyros and Bikakis 2011). The factors of Eq. (23.2) are (Tsamasphyros and Bikakis 2009):

$$P_0 = 4M_{xy}, K_b = \left[\begin{array}{l}\left(3.318 + 2.906\ln\alpha\right)\left(D_{11} + D_{22}\right) + \left(-8.124 + 1.938\ln\alpha\right)D_{12} \\ + \left(14.758 + 3.876\ln\alpha\right)D_{66}\end{array}\right]\dfrac{1}{\alpha^2}$$

$$K_{m1} = 0.576\left(N_x + N_y\right) + 0.734N_{xy}, K_{m2} = \left[0.62\left(A_{11} + A_{22}\right) + 0.412\left(A_{12} + 2A_{66}\right)\right]\dfrac{1}{\alpha^2}$$

in which the extension, bending stiffnesses of the composite, and bending, membrane stiffnesses of the FML are designated by A_{ij}, D_{ij} and K_b, K_m, discretely. In-plane forces N_x, N_y, N_{xy} and moments M_x, M_y, M_{xy} of titanium layers are calculated by assuming its fully plastic membrane and pure bending stress distributions (Sharma et al. 2021b). At first, as the plate exhibits a projectile initial velocity v, the initial conditions are:

$$w_0(0) = 0, \qquad \dot{w}_0(0) = v = \sqrt{\frac{2E_k}{M_0}} \tag{23.3}$$

The second stage starts with a subsequent deflection to composites delamination w_0^d and settles at the maximum plate deformation $w_0^{\max}$. The differential Eq. (23.1) is also effective during this stage. The plate deforms initially and shows a condensed velocity following delamination $\dot{w}_{02}$. The preliminary conditions are:

$$w_0(0) = w_0^d, \dot{w}_0(0) = \dot{w}_{02} \tag{23.4}$$

The third stage initiates from $w_0^{\max}$ and accomplishes when the load becomes zero. The equivalent differential equation of motion using Eq. (23.2) and (Tsamasphyros and Bikakis 2011) is:

$$\left(M_0 + m_e\right)\ddot{w}_0 + P_0 + \left(K_b + K_{m1}\right)\left(2w_0 - w_0^{\max}\right) + K_{m2}w_0^3 = 0 \tag{23.5}$$

wherein the load through unloading $P_U(w_0)$ is:

$$P_U(w_0) = P_0 + \left(K_b + K_{m1}\right)\left(2w_0 - w_0^{\max}\right) + K_{m2}w_0^3 \tag{23.6}$$

Since the plate displays a zero velocity at maximum deformation, the initial conditions are:

$$w_0(0) = w_0^{\max}, \qquad \dot{w}_0(0) = 0 \tag{23.7}$$

Differential Eqs. (23.1) and (23.5) and the initial conditions by Eqs. (23.3), (23.4), and (23.7) enable three distinct initial value problems, representing the physical impact case. Initially, $w_0^d, \dot{w}_{02}, w_0^{\max}$, and differential equations of motion are to be assessed. Subsequently, $(w_0, t), (P, t), \left(\dfrac{dw_o}{dt}, t\right)$ curves can be obtained.

23.4   EXPERIMENTAL METHOD

Four distinct FML layups consisting of titanium alloy Ti-6Al-4V sheets and unidirectional (UD) E-GFRP layers are manufactured by hand layup technique and compression molding, displaying comparable total metal layer thickness in Table 23.1 (Sharma et al. 2021a). Here, T3, T4, and T6 represent 0.3 mm, 0.4 mm, and 0.6 mm thick titanium sheets, and 0 and 90 indicate unidirectional composite layers along 0° and 90° directions, separately. Moreover, four different FML layups consisting of aluminum alloy 2024-T3 sheets and UDGFRP layers are manufactured, exhibiting analogous total metal layer thickness as titanium-based FMLs in Table 23.1 (Sharma and Velmurugan 2022). Here, A3, A4, and A6 denote 0.3 mm, 0.4 mm, and 0.6 mm thick aluminum sheets. The manufacturing procedure and arrangement of FML layups are detailed elsewhere (Sharma et al. 2021a). The areal weight densities of the FMLs are comparable in Table 23.1. Square FMLs of 100 by 100 mm² with 70 mm diameter aperture are clamped circumferentially, with a speckle pattern of random intensity applied on the back surface. The impact tests are executed using a pressed nitrogen gas gun setup. A barrel of 12.5 mm diameter and 1 m long and a hardened maraging steel hemispherical-nosed projectile having a mass, diameter, and total span of 16.7 g, 12.47 mm, and 19

TABLE 23.1

Specifics of Titanium- and Aluminum-Based FMLs

FMLs	Layups Ti FML	Al FML	Ti/Al thickness (mm)	Total thickness (mm)	Areal weight (kg/m²) Ti FML	Al FML
2/1-0.6	[T6/0/90/90/0/T6]	[A6/0/90/90/0/A6]	0.6 +0.6	3.10	8.5	6.5
3/2-0.3(O)	[T3/0/90/T6/90/0/T3]	[A3/0/90/A6/90/0/A3]	0.3 +0.6 +0.3	3.37	9.0	7.0
3/2-0.4	[T4/0/90/T4/90/0/T4]	[A4/0/90/A4/90/0/A4]	0.4 +0.4 +0.4	3.30	8.9	6.9
4/3-0.3	[T3/0/T3/90/90/T3/0/T3]	[A3/0/A3/90/90/A3/0/A3]	0.3 +0.3 +0.3 +0.3	3.41	9.1	7.1

mm are utilized. The projectile incident velocity is measured by a set of two infrared LED sensors-photodiode circuit assembly. The optics and cameras are conserved from projectile rebound by a steel protective inclusion, exhibiting two acrylic windows in the rear with an included angle of about 152.5° and four windows on the edges. A speckle pattern is a stereo imaged by two Photron SA 1.1 high-speed cameras, and two Lowell Pro-light lamps are used to lighten the specimen. The out-of-plane deflection response of aluminum-based FMLs at 30 J, 45 J, 60 J, and 75 J energies and equivalent velocities of 60 m/s, 73 m/s, 85 m/s, and 95 m/s to high-velocity normal impact is measured by correlating the successive images using Vic-3D software. The particulars of the experimental setup, fixing mechanism, calibration process of the stereo rig, and camera variables are reported by Sharma and Khan (2018).

23.5 RESULTS AND DISCUSSION

23.5.1 DYNAMIC PROPERTIES OF FML CONSTITUENTS

The equations of motion ahead of the composite failure presented in section 23.3 are solved by a reiteration scheme. The projectile's initial velocity is supposed to evaluate the projectile and the panel deflection. The tensile strain rate of glass/epoxy laminate and titanium sheets is calculated from maximum strain and corresponding time, approaching near 500/s and 700/s, independently. Here, a high-speed digital image correlation technique is applied to measure strain right on the specimen gauge area (Sharma and Velmurugan 2023). The tensile properties of FML constituents are described. In the case of unidirectional glass/epoxy laminate, a 50% increase with in-plane normal and shear stiffnesses is observed over static values while Poisson's ratio is akin at nearly 500/s (Sharma et al. 2021b). Also, fracture strain and static failure energy density increase to a small extent, $\varepsilon_f = 2.6\%$, and 50%, $e_t = 6$ MPa (Sharma et al. 2021b). A comparable fracture toughness with mode II and interlaminar shear strength is supposed under dynamic and static loads, $G_{IIC}^{G/E} = 2\,kJ/m^2$ and $(ILSS)_{G/E} = 20\,MPa$ (Sharma et al. 2021b). For titanium alloy Ti-6Al-4V sheet, approximately 7% and 56% increases in elastic modulus and flow stress are observed, separately, and about 25% decrease in failure strain at nearly 700/s (Sharma et al. 2021b). In the case of aluminum alloy 2024-T3 sheet, the elastic modulus, yield, ultimate stresses, and failure strain are observed to be near rate insensitive (Hodowany et al. 2000).

23.5.2 Transient Behavior of FML Plate

Characteristic histories of titanium-based FMLs such as transient deflection, contact force, and velocity with membrane and bending energies at 30 J impact energy are shown in Figure 23.1. Pre- and post-delamination phases of FML 2/1-0.6 are indicated by I(a) and I(b), separately in Figure 23.1a. Such phases are also apparent in Figure 23.1b and c. Comparable phases of the other three FMLs can be shown as well. For deflection to composites delamination, FML 2/1-0.6 displays a lower approximation whereas the other three FMLs show higher and comparable values in Figure 23.1a. However, FMLs 4/3-0.3, 2/1-0.6, and both 3/2s display greater, smaller, and intermediate maximum deflection, respectively, indicating rebound initiation. In the case of FMLs force history shown in Figure 23.1b, initially, it is clear that the central deflection of FMLs is zero as the plate does not experience any deformation. Afterward, the deflection rises until it attains the maximum force. The corresponding time of FMLs 2/1-0.6, 3/2-0.3(O), 3/2-0.4, and 4/3-0.3 is 153 μs, 183 μs, 208 μs, and 189 μs, separately. From this location, central deflection begins to reduce, showing rebound initiation and continuing until the impact event is completed. Moreover, in the beginning, the contact force of FMLs surges until composites develop delamination, at which titanium sheets yield. Successively, the hardening region with altered slope continues until reaching the maximum force. This is owing to the FMLs 2/1-0.6 and both 3/2s, in which 0° and 90° composites are arranged together. This causes the bonding of matrix cracks of the 90° layer by the adjoining 0° layer. This results in a reduction of force

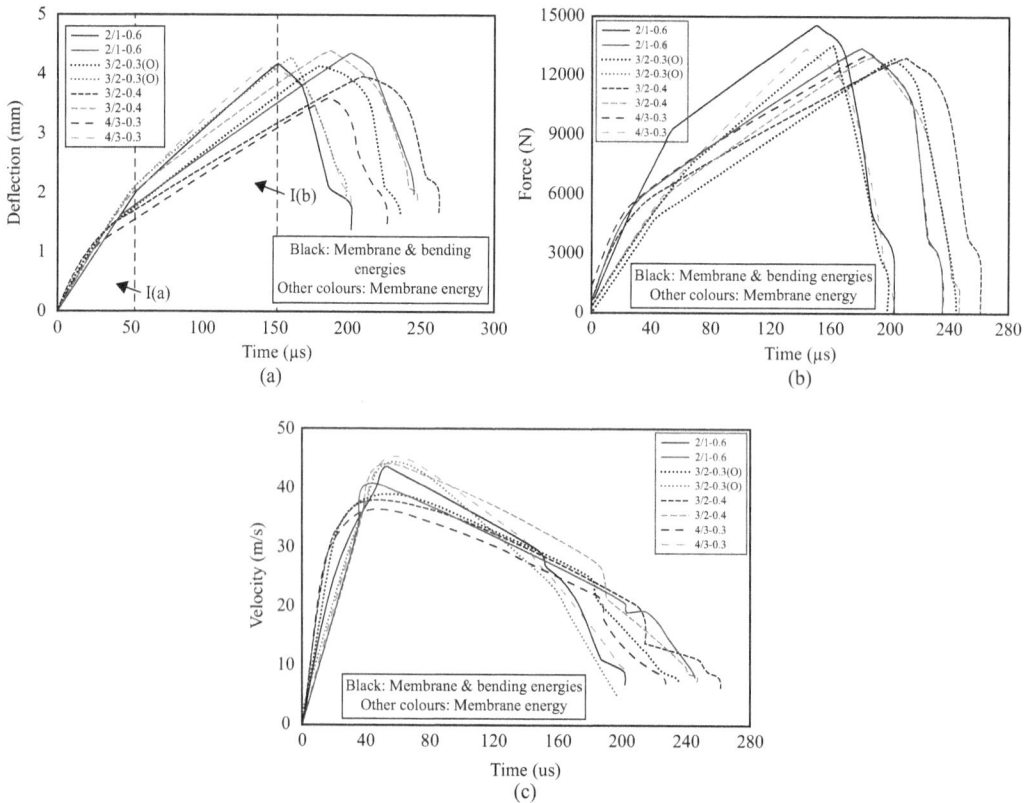

FIGURE 23.1 (a) Central deflection, (b) contact force, and (c) velocity of titanium-based FMLs 2/1-0.6, 3/2-0.3(O), 3/2-0.4, and 4/3-0.3, indicated by [T6/0/90/90/0/T6], [T3/0/90/T6/90/0/T3], [T4/0/90/T4/90/0/T4], and [T3/0/T3/90/90/T3/0/T3] layups, discretely at 30 J impact energy. T3, T4, and T6 denote 0.3 mm, 0.4 mm, and 0.6 mm thick titanium alloy sheets and 0 and 90 represent composite layers with fibers along 0° and 90° directions, respectively.

of all 0° layers and therefore FML. Constant slope reduction till achieving the maximum force is apparent due to matrix cracks bonding and yielding of titanium sheets. However, in the case of FML 4/3-0.3, 0° and 90° composites are disconnected by a metallic layer, resulting in a non-perpetual force decrease for 0° layers. In this viewpoint, the force history probably represents a linear slope until the maximum force more than additional FMLs. A higher, halfway (and analogous), and lower maximum force are shown by FMLs 2/1-0.6, both 3/2s, and 4/3-0.3, correspondingly and on the contrary for contact duration. This is enabled by the organization of composites within the FMLs. In the case of FML 2/1-0.6, all composites are structured together, resulting in ricocheting the laminate at a definite time, while for both FML 3/2s, [0/90] laminate is isolated by a metallic layer. This causes ricocheting off one of the [0/90] laminates at a specific time initially. Afterward, other [0/90] laminate ricochets at an augmented time. In the case of FML 4/3-0.3, 0° and 90° composites are detached by metallic layers. This causes ricocheting off one of the 0° layers at a precise time. Successively, other 0° layer ricochets in addition to [90]$_2$ laminate, each at a different amplified time. This is facilitated by the extent of damage evolution within the specimens, seeming greater for FML 4/3-0.3 because of more metallic layers. This is followed by both FML 3/2s presenting halfway and comparable damage extents and FML 2/1-0.6. However, FML 4/3-0.3 displays more spreading of damage than both FML 3/2s and 2/1-0.6, demonstrating disastrous damage. An enhanced ricochet and so the contact duration is shown by both FML 3/2s and 4/3-0.3 for separate [0/90] laminate and 0° layer than by FML 2/1-0.6 for [0/90]$_s$ laminate as elucidated above. Therefore, a more stable ricochet of composite layers succeeding maximum force is observed for FML 4/3-0.3, followed by both FML 3/2s signifying akin ricochet and FML 2/1-0.6. The corresponding plastic and elastic deformations of the titanium and composite layers continue until the falling load reaches zero. A residual deformation of FML 4/3-0.3 seems to be greater due to more metallic layers. This is followed by both FML 3/2s and 2/1-0.6 exhibiting a small number of thicker, thinner, less thick, and a few thicker metallic layers, illustrated in Figure 23.1a. Also, Figure 23.1c presents the velocity history of FMLs acquired from the ratio of deflection to equivalent time. FML 4/3-0.3 shows the maximum impact point velocity, succeeded by both FML 3/2s and 2/1-0.6. Thereafter, the velocity rebounds elastically at its natural frequency through the loading phase, succeeded by unloading. FML 4/3-0.3 seems to rebound by a lesser degree, trailed by both FML 3/2s and 2/1-0.6. Hence, a velocity seems to exhibit a comparable behavior to the deflection of FMLs. FML 4/3-0.3 displays lesser maximum force, greater energy absorption, deflection, velocity, and contact duration, and on the contrary for FML 2/1-0.6. Both FML 3/2s display the above features to be middle of the other FMLs. Thus, dispersing titanium layers across the thickness of FMLs predominantly affects the former features at 30 J.

Furthermore, in the case of FMLs with membrane energy only, analogous behavior is exhibited by deflection, contact force, the velocity to composites delamination, maximum force, and projectile rebound at 30 J as FMLs with membrane and bending energies in Figure 23.1. Specifically, a lesser initial slope of force history and maximum force is observed for FMLs with membrane energy only than FMLs with membrane and bending energies. The former aspects appear to be contrary to the deflection and velocity histories of FMLs. FMLs with membrane energy only display higher time, deflection, and velocity to composites delamination, maximum force, and impact closure events than FMLs with membrane and bending energies and in contrast for energy absorption. Hence, FMLs with membrane energy seem to contribute more substantially than bending energy.

23.5.3 PERFORMANCE PARAMETERS

The different performance parameters of titanium-based FMLs are predicted with membrane and bending energies and membrane energy only at 30 J, 45 J, 60 J, and 75 J impact energies (Sharma et al. 2021a). The maximum deflection of FMLs escalates with impact energies in Figure 23.2a, exhibiting higher maximum deflection for FML 4/3-0.3 and akin and lower values for the additional three FMLs.

Also, the maximum velocity of FMLs rises with impact energies in Figure 23.2b, displaying greater, lesser, and in-between maximum velocity for FMLs 4/3-0.3, 2/1-0.6, and both 3/2s, separately, following the behavior of maximum deflection of FMLs. The maximum force of FMLs increases initially with energies with higher and lower maximum force for FMLs 2/1-0.6, both 3/2s, and 4/3-0.3, independently in Figure 23.2c, exhibiting a contradictory behavior of the maximum deflection of FMLs. Moreover, the energy absorption of FMLs 2/1-0.6, both 3/2s, and 4/3-0.3 seem to be lower, midway (and akin), and higher, respectively and so are the metal/composite interfaces. This results in corresponding delamination energy and so the energy absorption. Matrix cracks, elastic deformation of fiber, inter-layer delamination, and plastic deformation of metal layers absorb energy. Thus, the energy absorption of FMLs increases by dispersing metallic layers as presented in Figure 23.2d, displaying a comparable behavior to the maximum deflection and maximum velocity of FMLs (Sharma et al. 2021a). Also, the above parameters of FMLs are affected considerably by spreading metallic layers, exhibiting analogous behavior with membrane and bending energies and membrane energy only. This recommends a substantial contribution of membrane energy over bending energy.

23.5.4   Validation of the Analytical Model

The predicted out-of-plane transient deflection history up to maximum deflection of aluminum-based FMLs 2/1-0.6, 3/2-0.3(O), 3/2-0.4, and 4/3-0.3 with membrane and bending energies and membrane energy alone at 30 J impact energy is compared with experiments in Figure 23.3 (Sharma and Khan 2018). Here, time is assessed by numerically integrating velocity and deflection. In experiments, the deflection is initially zero and subsequently increases until titanium layers yield. Up to this stage, titanium and composite layers are linearly elastic. Successively, the deflection history

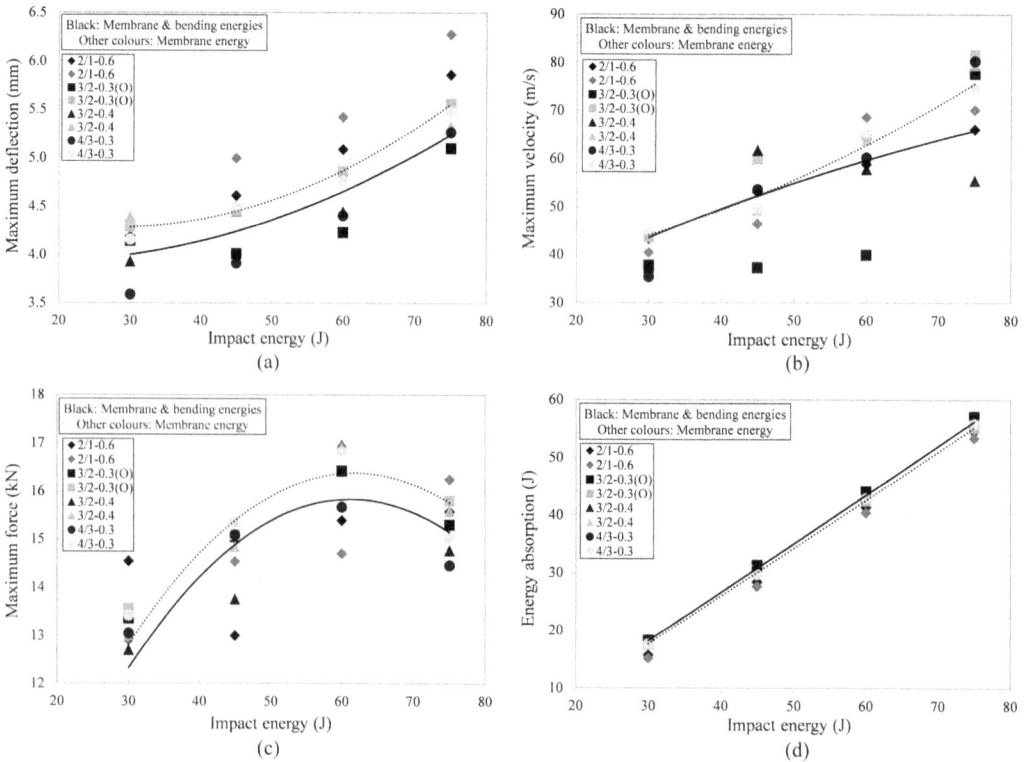

FIGURE 23.2   (a) Maximum deflection, (b) maximum velocity, (c) maximum force, and (d) energy absorption of titanium-based FMLs at various impact energies.

hardens until attaining its peak value. FMLs with membrane and bending energies and membrane energy alone exhibit elastic and plastic behaviors with two different slopes with time ranges of about 49–77 μs, 81–89 μs, and 77–139 μs, 89–158 μs up to peak deflection, respectively. The predicted maximum deflection, and time close to maximum deflection of the above FMLs sequence with membrane and bending energies and membrane energy only is to be within 7%, 13%, 8%, 23%; 10%, 4%, 6%, 10%; 32%, 32%, 30%, 20%; and 16%, 29%, 16%, 27% of experiments, discretely. Thus, the predictions of FMLs with membrane energy alone work relatively better with experiments than FMLs with membrane and bending energies in Figure 23.3, validating the analytical model. Further, membrane energy contributes primarily to bending energy.

Additionally, the predicted parameters of aluminum-based FMLs at different impact energies are associated with experiments in Figure 23.4 (Sharma and Khan 2018). The average difference between the estimated maximum deflection of FMLs 2/1-0.6, 3/2-0.3(O), 3/2-0.4, 4/3-0.3 with membrane and bending energies, and membrane energy alone at 30 J, 45 J, 60 J, 75 J is to be within 14%, 19%, 30%, 27%, and 9%, 9%, 22%, 21% of the measured data, individually. Lesser and greater (and comparable) maximum deflection is shown by FML 2/1-0.6 and an additional three FMLs, discretely, with akin behavior to titanium-based FMLs in Figure 23.2a. Moreover, the average variance between the projected maximum velocity and energy absorption of the above FMLs order with membrane and bending energies and membrane energy alone at the same energies is to be within 22%, 13%, 14%, 17%; 17%, 14%, 14%, 26%; 4%, 6%, 5%, 2%; and 17%, 6%, 4%, 6% of the measured data, separately. As a whole, this suggests that the predictions of FMLs with membrane energy alone better follow the experiments than FMLs with membrane and bending energies. Also, the above parameters of FMLs are affected by spreading metallic layers. The behavior of maximum deflection, velocity, and energy absorption of aluminum-based FMLs seems to be analogous, while the extent of the former and latter parameter sets is higher and lower than titanium-based FMLs in Figure 23.2a, b, and d. This recommends that the predicted deflection history up to its peak value and typical parameters of titanium-based FMLs with membrane and bending deformation energies at various impact energies are also anticipated to be a good match with experimental data.

23.6 SUMMARY

The high-velocity impact response of titanium-based FMLs at various impact energies before the first composite failure is estimated using mass-spring organization. Titanium-based FMLs with

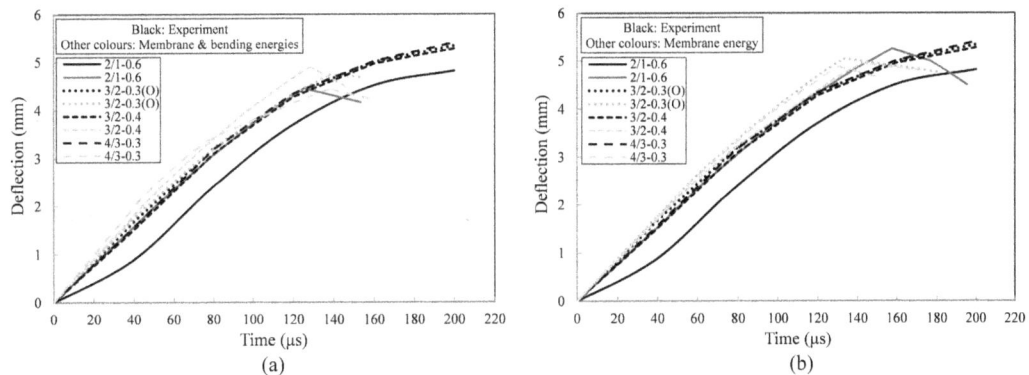

FIGURE 23.3 Transient out-of-plane deformation histories of aluminum-based FMLs 2/1-0.6, 3/2-0.3(O), 3/2-0.4, and 4/3-0.3, indicated by [A6/0/90/90/0/A6], [A3/0/90/A6/90/0/A3], [A4/0/90/A4/90/0/A4], and [A3/0/A3/90/90/A3/0/A3] layups, individually at 30 J impact energy. A3, A4, and A6 designate 0.3 mm, 0.4 mm, and 0.6 mm thick aluminum alloy sheets and 0 and 90 denote composite layers with fibers along 0° and 90° directions, respectively.

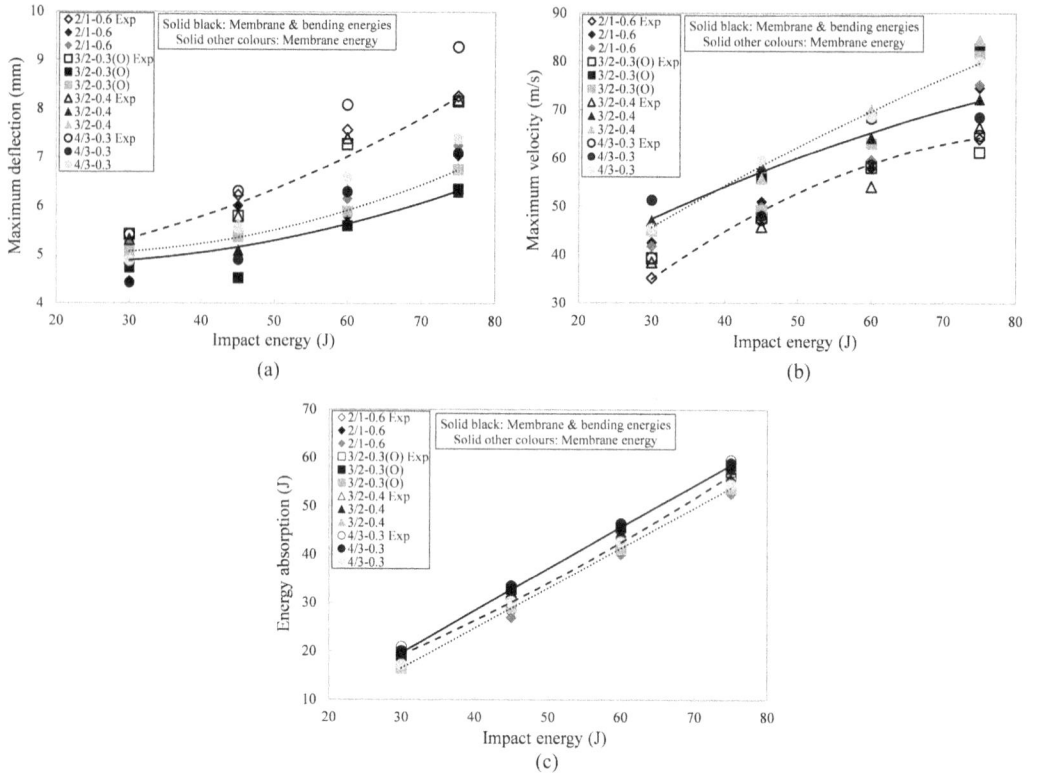

FIGURE 23.4 (a) Maximum deflection, (b) maximum velocity, and (c) energy absorption of aluminum-based FMLs at various impact energies.

more, less, and outer metallic layers show smaller, intermediate, and larger maximum force and vice versa for time to delamination, time to maximum force, maximum deflection, total duration, and composites recovery, respectively. FMLs with more metallic layers show more metal/composite connections, resulting in more delamination energy and so more total energy absorption. This is succeeded by FMLs with lesser and outer metallic layers, exhibiting higher delamination energy. Also, the estimated out-of-plane deformation history up to its maximum value, and the maximum deflection, velocity, and energy absorption of aluminum-based FMLs with membrane energy alone at 30 J, and at 30 J, 45 J, 60 J, and 75 J follow experiments better, respectively, than FMLs with membrane and bending energies. This validates the analytical model. The behavior of total energy absorption, and time to delamination, time to maximum deflection, maximum deflection, total duration of aluminum-based FMLs at various energies seem to be equivalent to titanium-based FMLs. However, the degree of previous and later parameter sets seems to be inferior, and greater for previous than later FMLs. This directs a higher impact resistance of titanium- than aluminum-based FMLs.

REFERENCES

Abdullah, M. R., and W. J. Cantwell. The high-velocity impact response of thermoplastic–matrix fibre–metal laminates. *The Journal of Strain Analysis for Engineering Design* 47(7), 432–443, 2012.

Burianek, D. A., and S. M. Spearing. Fatigue damage in titanium-graphite hybrid laminates. *Composites Science and Technology* 62(5), 607–617, 2002.

Chen, Y., B. Pang, W. Zheng, and K. Peng. Experimental investigation on normal and oblique ballistic impact behavior of fiber metal Laminates. *Journal of Reinforced Plastics and Composites* 32(23), 1769–1778, 2013.

Hodowany, J., G. Ravichandran, A. J. Rosakis, and P. Rosakis. Partition of plastic work into heat and stored energy in metals. *Experimental Mechanics* 40, 113–123, 2000.

Hoo Fatt, M. S., C. Lin, D. M. RevilockJr., and D. A. Hopkins. Ballistic impact of GLARE™ fibre-metal laminates. *Composite Structures* 61(1–2), 73–88, 2003.

Jakubczak, P., J. Bienia's, and M. Dro'zdziel. The collation of impact behaviour of titanium/carbon, aluminum/carbon and conventional carbon fibres laminates. Thin-Walled Structures 155, 106952, 2020.

Krishnakumar, S. Fiber metal laminates-the synthesis of metals and composites. *Materials and Manufacturing Processes* 9(2), 295–354, 1994.

Li, X., M. Y. Yahya, A. B. Nia, Z. Wang, and G. Lu. Dynamic failure of fibre-metal laminates under impact loading – Experimental observations. *Journal of Reinforced Plastics and Composites* 35(4), 305–319, 2016.

Ramadhan, A. A., A. R. Abu Talib, A. S. Mohd Rafie, and R. Zahari. High velocity impact response of Kevlar-29/epoxy and 6061-T6 aluminum laminated panels. *Materials & Design* 43, 307–321, 2013.

Sharma, A. P., and R. Velmurugan. Analytical modelling of low-velocity impact response characterization of titanium and glass fibre reinforced polymer hybrid laminate composites. *Thin-Walled Structures* 175, 109236, 2022.

Sharma, A. P., and R. Velmurugan. Damage and energy absorption characteristics of glass fiber reinforced titanium laminates to low-velocity impact. *Mechanics of Advanced Materials and Structures* 29(27), 6242–6265, 2022.

Sharma, A. P., and R. Velmurugan. Effect of high strain rate on tensile response and failure analysis of titanium/glass fibr reinforced polymer Composites. *Journal of Composite Materials* 55(24), 3443–3470, 2021a.

Sharma, A. P., and S. H. Khan. Influence of metal layer distribution on the projectiles impact response of glass fiber reinforced aluminum laminates. *Polymer Testing* 70, 320–347, 2018.

Sharma, A., and R. Velmurugan. Low-velocity impact perforation response of titanium/composite laminates: Analytical and experimental investigation. *Mechanics Based Design of Structures and Machines* 51(9), 5179–5212, 2023.

Sharma, A. P., R. Velmurugan, K. Shankar, and S. K. Ha. High-velocity impact response of titanium-based fiber metal laminates. Part I: experimental investigations. *International Journal of Impact Engineering* 152, 103845, 2021a.

Sharma, A. P., R. Velmurugan, K. Shankar, and S. K. Ha. High-velocity impact response of titanium-based fiber metal laminates. Part II: Analytical modelling. *International Journal of Impact Engineering* 152, 103853, 2021b.

Sinmazçelik, T., E. Avcu, M. Ö. Bora, and O. Çoban. A review: Fibre metal laminates, background, bonding types and applied test methods. *Materials & Design* 32(7), 3671–3685, 2011.

Tsamasphyros, G. J., and G. S. Bikakis. Analytical modeling of circular glare laminated plates under lateral indentation. *Advanced Composites Letters* 18(1), 11–19, 2009.

Tsamasphyros, G. J., and G. S. Bikakis. Dynamic response of circular GLARE fiber–metal laminates subjected to low velocity impact. *Journal of Reinforced Plastics and Composites* 30(11), 978–987, 2011.

Xu, M.-M., G.-Y. Huang, Y.-X. Dong, and S.-S. Feng. An experimental investigation into the high velocity penetration resistance of CFRP and CFRP/Aluminium laminates. *Composite Structures* 188, 450–460, 2018.

24 Transient Dynamic Response of Clay Brick Masonry Walls of Varying Aspect Ratios against Low-Velocity Impact Load

K. Senthil, Ankush Thakur, S. Rupali,
A. P. Singh and M. A. Iqbal

24.1 INTRODUCTION

Boundary walls of commercial and residential areas are generally made up of masonry walls. These walls, when constructed near roadsides or congested road areas, are subjected to vehicular crashes. It was observed from past cases that a wall exhibits local damage during impact events either due to the low tensile strength of the brick or because the impact energy increases the fracture energy of the material. As a mitigation measure, the speed of vehicles is restricted in residential areas or near public buildings; however, even at low speeds, local damage appears to be significant (Gilbert et al., 2002). Other than this, parameters such as aspect ratio (h/b), impact location, strength of brick and mortar, etc., also influence the wall response. Matsumura (1998) found that with an increase in aspect ratio, the shear strength was found to decrease in hyperbolic function. Further, Asad et al. (2020) concluded that with an increase in aspect ratio, the failure mode changes from two-way to one-way bending and damage progresses towards the corners of walls. In another study by Asad et al. (2021), the damage index was calculated as the ratio of the wall damage area to the area of the impactor (height × width) and it was concluded that with an increase in impact velocity, the damage index value keeps on decreasing. Further, out-of-plane load-carrying capacity as well as lateral stiffness were found to be affected by the in-plane aspect ratio of walls (Panto et al., 2019). Khuda et al. (2016) numerically studied the influence of aspect ratio on the out-of-plane behaviour of walls. It was concluded that a higher aspect ratio wall has the highest load-carrying capacity but the least ductility. The increase in load-carrying capacity for higher aspect ratio walls is due to the contribution of the torsional behaviour of joints (Chang et al., 2021). Janaraj and Dhanasekar (2016) presented design expression for unreinforced masonry walls which includes aspect ratio due to significant contribution to failure mode and capacity in confined masonry walls. Risi et al. (2019) studied the influence of aspect ratio on the in-plane and out-of-plane interaction of the masonry walls. It was concluded that square walls were more effective as compared to rectangular in terms of damage index. Similar results were reported by Agnihotri et al. (2013) where it was concluded that with an increase in aspect ratio, out-of-plane load-carrying capacity and shear strength of masonry walls were found to decrease.

Based on the literature survey, it was observed that aspect ratio affects the response of walls significantly; however, literature pertaining to the impact response of clay brick walls was found to be limited. Therefore, the present investigations are focused on estimating the transient dynamic response of clay brick masonry walls of varying aspect ratios under impact load through a comprehensive experiment and numerical simulations. The response of walls subjected to impact load was studied experimentally (see section 24.2) and validated numerically using elastoplastic constitutive

DOI: 10.1201/9781003352358-24

models (see section 24.3). Further, a brief numerical modelling strategy along with a mesh sensitivity study is reported in section 24.4. The results are compared with the experimental observation in terms of peak force and residual displacements in section 24.5. Lastly, major findings of the present work along with concluding remarks are given in section 24.6.

24.2   EXPERIMENTAL INVESTIGATION

The experiments were performed on clay brick masonry walls with aspect ratios of 1.00, 1.36, 2.11, 3.33, and 4.80, which were designated as W1, W2, W3, W4, and W5, respectively. The clay brick wall of dimension $1.2 \times 1.2 \times 0.110$ m has an aspect ratio of unity and is subjected to a 60 kg hemispherical nose-shaped impactor. The impactor comprises hard steel with a length of 630 mm and an impact head radius of 80 mm[1]. The test was carried out using a pendulum impact testing machine and the pendulum angle was between 28° and 30° (equivalent to 1.98 ms⁻¹ impact velocity). The response of walls of varying aspect ratios was studied under low-velocity impact loading. The results were presented in terms of force versus time and failure pattern and are discussed in this section.

24.2.1   Force-Time Response

The force versus time response on W1, W2, W3, W4, and W5 was studied for low-velocity impact loading. The peak force was 16.51, 17.5, 22.13, 16.29, and 15.22 kN, respectively.

The maximum peak force was observed for the wall with an aspect ratio of two, whereas beyond that, peak force was found to be decreasing. Also, instant decrement in peak force was observed for walls with aspect ratios of 3.33 and 4.80 as compared to walls with an aspect ratio of 2.00. The reason may be due to the propagation of stress waves within the wall as the number of joints was limited which made it more vulnerable. The peak force was found at 2 ms for W3 and W4, whereas the peak force was attained within 1.5 ms for W1, W2, and W5; see Figure 24.1.

24.2.2   Failure Mechanism

In the case of W1, vertical fracture cracks were observed near the impact region which passes through the bed joints of the walls. Further, it was observed that brick units were free from any crack; however, minor chipping was observed at the front face of the wall; see Figure 24.2(a–i). For W2, multiple vertical bed joint failures were observed on the wall which propagated along the full length and width of the brick joints. Similar to W1, no crack in the brick unit was observed; see Figure 24.2(b). Local shear sliding between brick joints was observed as stresses transmitted to bed joints provide a non-zero moment of resistance to impact load. Minor scabbing of mortar and brick material was also observed on the rear face of the wall. For W3, single vertical and horizontal fracture cracks were observed on both the front and rear sides of the wall; see Figure 24.2(c). Local shear sliding between the brick and mortar results in a vertical fracture line. It may be stated that impact load results in the formation of plastic hinges at the impact point which causes the formation of horizontal fracture cracks in the wall. In contrast to W1 and W2, the crack travels through the bricks as well as the mortar for W3. W4 and W5 also follow the same trend as W3, i.e., failure through the brick units and mortar rather than interface failure alone. However, the ejection of brick units with pure interface failure was observed on the right side of W5 (see Figure 24.2(d)) whereas cracking of brick units was observed on the left side. Also, a similar phenomenon was observed except ejection of brick units in the case of W6; see Figure 24.2(e). It was concluded that as the aspect ratio decreases, the mode of failure also changes from shear sliding alone to combined action of shear sliding and rocking failure.

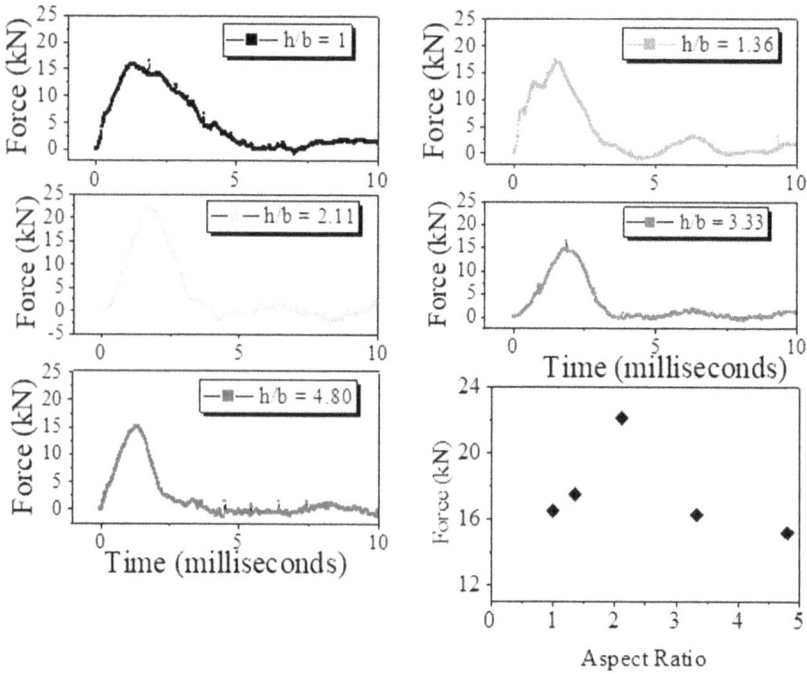

FIGURE 24.1 Influence of varying aspect ratios on peak force response of wall.

24.2.3 ENERGY ABSORPTION

The energy absorption by W1 and W2 has been calculated using the analytical expression shown in Eqn (24.1).

$$E_{ab} = \frac{1}{2} M \left(2gH\eta^2 - \left[\eta\sqrt{2gH} - \frac{1}{M} \int p(t)dt \right]^2 \right) \tag{24.1}$$

Where M is the hammer mass, g is the acceleration due to gravity, H is drop height, η is the efficiency of the system, in this case 98%, and $p(t)dt$ is the impulse value measured experimentally. The impulse $p(t)dt$ was measured by taking the area under the force-time history. The calculated impulse for W1, W2, W3, W4, and W5 was 46.98, 32.79, 38.94, 24.72, and 22.58 Ns, respectively. The results do not indicate any difference between the measured impulses. It has been observed that the energy absorption for W1 and W2 was 74.90 and 73.50 Joules, respectively. The impact duration was found to be increased on higher aspect ratio walls as compared to lower aspect ratio walls. The reduction in impulse on W2 was found to be reduced by 30% as compared to W1. However, variation is observed for W3 where impulse was found to increase by 18% as compared to W2. The highest reduction in the impulse was on W5 which was found to be reduced by 52% as compared to W1. Overall, the duration of impact played a significant role in determining the response of walls; however, the mass of the impactor is constant (60 kg) in all the tests. It has been observed that maximum energy is absorbed by W1 which is equivalent to 73.51 J. The energy loss is calculated based on the impact energy, and the energy absorbed by the wall is shown in Table 24.1. Despite having the same testing conditions, i.e., impactor contact area, wall base frictions, and support conditions for each wall, the loss is classified as the energy lost in damaging the brick walls. For instance, maximum damage has been observed in W5 with a maximum loss of energy equivalent to 63% though all the walls have mortar strength of 4.04 MPa.

Front face   (a)   Rear face

Front view   (b)   Back view

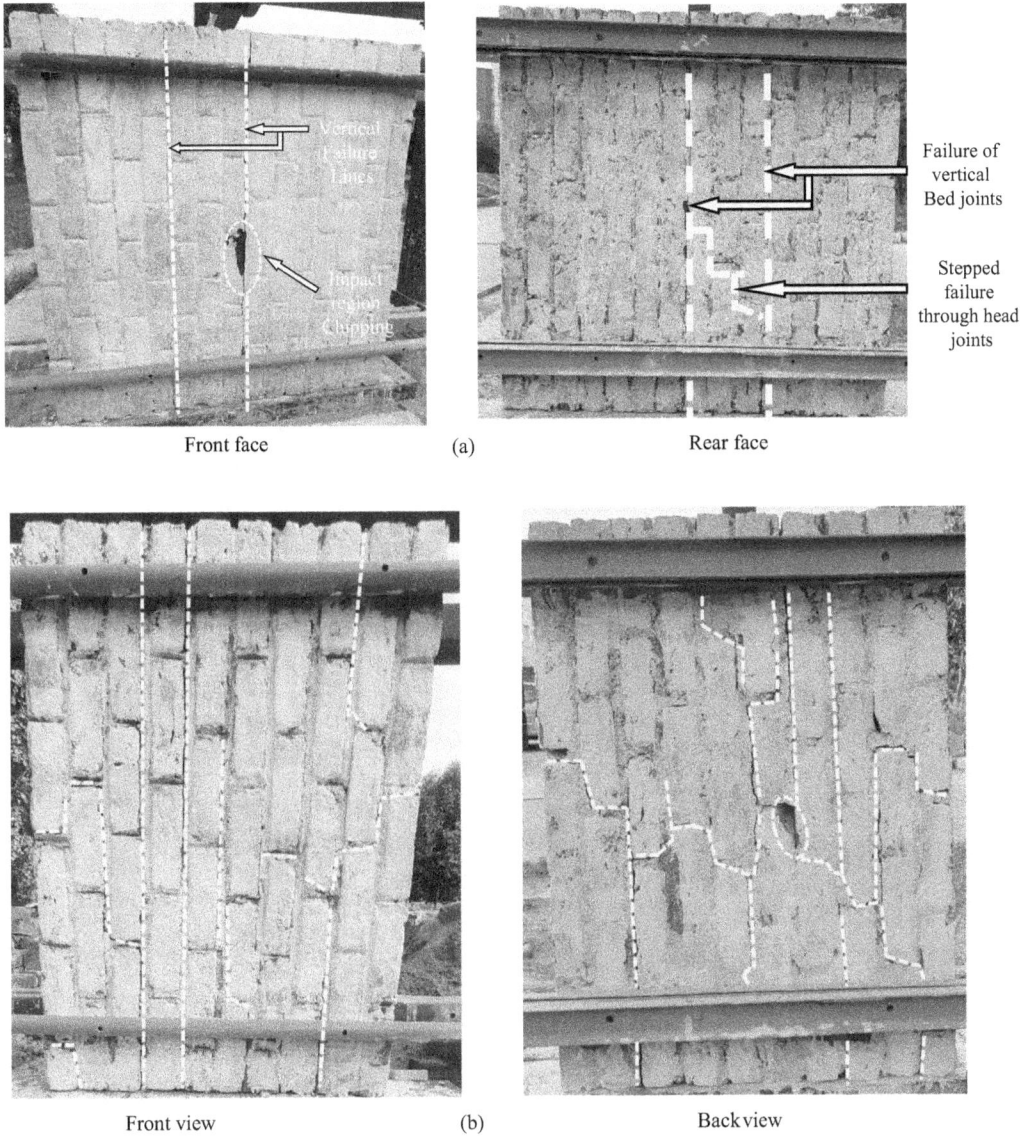

FIGURE 24.2 Deformation of (a) W1, (b) W2, (c) W3, (d) W4, and (e) W5 walls at right, left, and rear side face.

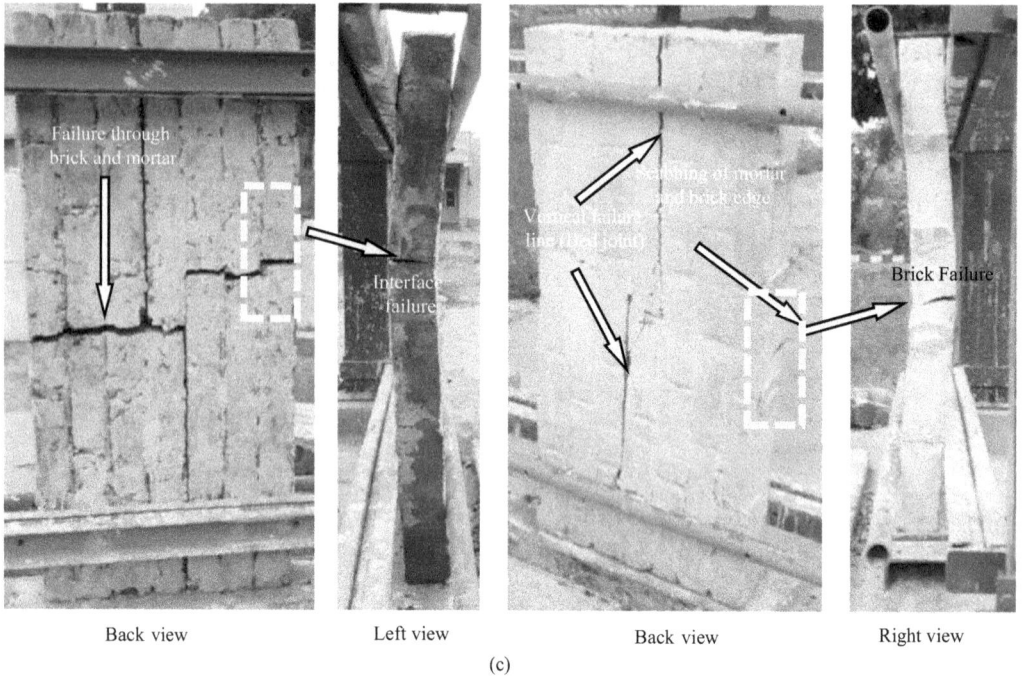

Back view   Left view   Back view   Right view

(c)

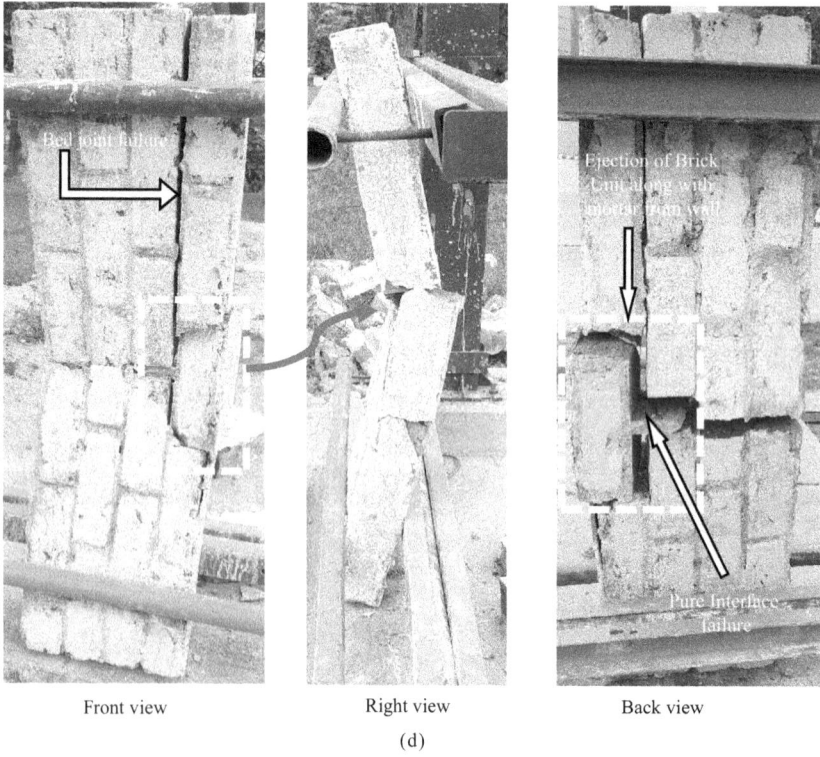

Front view   Right view   Back view

(d)

FIGURE 24.2 Continued

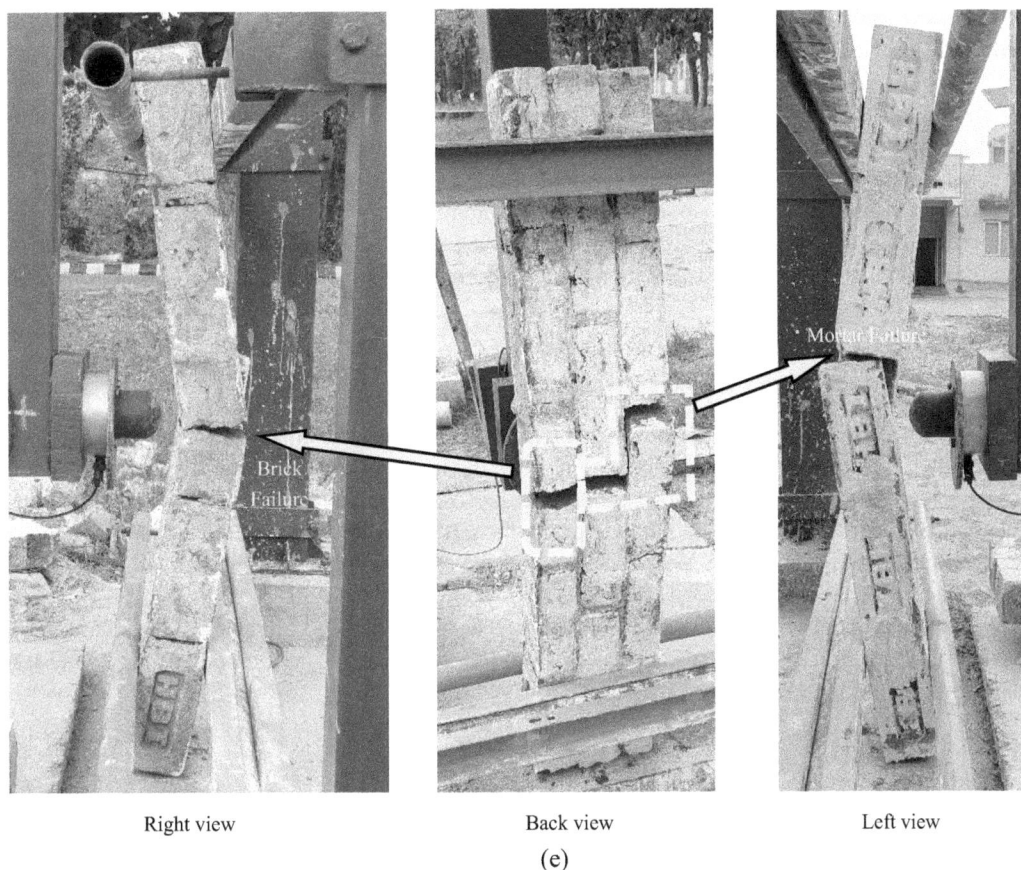

Right view   Back view   Left view

(e)

FIGURE 24.2 Continued

TABLE 24.1

Energy Absorption Capacity Calculated Using Eqn (24.1) on the Walls

Wall no.	Input energy (mgh)	Energy absorption capacity calculated by Eqn (24.1)	Energy loss (%)
W1	117.61	73.50	37%
W2	117.61	55.18	53%
W3	117.61	63.54	46%
W4	117.61	47.55	59%
W5	117.61	43.26	63%

24.3 CONSTITUTIVE BEHAVIOUR

The Drucker-Prager (DP) model was used to predict the response of masonry walls and is readily available in the ABAQUS materials library, whereas traction separation law was used to model the interface between brick and mortar and is discussed in this section.

24.3.1 DRUCKER-PRAGER MODEL

The elastoplastic response of clay brick walls was simulated using the Drucker-Prager (DP) model. The DP model is a hydrostatic stress-dependent model that allows isotropic hardening and softening of materials under compression, tension, and shear. Additionally, this model has been success-fully used in materials in which compressive strength was more than the tensile yield strength of the material (ABAQUS, 2019). Clay bricks satisfy both conditions; hence, the model is suitable for simulating masonry assemblage's stress-strain behaviour. Damage was defined based on uni-axial compressive strength (σ_c) combined with flow stresses, cohesion, and dilatation angle. The Drucker-Prager linear yield criterion was adopted in the present study.

Material properties used to simulate masonry compression assemblage are given in Table 24.2. Properties such as density, elastic modulus, and friction angle were determined from lab tests, whereas flow stress ratio (K) and dilatation angle (φ) were taken from ABAQUS (2019) and Mosalam et al. (2019) respectively.

24.3.2 BRICK–MORTAR INTERFACE BEHAVIOUR

Cohesive zone modelling (CZM) between brick units was used to model the interface between brick units. These were confined to form along with the layers of the bed and head joints which are basic features of the cohesive contact modelling approach. Traction separation law readily avail-able in ABAQUS was used to simulate the mechanical behaviour of brick joints, including damage initiation and damage evolution. Before the damage, it is assumed that cohesive behaviour follows a linear traction separation law; see Figure 24.3.

The stiffness coefficients shown in Figure 24.3 (K_{nn}, K_{ss}, and K_{tt}) are calculated based on the elastic modulus of brick (E_u) and mortar (E_m) and the thickness of the joint (t_m) using Eqn (24.2) and Eqn (24.3).

$$K_{nn} = \frac{E_u E_m}{t_m \left(E_u - E_m\right)};$$

(24.2)

TABLE 24.2

Drucker-Prager Model Properties for Clay Brick Masonry Walls

Density (γ)	Poisson ratio (υ)	Elastic modulus (E)	Friction angle (ϕ)	Flow stress ratio (K)	Dilatation angle (φ)
Kg/m³	–	MPa	°	–	°
1,980	0.15	2,075	38	1	11

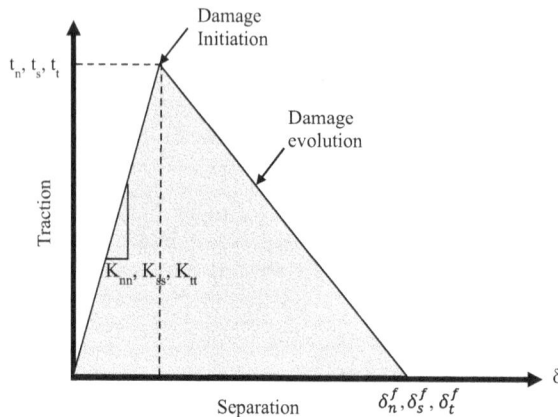

FIGURE 24.3   Traction separation response of brick interface. Source: Thakur et al. (2020).

$$K_{ss} = \frac{G_u G_m}{t_m \left(G_u - G_m \right)} \tag{24.3}$$

For damage, the quadratic traction criteria (see Eqn 24.4), which initiates damage upon user-defined parameters, was used. The normal (t_n) and shear stress (t_s and t_t) vectors represent Mode-I, Mode-II, and Mode-III failure modes. Other properties, such as cohesion, shear strength, tensile strength, friction coefficient, and fracture energy of joints are given in Table 24.3.

$$\left(\frac{t_n}{t_n^{max}} \right)^2 + \left(\frac{t_s}{t_s^{max}} \right)^2 + \left(\frac{t_t}{t_t^{max}} \right)^2 = 1 \tag{24.4}$$

24.4 NUMERICAL MODELLING AND MESH CONVERGENCE

The numerical modelling technique along with the mesh convergence study has been published elsewhere by authors (Senthil et al., 2022); however, key points are withdrawn and mentioned here. Simplified micro-modelling (SMM) technique was used to visualize the crack patterns in the brick–mortar interfaces. The elastoplastic response of the walls was incorporated by constitutive models discussed in the previous section. For simulation, the size of the brick considered was 230 mm, 110 mm, and 80 mm (length, breadth, height). It should be noted that the length and height of the brick are increased by 5 mm to compensate for the mortar thickness as the interface between the brick-and-mortar interface. Further cohesive behaviour was introduced between the head and bed joints of brick units using traction separation law as discussed above. The cohesive interface properties that were discussed are shown in Table 24.3. The velocity of 1.98 m/s was incorporated using a pre-defined option. The contact between impactor head and wall face is employed using surface-to-node (explicit). For simulating tangential behaviour, the coefficient of friction between these bodies was assumed to be 0.03 due to the low velocity of the impactor. Normal behaviour is simulated using the penalty contact option with a stiffness of 10^7 N/mm (Burnett et al., 2023; Thakur et al., 2023). In all the simulations, the impactor was considered the master surface and the wall the node-based slave surface. The hourglass control is kept default for all the 3D brick elements. Lastly, based on the mesh convergence study, it was concluded that a mesh size of 32 mm is suitable to carry out a simulation as it matched well with the experimental force-time history curves (Senthil et al., 2022).

24.5 COMPARISON OF EXPERIMENT AND FE SIMULATION RESULTS

The simulations were carried out on W1, W2, and W3 to verify the capability of the proposed model for simulating low-velocity impact behaviour. The simulations have been performed using the properties shown in Tables 24.2 and 24.3. The walls were simulated using a simplified micro-modelling approach as presented herein, by considering discrete rigid elements at the back and front faces of walls to act as boundary conditions similar to the experimental setup. The numerical results thus

TABLE 24.3

Interface Properties for Simulating Brick Unit Joints

Contact

Tangential behaviour	Normal behaviour				Cohesive behaviour				
		Traction separation behaviour					Damage		
		Stiffness coefficients (N/mm³)			Initiation (N/mm²)			Evolution (N/mm)	
Friction coefficient		K_{nn}	K_{ss}	K_{tt}	Normal	Shear I	Shear II	G_{Ic}	G_{IIc}
0.80	Hard contact	144	62	62	4.04	0.24	0.24	0.012	0.04

obtained were compared with the experimental results in terms of stress, force-time, and failure patterns, discussed in this section.

24.5.1 Force-Time Response

For W1, the response of walls in terms of peak force has been rather well predicted by the model with a deviation of 7% (overpredicted) from experimental results, whose peak force against time is depicted in Figure 24.4(a). Similarly, the response of W2 was well predicted in terms of peak contact force and the measured and predicted peak force was found to be 17.5 and 17.43 kN, respectively for the first impact. The drop of the force was achieved at 6 ms in the simulations, whereas the same was observed almost at 4 ms in the experiment; however, the rising part of the contact force was found to match well with the experiment; see Figure 24.4(b). Subsequently, the numerical response for W4 seems slightly divergent from the experimental results in terms of peak force. The measured and predicted peak contact force was 22.13 and 18.99 kN, respectively at first hit; see Figure 24.4(c). It was observed that the simulation underpredicted the peak force by 14% as compared to the experimental results; however, the rise and drop trend was found to match well with the experimental results until 3.55 ms.

24.5.2 Failure Pattern

The deformations of the simulated walls are shown in Figure 24.5. For W1, failure was mainly characterized by vertical joint failure, and the experimental and numerical crack patterns are shown in Figures 24.2(a) and 24.5(a), respectively. The residual displacement on the wall is found to be 7.56.

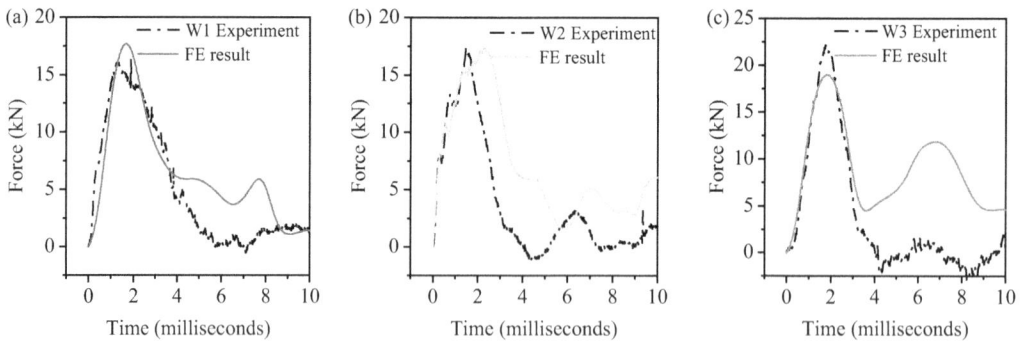

FIGURE 24.4 Comparison between experiment and FE result in terms of peak force-time for (a) W1, (b) W2, and (c) W3.

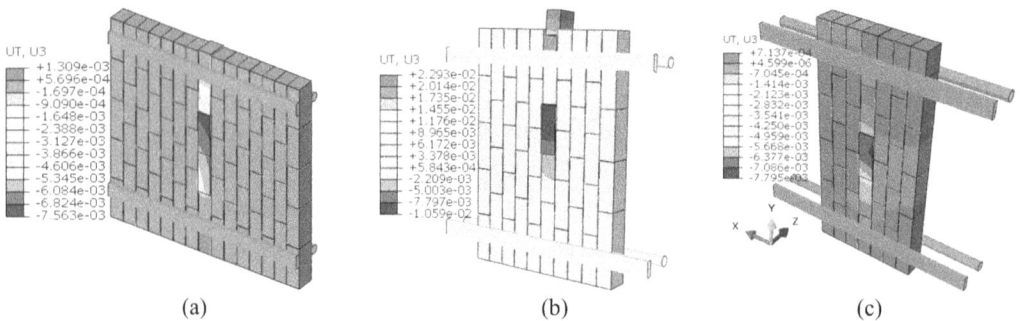

FIGURE 24.5 Simulated deformation pattern for (a) W1, (b) W2, and (c) W3.

In the case of W2, the stress was mostly concentrated on the bricks adjacent to the impact point and displacement reached a value of 10 mm; see Figure 24.5(b). Overall, it is concluded that the damage pattern is well predicted by the model. For the damage in the case of W3, the predicted failure pattern matched well with experimental results; however, cracks in the brick unit cannot be simulated. In order to verify the behaviour, the stress wave propagation in W3 at varying time steps is shown in Figure 24.6. The predicted damage in W3 is mainly characterized as interface failure rather than combined action of brick and interface failure which is observed in experiments and discussed in section 24.2. Also, the displacement contours showing failure at the end of the test are shown in Figure 24.5(c). The crack in the bricks cannot be modelled; the distribution of stress on matching with the force-time response and displacement contour revealed the expected crack pattern within the brick; however, it can still be regarded as a limitation of the proposed model. Conversely, it is worth mentioning that the simulation of considered tests is challenging given the existence of multiple failure mechanisms such as interface failure and brick unit failure; however, the proposed model is very well suited to determine the resistance offered by walls during impact events considering the simplicity of the model.

24.5.3 ENERGY ABSORPTION

The energy absorption in the case of simulated walls has been calculated using the applied and residual kinetic energy of the impactor; see Figure 24.7. An initial velocity of 1.98 m/s was able to produce a kinetic energy of 117.61 J. It was observed that all the simulated walls were able to sustain the first hit in simulation; however, the nature of energy absorption varies with time. Also, the rebounding of the impactor was observed to be similar to the experimental observations. For W1, the rebound of the hammer occurred after 0.05 seconds; however, for W2 and W3, the rebound of

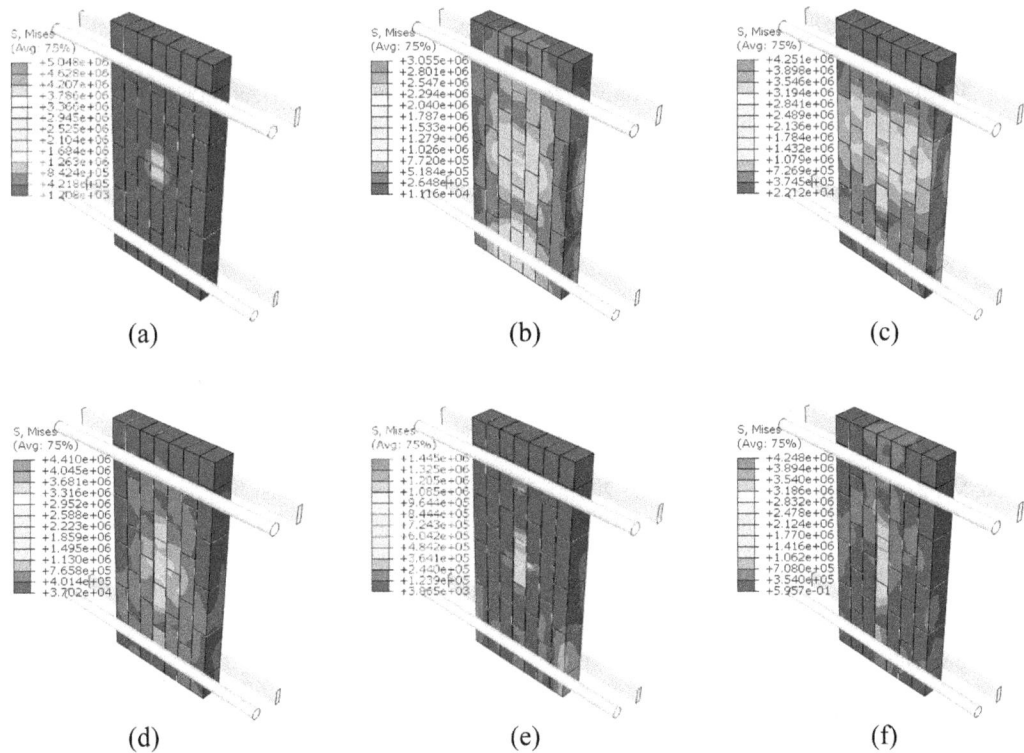

FIGURE 24.6 Stress wave propagation for W3 at (i) 2, (ii) 4, (iii) 6, (iv) 8, (v) 10, and (vi) 120 ms.

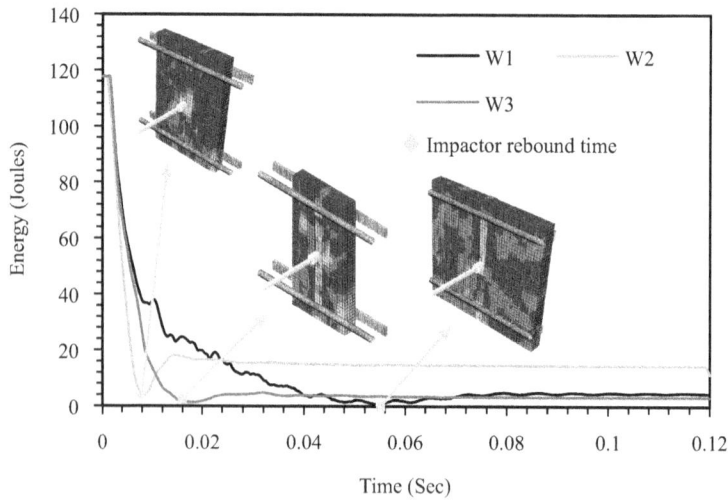

FIGURE 24.7 Energy absorption with time in simulated walls.

the hammer occur before 0.02 s. Considering this, it may be stated that the ductility of square walls (W1) is more compared to that of rectangular walls (W2 and W3).

24.6 CONCLUSION

Experimental and numerical investigations have been carried out on brick masonry walls with varying aspect ratios against low-velocity impact load. Based on the results, the following conclusions were drawn:

- It was observed that the aspect ratio is found to have a marginal effect on the peak force; however, it changes the failure pattern significantly.
- It was observed that the energy absorption capacity of walls with lower aspect ratios is more compared to the higher aspect ratio walls.
- The proposed model is very well suited to determine the peak force resistance offered by walls during impact events considering the simplicity of the model.

REFERENCES

Agnihotri, P., Singhal, V., & Rai, D. C., 2013. Effect of in-plane damage on out-of-plane strength of unreinforced masonry walls, *Eng. Struct.*, 57, 1–11.

Asad, M., Dhanasekar, M., Zahra, T., & Thambiratnam, D., 2020. Failure analysis of masonry walls subjected to low velocity impacts, *Eng. Fail. Anal.*, 116, 1–24.

Asad, M., Zahra, T., & Thambiratnam, D., 2021. Failure of masonry walls under high velocity impact – A numerical study, *Eng. Struct.*, 238, 112009.

Burnett, S., Gilbert, M., Molyneaux, T., Beattie, G., & Hobbs, B., 2007. The performance of unreinforced masonry walls subjected to low-velocity impacts: Finite element analysis, *Int. J. Impact Eng.*, 34(8), 1433–1450.

Chang, L. Z., Rots, J. G., & Esposito, R., 2021. Influence of aspect ratio and pre-compression on force capacity of unreinforced masonry walls in out-of-plane two-way bending, *Eng. Struct.*, 249, 113350.

De Risi, M. T., Di Domenico, M., Ricci, P., Verderame, G. M., & Manfredi, G., 2019. Experimental investigation on the influence of the aspect ratio on the in-plane/out-of-plane interaction for masonry infills in RC frames, *Eng. Struct.*, 189, 523–540.

Gilbert, M., Hobbs, B., & Molyneaux, T. C. K., 2002. The performance of unreinforced masonry walls subjected to low-velocity impacts: Mechanism analysis. *Int. J. Impact Eng.*, 27, 253–275.

Janaraj, T., & Dhanasekar, M., 2016. Design expressions for the in-plane shear capacity of confined masonry shear walls containing squat panels, *J. Struct. Eng.*, 142, 04015127.

Matsumura, A., 1998. Shear strength of reinforced masonry walls, in: 9th World Conf. Earthq. Eng., Tokyo, 121–126.

Mosalam, K., Glascoe, L., & Bernier, J., 2009. Mechanical properties of unreinforced brick masonry, Section1, 1–26.

Noor-E-Khuda, S., Dhanasekar, M., & Thambiratnam, D. P., 2016. An explicit finite element modelling method for masonry walls under out-of-plane loading, *Eng. Struct.*, 113, 103–120.

Pantò, B., Silva, L., Vasconcelos, G., & Lourenço, P. B., 2019. Macro-modelling approach for assessment of out-of-plane behavior of brick masonry infill walls, *Eng. Struct.*, 181, 529–549.

Senthil, K., Thakur, A., & Singh, A. P., 2022. Multi-hit impact resistance of masonry walls under large mass, in: D. Maity, P. K. Patra, M.S. Afzal, R. Ghoshal, C. S. Mistry, P. Jana, D. K. Maiti (eds.), *Recent Adv. Comput. Exp. Mech. Vol—I*, Springer Singapore, 177–188.

Simulia, ABAQUS User Guide (2019).

Thakur, A., Senthil, K., & Singh, A. P., 2023. Influence of constitutive models on the behaviour of clay brick masonry walls against multi hit impact loading, in: *Recent Trends in Wave Mechanics and Vibrations, Vol—125*, Springer, Cham, 77–86.

Thakur, A., Kasilingam, S., Singh, A. P., & Iqbal, M. A., 2020. Prediction of dynamic amplification factor on clay brick masonry assemblage, *Structures*, 27, 673–686.

25 A 3D Shear Deformation Theory for the Dynamic Response of 3D Braided Composite Shells under Low-Velocity Impact

Pabitra Maji and Bhrigu Nath Singh

25.1 INTRODUCTION

Three-dimensional (3D) braided composites are progressively being used in miscellaneous fields such as solid rocket nozzles, aircraft structures, civil structures, etc., for their different characteristics. In addition, some fibers in the 3D braided composite are oriented in the thickness direction, which generally works against delamination. The low-velocity impact commonly happens in real-life applications due to hail, birds, stones or tool-drop, debris particles, etc. The structural components are to be designed in such a way that the structure can work against low-velocity impact. The 3D braided geometric configuration is a complex type. It is constituted of two parts, such as yarn and matrix, and the 3D braided has exceptional material properties; it usually depends on configurations of the yarn and braided angle. The 3D braided geometric configuration is very complex. It is really challenging to calculate the accurate elastic properties of braided composites using theoretical models.

Wang and Wang [1] studied the thermo-elastic properties of 3D braided textile composites. In this study, the braided composites were manufactured based on the four-step braiding procedure. Sun et al. [2] developed a new mathematical model to calculate the effective elastic moduli of 3D braided composites. Further, based on multiscale modeling, Xu et al. [3] inspected the elastic moduli of 3D 4-step braided composites. Gao et al. [4] computed the material properties of 3D braided composites by theoretical prediction considering voids defects. Further, Zhai et al. [5] studied the mechanical behavior of 3D braided composites based on the multiscale finite element method under the thermal environment. Based on the mesh-free methods, Li et al. [6] utilized the unit cell model to predict the elastic moduli of 3D braided composite for the various configurations. Further, Shokrieh and Mazloomi [7] computed the mechanical properties of 3D braided composites with the theoretical prediction, and the results were also compared with the experimental predictions. Yang and Haung [8] examined the 3D braided composite plates with piezoelectric face sheets to scrutinize the dynamic stability and vibration characteristics. Haung et al. [9] performed the nonlinear modal and dynamic response of plates made of 3D braided composites using higher shear deformation theory. Further, Singh and Singh [10] worked out the modal and buckling characteristics of 3D braided plates using a new higher-order shear deformation theory. Recently, Maji and Singh [11, 12] and Maji et al. [13] explored the free vibration response of 3D braided composite panels based on the third-order shear deformation theory (TSDT) by finite element methods (FEM).

DOI: 10.1201/9781003352358-25

Sun and Chen [14] studied the low-velocity impact (LVI) of laminated composite plates based on the modified Hertzian contact law using the FEM and also evaluated contact force, shell displacement, and impactor displacement. They used Newmark beta time integration to solve the low-velocity impact. Chun and Lam [15] examined the dynamic characteristics of clamp-supported composite plates under low-velocity impact using higher-order theory. Pan et al. [16] examined the in-plane failure of 3D braided composites under various compression loads. Further, Hu et al. [17] studied the multiple impact response of 3D braided composites beam based on finite element analyses (FEA). Further, Shi et al. [18] studied the transverse impact behavior of biaxial/uniaxial braided tubes. Recently, Maji and Singh [19, 20] investigated the low-velocity impact of a 3D plate made of braided composites with TSDT at various locations on the plate.

To the best of the authors' knowledge, there is no research article related to the dynamic response of 3D braided composite spherical shells subjected to low-velocity impact. In this present investigation, results are obtained using FEM based on the TSDT theory. In the present work, Newmark's beta time integration algorithm is used to execute the low-velocity impact response of 3D braided spherical shells. The time-dependent contact force of the spherical shells is obtained based on nonlinear modified Hertzian contract law. To judge the accuracy and correctness of the present numerical models, various comparison studies are observed. In this study, time histories contact force, central deflection, impactor displacement, and impactor velocity are observed by varying fiber volume fractions, braided angles, boundary conditions, impactor radius, impactor velocity, thickness ratios, etc.

25.2 THEORETICAL FORMULATION

The schematic diagram of the 3D braided spherical shells (SPH) is shown in Figure 25.2 with a dimension of a × b × h and principal radius of curvature (R_x and R_y) in the Cartesian coordinate system. The mid-plane of the spherical shells is taken as a reference plane. The radius of curvature of the spherical shells is $R = R_x = R_y$, $R_{xy} = \infty$.

In this present study, a C° FEM is used, based on the 3D TSDT, incorporating twelve degrees of freedom per node. The displacement fields are given as follows:

$$u(x, y, z) = u_1(x, y) + zu_2(x, y) + z^2 u_3(x, y) + z^3 u_4(x, y), \tag{25.1}$$

$$v(x, y, z) = v_1(x, y) + zv_2(x, y) + z^2 v_3(x, y) + z^3 v_4(x, y), \tag{25.2}$$

$$w(x, y, z) = w_1(x, y) + zw_2(x, y) + z^2 w_3(x, y) + z^3 w_4(x, y). \tag{25.3}$$

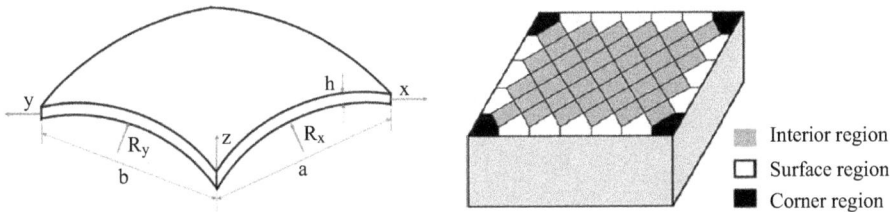

FIGURE 25.1 Schematic diagram of the 3D braided (a) spherical shell and (b) various regions of braided composites. Source: [7].

25.2.1 Strain-Displacement Relationship

The linear strain-displacement fields of the spherical shell can be expressed as,

$$\{\varepsilon\}_{xyz} = [B_e]\{d_e\} \tag{25.4}$$

Where $[B_e]$ defines the strain-displacement matrix.

25.2.2 Constitutive Equation

Based on Hooke's law, the linear stress-strain relationship for the spherical shell can be expressed as,

$$\{\sigma\}_{xyz} = [\bar{Q}]\{\varepsilon\}_{xyz} = [T][Q][T]^T \{\varepsilon\}_{xyz} \tag{25.5}$$

$[\bar{Q}]$ is defined as the elastic constant matrix of the 3D braided spherical shell.

An eight-noded isoparametric spherical shell element is utilized in the present finite element formulations. The elemental stiffness $[K_e]$ and mass matrices $[M_e]$ of the spherical shell in the natural coordinates system are expressed as,

$$\left[K_e\right] = \int_{-1}^{+1}\int_{-1}^{+1} [B_L]^T [D_e][B_L] |J| \, d\xi \, d\zeta \tag{25.6}$$

$$\left[M_e\right] = \int_{-1}^{+1}\int_{-1}^{+1} [N_e]^T [m][N_e] |J| \, d\xi \, d\zeta \tag{25.7}$$

The braided spherical shell is manufactured based on the four-step stable braiding process. For the material homogenization, the fibers are considered transversely isotropic, wherein the matrix is considered isotropic materials; the equivalent material properties of the 3D braided composite spherical shell are predicted based on the volume average method (VAM) using bridging techniques. $[\bar{S}]_{6X6}$ expresses the global compliance matrix.

The corresponding elastic properties of the 3D braided spherical shell are considered as,

$$E_{xx} = \frac{1}{\bar{S}_{11}}, E_{yy} = \frac{1}{\bar{S}_{22}}, E_{zz} = \frac{1}{\bar{S}_{33}}, \tag{25.8}$$

$$G_{xy} = \frac{1}{\bar{S}_{55}}, G_{xz} = \frac{1}{\bar{S}_{66}}, G_{yz} = \frac{1}{\bar{S}_{44}}, \tag{25.9}$$

$$v_{xy} = -\frac{\bar{S}_{12}}{\bar{S}_{11}}, v_{xz} = -\frac{\bar{S}_{13}}{\bar{S}_{11}} \ and \ v_{yz} = -\frac{\bar{S}_{23}}{\bar{S}_{22}} \tag{25.10}$$

25.2.3 Low-Velocity Impact (LVI)

The contact force (C_F) varies nonlinearly with the time-dependent indentation (φ). The time histories contact force of the spherical shell can be calculated based on the nonlinear modified Hertzian contact law as,

$$C_F(t) = K_c \varphi^{1.5}(t) \tag{25.11}$$

$$K_c = \frac{4\sqrt{1/r_{ip} + 2/R}}{3[(1-v_{ip}v_{ip})/E_{ip} + 1/E_{zz}]} \tag{25.12}$$

Where K_C defined the contact stiffness. r_{ip}, and E_{ip} are the radius and modulus of elasticity of the impactor. The equation of impactor motion is articulated as,

$$m_{ip}\ddot{w}_{ip}(t) + C_F(t) = 0; \ w_{ip}(0) = 0 \ and \ \dot{w}_{ip}(0) = V_i \tag{25.13}$$

The Newmark time integration algorithm is implemented for the low-velocity impact analyses; the governing equation of the motions is given as,

$$[M]\{\ddot{d}\}^{t+\Delta t} + ([K])\{d\}^{t+\Delta t} = \{C_F\}^{t+\Delta t} \tag{25.14}$$

25.3 RESULTS AND DISCUSSION

In this present numerical study, an eight-noded isoparametric 3D FEM using TSDT is implemented for the low-velocity impact of 3D braided composite spherical panels. Several judgment studies are performed to check the correctness of the numerical code. Table 25.2 shows the non-dimensional natural frequencies (NDFF) of laminated spherical shells with simply supported boundary conditions. The validation of the central displacement and contact force is illustrated in Figure 25.2. The NDFF of a 3D braided composite plate with simply supported with braided volume fractions and braid angles are presented in Table 25.3. It is clearly perceived that the present results are well suited

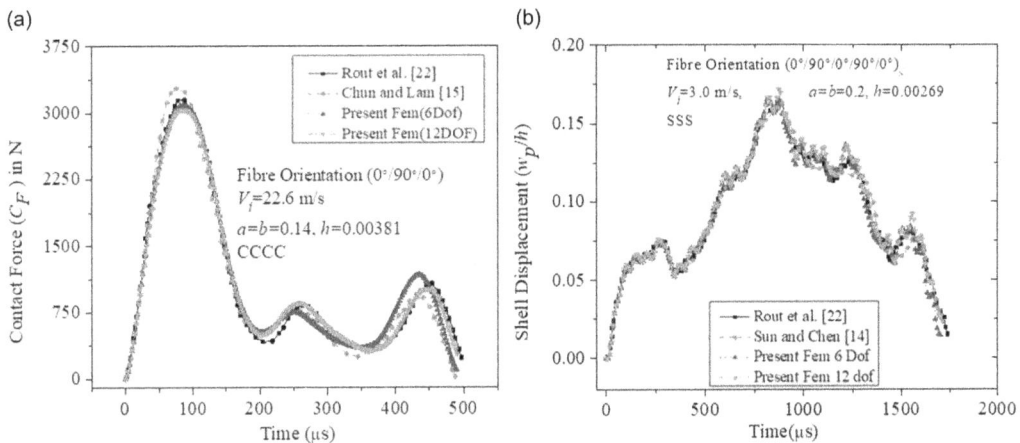

FIGURE 25.2 Centrally impacted, the time histories of (a) contact force and (b) central deflections of a square composite plate.

to the open literature results. The whole investigation is obtained with the mesh size of (16×16) and $\Delta t = 0.25$ μs time steps to reduce the computational cost. All displacements are in mm.

For CCCC cases:

$$x = 0, \ a; \ y = 0, \ b; u_1 = v_1 = w_1 = u_2 = v_2 = w_2 = u_3 = v_3 = w_3 = u_4 = v_4 = w_4 = 0. \quad (25.15)$$

In the SSSS cases,

$$x = 0, \ a; v_1 = w_1 = v_2 = w_2 = v_3 = w_3 = v_4 = w_4 = 0, \quad (25.16)$$

$$y = 0, \ b; u_1 = w_1 = u_2 = w_2 = u_3 = w_3 = u_4 = w_4 = 0. \quad (25.17)$$

25.3.1 Effect of Braided Impactor Mass

The effect of the impactor radius (R_i) on the low-velocity impact response of 3D braided composite spherical shells with fully simply supported (SSSS) boundary conditions (BCs) is illustrated in Figure 25.3. It is observed that contact force (C_F), central displacement (w_P), re-bound velocity (V_0), and impactor displacement (w_i) increase with the increase in impactor radius. The impactor mass (m_i) is directly related to the impactor radius (R_i). It is also identified that the contact time and peak

TABLE 25.1

Properties of Materials Used in the Entire Investigation

Materials	GPa					GPa	
	E_{11f}	E_{22f}	G_{12f}	G_{23f}	v_{12f}	E_m	v_m
	230.0	40.0	24.3	14.3	0.35	2.94	0.35

Source: [8].

TABLE 25.2

The NDFF of a Simply Supported Laminated Spherical Shell

R/a		0°/90°	0°/90°/0°
5	Present FEM	28.8364	31.0096
	Reddy and Liu [21]	28.840	31.020

$a/b = 1.0$, $a/h = 100$, $E_1/E_2 = 25$, $G_{12} = G_{13} = 0.5E_2$, $G_{23} = 0.2E_2$, $v_{12} = v_{13} = v_{23} = 0.25$.

TABLE 25.3

Variation of NDFF of the Simply Supported 3D Braided Composite Plate

Volume fraction (V_f)	Braid angle (α)	Yang and Haung [8]	Singh and Singh [10]	Present FEM
0.4	20	16.272	16.1892	15.6445
	30	16.237	16.1464	15.3837
0.45	20	16.523	16.4816	16.6317
	30	16.452	16.3748	15.9131

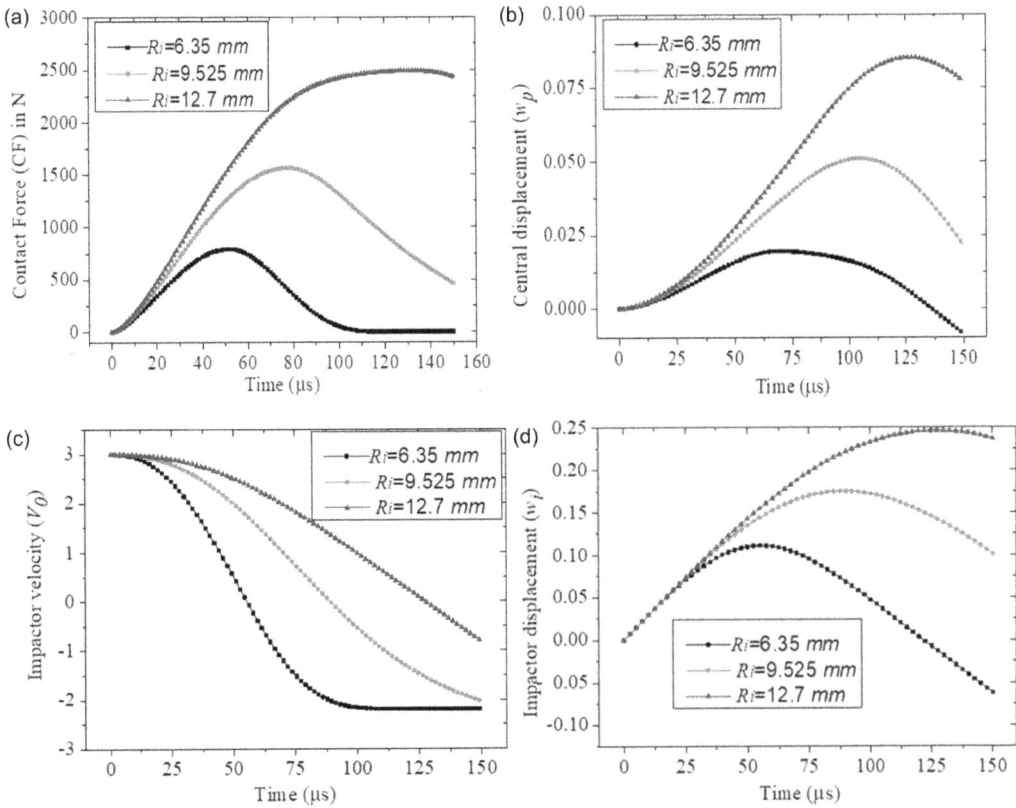

FIGURE 25.3 The influence of the impactor radius (R_i) on the time histories (a) contact force (C_F), (b) central deflection (w_P), (c) impactor re-bound velocity (V_o), and (d) impactor deflection ($a = b = 0.1$ m, $h = 0.015$ m, $Rx = Ry = 1.0$ m, $V_i = 3.0$ m/s, $V_f = 0.45$, $\alpha = 20°$).

value of contact force (C_F), central deflection (w_P), and impactor deflection (w_i) are shifted towards the right with the increase in the impactor radius (R_i).

25.3.2 Effect of Initial Impactor Velocity

The influence of the varying initial impactor velocity ($V_i = 1.5$, 2.0, 2.5, and 3.0 m/s) on the impact response of 3D braided composite spherical shell with SSSS BCs is demonstrated in Figure 25.4. It is evident that contact force (C_F) and central displacement (w_P) increase with the rise in impactor velocity (V_i). It may be noted that the time reaching the maximum values of the contact force (C_F) becomes shorter for the rise in initial impactor velocity; for the same investigations, the time reaching central displacement (w_P) becomes larger.

25.3.3 Effect of Braided Volume Fractions (V_f)

The time-dependent contact force (C_F) and central displacement (w_P) of the 3D braided square spherical shell with SSSS BCs at various braided volume fractions (V_f) are revealed in Figure 25.5. It is identified that the contact force (C_F) of the spherical shell increases with the increase in braided volume fractions (V_f), and the contact span becomes shorter. Interestingly, in contrast, the central displacement (w_P) of the spherical shell falls with the rise in V_f. This is because the contact stiffness of the spherical shell increases with the increase in V_f, which is as expected.

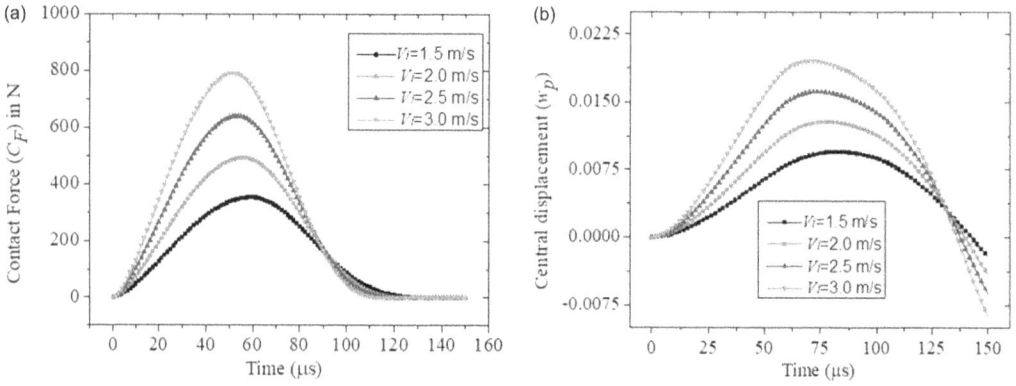

FIGURE 25.4 The influence of the initial impactor velocity (V_i) on the time histories of (a) contact force (C_F) and (b) central deflection (w_P) ($a = b = 0.1$ m, $h = 0.015$ m, $Rx = Ry = 1.0$ m, $R_i = 6.35$ mm, $V_f = 0.45$, $\alpha = 20°$).

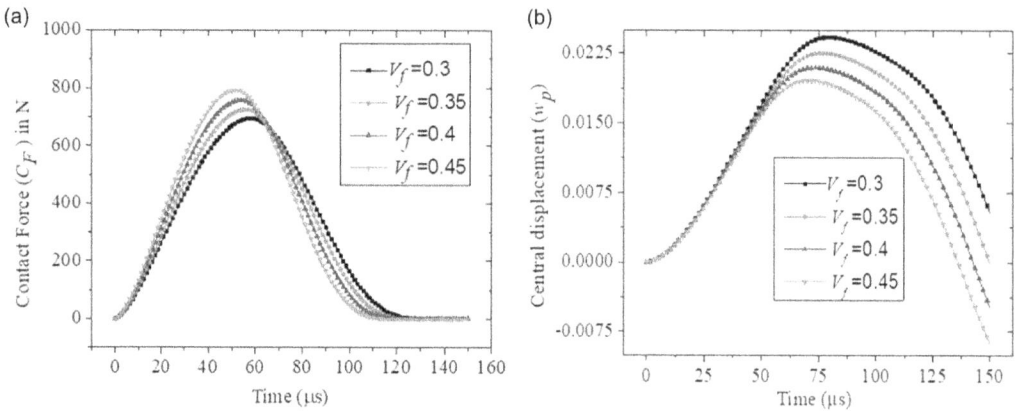

FIGURE 25.5 The influence of the braided volume fractions (V_f) on the time histories of (a) contact force (C_F) and (b) central deflection (w_P) ($a = b = 0.1$ m, $h = 0.015$ m, $Rx = Ry = 1.0$ m, $\Delta t = 0.25$ μs, $R_i = 6.35$ mm, $V_i = 3.0$ m/s, $\alpha = 20°$).

25.3.4 EFFECT OF BRAID ANGLE (α)

The time-dependent contact force (C_F) and the central displacement (w_P) of the 3D braided spherical shells with four sides CCCC BCs at different braided angles (α) are shown in Figure 25.6. It can be seen that the increase in braid angles leads to a higher contact force with small contact durations. Similarly, the higher braided angles also lead to a higher central displacement with greater contact durations.

25.4 CONCLUSION

In this study, the results are achieved based on the FEM using 3D TSDT. The modified nonlinear Hertzian contact law is implemented to calculate the contact force of the impact phenomenon. The Newmark beta time integration scheme is utilized to solve the time-dependent governing equations. In this study, to judge the accuracy of the present models, various parametric studies are obtained. It can be seen that impactor radius (R_i) has a significant influence on the contact force (C_F), central displacement (w_P), re-bound velocity (V_0), and impactor displacement (w_i). Further, impactor velocity

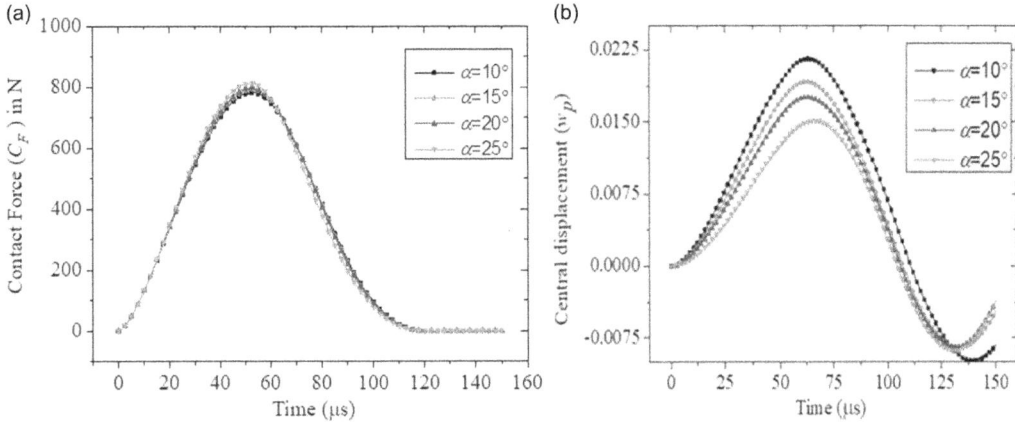

FIGURE 25.6 The influence of the braid angles on the time histories of (a) contact force (C_F) and (b) central deflection (w_P) ($a = b = 0.1$ m, $h = 0.015$ m, $Rx = Ry = 1.0$ m, $\Delta t = 0.25$ μs, $R_i = 6.35$ mm, $V_i = 3.0$ m/s, $V_f = 0.45$).

(V_i), braided volume fractions (V_f), and braided angles (α) also have remarkable performance on the low-velocity impact response. The clamp BCs have more influence than simply supported BCs.

REFERENCES

1. Y. Q. Wang and A. S. D. Wang, "Microstructure/property relationships in three-dimensionally braided fiber composites," *Compos. Sci. Technol.*, vol. 53, no. 2, pp. 213–222, 1995.
2. H. Sun, S. Di, N. Zhang, N. Pan, and C. Wu, "Micromechanics of braided composites via multivariable FEM," *Comput. Struct.*, vol. 81, no. 20, pp. 2021–2027, 2003.
3. Y. J. Xu, W. H. Zhang, and M. Domaszewski, "Microstructure modelling and prediction of effective elastic properties of 3D multiphase and multilayer braided composite," *Mater. Sci. Technol.*, vol. 27, no. 7, pp. 1213–1221, 2011.
4. X. Gao, L. Yuan, Y. Fu, X. Yao, and H. Yang, "Prediction of mechanical properties on 3D braided composites with void defects," *Compos. Part B Eng.*, vol. 197, no. November 2019, p. 108164, 2020.
5. G. Balokas, S. Czichon, and R. Rolfes, "Neural network assisted multiscale analysis for the elastic properties prediction of 3D braided composites under uncertainty," *Compos. Struct.*, vol. 183, no. 1, pp. 550–562, 2018.
6. L. Li and M. H. Aliabadi, "Elastic property prediction and damage mechanics analysis of 3D braided composite," *Theor. Appl. Fract. Mech.*, vol. 104, pp.102338 no. August, 2019.
7. M. M. Shokrieh and M. S. Mazloomi, "A new analytical model for calculation of stiffness of three-dimensional four-directional braided composites," *Compos. Struct.*, vol. 94, no. 3, pp. 1005–1015, 2012.
8. J. Yang and X. Huang, "Dynamic stability behavior of 3D braided composite plates integrated with piezoelectric layers," *J. Compos. Mater.*, vol. 43, no. 20, pp. 2223–2238, 2009.
9. X. Huang, X. Jia, J. Yang, and Y. Wu, "Nonlinear vibration and dynamic response of three-dimensional braided composite plates," *Mech. Adv. Mater. Struct.*, vol. 15, no. 1, pp. 53–63, 2008.
10. D. B. Singh and B. N. Singh, "New higher order shear deformation theories for free vibration and buckling analysis of laminated and braided composite plates," *Int. J. Mech. Sci.*, vol. 131–132, no. April, pp. 265–277, 2017.
11. P. Maji and B. N. Singh, "Free vibration responses of 3D braided rotating cylindrical shells based on third-order shear deformation," *Compos. Struct.*, vol. 260, no. July 2020, p. 113255, 2020.
12. P. Maji and B. N. Singh, "Shear deformation theory for free vibration responses of 3D braided pre-twisted conical shells under rotation," *Int. J. Comput. Methods Eng. Sci. Mech.*, pp. 1–20, 2021.
13. P. Maji, B. N. Singh, and D. B. Singh, "A third-order polynomial for the free vibration response of 3D braided curved panels using various boundary conditions," *Mech. Based Des. Struct. Mach.*, pp. 1–23, 2021.
14. C. T. Sun and J. K. Chen, "On the impact of initially stressed composite laminates," *J. Compos. Mater.*, vol. 19, no. 6, pp. 490–504, 1985.

15. L. U. Chun and K. Y. Lam, "Dynamic response of fully-clamped laminated composite plates subjected to low-velocity impact of a mass," *Int. J. Solids Struct.*, vol. 35, no. 11, pp. 963–979, 1998.

16. Z. Pan, W. Ouyang, M. Wang, J. Xiao, and Z. Wu, "In-plane compression failure mechanism of two-dimensional triaxial braided composite (2DTBC) material subjected to different load directions," *Mech. Mater.*, vol. 161, no. July, p. 104001, 2021.

17. M. Hu, J. Zhang, B. Sun, and B. Gu, "Finite element modeling of multiple transverse impact damage behaviors of 3-D braided composite beams at microstructure level," *Int. J. Mech. Sci.*, vol. 148, no. September, pp. 730–744, 2018.

18. L. Shi, Z. Wu, X. Cheng, Z. Pan, and Y. Yuan, "Transverse impact response of hybrid biaxial/uniaxial braided composite tubes," *Eng. Struct.*, vol. 244, no. July, p. 112816, 2021.

19. P. Maji and B. N. Singh, "A bridging model and volume averaging approach for the dynamic analyses of 3D braided plates based on 3D shear deformation theory under low- velocity impact," *Mech. Adv. Mater. Struct.*, pp. 1–22, 2022.

20. P. Maji and B. N. Singh, "Third-order shear deformation theory for the low-velocity impact response of 3D braided composite plates," in *Aerospace and Associated Technology*, 2022, pp. 106–111.

21. J. N. Reddy and C. F. Liu, "NASA contractor report 4056 a higher-order theory for a higher-order theory for geometrically nonlinear analysis of composite laminates," *NASA Contract. Rep. 4056*, p. 104, 1987.

22. M. Rout, S. S. Hota, and A. Karmakar, "Transient response of pretwisted delaminated stiffened shell under low velocity impact," *Int J Comput Methods Eng Sci Mech*, vol. 19, pp. 139–155, 2018.

26 Mechanical Characterization of Bio-Sandwich Structures with Composite Skins and Coconut Shell Powder-Filled Epoxy Core

Ankush P. Sharma, S. Manojkumar,
R. Velmurugan and K. Kanny

26.1 INTRODUCTION

A decent grouping of strength, modulus, toughness, and density is hard to achieve with several engineering materials (Rajesh et al. 2019). To conquer these limitations and to encounter the ever-growing call of contemporary day technology, composites are employed. The sandwich composite consists of high-strength and high-stiffness skins and lightweight core material. Light-weight materials unaffected by extreme loads are essential for current defense, aerospace, and naval usages. The projectile impact is one such circumstance. The former necessities are accomplished by composites (Paul, Velmurugan, and Gupta 2022). In the last 50 years, the design and manufacturing of light-weight and high-strength materials are majorly increasing. This is owing to the surge of polymer composites (Pradhan, Dwarakadasa, and Reucroft 2004). Ample combinations of biodegradable matrix and natural fillers are defined by several researchers. This endorses biodegradable composites of new classes with improved mechanical properties and helps to achieve low-cost products (Udhayasankar and Karthikeyan 2015). Numerous environmental rewards, viz. reduced need for non-renewable material sources, less pollution, and less greenhouse discharge, are shown by materials reinforced with natural fillers. Natural fillers have benefits over outdated ones, viz. low cost, high toughness, resistance to corrosion, low density, good properties with specific strength, and condensed wear of tools (Sarki et al. 2011).

Andezai, Masu, and Maringa (2020) have reported that composites with a rising weight percentage of CSP filler particles of both sizes exhibit a constant decrease in tensile strength. The stiffness of composites enhances beyond a critical volume fraction of reinforcing filler over a basic epoxy resin while elongation to break decreases. The predicted tensile strength and elastic moduli match well with experiments while the predicted elongation to break considerably differs from experiments. Sapuan, Harimi, and Maleque (2003) have reported that the tensile and flexural strengths of composites increase with CSP filler content in epoxy while corresponding strains decrease owing to materials becoming hard. Pradhan, Dwarakadasa, and Reucroft (2004) have investigated that for ultrahigh molecular weight polyethylene powder and CSP composite, good impact toughness is achieved at about 20 to 30 vol.% CSP. Oral, Kocaman, and Ahmetli (2022) have examined density, the velocity of both ultrasonic longitudinal and shear waves, elastic moduli, and Poisson's ratio of modified epoxy resin (MER)/modified CSP biocomposites, which are larger than those of MER/CSP biocomposites.

DOI: 10.1201/9781003352358-26

Salmah, Koay, and Hakimah (2012) have reported that the tensile strength and elongation at break of biocomposites reduce with infusing CSP, but the elastic modulus and thermal stability surge. The former properties of composites enhance with filler modified by acrylic acid while elongation at break reduces. Manjunatha Chary and Ahmed (2017) have observed that the tensile, flexural, and impact properties of CSP/epoxy composites decrease with rising particle size and volume fraction of filler. The hardness of CSP composites is higher than neat epoxy. This composite can be favorable for sectors, viz. automotive, aircraft, furniture, packing, barrier panels, etc. Deshpande and Rangaswamy (2014) have examined that hybrid composites filled with 10 vol.% CSP show maximum tensile stress. Composites infused with 15 vol.% CSP display maximum flexural stress, interlaminar shear stress, tensile modulus, and hardness. Mishra (2017) has observed that the tensile strength of composites surges up to 15 wt% of coconut shell dust, followed by shrinking to 20 wt%. Agunsoye, Odumosu, and Dada (2019) have investigated that the optimum hardness, tensile and flexural strengths of composites are achieved by infusing 25 wt% carbonized CSP to epoxy and optimum impact energy by that of 10 wt%. CSP can be used to produce automobile bumpers. Jagadeesh et al. (2020) have overviewed that the addition of natural fillers up to optimal weight percentage improves the mechanical and thermal properties of composites and further infusion may condense it. More studies on CSP composites are reported in the References (Rajesh et al. 2019; Udhayasankar and Karthikeyan 2015; Sarki et al. 2011; Nadzri et al. 2022).

It is observed that CSP composites have been characterized under mechanical loads by most of the above studies. While sandwich composites with bio-filler-filled epoxy as a core and synthetic fiber-reinforced epoxy as skins are yet to be investigated under tensile, flexural, and low-velocity single-impact perforation loads. Subsequently, in the current study, the sandwich panels with skins of glass fiber-reinforced polymer (GFRP) and carbon fiber-reinforced polymer (CFRP) and the core of the CSP-filled polymer matrix are fabricated using a co-curing technique. The influence of the addition of 0 wt%, 20 wt%, 30 wt%, and 40 wt% of CSP bio-filler within the matrix on the above properties and equivalent damage modes of sandwich panels has been brought out.

26.2 EXPERIMENTAL PROCEDURE

26.2.1 MATERIALS

Unidirectional (UD) E-glass and carbon fiber mats are procured from Poojan Fiber, Ahmedabad, India, and are employed as reinforcement materials. The areal density and volume density (measured by Archimedes method by weight) of UD glass fiber are 950 g/m^2 and 2.17 g/cc, separately. The former parameters of UD carbon fiber are 400 g/m^2 and 1.8 g/cc, separately. Matrix with epoxy LY 556 and hardener HY951 is procured from Huntsman India Private Ltd., and is used in the ratio of 100:10 parts by weight, discretely. The coconut shell powder (CSP) is acquired from Skylark Exporter, Rasipuram, India, and has a volume density of 0.65 g/cc (Udhayasankar and Karthikeyan 2015).

26.2.2 MANUFACTURING PROCEDURE

A co-curing process is used to make three layered sandwich panels, in which preparation of each layer is performed before the former layer is entirely cured. Initially, the bottom skin with a unidirectional glass fiber mat/epoxy layer is prepared. When the skin starts reaching a solid state, the coconut shell powder-reinforced epoxy is prepared, which is drizzled over the skin using a rubber mold, facilitating the desired shape. The core, which is three times thicker than the skin, is prepared by distributing the coconut shell powder in epoxy resin in pre-set quantities through a mechanical stirrer. The blend is heated at about 50 to 60° C. This diminishes the viscosity and attains an unvarying dispersal. When the core reaches an approximately solid state, the top skin is prepared above it analogous to the bottom skin. The complete specimen is cured for 24 h at room temperature. Similarly, the sandwich panel is prepared with a uni-directional carbon fiber mat/epoxy layer as top and bottom skins with CSP/epoxy as a core as the sandwich panel is based on glass fiber. Subsequently, the specimens are cut to the desired dimensions for experiments. The weight percentage of CSP within the epoxy is altered to 0, 20, 30, and 40% of the weight of the epoxy. The thickness of the core is maintained at 3 mm. In the case of sandwich panels based on glass fiber, the symbol G before and following the core denoted by C indicates a single layer of GFRP in the front (top) and back (bottom) skins, separately, which is designated as G/C/G-XX. Here, XX represents the weight percentage of CSP within the epoxy (core). For example, G/C/G-20 signifies a weight percentage of CSP of 20%. Similarly, for sandwich panels based on carbon fiber, the symbol C before and after the core represents a sole layer of CFRP in the front and back skins, discretely, which is designated as C/C/C-XX. The thickness of each skin and the total thickness of the specimen are about 0.5 mm and 4 mm, separately.

26.2.3 TESTING

The sandwich panels based on glass and carbon fiber-reinforced epoxy as skins and coconut shell powder of 0, 20, 30, and 40 weight percentages reinforced with epoxy as a core are tested under tensile and three-point flexural loads as per the standards ASTM D3039/D3039M-08, and ASTM D790-10, respectively. The specimens are cut using a water jet machine. Quasi-static tensile tests are carried out on a rectangular-shaped specimen using an FIE hydraulic universal testing machine (UTM) of 400 kN capacity with friction wedge grips. The gauge length, width, and total length of the specimen used are 100 mm, 25 mm, and 300 mm, individually. The nominal strain rate of the panels obtained is about 0.000051/s. The displacement data of the specimen is measured by a clip-on extensometer with a 50 mm gauge length. The equivalent output, from the load cell and the extensometer, is recorded by UTM software. The displacement obtained from the extensometer is divided by its gauge length to facilitate the engineering strain of the specimen. The stress of the specimen is obtained from the ratio of load provided by the load cell and the equivalent cross-sec-tional area. Moreover, three-point flexural tests are performed on rectangular beam specimens until failure using a Kalpak universal testing machine of 20 kN capacity. The loading span, width, and total length of the specimen used are 50 mm, 13 mm, and 120 mm, respectively. The rate of cross-head displacement used is 2 mm/min. The former displacement and the equivalent applied load are recorded by the software provided with UTM. The stress-strain curves presented in the resulting figures use engineering strain. Tensile and flexural modulus, strength, failure strain, and toughness are considered to examine the sandwich panels. Also, low-velocity single-impact perforation tests are carried out on specimens of 150 mm in length and 100 mm in width at its center at 18 J, 27 J, and 33 J impact energy levels. An in-house drop weight tower is used for impact testing as per the stan-dard ASTM D7136. The equivalent impact velocities are 2.27 m/s, 2.78 m/s, and 3.1 m/s, separately, which are calculated from the specified impact energy and mass of the impactor. The specimens are clamped within the fixture using four C clamps. The total weight of the impactor assemblage is about 7 kg and the tip diameter of the hemispherical steel impactor is 12.7 mm. The contact force-time history is acquired from a load cell attached to the impactor base using a data acquisition sys-tem supplied by NI. The above-mentioned data and mass of the impactor are employed to facilitate the velocity, displacement, and energy absorption of the impactor and the specimen (ASTM D7136/D7136M-15). The specimens are photographed from the top (impacted) and bottom (non-impacted) surfaces after the test to visually examine damages. This is also executed for specimens following tensile and flexural tests. All tests are performed on a minimum of three specimens of each type at room temperature with fibers oriented along the length.

26.3   RESULTS AND DISCUSSION

The density of the entire composite predicted from the rule of mixture decreases substantially with increased content of filler, as can be seen in Table 26.1.

26.3.1   TENSILE BEHAVIOR

The tensile strength, modulus, failure strain, and toughness of GFRP and CFRP sandwich composites are displayed in Figure 26.1. The tensile stress-strain curve of composites at different filler contents of 0 wt%, 20 wt%, 30 wt%, and 40 wt% is shown in Figure 26.1a. It is observed that the stress-strain curve and the corresponding strength, modulus, failure strain, and toughness of carbon composites decrease with increasing filler content from 20 to 40 wt%, in a contrast to glass composites (Livingston, Athijayamani, and Alavudeen 2021). The aforementioned parameters of different composites have been

TABLE 26.1
The Volume Densities of the Fabricated Composites

Composites	G/C/G-0	G/C/G-20	G/C/G-30	G/C/G-40	C/C/C-0	C/C/C-20	C/C/C-30	C/C/C-40
Theoretical density (kg/m³)	1,321	1,239	1,197	1,156	1,275	1,192	1,151	1,110

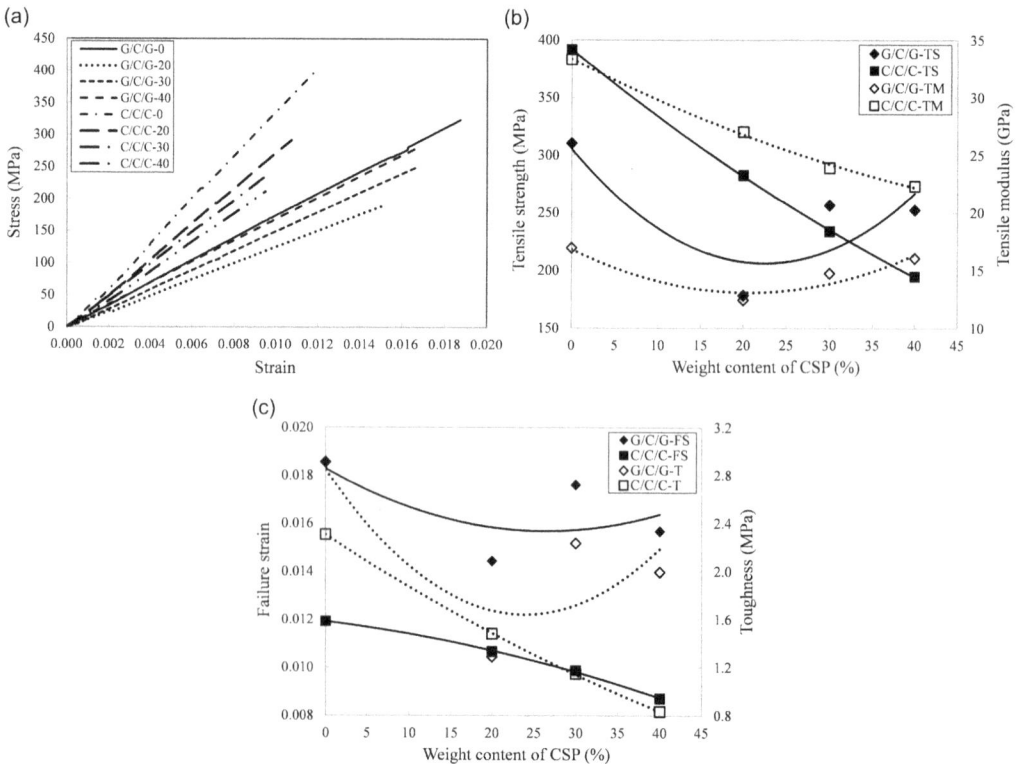

FIGURE 26.1   Tensile (a) stress-strain curve, (b) strength and modulus, and (c) failure strain and toughness of GFRP and CFRP composites. In the case of sandwich panels G/C/G-0, G/C/G-20, G/C/G-30, and G/C/G-40, G before and after the core designated by C indicates a single layer of GFRP in the front (top) and back (bottom) skins, individually, whereas 0, 20, 30, and 40 represent the weight percentage of CSP within the epoxy (core). Similarly, for sandwich panels C/C/C-0, C/C/C-20, C/C/C-30, and C/C/C-40, C before and after the core represents a single layer of CFRP in the front and back skins, separately.

plotted to appraise the influence of distinct filler content of CSP. From the stress-strain curves of composites shown in Figure 26.1a, it is observed that initially the core and the skin layers behave linear-elastic followed by an interfacial debonding between the matrix and the CSP filler which causes their cracking. This indicates a stress-strain curve with a slight slope change. The former event of cracking takes place at multiple locations within the core, resulting in delamination between the core and the skin. This keeps on increasing with loading until achieving the maximum stress of the specimen. Following the crack initiation of the core, the high-strength fiber layers carry the bulk of the load until the failure of the whole composite system. In this case, the fibers exhibit longitudinal splitting throughout the gauge length of the specimen which spans across the width. This event eventually leads to a fibers fracture and the subsequent widespread delamination between the skin and the core. The photographs of glass and carbon composites retrieved after the tests presented in Figure 26.2 show the former damage mechanisms. The cracking within the core can be seen with side views of G/C/G-20 and G/C/G-30 specimens. The carbon composites seem to exhibit lower damage than glass composites. The CFRP composites exhibit higher tensile properties except for failure strain and toughness compared to GFRP ones as anticipated. The data show that composite with 20 and 40 wt% filler contents withstands the highest tensile properties for CFRP and GFRP, separately. The reduction in properties of carbon composites from 0 to 40 wt% may happen due to weak interfacial bonding between the hydrophilic coconut shell particles and the hydrophobic polymer matrix. This leads to an ineffective transfer of load from CSP filler to the matrix and thus from core to skin. Also, the possibility of agglomeration and stress concentration of fiber is augmented by greater loading of filler, resulting in less energy required for crack propagation and material failure (Islam et al. 2017). The tensile behavior of composites exhibiting reductions with augmented CSP content is reported in the References (Manjunatha Chary and Ahmed 2017; Livingston, Athijayamani, and Alavudeen 2021; Islam et al. 2017; Bhaskar and Singh 2013; Somashekhar et al. 2018).

26.3.2 Flexural Behavior

The flexural strength, modulus, failure strain, and toughness of GFRP and CFRP composites are presented in Figure 26.3. The flexural stress vs. strain curve of composites at distinct filler contents of 0 wt%, 20 wt%, 30 wt%, and 40 wt% is shown in Figure 26.3a. It is observed that at first the core

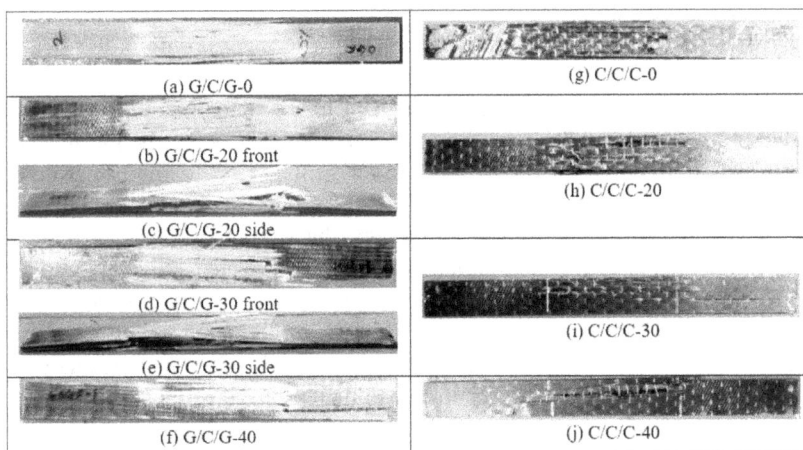

FIGURE 26.2 Surface of GFRP and CFRP composites following tensile load. In the case of sandwich panels G/C/G-0, G/C/G-20, G/C/G-30, and G/C/G-40, G before and after the core designated by C indicates a single layer of GFRP in the front (top) and back (bottom) skins, independently, whereas 0, 20, 30, and 40 represent the weight percentage of CSP within the epoxy (core). Similarly, for sandwich panels C/C/C-0, C/C/C-20, C/C/C-30, and C/C/C-40, C before and after the core denotes a single layer of CFRP in the front and back skins, separately.

(a)

(b)

(c)

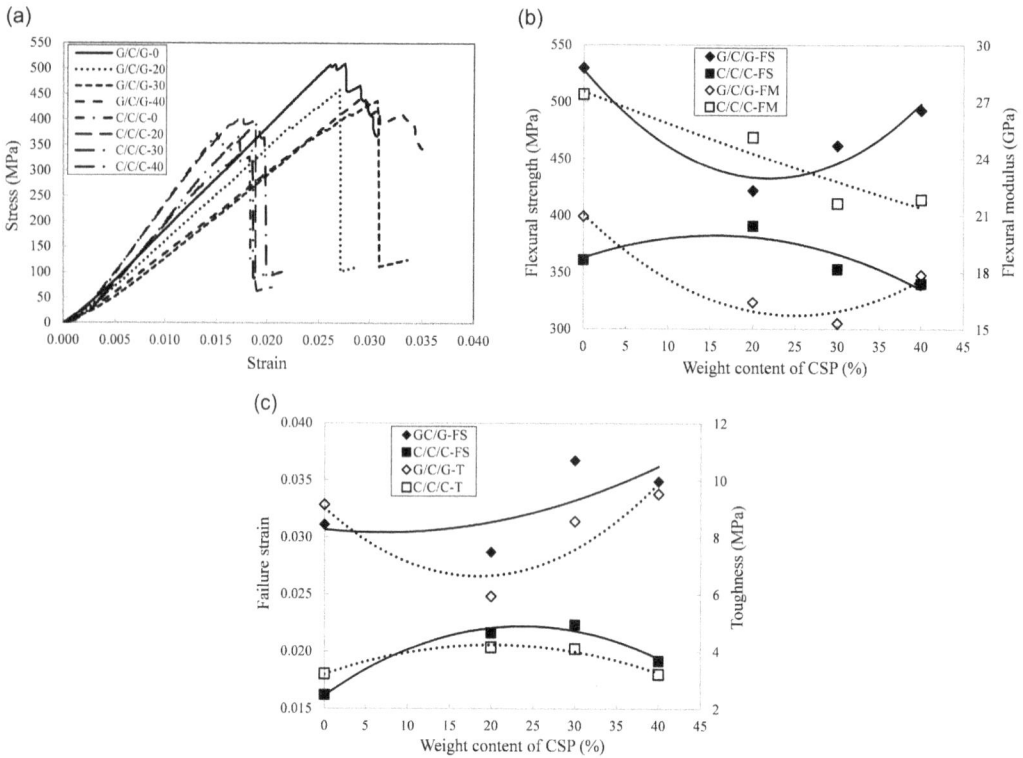

FIGURE 26.3 Flexural (a) stress-strain curve, (b) strength and modulus, and (c) failure strain and toughness of GFRP and CFRP composites. For sandwich panels G/C/G-0, G/C/G-20, G/C/G-30, and G/C/G-40, G before and after the core designated by C specifies a single layer of GFRP in the front (top) and back (bottom) skins, exclusively, while 0, 20, 30, and 40 signify the weight percentage of CSP within the epoxy (core). Similarly, for sandwich panels C/C/C-0, C/C/C-20, C/C/C-30, and C/C/C-40, C before and after the core denotes a single layer of CFRP in the front and back skins, separately.

and the skins exhibit linear elastic behavior. Successively the shear failure in the form of cracking of matrix and CSP filler occurs close to the location of the loading roller. The former event causes a trivial slope change with the stress-strain curve. This results in delamination between the core and the skin around the shear crack. The top skin loaded in compression by the top roller experiences failure along the fiber direction, especially close to the center. The existing shear crack with new small cracks keeps on developing with loading, causing successive delamination between the core and the adjacent skin at different locations. Ultimately, the bottom skin loaded in tension exhibits more damage with fibers along the longitudinal direction and consequent widespread delamination between the skin and the core compared to that between the top skin and the core. This leads to a drop in maximum stress. The aforementioned damages seem to increase within the composites to a small extent with increased loading of CSP from 0 to 40 wt%. Carbon composites exhibit local damage with skins especially close to the central loading region, signifying less damage than glass composites. Therefore, the former composites withstand higher bending loads than the latter ones. Skins that are stiffer support high bending load whereas core that is less stiff sustains more shear load. A thicker core (up to three times) of both skins resists more shear loads than the thinner core. Most of the interfacial region between the skin and the core seems to be intact even after the loading, indicating good bonding between the skin and the core. However, the specimens exhibit a small extent of permanent deformation with increased content of CSP from 0 to 40 wt%. This is observed to be lower for carbon composites than glass ones. These failure modes can be seen in photographs of glass and carbon composites presented in Figures 26.4 and 26.5, separately. It is noticed that the

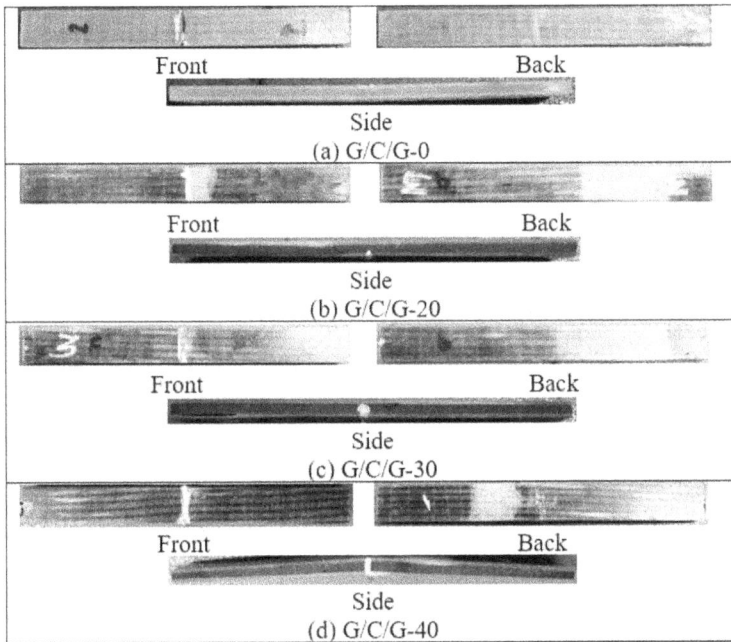

FIGURE 26.4 Surface of GFRP composite following flexural load. For sandwich panels G/C/G-0, G/C/G-20, G/C/G-30, and G/C/G-40, G before and after the core designated by C indicates a single layer of GFRP in the front (top) and back (bottom) skins, individually, while 0, 20, 30, and 40 represent the weight percentage of CSP within the epoxy (core).

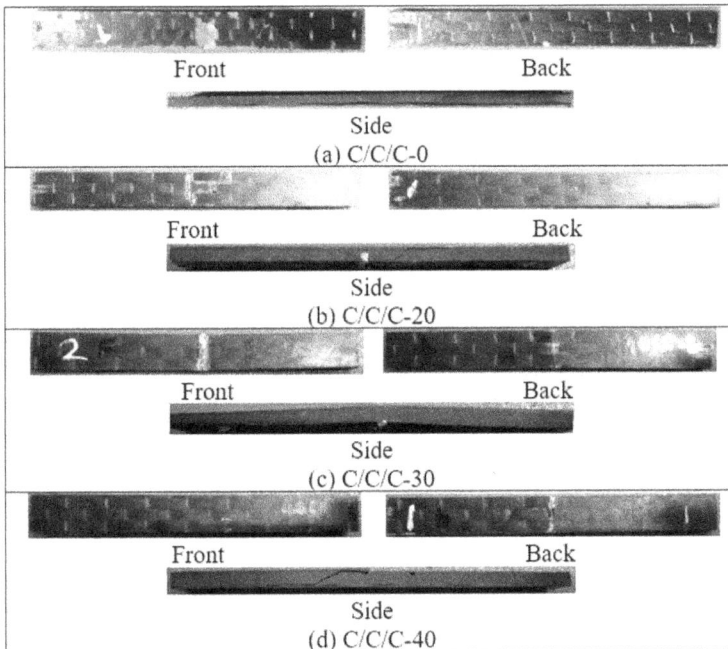

FIGURE 26.5 Surface of CFRP composite following flexural load. For sandwich panels C/C/C-0, C/C/C-20, C/C/C-30, and C/C/C-40, C before and after the core designated by C indicates a single layer of CFRP in the front (top) and back (bottom) skins, discretely, whereas 0, 20, 30, and 40 represent the weight percentage of CSP within the epoxy (core).

flexural strength, modulus, failure strain, and toughness of composites decrease with increased content of filler from 0 to 40 wt%. Failure strain and toughness of GFRP composite show an increasing trend from 20 to 30 wt%. The flexural stress-strain curve exhibits slope reductions with the inclusion of CSP filler from 0 to 40 wt%. Composites with 20 wt% of CSP content register higher bending properties. The reduction in flexural properties with augmented content of CSP filler appears to be due to weak interfacial bonding between CSP particles and polymer matrix. This results in a low bending load to be contributed by the matrix modified with CSP. Composites with condensed flexural behavior with amplified CSP content are reported in the References (Manjunatha Chary and Ahmed 2017; Jagadeesh et al. 2022; Livingston, Athijayamani, and Alavudeen 2021; Islam et al. 2017; Somashekhar et al. 2018; Abdul Khalil et al. 2017). The modulus of CFRP composite seems to be higher than GFRP composite while it is vice versa for strength, failure strain, and toughness. The flexural behavior of composites is observed to be similar under tensile loads.

26.3.3 Low-Velocity Impact Behavior

26.3.3.1 Force-Displacement Behavior and Damage

In this section, the analysis of composites is carried out under a low-velocity single-impact perforation event. The force-displacement behavior of glass composites at different filler contents of 0 wt%, 20 wt%, 30 wt%, and 40 wt% is presented in Figure 26.6a. It is observed that force shows a more gradual decrease for 20 wt%, 30 wt%, and 40 wt% composites following maximum forces of around 1,853 N, 1,994 N, and 1,903 N, discretely with increasing displacement compared to baseline composite (0 wt%). The images of the top (impact) and bottom (non-impact) damage surfaces of glass composites are shown in Figure 26.7. The damage mechanisms can be observed visually. The entire damage area exhibits local indentation, interfacial debonding between the matrix and the CSP, crushing of matrix and CSP within the core, delamination between the skin and the core, and splitting and fracture of fibers. The extent of the damage area of the composite seems to increase to a small extent with increased content of CSP filler from 20 to 40 wt% compared to the baseline composite. This is because of the agglomeration of CSP within the epoxy at higher weight percentages, viz. 20, 30, and 40, resulting in a weak interfacial bonding between the CSP and the matrix for composites under tensile and flexural loads. The baseline composite shows bottom face damage with two-leaf clovers with an identical extent of peanut-shaped damage area developed at the

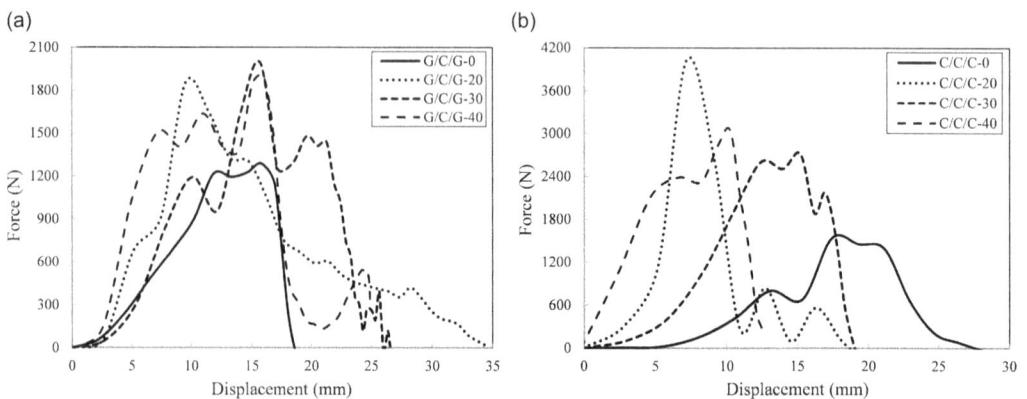

FIGURE 26.6 Impact force-displacement response of (a) GFRP and (b) CFRP composites. For sandwich panels G/C/G-0, G/C/G-20, G/C/G-30, and G/C/G-40, G before and after the core designated by C indicates a single layer of GFRP in the front (top) and back (bottom) skins, individually, whereas 0, 20, 30, and 40 signify the weight percentage of CSP within the epoxy (core). Similarly, for sandwich panels C/C/C-0, C/C/C-20, C/C/C-30, and C/C/C-40, C before and after the core denotes a single layer of CFRP in the front and back skins, separately.

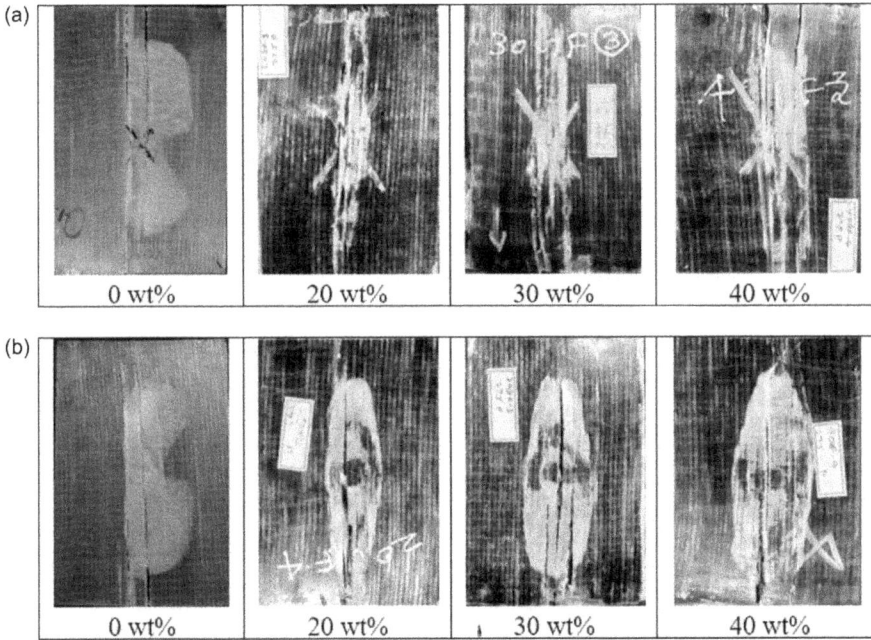

FIGURE 26.7 Impact-induced damage patterns of GFRP composites. 0 wt%, 20 wt%, 30 wt%, and 40 wt% represent the weight percentage of CSP within the epoxy (core) of sandwich panels G/C/G-0, G/C/G-20, G/C/G-30, and G/C/G-40, discretely. Here, G before and after the core designated by C indicates a single layer of GFRP in the front (top) and back (bottom) skins, individually.

interface between the skin and the core (Sharma, Khan, and Velmurugan 2019). The bottom layer of CSP-filled composites shows the separation of a vertical patch of fibers increasing to a small degree, majorly propagating along the length of the specimen, and reaching its boundary compared to the baseline composite. The force-displacement behavior of composites following maximum force well represents the aforementioned failure event. The bottom face of laminates exhibits more damage owing to higher bending stresses compared to the top face. Continuous regions with just about open curves following incessant reduction of forces are unveiled by force-displacement curves of composites, facilitating portions with widespread unstiffening. This appears to be owing to energy dissipation by friction up until discontinuing the impactor motion. Composites with 20 wt%, 30 wt%, and 40 wt% of CSP show this plateau following displacements of around 30.6 mm, 24.3 mm, and 20.2 mm, correspondingly, which do not appear to be physical ones. The ultimate displacement of the impactor for the aforementioned composites ranges to be around 34.5 mm, 26.6 mm, and 24.9 mm, respectively. This is equivalent to the perforation of composites which is defined as a small rise of energy away from the load-bearing capability of the plate. Such a plateau does not appear for the baseline composite. Composites infused with 20 to 40 wt% of bio-fillers exhibit higher maximum force and bending stiffness (slope of the force-displacement curve in the initial linear region), displacement, and energy absorption indicating its higher impact resistance than baseline composite. The force-displacement behavior of carbon composites shown in Figure 26.6b seems to exhibit comparable behavior to glass composites. However, the bending stiffness, maximum force, energy absorption, and displacement of the former composites with 20 wt%, 30 wt%, and 40 wt% appear to be higher, and lower than the latter ones, distinctly. Also, the damages on the top and bottom surfaces of carbon composites presented in Figure 26.8 appear to be similar; however, the extent of them seems to be lower than that of glass composites shown in Figure 26.7. This indicates a higher perforation resistance of carbon composites than glass composites.

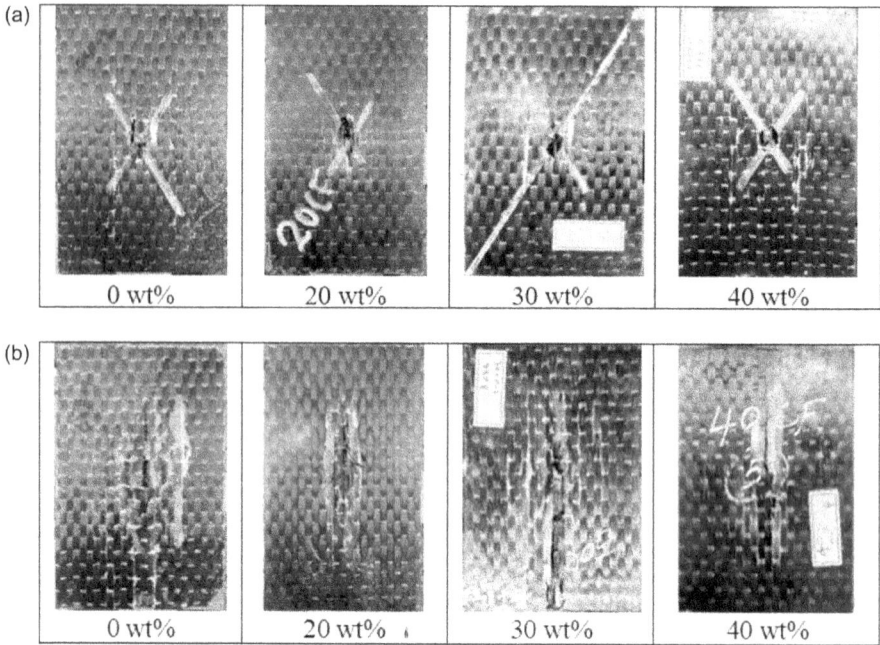

FIGURE 26.8 Impact-induced damage patterns of CFRP composites. 0 wt%, 20 wt%, 30 wt%, and 40 wt% represent the weight percentage of CSP within the epoxy (core) of sandwich panels C/C/C-0, C/C/C-20, C/C/C-30, and C/C/C-40, respectively. Here, C before and after the core designated by C represents a single layer of CFRP in the front (top) and back (bottom) skins, separately.

26.3.3.2 Performance Parameters

In this section, the effect of impact performance parameters on different content of CSP composites is brought out. The maximum force of glass composite with 20, 30, and 40 wt% of CSP fillers exhibits similar and higher extents compared to that of pristine composite as can be seen in Figure 26.9a. Maximum displacement and contact duration of glass composites with 20 to 40 wt% of CSP filler content plotted in Figure 26.9a and b also exhibit comparable behavior as the maximum force of

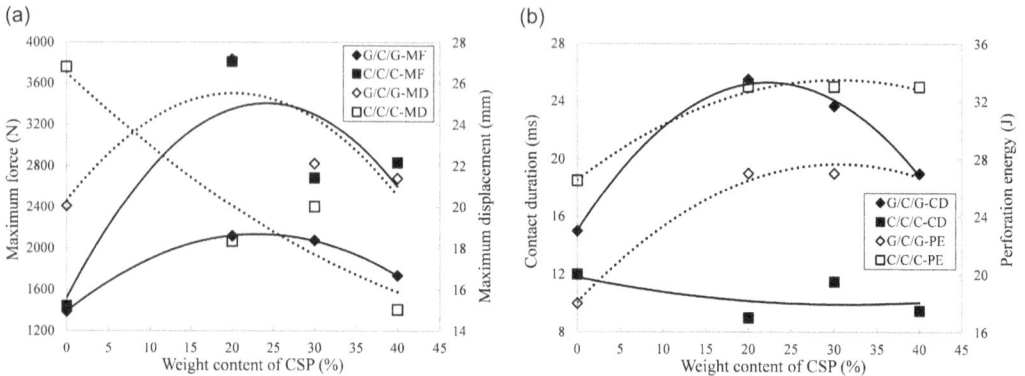

FIGURE 26.9 Variation of (a) maximum force and maximum displacement and (b) contact duration and perforation energy of GFRP and CFRP composites. In the case of sandwich panel G/C/G, G before and after the core designated by C indicates a single layer of GFRP in the front (top) and back (bottom) skins, individually. For sandwich panel C/C/C, C before and after the core represents a single layer of CFRP in the front and back skins, separately.

glass composites. Moreover, glass composite with bio-fillers unveils higher perforation energy than that of pristine composite in Figure 26.9b. This facilitates higher impact perforation resistance of glass composites with CSP than of the baseline composite. The carbon composites show similar behavior as glass composites. However, higher maximum force and perforation energy are exhibited by carbon composites than glass composites and vice versa for maximum displacement and contact duration. Overall, it can be stated that the impact perforation behavior of composites seems to be unaffected by varying content of CSP filler from 20 to 40 wt% within the matrix. This is owing to the effects of local penetration which seem to be predominant in impact perforation scenarios.

26.4 SUMMARY

Sandwich composites having epoxy infused with different content of coconut shell powder filler, viz. 0 wt%, 20 wt%, 30 wt%, and 40 wt%, as a core and glass fiber-reinforced epoxy and carbon fiber-reinforced epoxy as skins are evaluated under tensile, flexural, and low-velocity single-impact loading. A co-curing technique employed to prepare sandwich composites postpones skin-core debonding, resulting in advancing their bending performance. The tensile strength, modulus, failure strain, and toughness of composites diminish with amplified content of CSP filler from 0 to 40 wt%. This is believed to be due to the clustering of CSP fillers within the matrix leading to an ineffective transfer of load from CSP to the matrix and thereby from core to skin. This also happens with corresponding flexural properties of composites on account of weak interfacial bonding between CSP and the matrix. This results in a low bending load experienced by the matrix modified with CSP. However, the impact resistance of composites with 20 to 40 wt% of CSP fillers seems to be higher than the baseline composites. The impact parameters, viz. maximum force, displacement, contact duration, and perforation energy, are found to be almost independent of the above CSP contents. This happens due to the predominant local penetration effects in impact perforation cases. CFRP composite facilitates higher tensile, flexural, and impact properties than GFRP composite. The failure modes of composites include CSP-matrix interfacial debonding and cracking, skin-core delamination, longitudinal splitting, and breaking of fibers under tensile and impact loads. Meanwhile, core shear failure, skin-core delamination, and fiber damage on top and bottom skins around the loading region are experienced under bending load. Overall, the addition of CSP fillers to the polymeric core of sandwich composites decreases their tensile and flexural properties. However, it improves their energy absorption and resistance induced by impact. Sandwich composites with bio-filler facilitate lightweight and significantly durable material. Identical dispersal of filler particles in the matrix along with good affinity and linkage between skin and core are expected to occur on adding low content of CSP filler, viz. 5 wt%, 10 wt%, and 15 wt%, into the matrix. This would result in improved tensile and flexural properties of composites.

REFERENCES

Abdul Khalil, H. P. S., M. Masri, C. K. Saurabh, M. R. N. Fazita, A. A. Azniwati, N. A. Sri Aprilia, E. Rosamah, and R. Dungani. Incorporation of coconut shell based nanoparticles in kenaf/coconut fibres reinforced vinyl ester composites. *Materials Research Express* 4(3), 119501, 2017.

Agunsoye, J. O., A. K. Odumosu, and O. Dada. Novel epoxy-carbonized coconut shell nanoparticles composites for car bumper application. *The International Journal of Advanced Manufacturing Technology* 102, 893–899, 2019.

Andezai, A. M., L. M. Masu, and M. Maringa. Investigating the mechanical properties of reinforced coconut shell powder/epoxy resin composites. *International Journal of Engineering Research and Technology* 13(10), 2742–2751, 2020.

ASTM D7136/D7136M-15. Standard test method for measuring the damage resistance of a fiber reinforced polymer matrix composite to a drop-weight impact event. *ASTM International*, West Conshohocken, 2015.

Bhaskar, J., and V. K. Singh. Physical and mechanical properties of coconut shell particle reinforced-epoxy composite. *Journal of Materials and Environmental Sciences* 4(2), 227–232, 2013.

Deshpande, S., and T. Rangaswamy. Effect of fillers on E-glass/jute fiber reinforced epoxy composites. *International Journal of Engineering Research and Applications* 4(8), 118–123, 2014.

Islam, Md. T., S. C. Das, J. Saha, D. Paul, M. T. Islam, M. Rahman, and M. A. Khan. Effect of coconut shell powder as filler on the mechanical properties of coir-polyester composites. *Chemical and Materials Engineering* 5(4), 75–82, 2017.

Jagadeesh, P., M. Puttegowda, Y. G. T. Girijappa, S. M. Rangappa, and S. Siengchin. Effect of natural filler materials on fiber reinforced hybrid polymer composites: An Overview. *Journal of Natural Fibers* 19(11), 4132–4147, 2022.

Livingston, T., A. Athijayamani, and A. Alavudeen. Evaluation of mechanical properties of coconut shell particle/vinyl ester composite based on the untreated and treated conditions. *Materials Research Express* 8, 035309, 2021.

Manjunatha Chary, G. H., and K. S. Ahmed. Experimental characterization of coconut shell particle reinforced epoxy composites. *Journal of Materials and Environmental Sciences* 8(5), 1661–1667, 2017.

Mishra, A. Mechanical properties of coconut shell dust, epoxy-fly ash hybrid composites. *American Journal of Engineering Research* 6(9), 166–174, 2017.

Nadzri, S. N. I. H. A., M. T. H. Sultan, A. U. M. Shah, S. N. A. Safri, A. R. A. Talib, M. Jawaid, and A. A. Basri. A comprehensive review of coconut shell powder composites: Preparation, processing, and characterization. *Journal of Thermoplastic Composite Materials* 35(12), 2641–2664, 2022.

Oral, I., S. Kocaman, and G. Ahmetli. Preparation and ultrasonic characterization of modified epoxy resin/coconut shell powder biocomposites. *Journal of Applied Polymer Science* 139(11), 51772, 2022.

Paul, D., R. Velmurugan, and N. K. Gupta. Experimental and analytical studies of syntactic foam core composites for impact loading. *International Journal of Crashworthiness* 27(1), 299–316, 2022.

Pradhan, S. K., E. S. Dwarakadasa, and P. J. Reucroft. Processing and characterization of coconut shell powder filled UHMWPE. *Material Science and Engineering: A* 367(1–2), 57–62, 2004.

Rajesh, D. H., K. C. Ramachandra, G. R. Ravikumar, and K. K. Pavankumar. Study the mechanical properties of E-glass fiber and coconut shell particles in epoxy resin. *International Research Journal of Engineering and Technology* 6(5), 113–116, 2019.

Salmah, H., S. C. Koay, and O. Hakimah. Surface modification of coconut shell powder filled polylactic acid biocomposites. *Journal of Thermoplastic Composite Material* 26(6), 809–819, 2012.

Sapuan, S. M., M. Harimi, and M. A. Maleque. Mechanical properties of epoxy/coconut shell filler particle composites. *The Arabian Journal for Science and Engineering* 28(2B), 171–181, 2003.

Sarki, J., S. B. Hassan, V. S. Aigbodion, and J. E. Oghenevweta. Potential of using coconut shell particle fillers in eco-composite materials. *Journal of Alloys and Compounds* 509(5), 2381–2385, 2011.

Sharma, A. P., S. H. Khan, and R. Velmurugan. Effect of through thickness separation of fiber orientation on low velocity impact response of thin composite laminates. *Heliyon* 5(10), e02706, 2019.

Somashekhar, T. M., P. Naik, V. Nayak, Mallikappa, and S. Rahul. Study of mechanical properties of coconut shell powder and tamarind shell powder reinforced with epoxy composites. *IOP Conference Series: Materials Science and Engineering* 376, 012105, 2018.

Udhayasankar, R., and B. Karthikeyan. A review on coconut shell reinforced composites. *International Journal of ChemTech Research* 8(11), 624–637, 2015.

27 Numerical Investigation of the Influence of the Design Parameters on the Blast Mitigation Response of Steel Plate Subjected to Free-Air Blasts

A. Prakash, A. V. S. Siva Prasad and Raguraman Munusamy

27.1 INTRODUCTION

The behaviour of structures exposed to impulsive loads is predicted by the different modes of failure. Menkes et al. [1] conducted an experiment on fully clamped beams and proposed three modes of failure, i.e., mode I (inelastic deformation), mode II (tensile failure at the support) and mode III (transverse shear failure at the support). Several theoretical and experimental studies are conducted to predict the behaviour of plates under impulsive loads. Teeling-Smith [2] conducted an experiment on fully clamped thin circular plates under impulsive loads and observed modes of failure similar to the response of fully clamped beams [1]. Furthermore, in the mode II failure, partial tearing, complete tearing with increasing mid-deflection and complete tearing with decreased mid-deflection were observed by Nurick et al. [3] from the experiment on a clamped square plate under blast load. Shen et al. [4] theoretically predicted the failure response of a circular plate, considering strain-sensitive material, by adopting the failure criterion [5]. Since ductile materials undergo large plastic deformations under the dynamic impulsive load, a rigid plastic analysis on plates was conducted with a strain yield interface. The strain rate sensitivity of the materials should be considered due to the time-dependent deformation of the plate under dynamic impulsive load. Wierzbicki et al. [6] experimentally validated the viscoplastic theory of plates by correlating the effects of large deflection with the rate-dependent permanent deflection in the rigid plastic analysis. Olson et al. [7] numerically predicted the mode I and II response of the clamped square plate under impulsive load with finite element analysis. The numerical results also captured the plate rupture and are in good agreement with the experimental results. The development of constitutive material models exploited the computational efficacy by numerical modelling and validation of the material responses to impulsive loading. Impulsive loadings were numerically modelled as pressure pulses of different shapes and explosive detonations. However, the shape of the pressure pulse affects the material responses due to arbitrary approximations and the explosive detonation modelling requires fluid-structure interaction that reflects the exact physics of blast wave propagation. In this numerical study, the steel plate exposed to a free-air blast is simulated to capture its response. For this purpose, the Eulerian-Lagrangian coupled numerical model is simulated in ANSYS AUTODYN and validated with the existing experimental result. Further, the validated numerical approach is used to study the effect of plate geometry, charge mass and stand-off distance on the plate's response to an

DOI: 10.1201/9781003352358-27

explosion. To investigate the influence of plate geometry, a three-dimensional analysis is conducted to capture the mid-deflection of fully clamped circular and square plates. The numerical result of three-dimensional and axisymmetric simulations for the circular plate is also compared to obtain result accuracy. Since the axisymmetric analysis is in good agreement with three-dimensional analysis, it is adopted to capture the modes of failure for varying masses of 500 g, 700 g and 1,000 g, at various stand-off distances of 10 mm, 50 mm and 100 mm.

27.2 METHODOLOGY

27.2.1 BLAST WAVE

In the three-dimensional analysis, the blast wave is induced from the explosion of pentolite filled in the 1D wedge air model. An axial symmetry 1D wedge filled with Eulerian multi-material air is modelled for a radius of 60 mm with an element size of 0.5 mm. The pentolite explosive is filled into the air multi-material grid for a radius of 41.5 mm representing a mass of 500 g of spherical charge. At the end radius of the wedge, a flow-out boundary condition is applied to eliminate reflected pressure waves in the simulation. Gauges to measure output data were positioned at 51.5 mm, 55 mm and 60 mm. The detonation point for the explosive is fixed at the origin of the wedge as shown in Figure 27.1.

For the two-dimensional analysis, the Eulerian 2D multi-material element is used for modelling the air, and a semi-circle of diameter 83 mm pentolite equivalent to the mass of 500 g is filled in the grid as shown in Figure 27.2. In both analyses, the diameter of pentolite is filled to the Eulerian multi-material element based on the charge mass.

27.2.2 KINNEY AND GRAHAM EQUATION

The blast wave overpressure for a chemical explosion is given by [8]:

$$P_{so} = P_0 \frac{808\left[1+\left(\dfrac{Z}{4.5}\right)^2\right]}{\sqrt{1+\left(\dfrac{Z}{0.048}\right)^2}} \times \frac{1}{\sqrt{1+\left(\dfrac{Z}{0.32}\right)^2}} \times \frac{1}{\sqrt{1+\left(\dfrac{Z}{1.35}\right)^2}} \qquad (27.1)$$

Where P_{so} is the peak overpressure, P_o is the ambient pressure and Z is the scaled distance.

27.2.3 HOPKINSON-CRANK LAW

A dimensionless scaled distance is given by

$$Z = \frac{R}{\sqrt[3]{W}} \qquad (27.2)$$

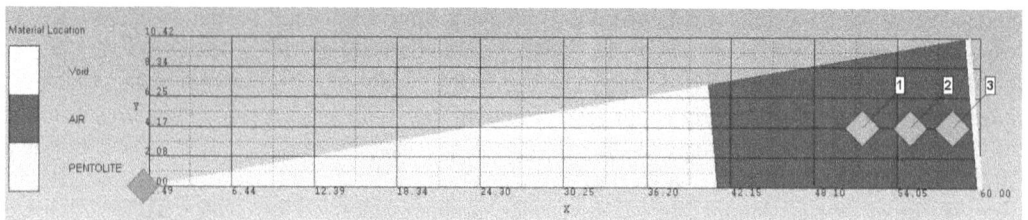

FIGURE 27.1   1D wedge model of air and pentolite.

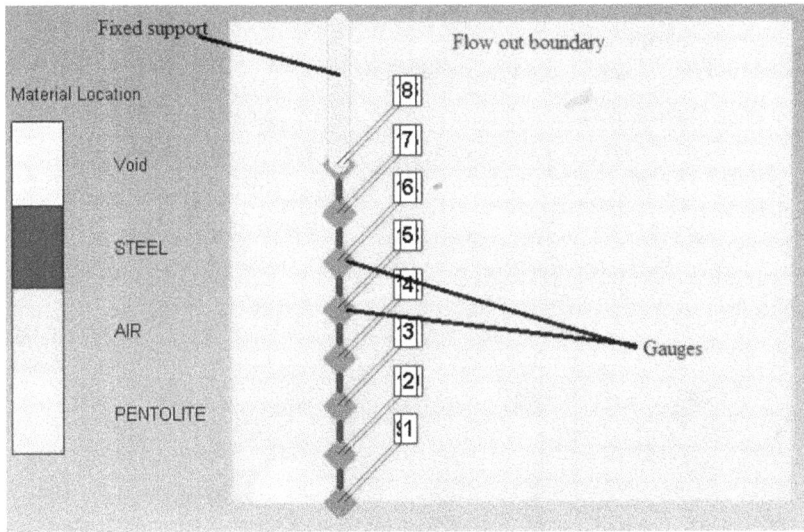

FIGURE 27.2 2D Eulerian-Lagrangian model.

where R is the distance from the centre of explosives and W is the explosive mass in the TNT equivalent.

27.2.4 TNT EQUIVALENT

Equivalent TNT weight for the chosen explosive weight is given by [8]:

$$W_e = W_{exp} \frac{H_{exp}}{H_{TNT}} \tag{27.3}$$

Where W_e is the TNT equivalent weight, W_{exp} is the weight of the actual explosive, H_{exp} is the heat of detonation of the actual explosive and H_{TNT} is the heat of detonation of TNT.

27.2.5 ANALYTICAL EVALUATION

The parameters for the calculation of peak overpressure are given in Table 27.1. By Kinney and Graham's equation (27.1), the overpressure at 200 mm for the detonation of pentolite of mass 500 g is found to be 1.37×10^4 kPa (137 bar). The calculated overpressure is in good correlation with the numerical result as shown in Figure 27.3.

27.2.6 EQUATION OF STATE

ANSYS AUTODYN material library material properties were used to define the equation of state for both pentolite and air. For pentolite, the JWL equation of state [9] is used in the numerical simulation. The relationship between the volume, energy and pressure of detonation products is defined by the following equation.

$$P = A\left(1 - \frac{w}{R_1 V}\right)e^{-R_1 V} + B\left(1 - \frac{w}{R_2 V}\right)e^{-R_2 V} + \frac{wE}{V} \tag{27.4}$$

where A and B are the pressure coefficients, R_1 and R_2 are the principal and secondary eigenvalues to depict the short- and long-range detonations and ω is the fractional part of the energy E contributing

TABLE 27.1

Parameters for Blast Calculation

Mass of pentolite	0.5 kg
TNT equivalent of pentolite mass	0.717 kg
Distance from the detonation centre, R	0.2 m
Scaled distance, Z	0.2235 $m/\sqrt[3]{kg}$
Ambient pressure, P_a	1×10^5 Pa

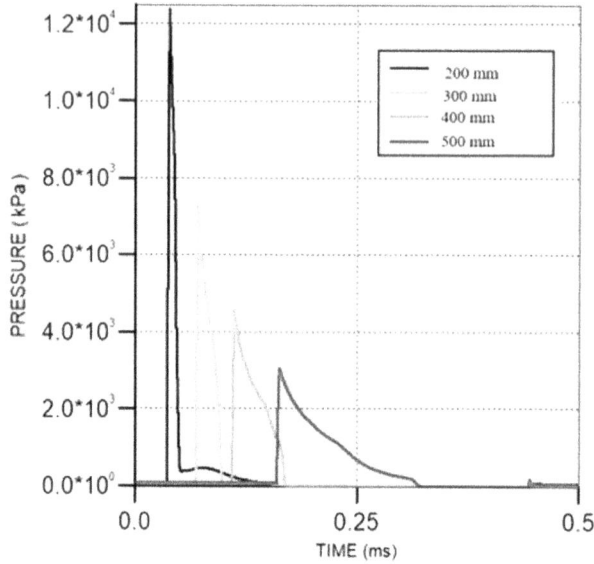

FIGURE 27.3   Shockwave attenuation.

to the pressure P. And for air material, the ideal gas equation of state is adopted in the simulation with an internal energy of $2.068 \times 10^5 mJ/mm^3$ to provide the ambient pressure of 101.3 kPa [10].

27.2.7   MATERIAL MODEL

The mild steel material is considered for the numerical study. The material model is carefully adopted for the ductile material to exhibit rate-dependent behaviour due to the time-sensitive blast pressure. For this purpose, the Johnson-Cook strength and failure model is adopted. The material strength and failure model parameters were adopted from the existing results [11].

Johnson-Cook strength model:

$$\sigma = \left[A + B\epsilon^n \right]\left[1 + Cln\dot{\epsilon}^* \right]\left[1 - T^{*m} \right] \tag{27.5}$$

$$T^{*m} = \left[\frac{T - T_r}{T_{melt} - T_r} \right] \tag{27.6}$$

where A is the yield stress (MPa), B is the hardening constant (MPa), n is the hardening exponent, C is the strain rate constant, $\dot{\epsilon}^*$ is the reference strain rate(/s^{-1}), m is the thermal softening constant, T_{melt} is the melting temperature and T_r is the room temperature.

The Johnson-Cook failure model depends on stress triaxiality, strain rate and temperature, and is given by

$$\varepsilon_f = \left[D_1 + D_2 exp\left(D_3 \frac{\sigma_m}{\bar{\sigma}} \right) \right]\left[1 + D_4 ln\left(\frac{\dot{\bar{\sigma}}^{pl}}{\dot{\varepsilon}_0} \right) \right]\left[1 + D_5 \hat{T} \right] \qquad (27.7)$$

where D_1 to D_5 are the damage parameters, σ_m is the mean stress, $\bar{\sigma}$ is the equivalent von Mises stress, $\dot{\bar{\epsilon}}^{pl}$ is the equivalent plastic strain and $\dot{\varepsilon}_0$ is the reference strain rate.

27.2.8 MODELLING OF BLAST WAVE INTERACTION WITH STEEL PLATE

The dimensions for the square plate are 1,000 mm × 1,000 mm and the diameter for the circular plate is considered as 1,000 mm. For all the simulation cases, the same dimensions were adopted with a constant plate thickness of 6 mm. In the three-dimensional model, the quarter part of the plate is modelled using Lagrangian elements and the air is modelled using the Eulerian multi-material model to the dimensions of 500 mm × 500 mm × 500 mm. The application of symmetry and other boundary conditions is explained in section 27.2.9. The 1D wedge model as shown in Figure 27.1 is detonated and the induced blast pressure expansion is allowed until the stand-off distance of 10 mm. At this time, the blast pressures developed in the 1D simulation are remapped to the 3D Eulerian-Lagrangian model as shown in Figure 27.4. The two-dimensional model is modelled using the axisymmetry condition for the analysis of a circular plate of diameter 1,000 mm. For this model, the steel plate is modelled to the radius of 500 mm using Lagrangian elements and the air is modelled using the Eulerian multi-material model to the dimension of 500 mm × 500 mm. Later, pentolite is filled to the Eulerian multi-material model by the geometrical space method based on the mass of the charge.

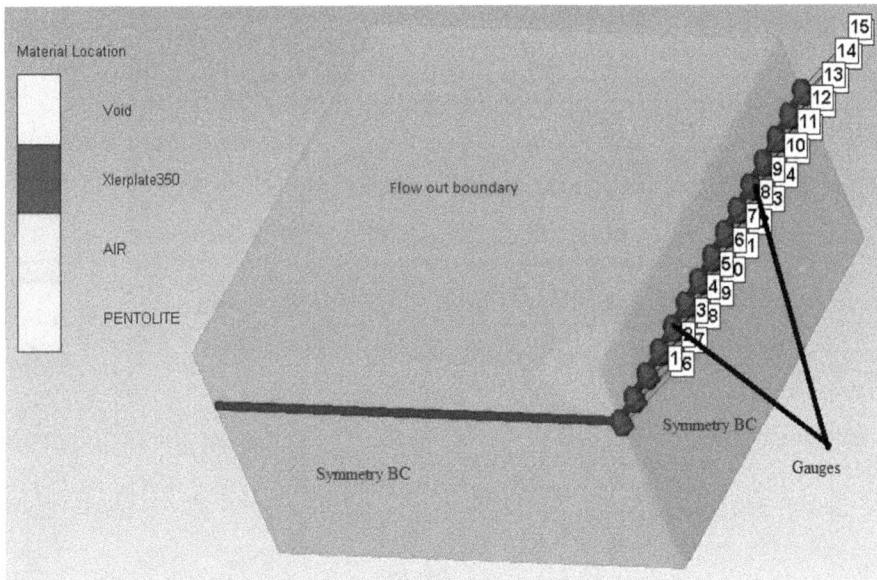

FIGURE 27.4 3D Eulerian-Lagrangian model.

27.2.9 BOUNDARY CONDITION

For the three-dimensional analysis, a symmetry boundary condition is applied to the quarter model in the X and Y direction. The plate is fully clamped at its boundary between the region 350 mm to 500 mm, so the nodes in the clamped region are constrained with fixed support conditions. To eliminate the reflection of pressure in the air domain, a flow-out boundary condition is applied as shown in Figure 27.4.

27.3 RESULTS

27.3.1 NUMERICAL APPROACH VALIDATION

The numerically modelled free-air blast simulation approach is validated with the centrally blast-loaded experimental result [12] by comparing the mid-deflection numerical result of the square plate. For the validation purpose, the free-air blast experimental setup from the work of Ackland et al. [12] is numerically simulated with pentolite of mass 500 g with 10 mm of stand-off distance from the plate. The mesh size of 5 mm is adopted for both steel plate and air from the results of the mesh resolution study. The obtained numerical result for the mid-deflection of the plate is compared with the experimental result, providing good agreement as shown in Figure 27.5.

27.3.2 EFFECT OF PLATE GEOMETRY

The three-dimensional free-air blast simulation of the circular plate is conducted for the pentolite of mass 500 g and the stand-off distance of 10 mm. The numerical result shows that the mid-deflection of the circular plate reaches a maximum of 136 mm, whereas the maximum mid-deflection of the square plate is 107 mm. From the comparison, the square plate geometry provides better blast resistance. An axisymmetric free-air blast simulation, subjected to the same mass and stand-off distance, is performed to capture the numerical accuracy of 2D with 3D analysis for the circular

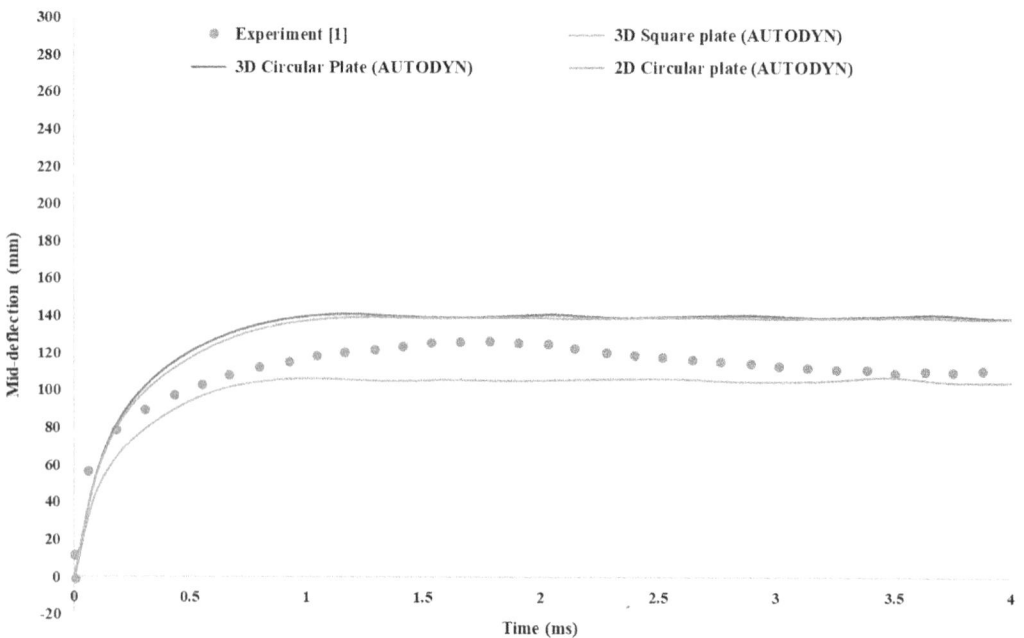

FIGURE 27.5 Comparison and validation of mid-deflection for mild steel plate.

plate. From the comparison, as shown in Figure 27.5, it is evident that 2D numerical results are in good agreement with three-dimensional analyses.

27.3.3 Effect of Mass and Stand-off Distance

The fully clamped mild steel circular plate of diameter 1,000 mm is subjected to the explosion of pentolite of varying masses 500 g, 700 g and 1,000 g for different stand-off distances to capture the mode of plates' response. By increasing the charge mass and reducing the stand-off distance in the axisymmetric Eulerian-Lagrangian simulations, the impulse on the plate is increased. The different combinations of charge mass with stand-off distance influence the plates' mode of responses as observed in Table 27.2. The charge mass and stand-off distance are varied to increase the impulse, thereby capturing the mode I (inelastic deformation at the centre) shown in Figure 27.6, mode II (partial tearing at the support) shown in Figure 27.7 and mode II (a) (complete tearing at the support) shown in Figure 27.8.

27.4 CONCLUSION

The dynamic behaviour of a mild steel square plate exposed to a chemical explosion is numerically simulated and the central mid-deflection of the plate is validated with the experimental result. Similarly, the central mid-deflection of a circular plate is captured by axisymmetric analysis which

TABLE 27.2
Mode of Responses for a Mild Steel Circular Plate under Free-Air Blast Explosion

Mass of pentolite (kg)	Stand-off distance (mm)	Mid-deflection (mm)	Modes of failure
500	10	136	Mode I – Inelastic deformation
	50	69.5	Mode I – Inelastic deformation
700	10	164	Mode II – Partial tearing
	50	84.33	Mode I – Inelastic deformation
1,000	50	–	Mode III – Transverse shear
	100	200	Mode II – Partial tearing

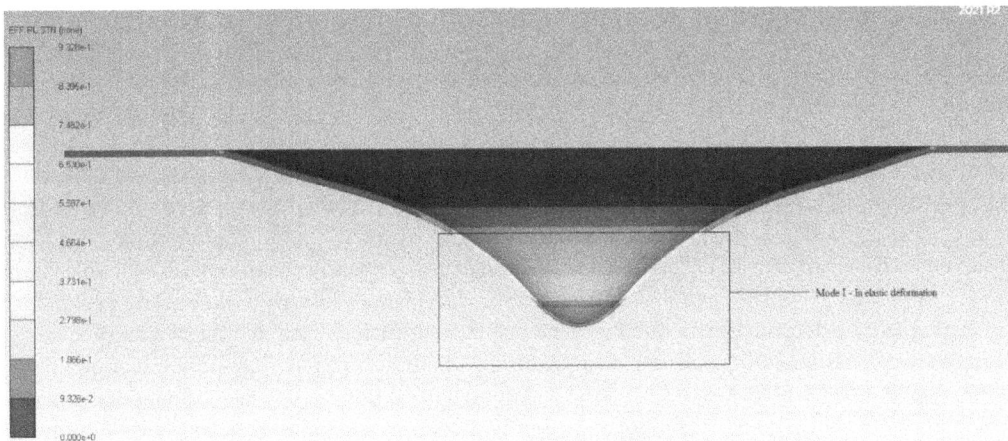

FIGURE 27.6 Mode I response of mild steel circular plate for the charge of mass 500 g at stand-off distance of 10 mm.

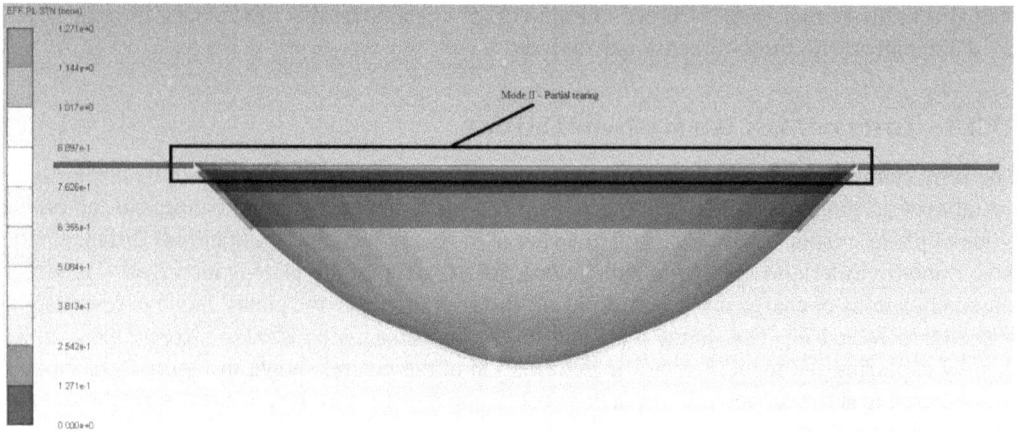

FIGURE 27.7 Mode II (partial tearing) response of mild steel circular plate for the charge of mass 1,000 g at stand-off distance of 100 mm.

FIGURE 27.8 Mode III (transverse shear failure) response of mild steel circular plate for the charge of mass 1,000 g at stand-off distance of 50 mm.

shows good agreement with the full-fledged three-dimensional analyses as shown in Figure 27.5. By using the validated numerical approach, further numerical studies were conducted to study the effect of plate geometry, charge mass and stand-off distance on the plates' behaviour. For the different combinations of charge mass and stand-off distance, axisymmetric analysis was performed on the circular plate with fully clamped boundary conditions to predict the different modes of plate failure. Thereby, mode I, mode II and mode III responses for the mild steel circular plate were observed as shown in Table 27.2. The transition of plates' response from mode I to mode III, as shown in Figures 27.6–27.8, exhibits the capacity of the plate to absorb the induced blast energy. The large inelastic deformation (mode I) occurs due to the high localisation of blast energy resulting from the narrowed shock front. The plates' clamping conditions did not have any effect on mode I failure due to the highly localised strains. The clamping condition of the plate influences the plates' behaviour in the mode II and III failure shown in Figures 27.7–27.8. The circular plate shows larger mid-deflection than the square plate for the same blast intensity, exhibiting the influence of plate geometry on the plates' behaviour.

REFERENCES

1. S Mendes and H Opat. Tearing and shear failures in explosively loaded clamped beams. *Exp. Mech*, 13:480–486, 1973.
2. RG Teeling-Smith and GN Nurick. The deformation and tearing of thin circular plate subjected to impulsive loads. Technical Report 1, 1991.
3. GN Nurick and GC Shave. The deformation and tearing of thin square plates subjected to impulsive loads—An experimental study. *International Journal of Impact Engineering*, 18(1):99–116, 1996.
4. WQ Shen and N Jones. Dynamic response and failure of fully clamped circular plates under impulsive loading. *International Journal of Impact Engineering*, 13(2):259–278, 1993.
5. WQ Shen and N Jones. A failure criterion for beams under impulsive loading. *International Journal of Impact Engineering*, 12(1):101–121, 1992.
6. T Wierzbicki and AL Florence. A theoretical and experimental investigation of impulsively loaded clamped circular viscoplastic plates. *International Journal of Solids and Structures*, 6(5):553–568, 1970.
7. MD Olson, GN Nurick, and JR Fagnan. Deformation and rupture of blast loaded square plates—Predictions and experiments. *International Journal of Impact Engineering*, 13(2):279–291, 1993.
8. GF Kinney and KJ Graham. *Explosive shocks in air*. Springer Science & Business Media, 2013.
9. EL Lee, HC Hornig, and JW Kury. *Adiabatic expansion of high explosive detonation products*. Technical report, Univ. of California Radiation Lab. at Livermore, Livermore, CA (United States), 1968.
10. TC Chapman, TA Rose, and PD Smith. Blast wave simulation using autodyn2d: A parametric study. *International Journal of Impact Engineering*, 16(5–6):777–787, 1995.
11. K Senthil, MA Iqbal, P Bhargava, and NK Gupta. Experimental and numerical studies on mild steel plates against 7.62 api projectiles. *Procedia Engineering*, 173:369–374, 2017.
12. K Ackland, C Anderson, and TD Ngo. Deformation of polyurea-coated steel plates under localised blast loading. *International Journal of Impact Engineering*, 51:13–22, 2013.

28 Numerical Simulation of Water Tank Used for Underwater Blast Testing

M. D. Goel and Nenshol Jayant Anand

28.1 INTRODUCTION

The idea of utilizing an explosive under the surface of the water has been around since the times of early nautical warfare, even though the resources to successfully use them were limited or did not exist then. This led to the requirement for understanding the underwater explosion (UNDEX) process, studying the consequences and in turn designing underwater blast-resistant structures that can withstand such a high dynamic loading. In spite of noteworthy efforts from various organizations like government, civil resources and military, it is broadly acknowledged that the properties of underwater explosion are still not well known [1]. Furthermore, most of the information is not in the civic domain because of tactical reasons and hence is not easily accessible to academic researchers. A water tank is a type of liquid storage tank that has become a crucial part of today's society, and due to its physical importance, these structures are being targeted by terrorists. In recent years, structural damage due to explosions has tremendously increased, mainly due to the result of terrorist activities, natural explosions due to chemical reactions or manmade accidents. Ramajeyathilagam and Vendhan [2] presented a study in which experimental and numerical underwater explosion was conducted on a rectangular mild steel plate submerged in the water. The experiment was carried out in a shock tank with a box model setup along with an air-backed state, and 10g of TNT was used. Numerical simulation was carried out using CSA/GENSA (DYNA3D). The Cowper-Symonds material model along with the elastoplastic material model and isotropic hardening for plasticity was used in their finite element (FE) analysis. Three different failure models were observed during analysis along with the rupture at the centre of the plate, with small tearing and shear and tensile failure [2]. Later on, Chang and Lin [3] presented a review of 242 storage tank accidents from all over the world and found that 85% of the accidents were due to fire and explosion. They also concluded that many of those accidents could have been avoided had good engineering designs been implemented. Afterwards, Jhung et al. [4] presented a numerical simulation for the impact effect of high-speed projectiles along with the effect of water on the impact analysis of the water tank structure. LS-DYNA [5] and ANSYS [6] FE software were used for the modelling and analysis of the water tank. It was concluded that fluid in the tank significantly reduces the impact effect. In the year 2013, Schiffer and Tagarielli [7] reported the ballistic response of a double-walled hull, and it was numerically analysed with the help of laboratory scale fluid-structure interaction (FSI) experiments and FE methods. The next year, Mittal et al. [8] presented a study on the investigation of mild steel cylindrical water tanks having open roofs under the blast loading. Maximum hoop stress, induced shear stress, peak sloshing of water and energy response were studied. Afterwards, Hsu et al. [9] imitated an experimental test of Ramajeyathilagam and Vendhan [2] with the help of FE software ABAQUS/Explicit [10]. The coupled acoustic structure (CAS) and coupled Eulerian-Lagrangian (CEL) models were used for the simulation of an underwater explosion on a mild steel plate in the shock tank. In the same year, Wang and Xiong [11] studied the simplified examination of the water tank against the blast load in order to

DOI: 10.1201/9781003352358-28

improve the structural resistance design. The numerical investigation focused on the generation of a pressure impulse (P-I) diagram in order to evaluate the damage to the water tank. Later on, Razic and Miralem [12] studied the effect of a real explosion on a reinforced concrete structure both underwater as well as in the air. From the results, a safe distance of 1 m from the epicentre of the explosion was formulated. They said that if detonation took place at this point, fragments would not reach the walls of the tank. Then, Yin et al. [13] analysed different configurations with the sacrificial coating on the thin hull and investigated the response of the hulls under blast loading. Hence, understanding the dynamic behaviour of liquid storage tanks towards blast loading through arduous numerical simulation is very essential due to limited experimental facilities. Based on the literature review and to the best of the authors' knowledge, it was noted that there has been no study available for the underwater explosion of a mild steel circular tank. It has also been noted from the literature that most of the work has been focused on ships, submarines and other common water structures and very little work has been carried out for sub-surface structures. Although bubble pulsation is an important process of underwater explosion, very few authors have focused on its monitoring and pulsation phenomena, mainly due to the complexity of the problem. Therefore, in this study, a mild steel water tank has been subjected to underwater blast loading and numerically analysed for various parameters along with the monitoring of bubble formation and pulsation process.

28.1.1 UNDERWATER EXPLOSION

When detonation of charge takes place, an extremely steep-fronted pressure, as can be seen in Figure 28.1, is developed at the point of detonation and propagates at very high speed in the covering fluid. These waves are termed shock waves and they propagate as spherical waves with speed more than that of sound. They are generally large discontinuous, compressive pressure waves, which are produced when a detonation wave reaches the surrounding fluid.

When the shock wave has covered a 2–3 charge radii distance, the speed of propagation holds to be constant and linear. Along with this, acoustic behaviour is considered, and within this distance, the propagation is highly nonlinear. The relationship between the shock wave pressure changes with respect to the time as per Eq. (28.1). In this equation, the relationship between peak shock wave pressure P_0 and shock wave sensitivity ($W^{1/3}/R$) is in power function. Along with this, a relationship for decay constant θ has also been reported [1].

$$P(t) = P_0 P(t) = P_0 e^{-\frac{t}{\theta}} \qquad 0 \leq t \leq \theta \tag{28.1}$$

FIGURE 28.1 Shock wave obtained from underwater explosion.

For TNT,

$$P_0 = 2.16 \times 10^4 \left(\left(\frac{W^{\frac{1}{3}}}{R} \right) \right)^{1.13}$$

$$\theta = 0.06 \left(W \right)^{\frac{1}{3}} \left(\left(\frac{W^{\frac{1}{3}}}{R} \right) \right)^{0.18}$$

Eq. (28.2) provides shockwave energy (E) and shock wave impulse (I) which can be computed using (Eq. 28.3). Shock wave energy, also called energy flux density, is the energy per unit area while shock wave impulse is the impulse per unit area generated after the shock wave reaches the surface [1].

$$E = \frac{1}{\rho_0 C_0} \int_0^t P^2 (t) dt \tag{28.2}$$

For TNT

$$E = 2.44 \times 10^3 \left(W \right)^{\frac{1}{3}} \left(\left(\frac{W^{\frac{1}{3}}}{R} \right) \right)^{2.02}$$

$$I = \int_0^t P(t) dt \tag{28.3}$$

For TNT

$$I = 1.44 \left(W \right)^{\frac{1}{3}} \left(\left(\frac{W^{\frac{1}{3}}}{R} \right) \right)^{0.89}$$

The similitude equation for the time period T_n and maximum bubble radius $A_{max,n}$ of the nth pulse are represented by Eq. (28.4) and Eq. (28.5), respectively [1].

$$T_n = K_n \frac{W^{\frac{1}{3}}}{Z_{0,n}^{\frac{5}{6}}} \tag{28.4}$$

$$A_{max,n} = J_n \frac{W^{\frac{1}{3}}}{Z_{0,n}^{\frac{1}{3}}} \tag{28.5}$$

Here, $Z_{0,n}$ is the hydrostatic pressure head at the centre at the start of the nth bubble pulse and is calculated by Eq. (28.6), where d_n is the depth of the bubble centre obtained at the start of the nth pulse.

$$Z_{0,n} = d_n + 10 \tag{28.6}$$

The approximate time at which the gas bubble obtains its nth minimum and maximum radii can be estimated using Eq. (28.7) and Eq. (28.8), respectively [1].

$$T_{\min,n} = \left\{ {}_{i=1}^{n} T_i \right. \tag{28.7}$$

$$T_{\max,n} = \left\{ {}_{i=1}^{n-1} T_i + \frac{T_n}{2} \right._i \tag{28.8}$$

Further, the minimum radius of the gas bubble can be calculated using Eq. (28.9), where f is constant whose value for TNT is 0.113 m/kg$^{1/3}$.

$$A_{\min 1} = fW^{\frac{1}{3}} \tag{28.9}$$

It is important to note that the main characteristics of a gas bubble pulse are pulse velocity (u), maximum pressure of gas bubble pulse (P_{BP}), bubble pulse impulse (I_{BP}) and duration of the bubble pulse (τ_{BP}), and these are represented by Eq. (28.10), Eq. (28.11), Eq. (28.12) and Eq. (28.13), respectively [1].

$$u = \frac{\pi}{2}\left(A_{\max 1} - \left(\frac{W^{\frac{1}{3}}}{\rho_{TNT} \times g} \right)^{\frac{1}{3}} \right. \tag{28.10}$$

$$P_{BP} = K_{BP} Z_0^{\frac{1}{3}} \left(\frac{W^{\frac{1}{3}}}{R} \right) \tag{28.11}$$

$$I_{BP} = K_{IBP} Z_0^{-0.4} \left(\frac{W^{\frac{2}{3}}}{R} \right) \tag{28.12}$$

$$\tau_{BP} = \frac{I_{BP}}{P_{BP}} \tag{28.13}$$

28.2   GEOMETRICAL AND FINITE ELEMENT (FE) DETAILS

Herein, a mild steel water tank with a radius of 8 m and height of 10 m is numerically simulated. Water is filled up to a height of 9 m and a TNT charge of 3 kg is placed at 2 m from the top and at the centre of the tank; this complete setup is shown in Figure 28.2. This water tank has been subjected to an underwater explosion for investigation of the pressure contour on the surface of the mild steel tank and monitoring of the bubble formed during the underwater explosion process. Figure 28.2 shows the model of the water tank along with the placement of the explosive TNT charge of 3 kg mass inside the water surface at a depth of 2 m from the top of the tank. The tank is modelled as a shell element having a thickness of 10 mm. The water is modelled as Eulerian domain with a radius the same as that of the tank while the depth of the water is 9 m. The charge is placed at the centre of the tank at a distance of 8 m from the bottom of the tank and 1 m inside the water domain from the top. The water tank is modelled using FE package ABAQUS/Explicit [10], wherein the tank is modelled as a 3-D deformable shell with S4R (four-node, quadrilateral, stress/displacement shell element with reduced integration and large strain formulation) mesh element having a total of 18,673 mesh elements. The water is modelled as a 3-D Eulerian domain using EC3D8R (eight-node

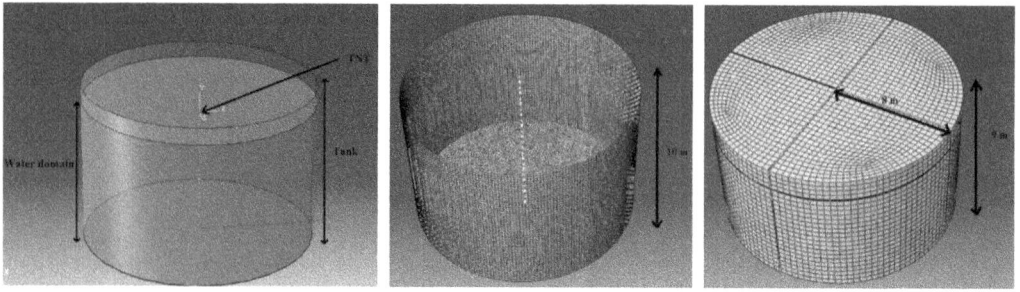

FIGURE 28.2 CAD and FE model of water tank for numerical simulation along with mesh details.

linear brick, multi-material, reduced integration with hourglass control) mesh element, having a total of 47,656 elements.

The general contact definition is one of the simplest contact definitions available in ABAQUS/ Explicit, which allows defining contact between surfaces with few limitations. General contact allows users to describe the contact amongst all or numerous areas of a model. The general contact system also implements contact among Eulerian materials and Lagrangian surfaces. This system automatically reimburses for mesh size disagreements to avoid infiltration of Eulerian material through the Lagrangian surface. Self-contact and contact between element-based surfaces are specified in the general contact definition. Contact force is equal to penalty stiffness multiplied by the penetration distance. ABAQUS/Explicit selects penetration distance such that the effect on time increment is minimum.

28.2.1 MATERIAL MODELS

During blast loading, the structure will undergo large deflection at a high strain rate accompanied by a large temperature variation and hence, for such simulations, the J-C (Johnson-Cook) material model is most suitable due to its effective formulation, simplicity and satisfyingly accurate results [14]. Johnson and Cook, in 1983, developed a constitutive model for material subjected to large strain, higher strain rate and temperature variations [14]. While developing this material model, various experiments were conducted on Armco iron, OFHC copper and 4340 steel with the help of a Hopkinson bar apparatus. These laboratory tests govern the properties of temperature, pressure on the strain to fracture and strain rates of the material. From the test data obtained, a cumulative-damage fracture model having separate properties of strain rate, temperature and pressure was modelled [14].

The hardening properties of the materials can be defined using the Johnson-Cook material law [14]. The flow stress of materials is stated in Eq. (28.14) as [14],

$$\sigma = (A + B\varepsilon^n)(1 + C \ln \dot{\varepsilon}^*)(1 - T^{*m}) \tag{28.14}$$

Here, σ represents equivalent stress and ε represents equivalent plastic strain. A, B, C, m and n are material constants. The components in the first bracket represent the strain hardening effect, the second bracket represents strain rate effects and the third bracket represents temperature effects. The material constants characterize the following physical phenomena:

A = yield stress at reference state, B = strain hardening constant, C = strengthening coefficient, n = strain hardening coefficient, m = thermal softening coefficient.

The $\dot{\varepsilon}^*$ represents the dimensionless strain rate as reported by Eq. (28.15) and T^* represents homologous temperature as reported by Eq. (28.16). T_m is melting temperature, T_{ref} is temperature at reference condition and $\dot{\varepsilon}_{ref}$ is strain rate at reference condition.

$$\varepsilon^{\dot{*}} = \varepsilon^{\dot{}}/\varepsilon^{\dot{}}_{ref} \tag{28.15}$$

$$T^* = \frac{T - T_{ref}}{T_m - T_{ref}} \tag{28.16}$$

Hence, in this study, the mild steel water tank is defined using the Johnson-Cook material model [14] and its properties are reported in Table 28.1. Equation of state (EOS) material properties are being used to define water properties in the FE model, and these properties are shown in Table 28.2. In this study, the Jones-Wilkens-Lee (JWL) EOS is used for defining explosive charge property because of its ease in hydrodynamic calculations and also due to its very close association with the experimental test results. The JWL EOS comprises factors presenting the relationships among volume, energy and pressure of detonation charge. The prime value of the JWL EOS lies in its ability to give a precise illustration of the Chapman-Jouguet (C-J) adiabat. The JWL EOS is represented in the form of Eq. (28.17), where V stands for the relative velocity, P refers to pressure and E to energy. Also, A and B refer to the pressure coefficients along with R_1 and R_2 which are the principal and secondary Eigenvalues to portray the short-range and long-range performance of the explosive charges correspondingly.

$$P = A\left(1 - \frac{\omega}{R_1 V}\right)e^{-R_1 V} + B\left(1 - \frac{\omega}{R_2 V}\right)e^{-R_2 V} + \left(\frac{\omega E}{V}\right) \tag{28.17}$$

Meanwhile, TNT material properties were defined using the JWL EOS and the same are reported in Table 28.3. By using the Geers-Hunter model, which defines an underwater explosion charge [1], the blast is modelled. Table 28.4 shows the charge parameters defined in the Geers-Hunter model as reported by Khani and Emamzadeh [15].

TABLE 28.1
Johnson-Cook Model Parameters for Mild Steel

Parameters	Values
Modulus of elasticity, E *(N/m²)*	203×10^9
Poisson's ratio, μ	0.33
Density, *(kg/m³)*	7,850
Yield stress constant, A *(N/m²)*	304.330×10^6
Strain hardening constant, B *(N/m²)* and N	422.007×10^6; 0.345
Constant, C	0.0156
Thermal softening constant, M	0.87
Reference strain rate (S⁻¹)	0.0001
Melting temperature	1,800
Transition temperature	293
Fraction strain constant	0.1152
D1	1.0116
D2D3D4	−1.7684
D5	−0.05279
	0.5262

Source: [16].

TABLE 28.2
Material Properties of Water

Parameter	Values
Density (kg/m³)	1,000
Velocity of sound through water (mm/sec)	1,500,000
Material constant (Gamm0)	0
Material constant (s)	0
Viscosity (N-s/mm²)	1×10^{-9}

Source: [8].

TABLE 28.3
JWL Equation of State Parameters for TNT

Material property	Values
Detonation wave speed	6,930 m/s
A	3.738×10^{11} Pa
B	3.747×10^{9} Pa
Ω	0.35
R_1	4.15
R_2	0.9
Detonation energy density	4.29×10^{6} J/kg

Source: [17].

TABLE 28.4
Geers-Hunter Model Parameters

Name of constant	Value
Charge constant1	5.21×10^{7}
Charge constant2	9×10^{-5}
Similitude spatial exponent	0.13
Similitude temporal exponent	0.18
Charge constant3	1.045×10^{9}
Ratio of specific heats for explosion gas	1.27
Charge material density (kg/m³)	1,600

Source: [15].

28.3   VALIDATION OF NUMERICAL SCHEME

Ramajeyathilagam and Vendhan [2] performed underwater blast explosion simulation experimentally as well as numerically on a mild steel plate placed inside a shock tank filled with water (refer to Figure 28.3). In order to validate the FE model, the setup of the test performed by [2] was modelled in ABAQUS/Explicit [10]. Ramajeyathilagam and Vendhan [2] have used PEK I (plastic explosive) as an explosive with a standoff distance of 0.15 m for generating an underwater blast load.

FIGURE 28.3 Experimental setup of Ramajeyathilagam and Vendhan (2004). Sources: [2]; www.sciencedi-rect.com/science/article/pii/S0734743X04000132.

The experiment was carried out in a 15 m × 12 m × 10 m tank by them. A mild steel plate of size 0.55 m × 0.45 m × 0.002 m with an exposed area of 0.30 m × 0.25 m was used for conducting the underwater blast experiment. Water of dimension 15 m × 12 m × 10 m was modelled as an Eulerian part instance, as the CEL formulation is to be used in the present FE modelling. The dimension of the water part is taken as bigger so that a proper node set of the nodal element of Eulerian and Lagrangian elements can be obtained while using the volume-fraction tool in the interaction property of ABAQUS/Explicit [10]. The shock tank is modelled as a 3-D deformable solid with acoustic material properties and AC3D8R (eight-node linear acoustic brick, reduced integration, hourglass control) mesh element having bulk modulus 160 GPa and mass density of 7,860 kg/m³. The acoustic element in ABAQUS/Explicit [10] allows the user to generate a non-reflecting surface condition, therefore acoustic element properties are defined for the tank. The mild steel plate was modelled as a 3-D deformable shell with S4R (four-node, quadrilateral, stress/displacement shell element with reduced integration and large strain formulation) mesh element having a total of 832 mesh elements. As in the numerical analysis of [2], strain-rate-dependent material properties were defined for mild steel plate; therefore, the Johnson-Cook model, having strain-rate-dependent properties, was used in this FE model as per [2]. The water part instance is modelled as a 3-D Eulerian EC3D8R (eight-node linear brick, multi-material, reduced integration with hourglass control) mesh element, having a total of 2,479,400 elements. EOS material properties are being used to define the water properties in the FE model, the same as those used by [2]. Based on the FE analysis, deformation behaviour with the distance from the edge of the tank is shown in Figure 28.4. It can be observed from this figure that both results are in good agreement, hence validating the numerical scheme. Peak deflection at the centre of the mild steel plate (m) has been reported as 0.06166 m by [2], wherein, in this numerical simulation, this peak defection is computed as 0.06187 m with a deviation of 0.34% only.

28.4 RESULTS AND DISCUSSION

Herein, a mild steel water tank having a radius of 8 m and height of 10 m is numerically simulated for underwater blast loading. The tank is filled with water up to a height of 9 m and a TNT charge of

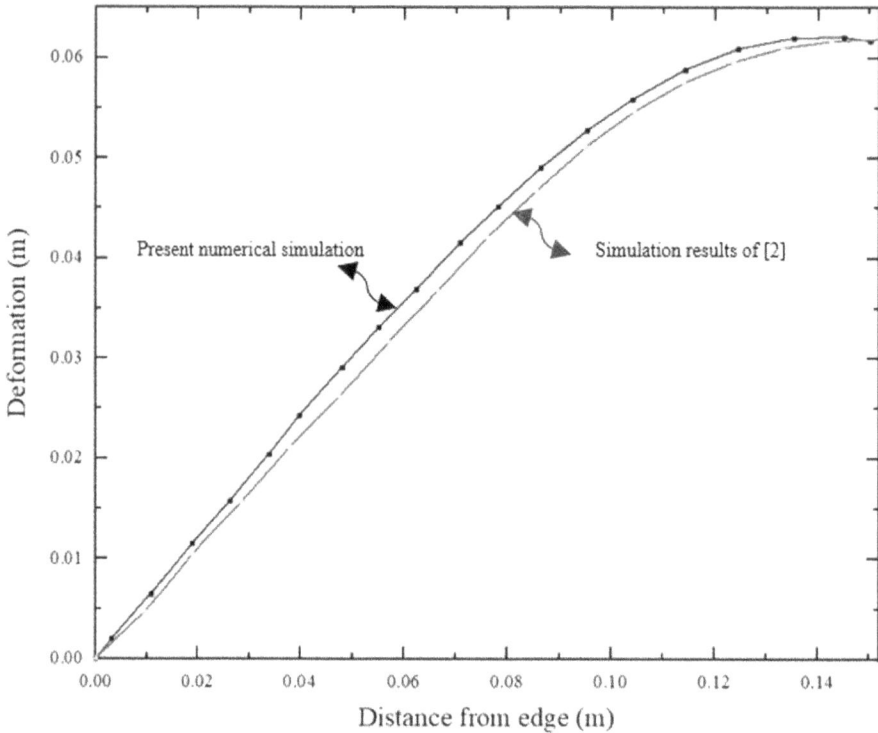

FIGURE 28.4 Comparison of deflection of present FE simulation with the numerical results of Ramajeyathilagam and Vendhan (2004). Sources: [2]; www.sciencedirect.com/science/article/pii/S0734743X04000132.

3 kg is placed at 2 m from the top and is located along the centre line of the tank. UNDEX is used to analyse the tank along with a CEL formulation of the whole underwater blast. It is to be noted that investigating the vigorous procedure of underwater explosion bubbles is a complicated process and it needs the consideration of various models and theories. The UNDEX technique numerically simulates the dynamic procedure of an underwater explosion gas bubble in three dimensions along with the simulation of the effect of bubble pulsation and shock wave generated during the underwater explosion process on the surface of the mild steel water tank. Figure 28.5 shows the process of bubble formation and shape up to the considered time instant. Initially, a non-spherical-shaped bubble is obtained due to the interaction of reflected waves from the water tank surface, but as the process continues, a completely spherical-shaped bubble is obtained. During the initial phase, the gas bubble expands outward rapidly with high pressure and high temperature in it, and this rapid expansion is captured in this simulation and has been presented at various time intervals in Figure 28.5. Some distortion can still be seen on the surface of the bubble, which signifies the need for providing a very fine meshing for the analysis. As this process requires much time for simulation, efforts were made for workable mesh refinement, based on the availability of computational facilities, but still, a more refined meshing may help to get a proper spherical gas bubble. The approximate maximum bubble radius (A_{max}) was calculated as 2.24 m using Eq. (28.5). The depth at which the explosion was carried out in the present study was 1 m below the surface of the water, so this implies that the first bubble will reach the surface of the water and the same will be observed in Figure 28.5. From this figure, it can be clearly observed that the first bubble touches the outer surface of the water in the proposed model. As the bubble pulsation continues until the whole energy inside the bubble is expended or the bubble reaches close to the object or surface, it can be concluded that only one bubble is observed in this analysis. Further, using Eq. (28.4), the time duration (T_n) for

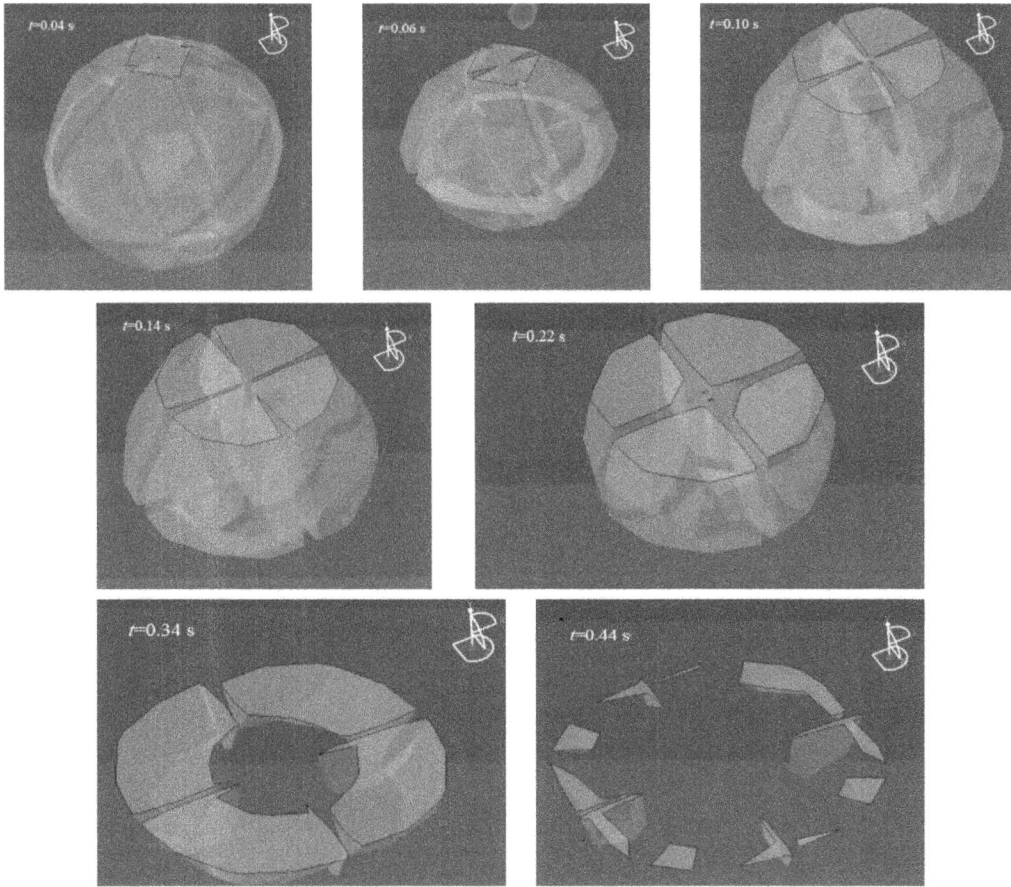

FIGURE 28.5 Bubble formulation, progress and collapse phenomena from an underwater explosion of 3 kg TNT in a mild steel water tank at different time instances.

the first bubble pulsation was calculated as 0.44 s. This time duration implies that the maximum bubble radius is to be obtained within this period and it can be observed that the bubble has reached the surface within this time period. Hence, it can be concluded that this numerical simulation follows the empirical formula result.

Figure 28.6 shows the variation of developed pressure with time for the present simulation. From this, it can be observed that the peak pressure developed on the wall of the tank was 5.7 MPa at the time instance of 0.04 s. As the stress generated on the water tank wall is quite a bit lower than the dynamic yield stress of steel, no deformation is observed on the water tank wall. Since the density of water and air is quite a bit less than that of mild steel, there is a rarefaction wave reflected at the interface of water and steel and similar to air and steel. This reflected wave led to the unloading of the intensity of the shock wave and for the fact that the density of air is lower than that of water, this unloading effect of the shock wave due to the reflected wave is less in the air than in the water. Therefore, peak pressure at the top portion of the tank above the water surface is observed. Further, along with the pressure from the shock wave, hydrostatic pressure due to water and gravity loading on the wall of the tank were also considered to create a realistic condition for the numerical simulation. Due to the fact that the base of the tank was considered rigid and fixed, the minimum pressure was observed at the base of the tank wall throughout the simulation. Figure 28.7 shows the pressure contours in the tank at various time instances. From this figure, it can be observed that the peak pressure was always at the top of the tank above the water surface while minimum pressure was observed at the base of the tank.

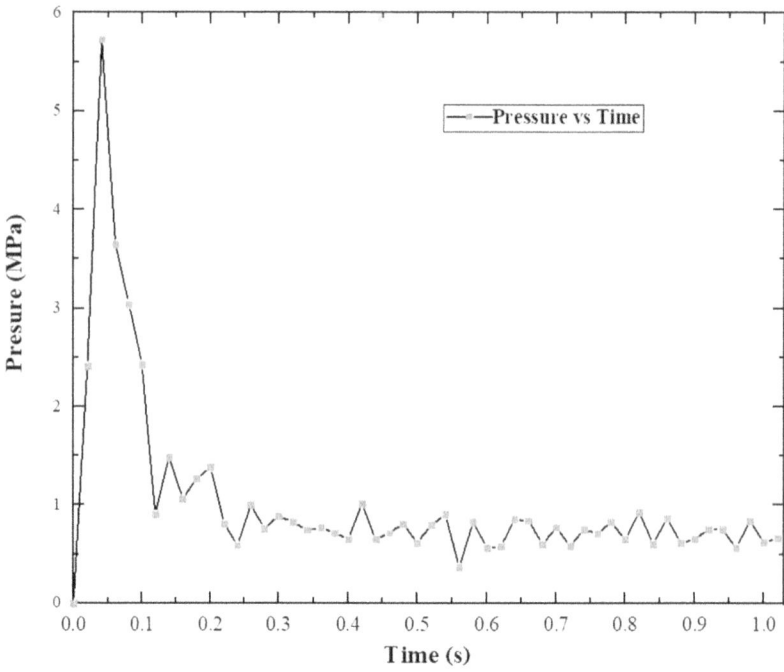

FIGURE 28.6   Pressure variation with time on the wall of the tank.

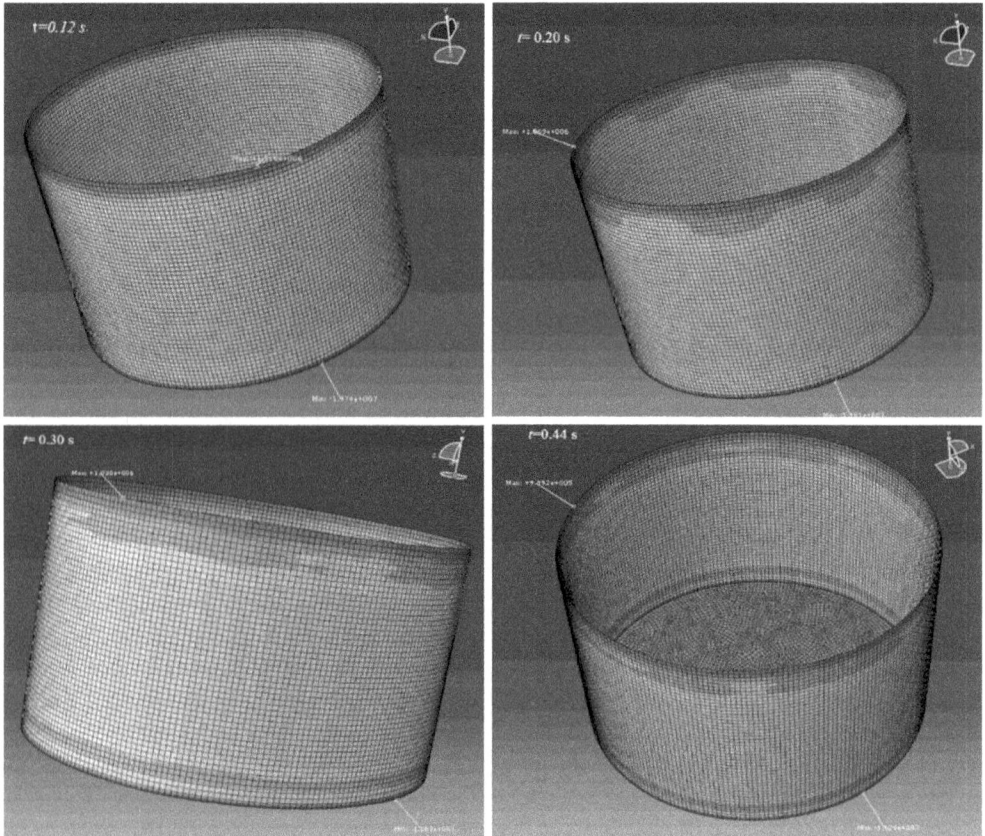

FIGURE 28.7   Position of maximum and minimum pressure on tank's wall for different time instances.

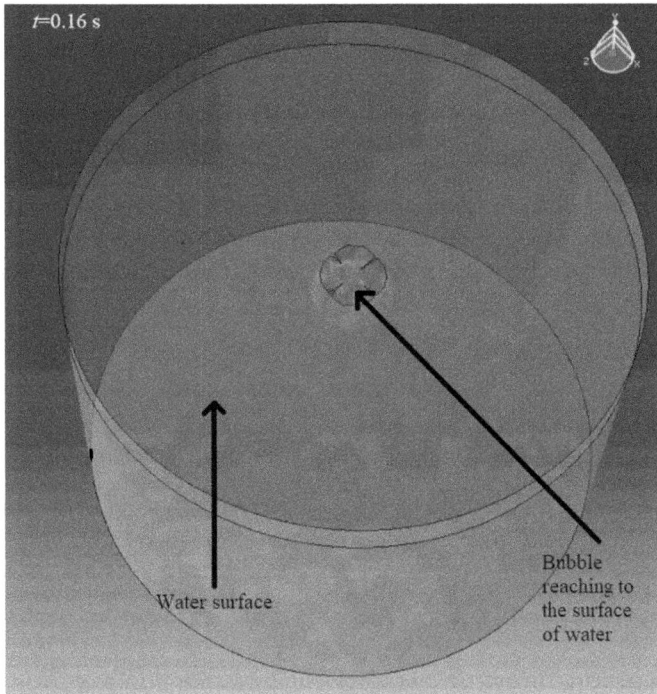

FIGURE 28.8   Bubble from underwater explosion reaching the surface of the water.

Further, the approximate maximum bubble radius was calculated as 2.24 m using Eq. (28.5). The depth at which the explosion was carried in the present study was 1 m below the surface of the water, so this implies that the first bubble will reach the surface of the water; this can be seen in Figure 28.8. In this figure, we can clearly see the first bubble touching the outer surface of the water in the proposed model. As the bubble pulsation continues until the whole energy inside the bubble is expended or the bubble reaches close to the object or surface, it can be concluded that only one bubble will be obtained from this analysis. Future, using Eq. (28.4), the time duration for the first bubble pulsation was calculated as 0.44 s. This time duration implies that the maximum bubble radius is to be obtained within this period, and as we can see, our obtained bubble has reached the surface within this time period, hence it can be concluded that this numerical simulation follows the empirical formula result.

28.5   SUMMARY AND CONCLUSION

In this numerical study, a TNT charge is modelled inside the Eulerian domain for an underwater explosion inside the tank using the CEL formulation available in ABAQUS/Explicit. The following conclusions are made from the investigation:

(i) Only one bubble is observed in the present analysis. Further, the bubble in its first cycle reaches the surface of the water tank and hence no further bubble is observed from the present analysis.

(ii) Obtaining a proper spherical bubble shape requires refined meshing along with a non-reflecting surface. Since when the reflected wave meets the gas bubble, the rarefaction wave is generated, and due to the superposition of shock waves and reflected waves, the pressure pitch of the flow field becomes extremely complicated, and this leads to the distortion of the bubble shape.

(iii) The peak pressure is observed at the top of the water tank above the water surface and minimum pressure is observed at the base of the tank for the parameters considered in the present study.

(iv) No deformation is observed at the surface of the tank wall as the dynamic yield stress of mild steel is much higher than the stress generated by the present underwater explosion. Thus, tank material is under the elastic range only.

(v) Based on this numerical simulation, comparatively higher pressure on the tank wall is observed as compared to the base of the tank. The reason for such behaviour may be attributed to the effect of a reflected wave from the base surface and the hydrostatic pressure of the water.

REFERENCES

1. R.H. Cole, *Underwater explosion*, 1948, Dover, New York, USA.
2. K. Ramajeyathilagam, C.P. Vendhan, Deformation and rupture of thin rectangular plates subjected to underwater shock, *Int. J. Impact Eng.*, 2004, 30(4), 699–719.
3. T.I. Chang, C. Lin, A study of storage tank accidents, *J. Loss Prev. Process Ind.*, 2006, 19(6), 51–59.
4. M.J. Jhung, S.J. Jeong, J.C. Jo, Impact analysis of water storage tank, *Nucl. Eng. Technol.*, 2006, 38(6), 681–688.
5. J.O. Hallquist, LS-DYNA® Keyword User's Manual Volume I. LSTC, 2009, Version, 971.
6. T. Stolarski, Y. Nakasone, S. Yoshimoto, *Engineering analysis with ANSYS software*, 2018, Butterworth-Heinemann, .U.K.
7. Schiffer, V.L. Tagarielli, The one-dimensional response of a water-filled double hull to underwater blast: Experiments and simulations, *Int. J. Impact Eng.*, 2013, 63(8), 177–187.
8. V. Mittal, T. Chakraborty, V. Matsagar, Dynamic analysis of liquid storage tank under blast using Coupled Euler-Lagrange formulation, *Thin-Walled Struct.*, 2014, 84(14), 91–111.
9. C. Hsu, C. Liang, A. Nguyen, T. Teng, A numerical study on underwater explosion bubble pulsation and the collapse process, *Ocean Eng.*, 2014, 81(14), 29–38.
10. Abaqus/Explicit User's Manual, Dassault Systèmes Simulia Corporation, Providence, Rhode Island, USA, 2011, Version 6.11.
11. Y. Wang, M. Xiong, Analysis of axially restrained water tank under blast loading, *Int. J. Impact Eng.*, 2015, 86(15), 167–178.
12. F. Razic, B. Miralem, Underwater explosion effects of 60mm H.E. mortar bomb on a cylindrical concrete structure-PIT, *Def. Technol.*, 2018, 15(19), 65–71.
13. C. Yin, Z. Jin, Y. Chen, H. Hua, Effects of sacrificial coatings on stiffened double cylindrical shells subjected to underwater blasts, *Int. J. Impact Eng.*, 2020, 136(10), 103–412.
14. G.R. Johnson, W.H. Cook, A constitutive model and data for metals subjected to large strains, high strain rates and high temperatures, Proceedings of the 7th International Symposium on Ballistics, The Hague, The Netherlands, April 1983.
15. N.H. Khani, S.S. Emamzadeh, The effects of underwater explosion on the fixed base offshore platform at 10 metres distance, *Specialty J. Eng. Appl. Sci.*, 2017, 3(2), 11–18.
16. M.A. Iqbal, K. Senthil, P. Bhargava, N.K. Gupta, The characterization and ballistic evaluation of mild steel, *Int. J. Impact Eng.*, 2015, 78(15), 98–113.
17. W.T. Lui, F.R. Ming, A.M. Zhang, X.H. Miao, Y.L. Liu, Continuous simulation of the whole process of underwater explosion based on Eulerian finite element approach, *Appl. Ocean Sci.*, 2018, 80, 125–135.

29 Performance of RC Plates Subjected to Explosive Loading

Vimal Kumar

29.1 INTRODUCTION

In the recent past, structures are seen to have faced an increased threat of internal and external blast loading during their service life [1]. Consequently, this area brought a wide concern among structural engineers on the strength and durability aspects of structures exposed to explosive loadings. To address the performance of plates against blast loading, some studies were carried out which include the performance of plain concrete, reinforced concrete [2–4], retrofitted RC members [1, 5], FRP-concrete composite [2], steel-concrete composite [6], ultra-high-performance concrete, [5] etc. Namely, two quantities were said to have a major influence on the behaviour of plates under explosion, which are the mass of the explosive and the standoff distance [7]. However, their effects are not completely known in closed-field explosions wherein the target is just supported at two opposite edges [8]. Such boundary conditions can fulfil the requirements of temporary military protection in adverse and forward military areas. This study presents a systematic experimental and numerical investigation of reinforced concrete plates exposed to close-in blasts. The span and thickness of the square plates were considered 1.0 m and 100 mm, respectively. Three different amounts of explosives were exploded from three detonation distances to study the damage resistance of the plates. The experimental results were reproduced through three-dimensional numerical simulations performed in ABAQUS software.

29.2 EXPERIMENTAL OVERVIEW

29.2.1 MATERIAL AND TEST SPECIMENS

The square-shaped plates were cast to explore their performance under explosive loading. The size of the plates was kept at 1.0 m, keeping the thickness of the plates uniformly 100 mm. The OPC (43-grade cement), tap water, natural river sand and crushed 4.75–10 mm basalt aggregate were used for the preparation of concrete mix following the procedure for weight-batching defined in IS 10262:2009, IS 383:2009 and IS 456:2000. The quantity of water was restricted to 40% of the cement mass. The average 28-day strength of the concrete cubes obtained under a compression testing machine was 46.80 N/mm². All the plates were reinforced using 5 mm radius steel bars which were placed at a spacing of 0.1 m centre to centre; see Figure 29.1(a). The plates thus cast were cured in water for 28 days and thereafter transported for testing under explosion.

29.2.2 EXPERIMENTAL APPROACH

A total of nine RC plates were tested under blast loading through the detonation of gelatine explosive. Three different masses of gelatine were detonated from three different distances. Thus, the resultant scaled distance was obtained for three quantities of gelatine, i.e. their equivalent TNT explosives, and three distances which varied from 0 to 0.527 m/kg$^{1/3}$. The gelatine was put at the centre of the plate at a suitable detonation distance varying from in-contact explosion to 0.5 m. The

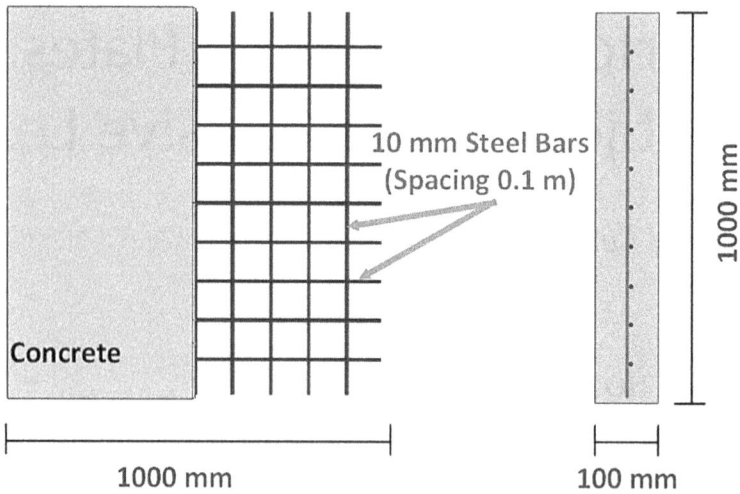

(a) Reinforcement in Plate

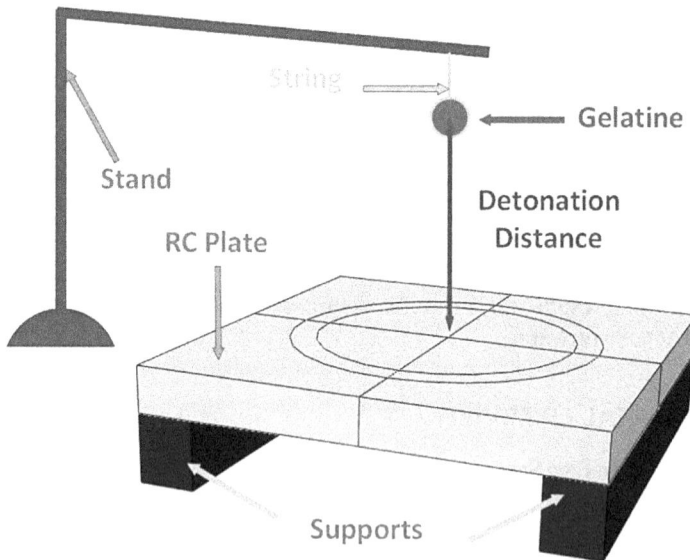

(b) Experimental Arrangement (not to scale)

FIGURE 29.1   RC plate showing (a) arrangement of bars and (b) experimental arrangement. Source: www .sciencedirect.com/science/article/abs/pii/S0141029619317432.

plate was just resting on the support at its two opposite edges to replicate the typical support condition in the field area as shown in Figure 29.1(b). The important findings from some of the tests are discussed in subsequent sections of this chapter.

29.3   FINITE ELEMENT MODELLING

The numerical simulation was carried out in ABAQUS software to replicate the experimental findings. The quarter model (0.5 m × 0.5 m × 0.1 m) of the RC plate was considered for the numerical study, taking into account the advantages of the symmetric loading and support conditions. Both the concrete and the steel were modelled as deformable materials such that the reinforcing bars were embedded inside the concrete [9]. The contact between the reinforcement and concrete was assumed a perfect bond without any slippage at the interface. The mesh size for these embedded steel bars/elements was uniformly considered 1 mm. The concrete was discretized into solid elements of varying sizes from 1.5 to 5 mm keeping finer mesh at the centre and coarser mesh at the edges of the plate. Eight nodded brick elements were considered at the centre and edges of the concrete. The mesh gradually varied from centre to edge. While two nodded link elements were considered for bar elements. A friction-based kinematic contact technique was used between the plate and support, considering the frictional coefficient as 0.3. The supports at their bottom were fixed against all six degrees of freedom. The finite element (FE) model contained approximately 1.685 million elements in the quarter model. In some other models, a three-millimetre-thick layer of CFRP [0/90/0] composite was also used on the surface of the concrete. The CFRP was uniformly divided into mesh sizes of 1–2.5 mm. Therefore, the FE model with the FRP laminate had approximately 0.121 million additional solid brick elements. The detail of the meshing used in the FE model is presented in Figure 29.2.

29.4 CONSTITUTIVE MODELLING

The blast was simulated by employing conventional weapon techniques in ABAQUS. The concrete, reinforcing steel and CFRP were modelled employing Holmquist-Johnson-Cook (HJC), metal

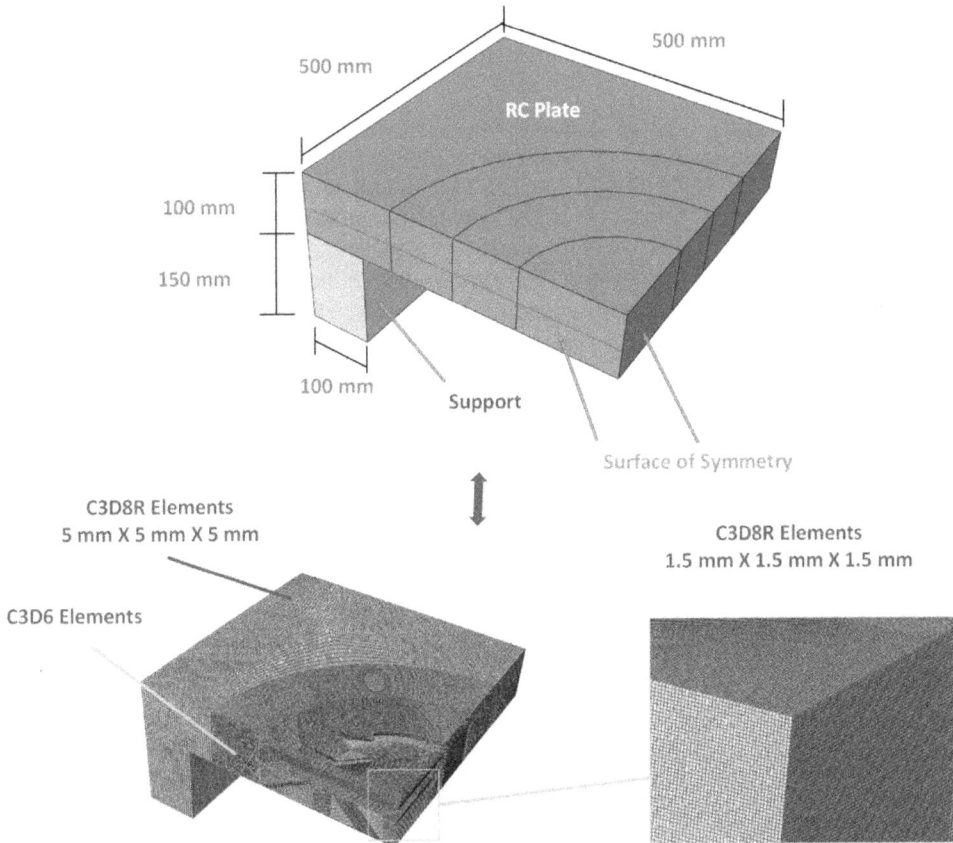

FIGURE 29.2 Detail of meshing in quarter FE model. Source: www.sciencedirect.com/science/article/abs/pii/S0141029619317432.

plasticity and Hashin models, respectively. The properties of these materials have been taken from available studies [10–12] which are mentioned in Tables 29.1 and 29.2.

29.5   RESULTS AND DISCUSSION

29.5.1   Damage Resistance

The damage that occurred in the typical reinforced concrete plates has been presented in Figures 29.3 and 29.4, respectively, for 1.37 and 3.25 kg gelatine. The corresponding scaled distance of the explosion was 0.5273 and 0.3968 m/kg$^{1/3}$, respectively, at a fixed detonation distance of 0.5 m. The characteristics of the blast pressure are significantly governed by the detonation distance such that it acts uniformly for larger values of detonation distances and non-uniformly for smaller detonation distances. The latter case usually deals with localized damage in concrete structures. For the higher value of scaled distance, i.e. 0.5273 m/kg$^{1/3}$, the plate only witnessed flexural deformation and cracking along with hairline cracks. As such, the plate did not witness spalling or crater and therefore can be rendered safe against the blast loading. The FE simulations also witnessed minor damage on the surfaces of the plate. The predicted damage observed was relatively higher compared to the actual damage due to the mesh size and limitation of the FE model because the damage was of the order of mesh size. On the other hand, the plate witnessed a splitting crack at the centre and increased cracking with a reduction in the scaled distance to 0.3968 m/kg$^{1/3}$. The reduced scale distance increased the intensity of the blast load on the surface of the plate. The plate witnessed one-way splitting and bending parallel to support rendering it unsafe against blast loading. Although, no scabbing was noticed on the plates. The FE simulations over-predicted the damage on the plates. However, CFRP or steel lamination can be suggested to improve the performance of the plate.

29.5.2   Blast Pressure

The blast pressure history for the reinforced concrete plates has been presented in Figure 29.5(a) and (b), respectively, for the scaled distances 0.3968 and 0.5273 m/kg$^{1/3}$. The pressure was predicted at the centre of these plates. Irrespective of the amount of explosives, the plates witnessed a sharp peak just after the blast due to the interaction of the shock wave with the plate. The pressure then reduced suddenly in the next fraction of a millisecond for both plates. Thus, the blast pressure response is more or less seen to have a triangular shape acting over a fraction of a millisecond. The magnitude

TABLE 29.1

The H-J-C Constitutive Parameters Used in the Present Study

HJC parameter	Value	HJC parameter	Value
ρ (kg/m^3)	2430	$D1$	0.04
A	0.79	$D2$	1
B	1.6	P_{crush} (GPa)	0.01567
N	0.61	P_{lock} (GPa)	0.8
C	0.007	μ_{crush}	0.001
S_{MAX}	7.0	G (MPa)	14,648
K_3 (GPa)	208	μ_{lock}	0.1
f_c' (MPa)	47	K_1 (GPa)	85
T (MPa)	3.80	K_2 (GPa)	−171

Sources: [10, 11].

TABLE 29.2
CFRP Material Parameters Used in the Present Study

Material parameter	Value
E_{xx} (GPa)	164
E_{yy} (GPa)	12
E_{zz} (GPa)	12
G_{xy} (GPa)	4.5
G_{xz} (GPa)	4.5
G_{yz} (GPa)	2.5
ν_{xy}	0.32
ν_{yz}	0.45
ν_{xz}	0.32
X_{xt} (MPa)	2,724
X_{xc} (MPa)	111
X_{yt} (MPa)	50
X_{yc} (MPa)	1,690
X_{zt} (MPa)	290
X_{zc} (MPa)	290
S_{xy} (MPa)	120
S_{xz} (MPa)	137
S_{yz} (MPa)	90
ρ (kg/m^3)	1,800

Source: [12].

of blast pressure was noticed as 43.7 N/mm^2 at the centre of the plate for a scaled distance of 0.3968 m/kg$^{1/3}$ which significantly reduced to about 38% (i.e. 16.6 N/mm^2) with an increase in the scaled distance to 0.5273 m/kg$^{1/3}$.

29.5.3 INFLUENCE OF CFRP ON THE PERFORMANCE OF THE PLATES

The effects of a CFRP layer were studied on the damage resistance of the plates for a scaled distance of 0.3968 m/kg$^{1/3}$. The protection layer was noticed to have significantly reduced the damage to the concrete. The mass of the concrete was measured at 243.1 and 119.3 kg, respectively, before and after the blast for the full plate without CFRP lamination as shown in Figure 29.6. The mass of the full plate is considered four times the mass of the quarter plate. For the plate with CFRP lamination at the back surface, the mass of the concrete plate was measured at 169.9 kg. On the other hand, the mass of the plate measured 187.19 kg when the front surface of the plate was laminated. The higher mass of the plate with CFRP was primarily a cause of reduced damage in the concrete as a result of the provided protective CFRP layer. This layer also showed some improvement in the failure mode by reducing the splitting damage in the concrete up to some extent. The CFRP at the back surface

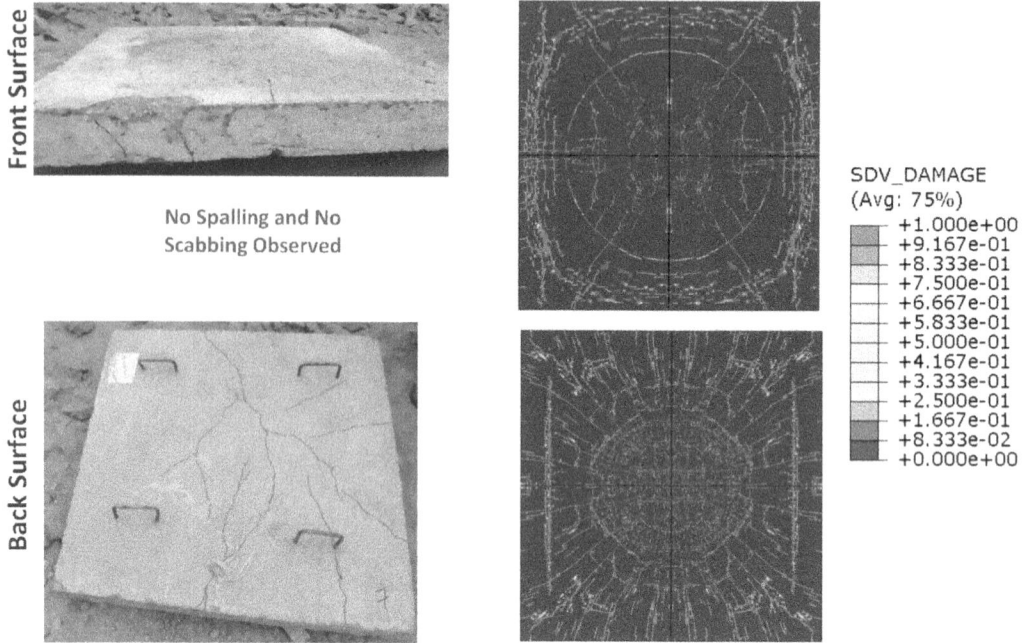

FIGURE 29.3   Damage in reinforced concrete plate against blast loading for scaled distance 0.5273 m/kg$^{1/3}$. Source: www.sciencedirect.com/science/article/abs/pii/S0141029619317432.

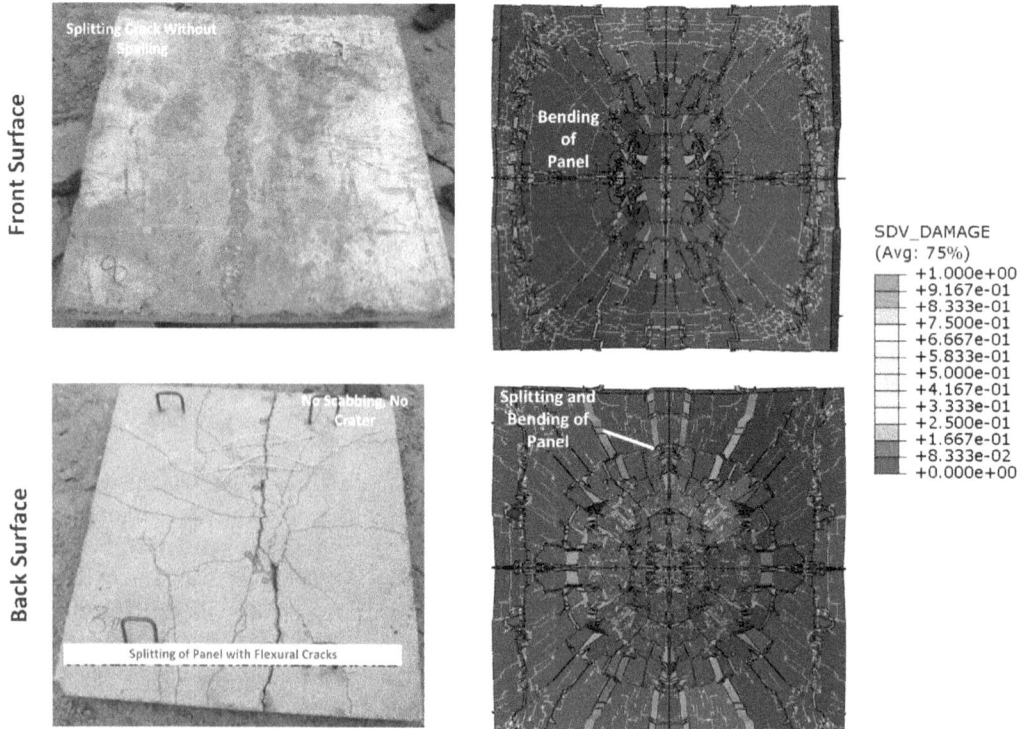

FIGURE 29.4   Damage in reinforced concrete plate against blast loading for scaled distance 0.3968 m/kg$^{1/3}$. Source: www.sciencedirect.com/science/article/abs/pii/S0141029619317432.

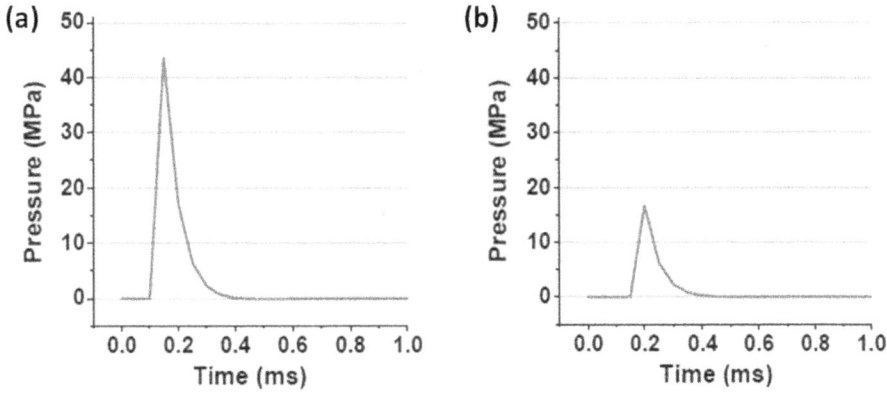

FIGURE 29.5   Blast pressure on plates against scaled distances (a) 0.3968 and (b) 0.5273 m/kg$^{1/3}$. Source: www.sciencedirect.com/science/article/abs/pii/S0141029619317432.

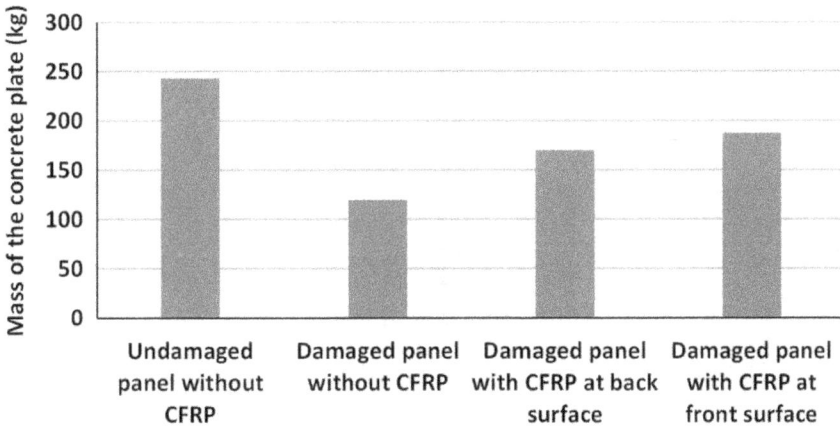

FIGURE 29.6   Level of damage in the reinforced concrete plate for scaled distance 0.3968 m/kg$^{1/3}$. Source: www.sciencedirect.com/science/article/abs/pii/S0141029619317432.

was seen to be more effective against splitting deformation of the plate. The predicted damage in the plates with and without CFRP has been presented in Figure 29.7.

29.6   CONCLUSION

The performance of square reinforced concrete plate (1 m × 1 m) was investigated under blast loading through the experimental and numerical approach. The behaviour of a square reinforced concrete plate was obtained for scaled distances 0.3968 and 0.5273 m/kg$^{1/3}$ and discussed. The damage in the plate was noticed to reduce with an increase in the scaled distance, i.e. with a reduction in the mass of the explosive or an increase in the detonation distance. The plate witnessed splitting damage for smaller scaled distances while it witnessed flexural deformation/crack for higher scaled distances. The simulations reasonably reproduced damage in the plates; however, their accuracy is governed by the mesh size. The blast pressure increased with a decrease in the standoff distance. The blast pressure was noticed to have an approximately triangular pulse. The CFRP laminate was noticed to have reduced damage in the concrete hence improving their performance. The CFRP at the back surface was seen to be more effective against splitting deformation of the plate.

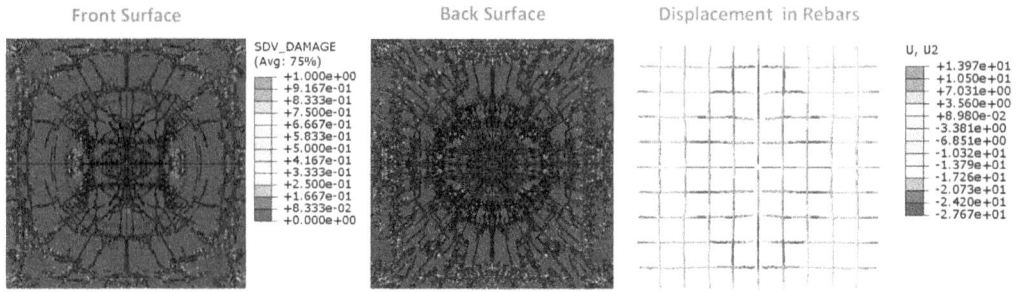

(a) Damage in panel without CFRP

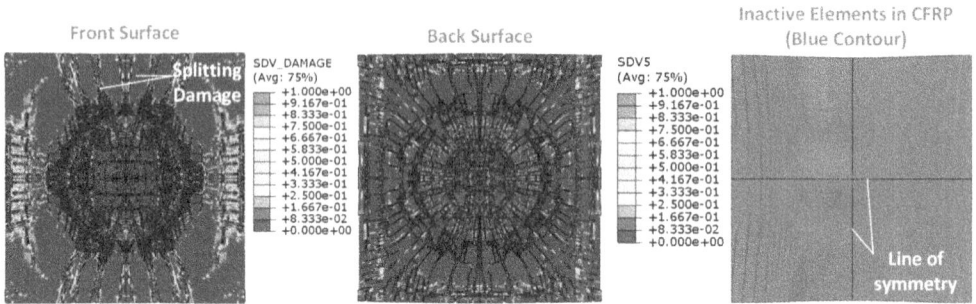

(a) Damage in panel with CFRP at top surface

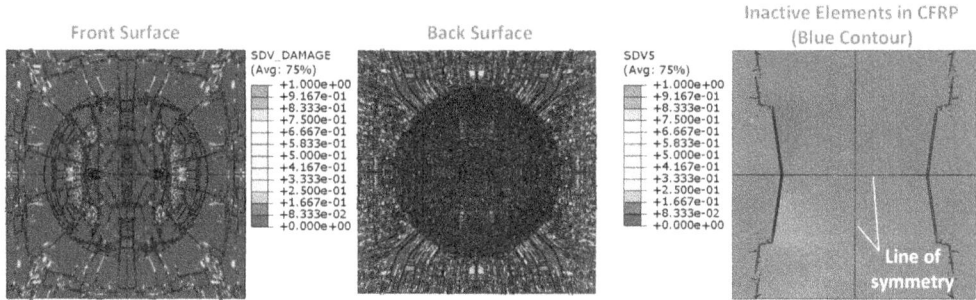

(c) Damage in panel with CFRP at bottom surface

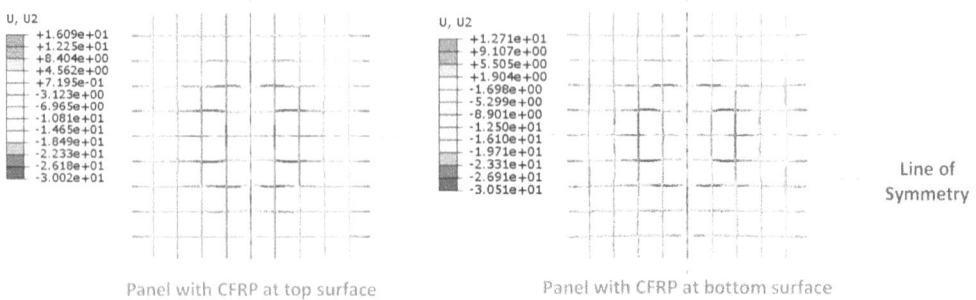

(d) Displacement contour in reinforcement

Time 0.004

FIGURE 29.7 Predicted damage in the plates with and without CFRP lamination (at t = 0.004 s).

ACKNOWLEDGEMENTS

The author is highly grateful to Dr Abhishek Rajput, IIT Indore, and Dr Venkatesan, CSIR-SERC Chennai, for helping to carry out simulations. The figures 29.1 to 29.6 are presented in original/modified form with the permission of Elsevier.

REFERENCES

1. H.M. Elsanadedy, T.H. Almusallam, H. Abbas, Y.A. Al-Salloum, S.H. Alsayed, Effect of blast loading on CFRP-retrofitted RC columns – A numerical study, *Lat. Am. J. Solids Struct.*, 8 (2011) 55–81.
2. C.P. Pantelides, T.T. Garfield, W.D. Richins, T.K. Larson, J.E. Blakeley, Reinforced concrete and fiber reinforced concrete panels subjected to blast detonations and post-blast static tests, *Eng. Struct.*, 76 (2014) 24–33.
3. G. Thiagarajan, A.V. Kadambi, S. Robert, C.F. Johnson, Experimental and finite element analysis of doubly reinforced concrete slabs subjected to blast loads, *Int. J. Impact Eng.*, 75 (2015) 162–173.
4. K.V. Kartik, Evaluation of reinforced concrete plates under blast loading, Thesis, Indian Inst. Technol. Roorkee, Roorkee India, 2016.
5. C. Wu, D.J. Oehlers, M. Rebentrost, J. Leach, A.S. Whittaker, Blast testing of ultra-high performance fibre and FRP-retrofitted concrete slabs, *Eng. Struct.*, 31 (2009) 2060–2069.
6. J.C. Bruhl, A.H. Varma, Summary of blast tests on steel-plate reinforced concrete walls, Struct. Congr. 2015, no. 1995, 151–159, 2015.
7. M.A. Basset, M. Abdelwahab, K.M. Abdelgawad, M.N. Fayed, Evaluating and improving the blast resistance capacity of the RC fences, *J. Mech. Civ. Eng.*, 17 (2020) 26–47.
8. V. Kumar, K.V. Kartik, M.A. Iqbal, Experimental and numerical investigation of reinforced concrete slabs under blast loading, *Eng. Struct.*, 206 (2020) 110125.
9. Abaqus 6.14 documentation.
10. T.J. Holmquist, G.R. Johnson, W.H. Cook, A computational consititutive model for concrete subjected to large strains, high strain rates, and high pressures, Fourteenth Intenational Symposium on Ballistics, Canada (1993) 1–10.
11. V. Kumar, M.A. Iqbal, A.K. Mittal, Study of induced prestress on deformation and energy absorption characteristics of concrete slabs under drop impact loading, *Constr. Build. Mater.*, 188 (2018) 656–675.
12. Y.B.S. Sastri, P.R. Budarapu, Y. Krishna, S. Devaraj, Studies on ballistic impact of the composite panels, *Theor. Appl. Fract. Mech.*, 72 (2014) 2–12.

30 Blast Response Analysis of Composite Column

A. S. Bhonge and M. D. Goel

30.1 INTRODUCTION

An increase in terrorist activities in the past decades has led to a copious number of causalities and significant economic loss. Generally, these types of attacks are carried out as explosive attacks targeting structures of high significance and crowded facilities. A study by the Global Terrorism Index (GTI) acknowledged the effect of such activities and indicated their impact on the economic loss of the country. Apparently, from such activities, civilians and structures are the most affected, and the consequences could be as bad as a fatal loss [1]. In such situations, structures remain the most vulnerable object along with the people residing in them. So, preventive measures need to be adopted to avoid these kinds of threats by mitigating the impact of blast load on structures through appropriate research in designing structures.

With the increase in the threat to structures, research has been made to improve the resistance of structures against the blast load. The usage of advanced material with higher strengths and cautious design of all structural elements and critical sections is suggested in the planning stage itself for minimizing the effect of blast load. Strategies to mitigate the effect of blast load using protective structures such as sacrificial blast walls and increased stand-off distance are necessary precautionary measures to minimize the effect of blast load on structures [2]. Rather than studying the complete structure, the study of the behaviour of individual elements of structures is also equally important. While many structural components are studied such as walls, slabs, etc., columns are termed as a key element by many researchers [3, 4]. This is because columns play a very important role in transferring and bearing the load of entire structures, and the failure of this single element may lead to the progressive collapse of the entire structure. Many studies suggest the use of composite material as key to improving the response of columns under the effect of blast load. Figure 30.1 shows a composite column subjected to blast load. Wu et al. recommended the usage of a composite column designed with a centrally embedded structural steel I-section within the reinforced concrete section of a column to improve the blast resistance using numerical investigation [5].

Lately, there has been a surge in the usage of concrete-filled steel tube (CFST) columns in the construction of high-rise buildings and bridge piers. Some research in recent times has been carried out to check the feasibility of CFST columns for resisting blast loads. Fujikura et al. studied a CFST column of a bridge pier and found that the column behaves in a ductile manner when subjected to blast load. This happened due to the introduction of a steel tube which provides additional ductility to the column, additionally minimizing the spalling of concrete inside the steel tube [6]. Remennikov et al. studied a steel tubular column section with and without concrete infill for near-field explosion using finite element (FE) analysis which found that the concrete plays an important role in minimizing the plastic deformation of steel tubes such as local buckling by providing resistance to the column [7].

Budzaik and Garbowski analyzed a CFST column against surface blast load and specified the least intensity of blast load required for the failure of the entire composite column. Various strengthening solutions could be adopted to reduce the effect of blasts on composite columns [8]. Although,

DOI: 10.1201/9781003352358-30

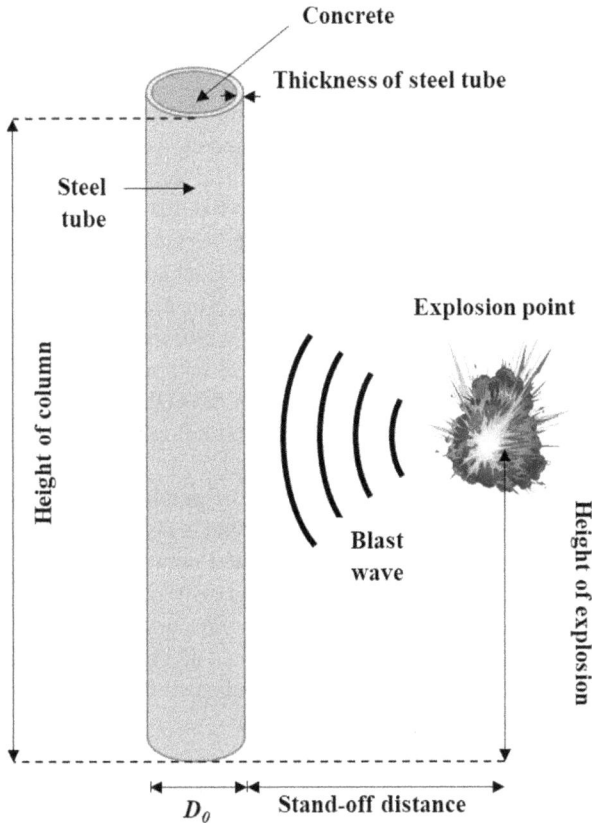

FIGURE 30.1 Representation of composite column subjected to blast load.

during the design of any structure or any of its elements, the intensity of attack cannot be practically found, therefore some hypothetical cases are considered for analysis and their effect produced. Thus, the present study emphasizes the behaviour of composite columns for various intensities of blast load in terms of scaled distances. Many researchers have ignored the use of steel reinforcement in CFST columns due to complex construction practices.

Therefore, in the present study, the response of the column is investigated based on the distinguished parameters such as the height of the explosion, design parameters such as the thickness of the outer steel tube, the grade of concrete and the steel reinforcement ratio in the column forming, a comprehensive analysis of composite columns subjected to blast load. A study of all parameters is carried out considering three distinct blast load cases as a single blast load case cannot express the effects and composite column and it behaves differently for higher loading conditions.

30.2 FINITE ELEMENT MODEL

A three-dimensional non-linear FE model of a composite column is modelled using ABAQUS/ Explicit software [9]. In the present study, a CFST as a composite column is used for analysis against blast loading. The FE model of the composite column is 6 m in height with a steel tube of diameter 406.4 mm having concrete and steel reinforcement inside it. The model for the concrete is created using a C3D8R, eight-noded brick element considering reduced integration and enhanced hourglass effect. M30 grade of concrete is used in the present numerical simulation. The steel tube is modelled using an S4R, four-noded shell element considering reduced integration and enhanced hourglass to avoid numerical uncertainties developed in the FE model.

The steel tube of grade S235 modelled in the current analysis has a circular cross-section with a constant thickness of 8 mm [8]. The steel reinforcement in the column is provided as 12 numbers of 16 mm diameter longitudinal bars and 8 mm diameter stirrups as transverse reinforcement with a varying spacing of 150 mm at one-third distance from both ends and 300 mm spacing at the remaining one-third distance. The reinforcements are modelled as T3D2, a two-noded truss element available in the ABAQUS/Explicit element library [9]. The steel used for rebars is of grade BS500 [8]. The boundary condition of the composite column is fixed at the base of the column and pinned at the head of the column. Contact between the concrete surface and the inner steel tube surface is a surface-to-surface contact which is defined as a "hard" contact, and for steel reinforcement inside the concrete, an embedded region constraint is adopted. The load applied on this column is a blast load applied through the ConWep programme of ABAQUS/Explicit [9].

Preliminary analysis is performed only on blast load without any axial load on columns. The complete assembly of the composite column modelled in ABAQUS/Explicit is shown in Figure 30.2. Element size for the entire assembly is kept to 10 mm based on mesh convergence. The analysis is carried out for 50 ms after the explosion of the blast load.

However, some of the results shown in the form of the graph are for a lower time scale to show the difference in peak displacement. Later on, an axial load is also applied along with a blast load on the column. The axial load capacity column is calculated numerically through ABAQUS/Explicit and it is applied with a gradual increase in the first phase of 10 ms, and after the application of axial load on the head of the column, blast load is induced for subsequent analysis. In the present analysis, blast load is assigned at the mid-height of the column, i.e., at 3 m from the base of the column, and three distinct blast load cases are considered. These load cases for the explosion are stated in terms of scaled distance as mentioned in Table 30.1.

30.3 VALIDATION OF FINITE ELEMENT MODEL

In the present study, ABAQUS/Explicit is used for validating the FE model, doing so by comparing the results from the current FE model to the results presented by Budzaik and Garbowski [8]. Budzaik and Garbowski analysed a 6 m high steel-concrete composite column having a diameter of 406.4 mm. They performed the analysis for a hemispherical surface blast of four distinct scaled distances, 0.1667, 0.2087, 0.2459 and 0.2894 $m/kg^{1/3}$, and provided strengthening solutions for the same. The simulation was carried out using a C20/25 grade of concrete using the Drucker-Prager

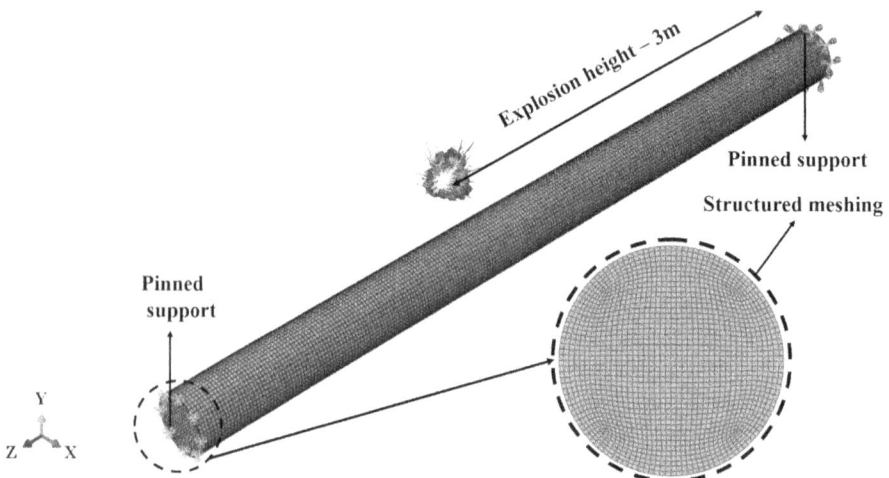

FIGURE 30.2 Comprehensive assembly of composite column modelled in ABAQUS/Explicit.

TABLE 30.1
Details of Blast Load Cases Applied to Composite Column

Scaled distance (m/kg$^{1/3}$)	Stand-off distance (m)	Charge weight (kg)
0.4	1	15.625
0.35	1	23.32
0.3	1	37

material plasticity model focusing on the tensile failure of concrete. The steel used in this column consisted of grade S235 structural steel for the outer tube and grade BS500 for reinforcing steel, and both were modelled using the elastoplastic material model in the present study. For comparison, the kinetic energy and internal energy of the present model are compared to those presented by [8] as shown in Figure 30.3. Figure 30.3 shows that the nature of graphs and their values are similar in both cases. Further, the percentages of error in peak kinetic energy value and final internal energy value are 3.4% and 1.24%, respectively. Since the error in comparison is less than 5%, the present FE model is equivalent to the FE model presented by Budzaik and Garbowski [8]. For further comparison, the vertical displacement in the column measured in the present model is 2.64 mm to that of Budzaik and Garbowski as 2.2 mm [8]. As the difference between displacement in the present simulation and the authors' results is a mere 0.44 mm, this value is within acceptable limits. Since the difference in results in the present FE model and the reference paper are of the same magnitude, the model is said to be validated.

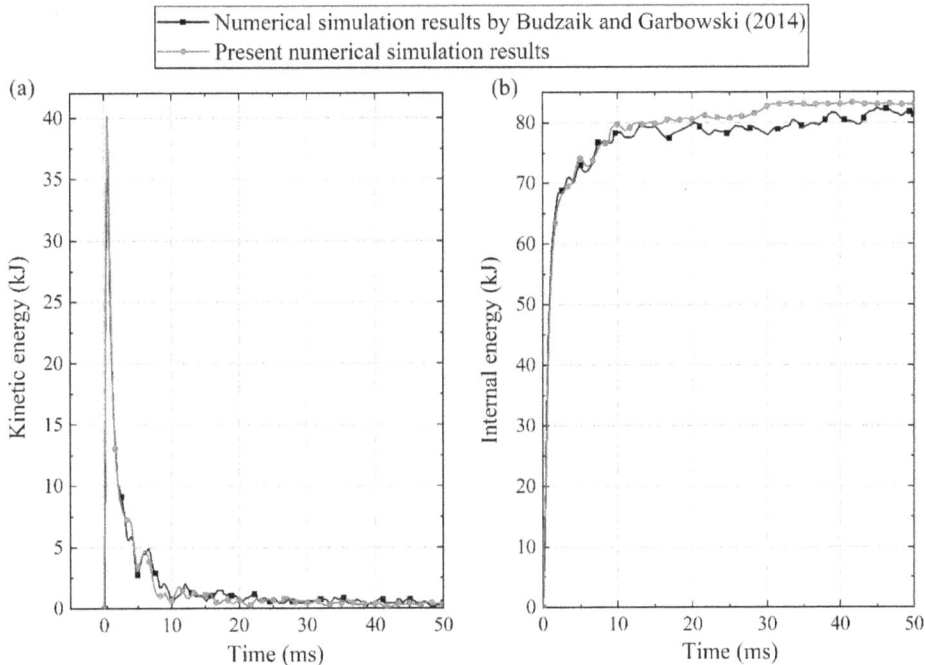

FIGURE 30.3 Comparison of results for validation of finite element model. Source: www.researchgate .net/publication/262972037_Failure_Assessment_of_Steel-Concrete_Composite_Column_Under_Blast _Loading.

30.4   RESULTS AND DISCUSSION

In the present analysis, a FE model of a CFST column is prepared using ABAQUS/Explicit and is investigated for air blast load. The results from the present analysis are mainly oriented towards the transverse displacement of the column and the amount of damage that occurred in the concrete. The parametric study associated with the varying height of explosion, the thickness of steel tube, grade of concrete and reinforcement ratio is performed to understand the effect of varying parameters on the response of composite column.

30.4.1   EFFECT OF VARYING HEIGHT OF EXPLOSION

Three different heights of explosion are adopted in this analysis for investigating the effect of blast load. These heights are at a one-third distance measured from both ends of the column and at the mid-height of the column. The schematic of the same is shown in Figure 30.4 where explosions take place individually on the column. The heights of the explosions are 2 m, 3 m and 4 m respectively.

Figure 30.5 signifies the displacement-time history of the composite column which shows the transverse movement, i.e., lateral movement of the column at different heights of explosion considered in the current analysis. From this result, it is obvious that in all three load cases, the transverse displacement of the column is maximum when the explosion is carried out at the mid-height of the column. The remaining two heights of explosion show a lower transverse displacement as these explosions take place near the supports of the column which provides resistance to the transverse displacement of the column.

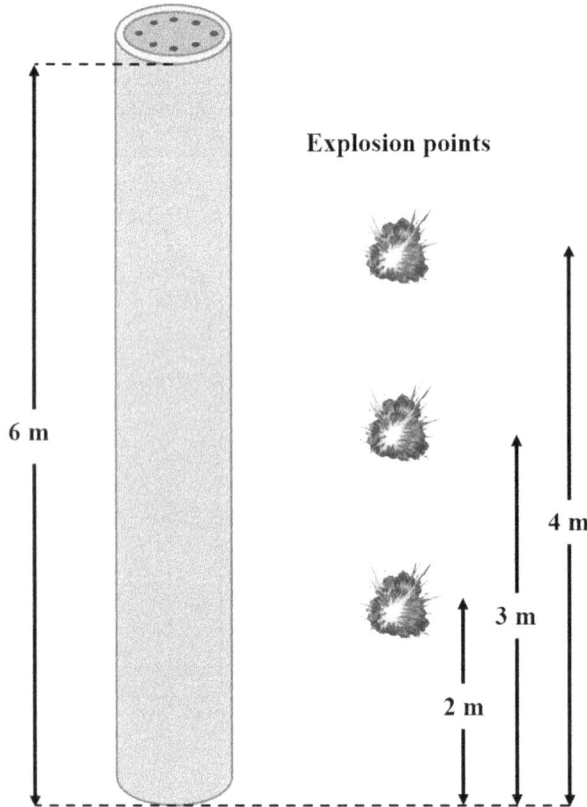

FIGURE 30.4   Schematic of FE model subjected to blast load.

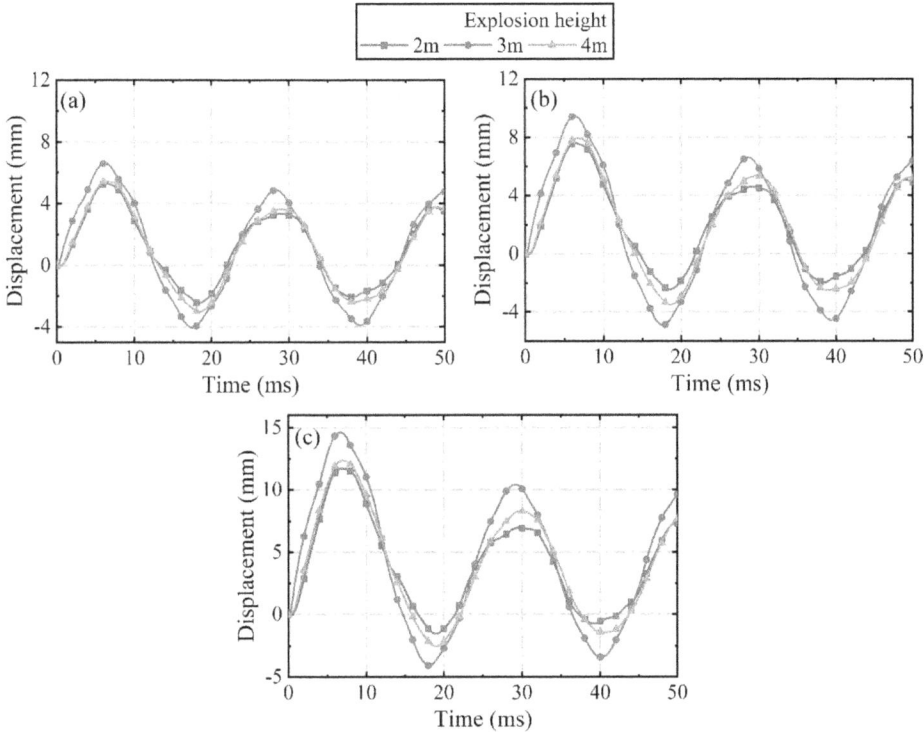

FIGURE 30.5 Comparison of displacement-time history for varying height of explosion for scaled distances of (a) 0.4 m/kg$^{1/3}$, (b) 0.35 m/kg$^{1/3}$ and (c) 0.3 m/kg$^{1/3}$.

30.4.2 Effect of Varying Thickness

For the present analysis, the variation in the thickness of the steel tube is based on the actual thicknesses available practically, and the Indian standards, IS 3601:2006 and IS 1161:1998, are used for geometric and material properties for different types of steel tubes available for engineering uses [10, 11]. Conferring to IS 3601:2006, four distinct thicknesses of the 406.4 mm diameter steel tube are considered, and these are 6.3 mm, 8.8 mm, 10 mm and 12.5 mm, respectively. The FE model of the composite column is modified according to the different thicknesses of the outer steel tube in such a way that the outer diameter of the column remains constant for all the thicknesses.

The result of this analysis in terms of displacement-time history is shown in Figure 30.6 for three scaled distances of 0.4 m/kg$^{1/3}$, 0.35 m/kg$^{1/3}$ and 0.3 m/kg$^{1/3}$, respectively. Based on the results from this analysis, it is seen that FE models with the smallest steel tube thickness, i.e., 6.3 mm, result in higher lateral deflection. From the results, a pattern can be constructed that the peak displacement at the mid-height of the column decreases with the increase in the thickness of the steel tube of the composite column.

A reason for this could be that the steel tube with more thickness can absorb a higher amount of energy, leading to reduced displacement. Technically, the area of concrete is reduced and the area of steel is increased whenever there is an increase in the thickness of the steel tube, and the strength of concrete compared to the steel tube is much lower, which further results in lower displacement. From the present study, it is evident that the peak displacement decreases with the increase in steel tube thickness and a further increase in the steel tube thickness will lead to comparatively lower displacement. More damage could have happened to the column while modelling using concrete alone. This can be attributed to the fact that concrete spalling plays a major role in damaging the column, whereas, while using the steel tube, damage to the column got significantly reduced due to

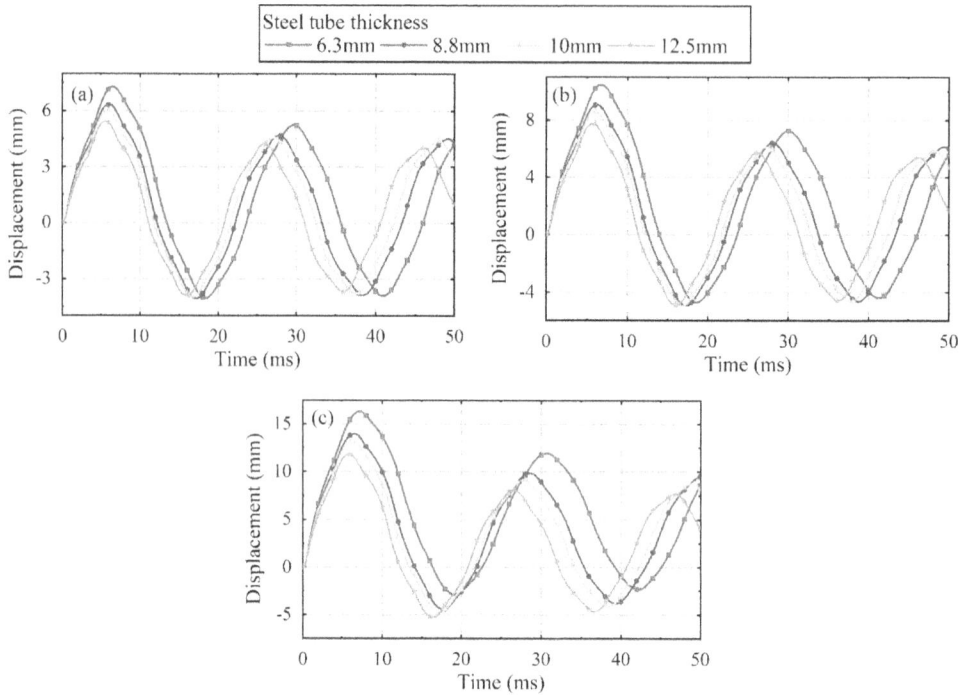

FIGURE 30.6 Comparison of displacement-time history for varying thickness of steel tube for scaled distances of (a) 0.4 m/kg$^{1/3}$, (b) 0.35 m/kg$^{1/3}$ and (c) 0.3 m/kg$^{1/3}$.

the restriction on spalling of the concrete because of the presence of the steel tube. Thus, the advantage of using the steel tube is that it acts as a protective covering to the concrete core, resulting in the steel tube absorbing more energy and saving the concrete from cracking and spalling, which would have been caused when directly exposed to blast load. Although there exists a relation that the peak deflection of the column reduces with increasing steel tube thickness, the amount of reduction in peak displacement degrades.

Graphically it is seen from Figure 30.6 that there exists a significant decrease in peak displacement when the thickness is increased from 6.3 mm to 8.8 mm, but when the thickness is increased from 10 mm to 12.5 mm, the difference in magnitude of peak displacement of the column is lower than previously, even though the difference in the magnitude of increase in the thickness of the steel tube is constant at 2.5 mm.

This behaviour is observed in all three scaled distances and a major reduction in peak deflection is observed for the scaled distance of 0.3 m/kg$^{1/3}$ where the peak deflection of the column is reduced by a magnitude of 2.36 mm when the thickness of the steel tube is increased from 6.3 mm to 8.8 mm, and a reduction of 1.38 mm in peak displacement is observed when the thickness of the steel tube is increased from 10 mm to 12.5 mm. Therefore, from the present parametric study, it is evident that the decrease in the magnitude of peak displacement reduces with the increase of steel tube thickness to a higher value, and a further increase in the steel tube thickness might lead to a lower displacement, but the reduction in the magnitude of deflection will be relatively lower.

Figure 30.7 shows the reduction in damage to concrete while using higher thicknesses of steel tube for a scaled distance of 0.3 m/kg$^{1/3}$. So, from this parametric study, it can be observed that the decrease in peak displacement reduces with the increase in steel tube thickness but up to a certain limiting thickness of the steel tube only. A further increase in the steel tube thickness will certainly lead to a reduced displacement but the reduction in the magnitude of displacement will be relatively lower.

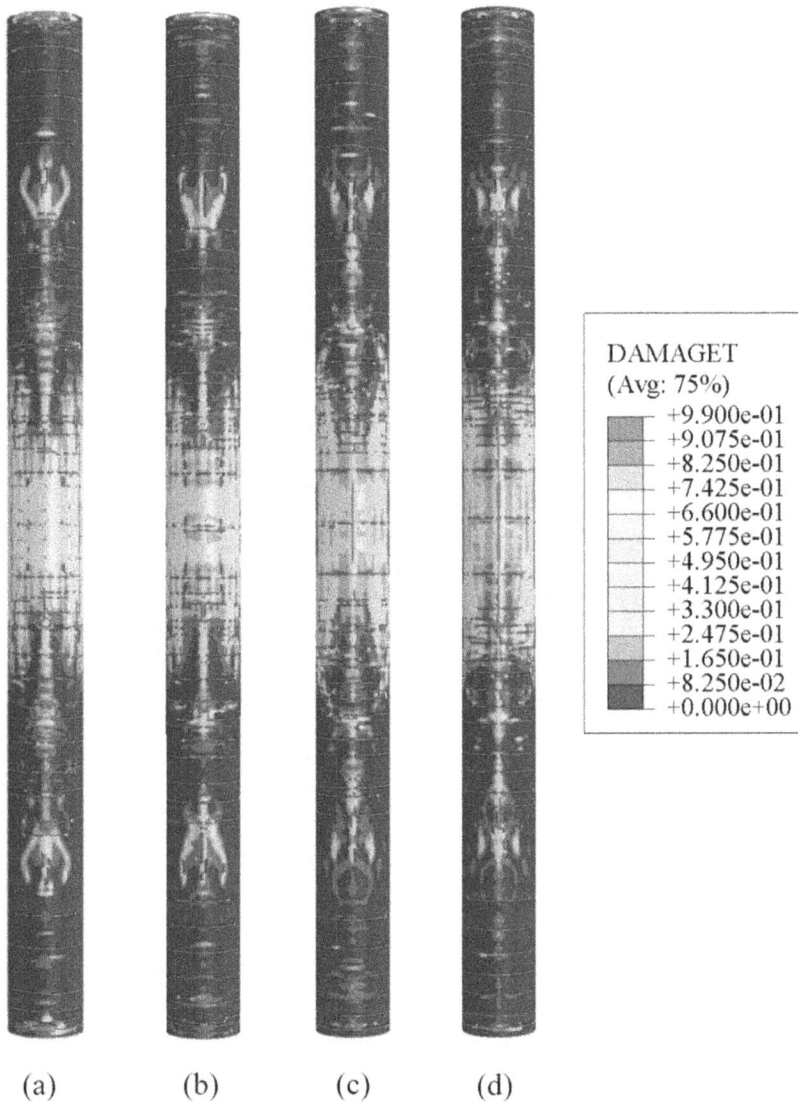

DAMAGET
(Avg: 75%)
+9.900e-01
+9.075e-01
+8.250e-01
+7.425e-01
+6.600e-01
+5.775e-01
+4.950e-01
+4.125e-01
+3.300e-01
+2.475e-01
+1.650e-01
+8.250e-02
+0.000e+00

(a)   (b)   (c)   (d)

FIGURE 30.7 Effect of varying steel tube on concrete for steel tube of thickness (a) 6.3 mm, (b) 8.8 mm, (c) 10 mm and (d) 12.5 mm.

30.4.3 EFFECT OF VARYING GRADES OF CONCRETE

For this parametric analysis, four distinct normal strength concrete (NSC) grades of concrete are used, namely, M20, M30, M40 and M50. The properties input in ABAQUS/Explicit for concrete include compressive and tensile behaviour and their damage parameters, which are according to Hafezolghorani et al. [12]. Figure 30.8 demonstrates the result of the increasing grade of concrete which shows the reduction in damage while using higher grades of concrete of a scaled distance of 0.3 m/kg$^{1/3}$. This particular scaled distance is exhibited as it shows a clear difference as compared to other load conditions. These damage contours shown in Figure 30.8 are at the mid-height of the column shown as a cross-section, and this failure is prolonged up to some distance from the mid-height of the column, which differs according to load cases. Results exhibit that there are a reduced number of tensile cracks in concrete for a higher grade of concrete, and compression damage is also reduced with the provision of a higher grade of concrete, resulting in lower concrete crushing.

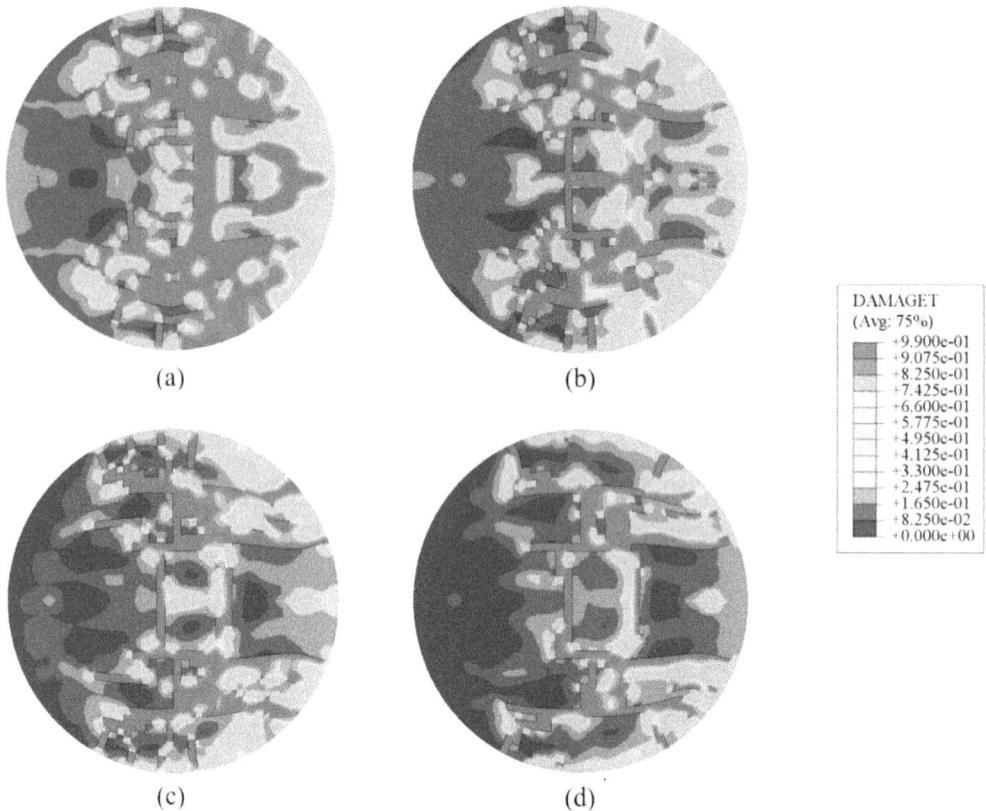

DAMAGET
(Avg: 75%)
+9.900e-01
+9.075e-01
+8.250e-01
+7.425e-01
+6.600e-01
+5.775e-01
+4.950e-01
+4.125e-01
+3.300e-01
+2.475e-01
+1.650e-01
+8.250e-02
+0.000e+00

(a)   (b)

(c)   (d)

FIGURE 30.8 Comparison of extent of damage in concrete in terms of tension damage for M20, M30, M40 and M50 grades of concrete.

Tensile damage, shown in Figure 30.8 as compression damage to the concrete because of the explosion, is not substantial as the steel tube provided a shield against blast load. Similar results are seen in all three scaled distances where the use of a higher grade of concrete results in lower damage to the concrete core inside the composite column. Further, it is to be noted that tensile damage in concrete is the main cause of damage and generally occurs at the face of the column opposite to an explosion as this face of the column experiences tensile stresses due to stretching.

Due to the low tensile strength of concrete, tension governs the failure in the concrete and in the present study, it occurs at the face of the column opposite to the explosion face. This is due to the reason that this side of the column experiences tensile stresses due to stretching, which is caused due to bending of the column and the reflection of a compressive wave into the tensile wave at the opposite face. The result of this analysis is also compared in terms of peak displacement, but there exists no significant decrease in peak displacement of the column with an increase in the grade of concrete. Besides, it is detected that there is a reduction of 1.06 mm in peak deflection when the grade is increased from M20 and M50 for a scaled distance of 0.3 m/kg$^{1/3}$. This reduction is much lower for the other two load cases, from which it can be inferred that using higher grades of concrete results in the reduction of displacement of concrete, but comparatively, reduction in the damage to concrete is much more significant.

30.4.4 Effect of Varying Reinforcement Ratio

The design of a CFST column can be carried out considering steel reinforcements as well as without considering the steel reinforcements as mentioned in Eurocode 4. In the past, many authors excluded the use of steel reinforcement in CFST columns but the present analysis is carried out by considering the reinforcement. This has been taken care of by considering the reinforcement ratio of 6% (in the absence of overlapping bars) and 4% (in the presence of overlapping bars) as suggested by the Indian standard code, IS 456:2000 [13].

In the current parametric analysis, the reinforcement ratio is increased by 1% each up to 6% starting from 2%. The FE model of the composite column is modified according to Table 30.2 which depends on the amount of reinforcement required to satisfy the criteria. The constant transverse reinforcement of a 6 mm diameter bar is provided as it satisfies all the standards after incrementing the longitudinal rebars.

Figure 30.9 exhibits the displacement-time history of all reinforcement ratios for all loading conditions and suggests that there could not be found a significant amount of impact while using a higher percentage of reinforcement. To make a clear picture, the displacement-time history is limited to 8 ms, which is 50 ms for the previous analysis. Results from this analysis show that there is an insignificant improvement due to steel reinforcement in the composite column.

TABLE 30.2
Details of Reinforcement in the Column Based on the Percentage of Reinforcement

Percentage of reinforcement	Area of steel required (mm²)	Reinforcement provided	Area of steel provided (mm²)	Actual percentage of reinforcement
2%	2,394.08	12 nos. – 16 mm	2,412.74	2.01%
3%	3,591.12	12 nos. – 20 mm	3,769.91	3.14%
4%	4,788.16	16 nos. – 20 mm	5,026.55	4.19%
5%	5,985.21	16 nos. – 22 mm	6,082.12	5.08%
6%	7,182.25	16 nos. – 24 mm	7,238.23	6.04%

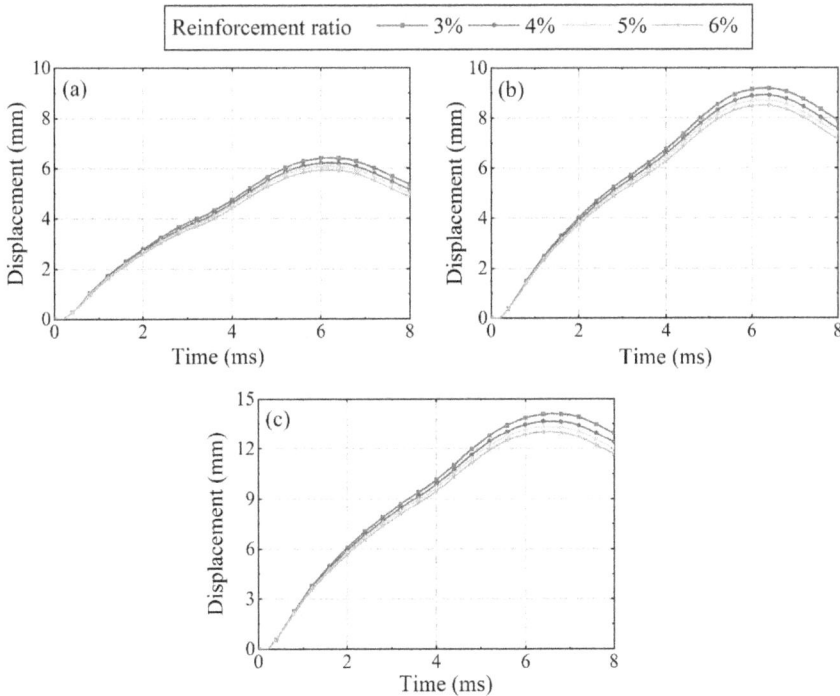

FIGURE 30.9 Comparison of displacement-time history for varying reinforcement ratio for scaled distances of (a) 0.4 m/kg$^{1/3}$, (b) 0.35 m/kg$^{1/3}$ and (c) 0.3 m/kg$^{1/3}$.

However, it does provide additional ductility to the composite column and provides strength to the system to minimize the lateral deflection, but in a very limited amount. For the record, in Figure 30.9(c) the variance in peak displacement between 3% and 2% of the reinforcement ratio is just 0.45 mm, which is not advisable for the use of reinforcement as a blast mitigation technique for the composite columns. This may be attributed to the fact that the primary purpose of steel reinforcement which includes the increase in strength, providing ductility, preventing shear failure, etc. is already satisfied by the introduction of the outer steel tube.

The results from this analysis show that there isn't a significant amount of impact after the increase in reinforcement ratio as shown in Figure 30.10 where the focus on increasing the reinforcement ratio is identical. The supplementary analysis is carried out on the CFST column without any steel reinforcement to check the impact of the reinforcement ratio on the column for a scaled distance of 0.3 $m/kg^{1/3}$, and the results of the same are described in Figure 30.10. From the successive result, it is noted that the initial reinforcement in the column contributes to a reduction in a substantial amount of peak displacement, but an additional increase in reinforcement is not suggested.

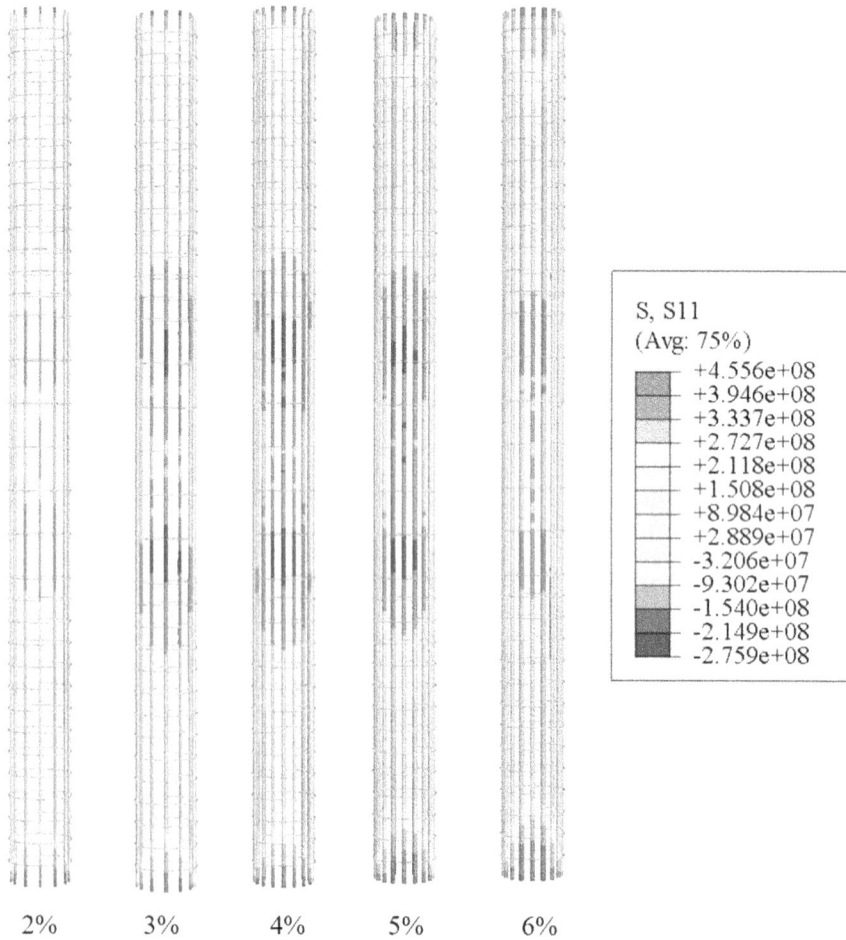

S, S11
(Avg: 75%)
+4.556e+08
+3.946e+08
+3.337e+08
+2.727e+08
+2.118e+08
+1.508e+08
+8.984e+07
+2.889e+07
-3.206e+07
-9.302e+07
-1.540e+08
-2.149e+08
-2.759e+08

| 2% | 3% | 4% | 5% | 6% |

FIGURE 30.10 Stress distribution in reinforcement for varied percentages under blast load of scaled distance 0.3 $m/kg^{1/3}$.

30.5   CONCLUSION

The maximum displacement of the composite column occurs when the explosion is at the mid-height of the column. This height of explosion is stated as critical height, and analysis of further parameters is performed considering mid-height explosion only. The most predominant parameter that helps in mitigating the effect of blast load is the thickness of the outer steel tube. The results show that the high energy absorption characteristic of steel causes a reduction in peak displacement of the column which further reduces the damage to the concrete inside the steel tube. Another effective parameter in the present study is the application of a high grade of concrete. A higher grade of concrete resulted in the reduction of tensile and compressive failure in concrete but was not significant enough in reducing the transverse displacement of the column. Using steel reinforcement in the CFST column is generally ignored by many authors; in the present study, it is found that the increase in reinforcement ratio beyond the design limit does not help in mitigating the effect of blast load.

REFERENCES

1. P.A. Shirbhate, M.D. Goel, A critical review of blast wave parameters and approaches for blast load mitigation, *Arch. Comput. Methods Eng.* 28 (2020) 1713–1730.
2. M.D. Goel, V.A. Matsagar, Blast-resistant design of structures, *Pract. Period. Struct. Des.* 19(2) (2014) 04014007.
3. J. Zhang, S. Jiang, B. Chen, C. Li, H. Qin, Numerical study of damage modes and damage assessment of CFST columns under blast loading, *Shock. Vib.* (2016) 3972791.
4. M.A. Galal, M. Bandyopadhyay, A.K. Banik, Progressive collapse analysis of three-dimensional steel–concrete composite building due to extreme blast, *J. Perform Constr. Facil.* 34(3) (2019) 04020021.
5. K.C. Wu, B. Li, K.C. Tsai, The effects of explosive mass ratio on residual compressive capacity of contact blast damaged composite columns, *J. Constr. Steel Res.* 67 (2011) 602–612.
6. S. Fujikura, M. Bruneau, D. Lopez-Garcia, Experimental investigation of multihazard resistant bridge piers having concrete-filled steel tube under blast loading, *J. Bridg. Eng.* 13(6) (2008) 586–594.
7. A.M. Remennikov, B. Uy, Explosive testing and modelling of square tubular steel columns for near-field detonations, *J. Constr. Steel Res.* 101 (2014) 290–303.
8. M. Budzaik, T. Garbowski, Failure assessment of steel-concrete composite column under blast loading, *Eng. Trans.* 62(1) (2014) 61–84.
9. Simulia D. S. C. ABAQUS 6.14 Analysis user's manual getting started with ABAQUS interactive edition. Dassault Systems, Rhode Island, 2014.
10. BIS (Bureau of Indian Standards). Steel tube for structural purposes, IS: 1161:1998, New Delhi, India, 1998.
11. BIS (Bureau of Indian Standards). Steel tubes for mechanical and general engineering purposes, IS: 3601:2006, New Delhi, India, 2006.
12. M. Hafezolghorani, F. Hejazi, R. Vaghei, M.S.B. Jaafar, K. Karimzade, Simplified damage plasticity model for concrete, *Struct. Eng. Int.* 27(1) (2017) 68–78.
13. BIS (Bureau of Indian Standards). Plain and reinforced concrete – Code of practice, IS: 456:2000, New Delhi, India, 2000.

31 Characterization of Near-Field Blast Response of Aluminum Honeycomb Using Finite Element Simulations

Rohit Sankrityayan, Anoop Chawla,
Sudipto Mukherjee and Devendra K. Dubey

31.1 INTRODUCTION

The basic function of honeycombs and other cellular structures is to absorb energy. Under varying loads, the manner of energy absorption differs. In the past, researchers have studied energy absorption under controlled loading circumstances such as quasi-static and impact scenarios [1, 2]. When a honeycomb structure is subjected to blast loading, the rate of deformation is high since the loading mainly comprises intense shock wave propagation. The deformation behavior of cell wall material under such conditions is difficult to predict accurately using quasi-static deformation mechanics theory. Many researchers have well explored the application of honeycombs for quasi-static to low-impact loads. However, the shock loading response of the honeycombs has not been thoroughly analyzed, particularly for the highly dynamic phenomenon (i.e., near-field blast). In addition, finite element (FE) modeling of honeycomb structures is crucial because of its complex geometrical aspects.

In the wake of this, a finite element tool was adopted to investigate the extreme loading response of honeycomb structures and to carry out a parametric study. In the current work, the near-field blast loading was performed on a honeycomb core sandwiched between two face sheets. A setup as shown in Figure 31.1(a) was used. The force transmitted to the steel rod and energy absorption by core compression were computed by changing honeycomb parameters. Figure 31.1(b) illustrates the deformation of the honeycomb subjected to a near-field blast.

Honeycomb structures are predominately subjected to dynamic conditions. However, more emphasis is given to the performance of honeycombs against quasi-static loading conditions. This could be due to the simplicity involved in the numerical and theoretical analysis under quasi-static conditions. According to many authors [3], the deformation profile of honeycombs is assumed to be independent of the loading rate. However, in the actual scenario, the response of honeycombs such as failure modes, maximum compressive load, maximum core displacement, and specific energy absorption is critically dependent on the loading rate.

Given the application of honeycomb as a potential energy-absorbing material, the current work explores the honeycomb response under extreme loading conditions using hydrocode-based numerical tools. The study focuses on finite-element modeling of the response of honeycomb structures under a shock environment. A precise and efficient numerical modeling approach for such cases has been presented in subsequent sections.

DOI: 10.1201/9781003352358-31

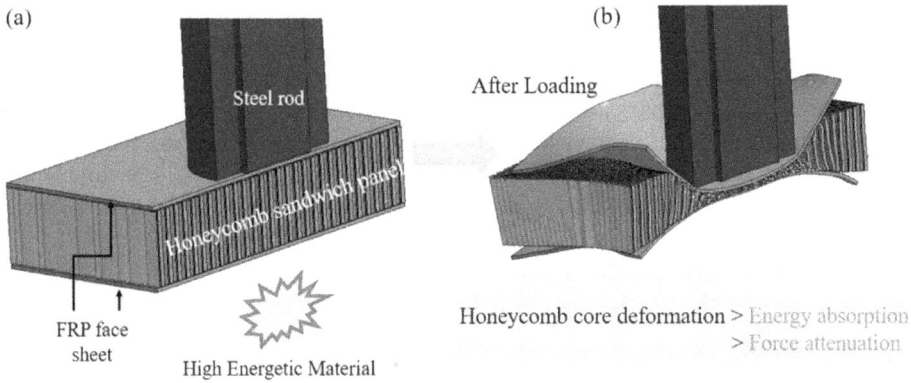

FIGURE 31.1 (a) Honeycomb sandwich under blast setup and (b) honeycomb deformation under blast loading.

31.2 LITERATURE REVIEW

Cellular structures, i.e., metallic honeycomb, have an excellent energy-absorbing capacity and have been found to mitigate or reduce the damaging effect of blast and impact. Honeycomb is the most prevalent cellular material used as energy-absorbing core material, followed by polymeric and paper honeycomb. Sun et al. [4] performed impact experiments on aluminum face sheets and aluminum honeycomb sandwich panels. It was found that the peak forces, energy absorption (EA), and specific energy absorption (SEA) increase when the face sheet thickness increases. Also, the SEA increases with the increased cell size of the honeycomb core. Chen et al. [5] reported an enhancement in the perforation resistance of sandwich panels by reducing honeycomb cell size and increasing face sheet thickness. Optimization of honeycomb design parameters and face sheet are required to reach the desired performance for impact resistance. Nurick et al. [6] and Karagiozova et al. [7] performed blast loading on aluminum honeycomb cores with steel face sheets. They found that the blast resistance of the core was better for uniform loading than for localized loading. It was also found that the load transferred through the panel depends upon the core thickness, load intensity, and flexibility of the sandwich structure. Shiqiang Li et al. [8] studied the effect of geometry on the type of deformation modes and blast response. They found that core arrangement affects the blast resistance and energy absorption capability; a core of a higher density arranged closer to the blast performs better. Theobald et al. [9] performed the blast experiment on sandwich panels of unbound aluminum foam and aluminum core with steel faces as face sheets. They observed that for both the foam and honeycomb core, face sheet thickness had a significant effect on the blast resistance performance. Also, it was found that as a core, metallic honeycomb performs better than metallic foam. Studies on the energy absorption/dissipation rate of honeycomb materials are limited to low-velocity impact and far-field blast loading conditions. There is no work done on characterizing the near-range blast damage of honeycomb materials. In this study, the blast response of aluminum honeycomb for near-field conditions is motivated by this major limitation.

31.3 FINITE ELEMENT SIMULATION STUDY

31.3.1 SIMULATION SETUP

A parametric study on aluminum honeycomb subjected to blast loading was carried out using the ConWep method (based on polynomial-derived empirical equations) [10]. The ConWep approach is computationally efficient and might be used for parametric studies; however, it lacks the actual effects of explosives and air modeling. The simulation setup was built on the near-range blast experimental scenario, which had been developed to evaluate the performance of the anti-mine boot by Karahan et al. [11, 12]. The finite-element arrangement replicated from the experimental setup is depicted in Figure 31.2. Table 31.1 shows the parameters used in the finite element setup.

The setup comprises a cylindrical steel surrogate to replicate the human leg. The sample material was aluminum honeycomb, sandwiched between face sheet material. Dyneema was chosen as a face sheet material because of its high tensile strength-to-weight ratio [13]. The sample material was placed in contact with the steel rod. Lumped mass of 35 kg was kept at the top of the rod to

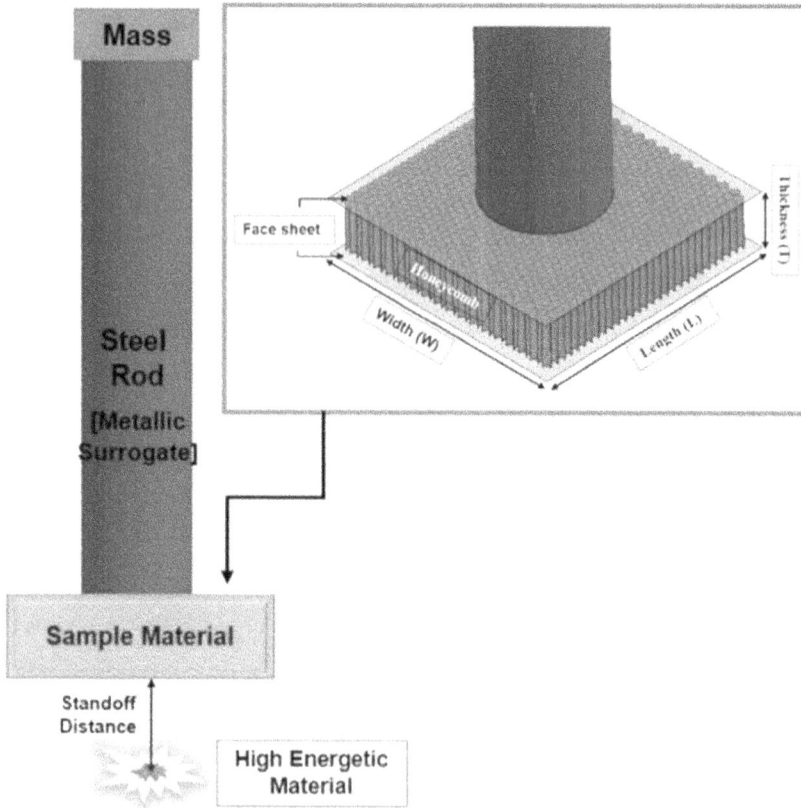

FIGURE 31.2   Parametric simulation setup in LS-Dyna.

TABLE 31.1
Parametric Simulation Parameter Details

FE setup parameters	Values
Steel rod diameter	50 mm
Steel rod length	600 mm
Face sheet material	Dyneema
Face sheet thickness	1.5 mm
Core dimensions	Length (L) = 100 mm, width (W) = 100 mm, and thickness (T) = 20 mm, 30 mm, 40 mm, 50 mm, 60 mm
Mass (kept on the top of the steel rod)	35 kg
Standoff distance	40 mm
Mass of the charge	10 g, 20 g, 30 g, 40 g, 50 g

represent half bodyweight. The explosive mass was varied from 10 g to 50 g based on the amount of charge used in explosives mines [14]. The standoff distance was kept constant in the current study.

31.3.2 CONTACT DEFINITION

The contact definition between the parts is shown in Table 31.2. Node-to-surface contact was employed between the honeycomb core and face sheet material. Node-to-surface contact was defined between the null and node sets of all parts. A null element layer was defined at the bottom of the sandwich panel, and the blast load was subjected through it. The null element segment set was effective in providing a stable contact between parts of different stiffnesses; it also does not affect the simulation results with regard to stress, strain, or energy absorption. Automatic single surface contact was used to capture the self-contact due to the folding of honeycomb cell wall material.

31.3.3 BASELINE MATERIAL MODELS

31.3.3.1 Aluminum Honeycomb

Aluminum honeycomb was modeled as elastic piecewise plastic using a Mat-024 material card available in LS-Dyna. The mechanical properties of the aluminum honeycomb core were taken for AL5052 aluminum grade [15] (as shown in Table 31.3). A stress-strain plot at different strain rates is shown in Figure 31.3.

31.3.3.2 Steel Rod

The steel rod is used in the simulation setup as a metallic surrogate and is modeled after the Johnson-Cook material model, whose parameters are given in Table 31.4 [16, 17].

31.3.3.3 Dyneema Face Sheet

Dyneema composite was used as a face sheet material as the sandwich configuration for the aluminum honeycomb core. The Dyneema was modeled using the enhanced composite damage (Mat-54/56) material model available in LS-Dyna, whose input parameters are given in Table 31.5 [13].

TABLE 31.2
Contact Definitions

Parts in contact	Contact type
Bottom FRP and aluminum honeycomb	Node to surface contact
Top FRP and aluminum honeycomb	
Top FRP and steel rod	
Null layer and all part set	
Aluminum honeycomb	Automatic single-surface contact

TABLE 31.3
Material Properties of Aluminum

Material	Density (g/cm³)	Young's modulus (GPa)	Yield stress (MPa)	Poisson's ratio
AL5052	2.68	69	180	0.33

Source: [15].

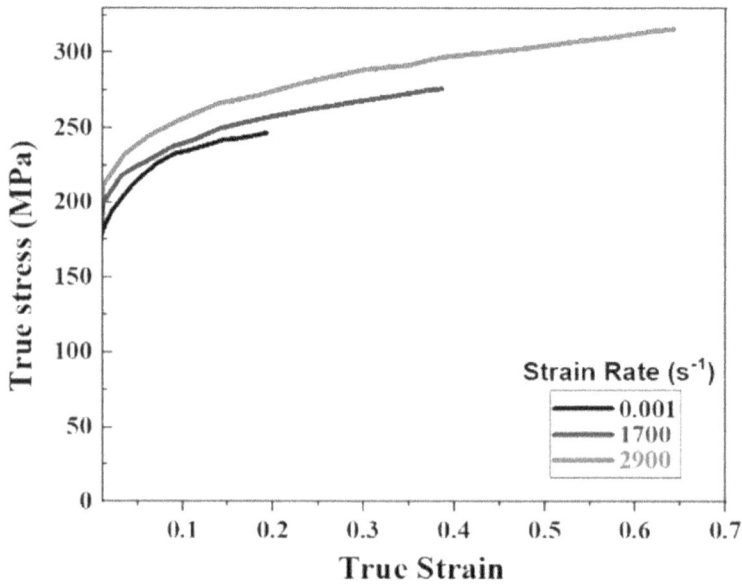

FIGURE 31.3   Stress-strain curve of aluminum 5052 at different strain rates. Source: [15].

TABLE 31.4
Johnson-Cook Material Model (Mat-015) Parameters of Hardened AISI 4340

Young's modulus (GPa)	207	A (MPa)	792	D1	0.050
Poission's ratio	0.3	B (MPa)	510	D2	3.440
Heat capacity (J. Kg^{-1} .K^{-1})	477`	C	0.014	D3	-2.120
Melting point (K)	1793	m	1.030	D4	0.002
ε_0	1	n	0.260	D5	0.610

Sources: [16, 17].

TABLE 31.5
Material Properties of Dyneema

Dyneema HB26

Parameters	Orthotropic
Ref. density (g/cc)	0.97
E11 (MPa)	34,257
E22 (MPa)	34,257
E33 (MPa)	3,260
Poisson's ratio 12	0
Poisson's ratio 23	0.013
Poisson's ratio 31	0.013
Shear modulus 12 (MPa)	173.8
Shear modulus 23 (MPa)	547.8
Shear modulus 31 (MPa)	547.8

Source: [13].

31.4 RESULTS AND DISCUSSION

31.4.1 HEXAGONAL ALUMINUM HONEYCOMB: A PARAMETRIC STUDY

Honeycomb core density is a function of the ratio of wall thickness (t) to cell edge size (L). The t/L ratio is a crucial parameter in determining the performance of honeycomb material under different loading scenarios. Hexagonal aluminum honeycomb has higher energy absorbing tendency in a direction along the cell axis due to progressive buckling of the cell wall. A parametric study was performed to investigate the response of aluminum honeycomb under extreme blast load. The response was analyzed in terms of the following parameters: (a) mesh convergence, (b) effect of scaled distance, (c) specific energy absorption, and (d) effect of t/L ratio on force transmission. The cell size of the honeycomb was varied from 9.6 mm to 1 mm and the wall thickness (t) is constant. The density and t/L ratio are mentioned for the corresponding cell sizes in Table 31.6.

31.4.2 MESH CONVERGENCE

Mesh convergence is essential for building a reliable numerical model. The meshed models were subjected to mesh sensitivity and convergence studies for mesh sizes ranging from 2 mm down to 0.25 mm. Peak force was found to start converging below 1 mm mesh size. However, below a mesh size of 0.8 mm, computational efficiency declined. Transition meshing was generated with a minimum element size of 0.25 mm, facing the blast wave, and a maximum size of 2 mm along the core thickness. Transition meshing had shown a close agreement with the convergence results and a substantial drop in the computational time (see Figure 31.4(a)).

TABLE 31.6
Honeycomb Cell Size Variation

Cell size (in mm)	Density (Kg/m3)	t/L
9.6	16.588	0.0054
6.4	24.88	0.0081
4.8	33.792	0.011
3.2	49.152	0.016
2.0	79.152	0.026
1.0	159.7	0.052

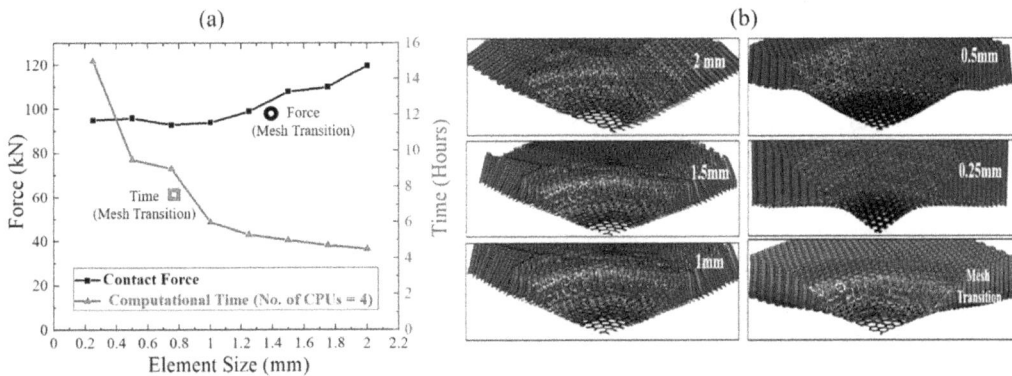

FIGURE 31.4 (a) Mesh convergence study and (b) final deformation of honeycomb with varying mesh sizes.

The smaller mesh size resulted in stable plastic folding of cell wall material rather than buckling. Transition meshing was cost-effective because it triggered the cell wall material folding due to the presence of small elements in the blast direction. For further analysis, a mesh size of 0.8 mm was chosen based on the mesh convergence study.

31.4.3 DEFORMATION PROFILE

Contact force was measured between the steel rod and the top Dyneema face sheet. To help visualize the phenomenon, Figure 31.5 depicts the deformation profile of a honeycomb of a cell size of 1 mm in relation to the contact force vs. time plot. At 0.0242 ms (1), the compression of the core material began, and the first peak was seen. This primary peak was caused by a reflected blast wave triggered by overpressure interacting with the structure. Following, the honeycomb was compressed, and the secondary peak appeared at 0.0476 ms (2) due to the plastic collapse of the cell wall material. Around 0.0758 ms (3), the central portion of the honeycomb approached densification, and the force progressively decreased. Force was almost constant after 0.2 ms (4).

31.4.4 EFFECT OF SCALED DISTANCE

The variation in energy absorption by the aluminum honeycomb core with the scaled distance was investigated by varying the amount of charge. The impulse transferred from the bottom to the top was computed. After the full collapse of the cell wall, energy absorption was nearly constant below the scaled distance, $Z = 0.2$ m/kg$^{1/3}$ (until the beginning of the densification process).

The generated pressure was extremely high for smaller-scaled distances, resulting in a large change in the shape of the shell element over a single timestep (Figure 31.6(b)). The deformation was extensive for the scaled distance of 0.1 m/kg$^{1/3}$, resulting in a decrease in timestep, while the deformation was stable for scaled distances of more than 0.25 kg/m$^{1/3}$ (see Figure 31.6(a)). When the scaled distance was less than 0.2 m/kg$^{1/3}$, the computing time increased substantially due to a decrease in FE timestep (see Figure 31.6(a)). To solve this, elements having a timestep of less than 9.0e-5 ms were eliminated. This helped in the successful completion of the simulation.

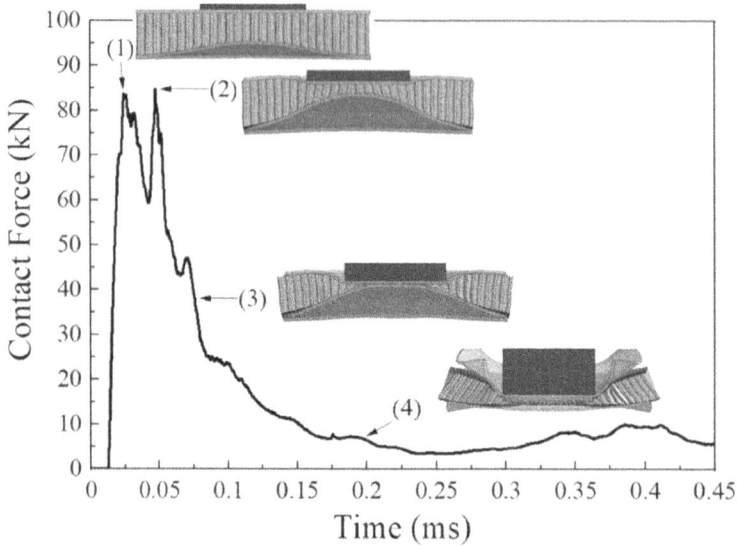

FIGURE 31.5 Aluminum honeycomb deformation w.r.t to the contact force vs. time (measured at the interface of the steel rod and top Dyneema).

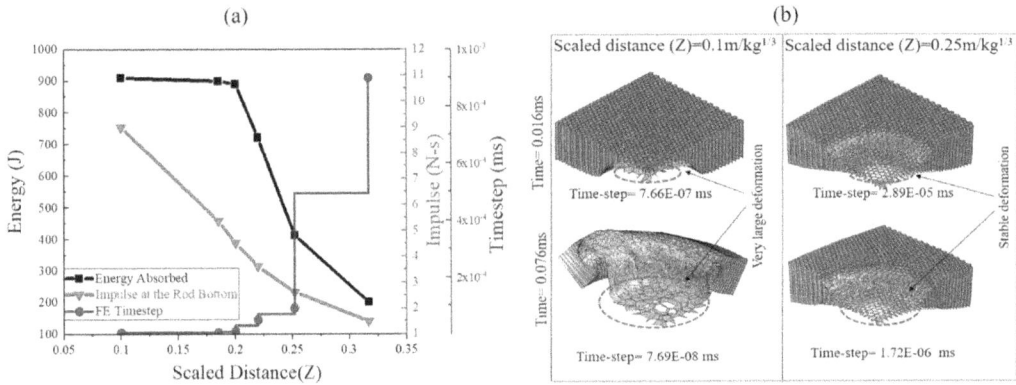

FIGURE 31.6 (a) Effect of scaled distance on energy and (b) deformation profile of aluminum honeycomb of cell size 1 mm for $Z = 0.1$ m/kg$^{1/3}$ and 0.25 m/kg$^{1/3}$.

31.4.5 SPECIFIC ENERGY ABSORPTION (SEA)

Energy absorption per unit mass and per unit volume is one of the prime requisites in the development of any protective equipment for impact and blast loading. Energy absorption per unit mass and unit volume are termed specific energy absorption (SEA). The efficiency of honeycomb core material under blast load was investigated using specific energy absorption evaluation techniques. Figure 31.7(a) shows energy absorption per unit mass by a honeycomb core for various cell sizes and core heights. With a decrease in core density (or increase in cell size), SEA (J/g) rises. For core thicknesses of 40 mm, 50 mm, and 60 mm, the SEA (J/g) was constant for cell sizes of 1 mm, 2 mm, 3.2 mm, 4.7 mm, and 6.4 mm. The maximum SEA (J/g) was found in a honeycomb of core thickness of 20 mm.

31.4.6 EFFECT OF t/L RATIO ON FORCE TRANSMISSION

The peak section of force transmission through the honeycomb sandwich core to the steel rod, just above the upper Dyneema, is shown in Figure 31.8, as a function of the t/L ratio and honeycomb core thickness. When building any piece of equipment that protects against extreme loading such as impact and blast loads, force transfer and energy absorption are some of the critical aspects to consider. With the core thickness variation, the honeycomb with t/L = 0.026 demonstrated the least force transmission. Unlike other cases, the t/L = 0.0054 curve exhibited

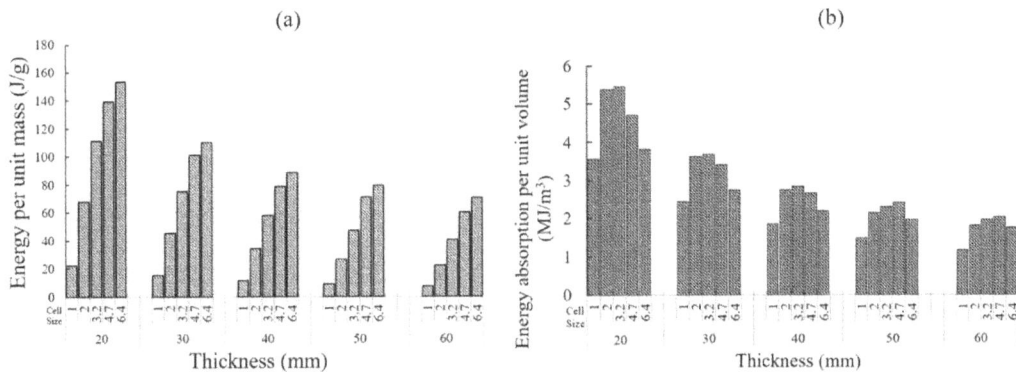

FIGURE 31.7 Specific energy absorption: (a) per unit mass and (b) per unit volume.

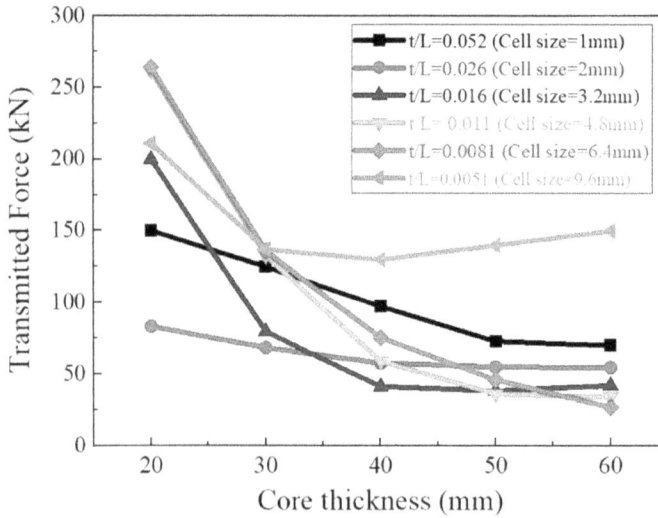

FIGURE 31.8   Effects of t/L ratio with varying core height on force transmission.

increasing force transmission with increasing thickness. The thickness of the core material is also essential, as most of the variations in force can be seen for thickness variations between 20 mm and 40 mm.

31.5   CONCLUSION

Under blast loading, a mesh convergence study on hexagonal aluminum honeycomb revealed a substantial difference in force transmission. The force varies wildly as the mesh size is increased. FE results show a close match (percentage difference below 3%) to a mesh size between 0.8 mm and 0.25 mm. The 0.8 mm mesh size was demonstrated to be more computationally efficient and has been used for further analysis for near-field blasts. Transition meshing is much more cost-effective. The amount of energy absorbed is determined by the impulse applied to the sandwich panel. The higher the impulse load, the higher the deformation, and thus the greater the energy absorption. The plastic collapse of cell wall material, on the other hand, controls the aluminum honeycomb core's energy absorption. After the honeycomb core is compacted, energy absorption due to the plastic collapse of cell wall material remains relatively constant. Global bending of a compacted honeycomb core, on the other hand, may result in an increase in finite-element energy calculation, if proper damage (or element deletion criteria) is not employed. The computation time is affected by the amount of overpressure applied to the sandwich panel. The geometry of the cell wall material changes rapidly at lower scale distances, resulting in a significant reduction in computing timestep. To solve this, the simulation was successfully completed using element elimination based on the element timestep. With the core thickness variation, a hexagonal aluminum honeycomb with t/l = 0.026 demonstrates the least force transmission. In cells of sizes of 2 mm and 3.2 mm, specific energy absorption per unit mass and volume is higher. Overall, the current study proposes efficient modeling of honeycomb structural materials to evaluate their efficacy in near-field explosion scenarios.

ACKNOWLEDGMENTS

The authors thankfully acknowledge the Defence Research and Development Organisation for funding, vide grant # DFTM/03/3203/M/01/JATC.

REFERENCES

1. *Cellular Solids: Structure and Properties*. 1997. [Online]. Available: doi: 10.1017/CBO9781139878326

2. A. Chawla, S. Mukherjee, D. Kumar, T. Nakatani, and M. Ueno, "Prediction of crushing behaviour of honeycomb structures," *International Journal of Crashworthiness*, vol. 8, no. 3, pp. 229–235, Jan. 2003, doi: 10.1533/ijcr.2003.0227.

3. T. Thomas and G. Tiwari, "Crushing behavior of honeycomb structure: A review," *International Journal of Crashworthiness*, vol. 39, no. 2, 2019, doi: 10.1080/13588265.2018.1480471.

4. G. Sun, D. Chen, X. Huo, G. Zheng, Q. Li, and Q. X. Li, "Experimental and numerical studies on indentation and perforation characteristics of honeycomb sandwich panels," *Composite Structures*, vol. 184, pp. 110–124, Jan. 2018, doi: 10.1016/j.compstruct.2017.09.025.

5. G. Sun, D. Chen, H. Wang, P. J. Hazell, Q. Li, and Q. X. Li, "High-velocity impact behaviour of aluminium honeycomb sandwich panels with different structural configurations," *International Journal of Impact Engineering*, vol. 122, pp. 119–136, Dec. 2018, doi: 10.1016/j.ijimpeng.2018.08.007.

6. G. N. Nurick, G. Langdon, Y. Chi, and N. Jacob, "Behaviour of sandwich panels subjected to intense air blast – part 1: Experiments," *Composite Structures*, vol. 91, no. 4, pp. 433–441, Dec. 2009, doi: 10.1016/j.compstruct.2009.04.009.

7. D. Karagiozova, G. N. Nurick, and G. Langdon, "Behaviour of sandwich panels subject to intense air blasts – part 2: Numerical simulation," *Composite Structures*, vol. 91, no. 4, pp. 442–450, Dec. 2009, doi: 10.1016/j.compstruct.2009.04.010.

8. S. Li *et al.*, "Sandwich panels with layered graded aluminum honeycomb cores under blast loading," *Composite Structures*, 2017, doi: 10.1016/j.compstruct.2017.04.037.

9. M. D. Theobald, G. Langdon, G. N. Nurick, S. Pillay, A. Heyns, and R. P. Merrett, "Large inelastic response of unbonded metallic foam and honeycomb core sandwich panels to blast loading," *Composite Structures*, vol. 92, no. 10, pp. 2465–2475, Sep. 2010, doi: 10.1016/j.compstruct.2010.03.002.

10. A. Erdik and V. Uçar, "On evaluation and comparison of blast loading methods used in numerical simulations," *Sakarya University Journal of Science*, vol. 22, no. 5, pp. 1385–1391, Oct. 2018, doi: 10.16984/saufenbilder.357629.

11. M. Karahan, E. A. Karahan, and Nevin Karahan, "Blast performance of demining footwear: Numerical and experimental trials on frangible leg model and injury modeling," *Journal of Testing and Evaluation*, 2017, doi: 10.1520/jte20160340.

12. R. K. Pinto, "A gun based test method to simulate mine blast against boots," http://dspace.lib.cranfield.ac.uk/handle/1826/10991.

13. M. K. Hazzard, R. S. Trask, U. Heisserer, M. Van Der Kamp, and S. R. Hallett, "Finite element modelling of Dyneema® composites: From quasi-static rates to ballistic impact," *Composites Part A-applied Science and Manufacturing*, 2018, doi: 10.1016/j.compositesa.2018.09.005.

14. R. M. Harris, M. S. Rountree, L. V. Griffin, R. A. Hayda, and T. Bice, "Final report of the Lower Extremity Assessment Program (LEAP 99–2). volume 2," 2000, doi: 10.21236/ada409059.

15. P. Song, W. Li, W. B. Li, and X. Wang, "A study on dynamic plastic deformation behavior of 5052 aluminum alloy," *Key Engineering Materials*, 2019, doi: 10.4028/www.scientific.net/kem.812.45.

16. M. Karahan, N. Karahan, and E. A. Karahan, "An investigation of blast performance of mine protective military boots," 2017[https://api.semanticscholar.org/CorpusID:212594841].

17. M. Karahan and N. Karahan, "Development of an innovative sandwich composites for the protection of lower limbs against landmine explosions," *Journal of Reinforced Plastics and Composites*, vol. 35, no. 24, pp. 1776–1791, Dec. 2016, doi: 10.1177/0731684416668261.

32 Parametric Study on Blast Response Mitigation Using Tube-Reinforced Honeycomb Sandwich Structures

Payal Shirbhate and M. D. Goel

32.1 INTRODUCTION

Due to the surge in terrorist acts occurring over the past few decades, the demand for increasing the protection level of public gathering places (e.g., shopping complexes, institutions, theaters, etc.), government organizations, and other important buildings has also increased. The blast-resistant design of structures includes efficient energy absorption mechanisms. The sandwich structure is found to be the most efficient energy-absorbing cellular composite structure. Thus, scientists and designers are consistently suggesting sandwich structures over conventional materials for mitigating the damaging effects that may occur due to blast loading. The general layout of a sandwich structure consists of facesheets separated by core material. The components, namely the core, facesheets, and material of these components, govern the energy absorption characteristics of the whole sandwich structure. Sandwich panels also tend to dissipate energy by undergoing plastic deformation. The sandwich structures have been found to be more effective in blast response mitigation than solid plates with equal areal densities [1, 2]. The provision of a core between the facesheets not only improves the sandwich structure's blast resistance capability but also improves energy absorption performance by systematically distributing stress [3]. There are numerous investigations available on the blast resistance behavior of conventional honeycomb sandwich structures. Some researchers experimentally investigated the dynamic response of honeycomb sandwich structures under blast loading [4–10]. The parameters that govern the blast resistance capabilities of honeycomb sandwich structures include cell wall thickness, core height, filling strategy in reinforced sandwich structures, and cell size [11]. In addition to these parameters, facesheet material, facesheet thickness, and core play an important role in blast response reduction as compared to the backsheet. Some researchers also proposed analytical solutions to investigate the response of sandwich panels subjected to blast loading [12].

To further improve the load transfer mechanism and energy absorption performance, various other ways are proposed by several researchers. Li et al. [13, 14] evaluated the behavior of a layered honeycomb sandwich structure under blast load. It was observed from this study that panels with the highest relative density placed near the blast impact zone showed low backsheet deformation. The response of a new type of configuration with a biomimetic layered honeycomb sandwich has been compared with that of a single-layer conventional honeycomb sandwich structure. The layered and staggered arrangement in biomimetic layered honeycomb structure absorbed more energy by sacrificing the layers than in conventional honeycomb [15]. Nayak et al. [16] carried out the process of design optimization of honeycomb sandwich panels to get the maximum reduction in response under the impact due to blast load. Goel et al. [17–19] in their studies proposed to provide stiffeners

DOI: 10.1201/9781003352358-32

to enhance the performance of sandwich structures for mitigation of blast loads. The strategic place-ment of stiffeners is necessary to achieve the maximum reduction in response under impulsive load conditions occurring due to blast. Thus, based on the literature survey, an attempt is made in the current study to perform a parametric analysis of a reinforced honeycomb sandwich structure under blast load.

32.2   FINITE ELEMENT MODELING

32.2.1   GEOMETRIC DETAILS OF SANDWICH STRUCTURE

A sandwich panel with dimensions of 300 mm × 300 mm is considered to evaluate the response of a reinforced honeycomb sandwich structure. The complete assembly of the sandwich panel with dimensional details is represented in Figure 32.1(a). The facesheet and backsheet thickness is 5 mm, whereas other geometric details of the honeycomb core are illustrated further. The cell size (c) of the honeycomb is 19 mm; the diameter of the tube reinforcement (d_r) is 12.7 mm; and the cell wall thickness of the honeycomb for a single wall (t_w) is 0.05 mm and a double wall ($2t_w$) is 0.1 mm. The wall thickness of tube reinforcement (t_r) is 0.05 mm. The depth of the honeycomb core (h_c) as well as tube reinforcement considered for finite element analysis is 50 mm. Adhesive is not modeled between honeycomb cells and reinforcement. The relative densities of a bare hexagonal honeycomb core and a tube-reinforced honeycomb core are obtained as per Equations (32.1) and (32.2) [11], where ρ^* and ρ^s represent the density of the honeycomb structure and the density of solid material, respectively.

$$\frac{\rho^*}{\rho^s} = \frac{2t/h}{(1+\sin\theta)\cos\theta} \tag{32.1}$$

$$\frac{\rho^*}{\rho^s} = \frac{4th + \pi dD}{(1+\sin\theta)2h^2\cos\theta} \tag{32.2}$$

32.2.2   MATERIAL MODEL

The honeycomb sandwich panel composed of facesheets and core is assumed to be made up of aluminum alloy due to its advantages of low density and plastic deformation capabilities. The hon-eycomb structure is composed of AA-3003, whereas the tube reinforcement and facesheets are

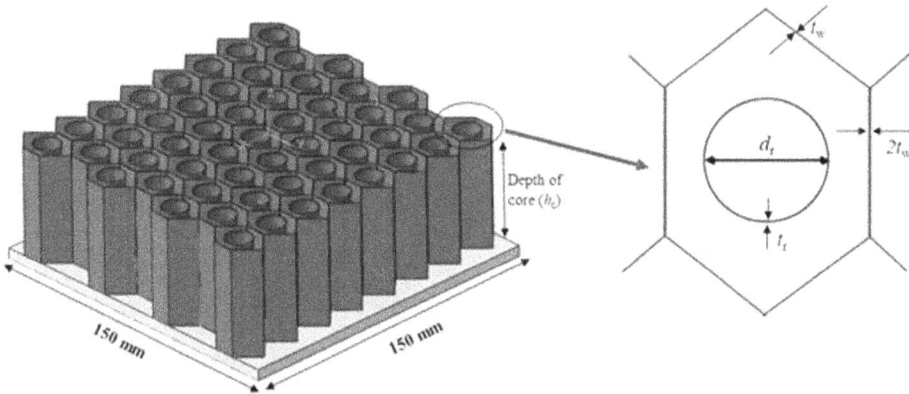

FIGURE 32.1   Tube-reinforced honeycomb sandwich structure's geometric specifications of core and reinforcement.

composed of AL-5052 and AL-1100, respectively. The mechanical properties of the aluminum alloys are represented in detail in Table 32.1. The material behavior of the honeycomb sandwich structure is modeled using the elasto-plastic material model available in ABAQUS.

32.2.3 NUMERICAL MODELING AND BLAST LOADING

The blast load resulting from a charge weight of 1 kg TNT at a 300 mm standoff distance from the facesheet is simulated using ABAQUS. Due to the symmetry, quarter geometry is modeled in the present analysis. The symmetric boundary conditions are applied to the sandwich structure as represented in Figure 32.2. The honeycomb cell is meshed with 30 elements through the thickness of the core, five elements in the vertical web, and seven elements in the inclined web of the honeycomb core, whereas the facesheet and backsheet are meshed using 30×30 elements and five elements in the thickness direction. The contact between facesheets and core is modeled using the TIE constraint option available in ABAQUS.

The reinforced honeycomb sandwich structure is analyzed based on the ConWep blast load analysis method developed by Kingery and Bulmash [20] as represented in Equation (32.3). In this method, the blast pressure is obtained as a combined effect resulting from incident pressure (Pi) and reflected pressure (Pr). The equation is valid for scaled distances ranging from 0.05 to 40 m/kg1/3.

TABLE 32.1
Mechanical Properties of Aluminum Alloy

Aluminum alloy	Young's modulus (GPa)	Yield stress (MPa)	Poisson's ratio	Density (kg/m³)
AA-3003	70	194	0.3	2,710
AL-5052	70	70	0.3	2,710
AL-1100	70	140	0.3	2,710

Sources: [5, 11].

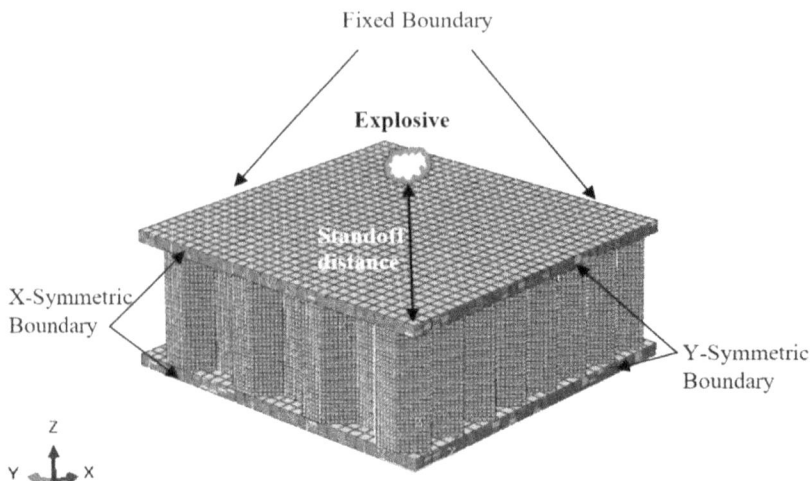

FIGURE 32.2 Finite element setup of a quarter model of a sandwich structure in commercial software ABAQUS.

$$P_t = P_r \cos^2 \theta + P_i \left(1 + \cos^2 \theta - 2\cos\theta \right) \tag{32.3}$$

$$Z = \frac{R}{W^{\frac{1}{3}}} \tag{32.4}$$

32.2.4 VALIDATION OF FE MODEL

The deformation behavior of sandwich plates has been investigated using ABAQUS to study the response of honeycomb sandwich plates subjected to ConWep blast loading. To validate the blast model, literature authored by Dharmasena et al. [21] is referred to. For this validation, a honeycomb sandwich plate of size 610 mm × 610 mm × 61 mm consisting of a square honeycomb core was considered as represented in Figure 32.3(a). The honeycomb core is 0.76 mm thick and 51 mm wide, with spacing between the core web of 30.5 mm. The size of both top and bottom plates was 610 mm × 610 mm × 5 mm. The 1 kg TNT was positioned 100 mm vertically above the top plate in the z-direction. The sandwich structure was made of high-ductility stainless steel alloy material. Assuming the symmetry of the solution, a one-quarter model of a full plate was used for simulation.

The sandwich panel model has been analyzed using ABAQUS and the results were compared with the experimental results of Dharmasena et al. [21]. The central point deflection of the back face with respect to distance from the center of the plate is represented in Figure 32.3(b). It can be observed from Figure 32.3(b) that the results of numerical simulation and experimental results are comparable. Peak pressure from experimental and FE results are 366 and 365.34 Pa, respectively, which are nearly equal with a percentage variation of 0.18%. However, there exists a small difference in displacement-time history with a maximum percentage deviation of 0.66% and 2.3% for facesheet and backsheet respectively. Thus, it can be said that the finite element model is validated and can be further used for the analysis.

32.3 RESULTS AND DISCUSSION

The current study represents a numerical investigation of the effect of reinforcement on the blast resistance performance of a sandwich structure. Further, parametric analysis is conducted to get the effective configuration for blast load mitigation. The detailed discussions on results obtained from the study are represented further.

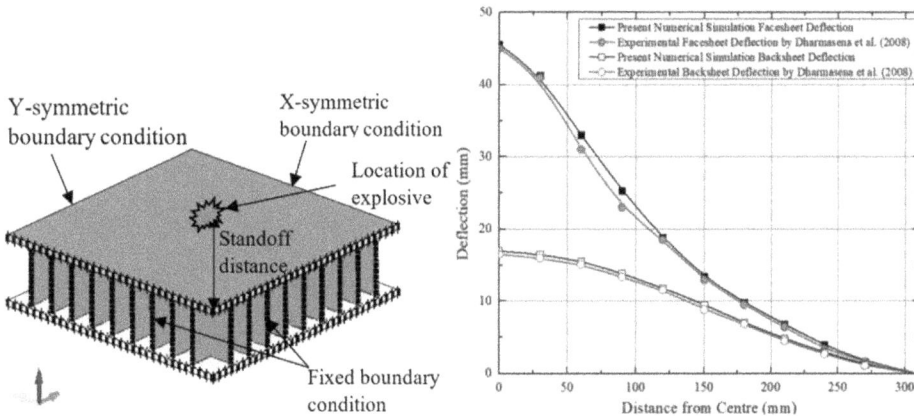

FIGURE 32.3 (a) One-fourth symmetrical FE model of the sandwich plate with explosive; (b) comparison of deflection versus distance from the center of the square honeycomb sandwich panel with experimental results.

32.3.1 Effect of Reinforcement

The effect of reinforcement on deformation modes and blast resistance performance of the honeycomb sandwich structure is discussed in detail in the subsequent section. Here, the response of the bare honeycomb core is compared with the tube-reinforced honeycomb core sandwich structure based on peak backsheet deflection. Initially, when the blast wave arrives at the sandwich structure, it interacts with the facesheet. Thereafter, due to the contact between the facesheet and the core, impulse energy is transferred to the core and backsheet. As the explosive charge is located centrally, the blast pressure is maximum at the center and reduces circumferentially. Figure 32.4 depicts the modes of deformation and variation of displacement with time at the center point of the backsheet of the sandwich panel. It can be noted that the provision of reinforcement in honeycomb cells resists the impulse transferred to the backsheet compared to conventional honeycomb. The maximum backsheet deformation for conventional and reinforced honeycomb sandwich structures is 35.43 mm and 32.38 mm, respectively, which indicates an enhancement in resistance to deflection. The relative density of conventional and reinforced honeycomb cores is 0.0214 and 0.0265 obtained using Equations (32.1) and (32.2), respectively. The relative densities of the reinforced honeycomb are 23.83% higher than the conventional honeycomb structure. With the compromise of a 23.83% increase in relative density, an 8.6% improvement in deformation resistance is observed. It can also be observed from Figure 32.4(b) that there is a reduction in kinetic energy of the overall sandwich structure when a reinforced honeycomb core is used instead of a general honeycomb configuration. The total plastic energy dissipated due to inelastic deformation for conventional and reinforced honeycomb sandwich structures is presented in Figure 32.4(c). The energy absorption capability

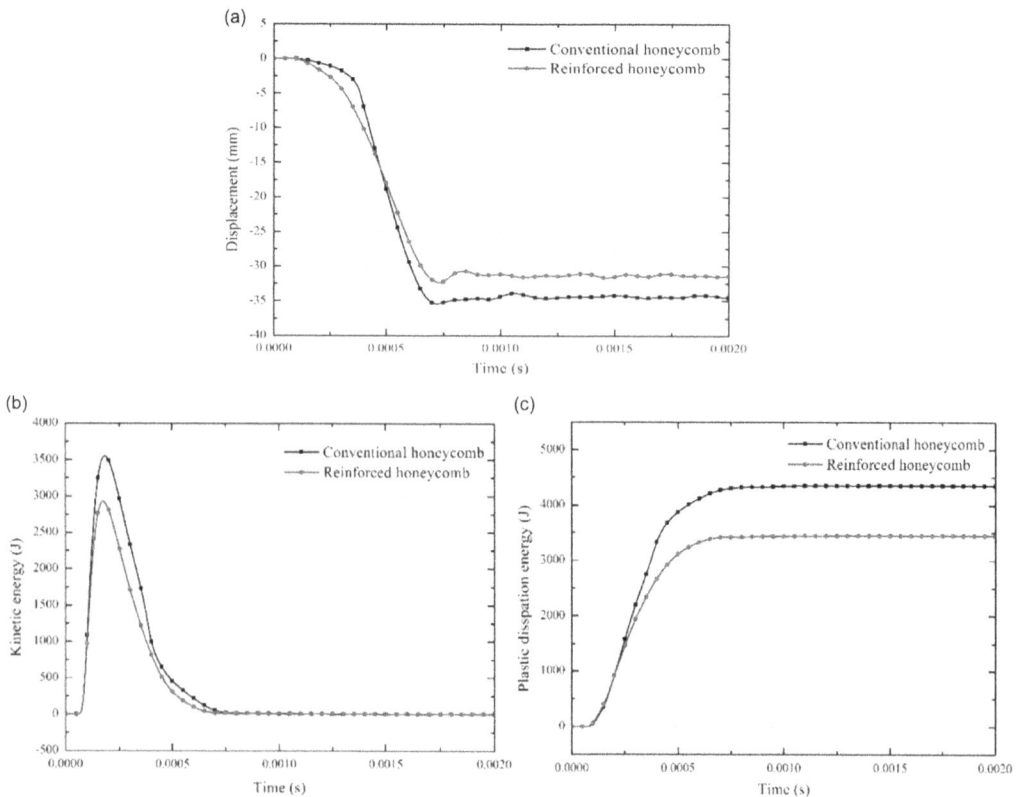

FIGURE 32.4 (a) Displacement-time response; (b) kinetic energy; (c) plastic dissipation energy of conventional and reinforced honeycomb sandwich structure.

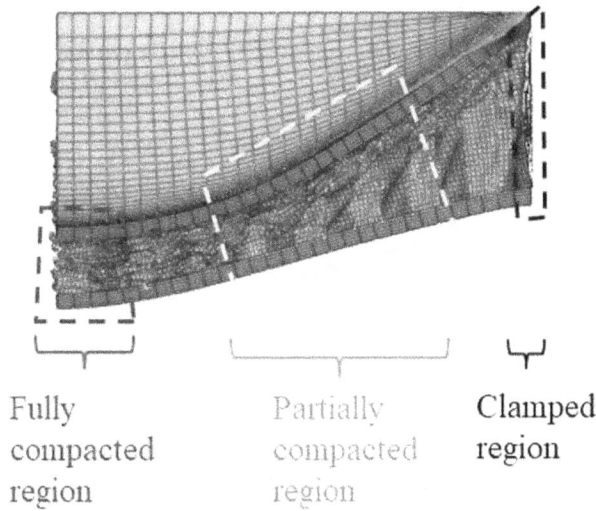

Fully
compacted
region

Partially
compacted
region

Clamped
region

FIGURE 32.5 Deformation mode of reinforced honeycomb sandwich structure.

of a honeycomb sandwich structure filled with tube reinforcement is higher. Thus, hereby it can be said that the purpose of providing a reinforced sandwich structure is fulfilled in the present case as compared to the bare honeycomb core sandwich structure.

The maximum deformation mode of the reinforced honeycomb sandwich structure is represented in Figure 32.5. The deformation of the sandwich structure is majorly divided into three regions, namely fully compacted, partially compacted, and clamped regions. As the explosive is located at the central point of the facesheet, the region near the central portion of the sandwich panel experienced maximum crushing. Due to the reduction in the intensity of the blast load away from the central point of the sandwich structure, this portion is partially folded with cell walls buckled. This portion of the sandwich structure is least affected as it is located close to the clamped boundary.

32.3.2 EFFECT OF CELL WALL THICKNESS

To study the effect of cell wall thickness, different configurations with varying wall thicknesses of reinforcement and honeycomb are considered. Here, H represents a honeycomb core, and T represents tube reinforcement followed by the wall thickness value. Four combinations of cell wall thicknesses of reinforcement and core are considered for the present investigation, namely H0.1T0.1, H0.1T0.05, H0.05T0.05, and H0.5T0.1. The cell size of 12 mm and core depth of 50 mm are kept constant during the investigation of the effect of cell wall thickness.

From the displacement-time curve shown in Figure 32.6, it can be noticed that the configuration having a cell wall thickness of the core of 0.1 mm and a tube reinforcement thickness of 0.05 mm resisted more blast load among all the configurations in the current analysis. Increasing the wall thicknesses of the reinforcement or core promotes the stiffness of the sandwich panel resulting in more backsheet deformation. It can also be noted from the curve represented in Figure 32.6 that the deformation responses of H0.05T0.05 and H0.05T0.1 do not have much difference. This is due to the fact that the material used for reinforcement is weaker compared to conventional honeycomb. Thus, reducing the wall thickness of the core and increasing the wall thickness of the reinforcement does not enhance the blast resistance capability. Hence, it can be concluded from this parametric study that a stronger core with weak reinforcement is the ideal combination for achieving maximum output from the reinforced honeycomb sandwich configuration.

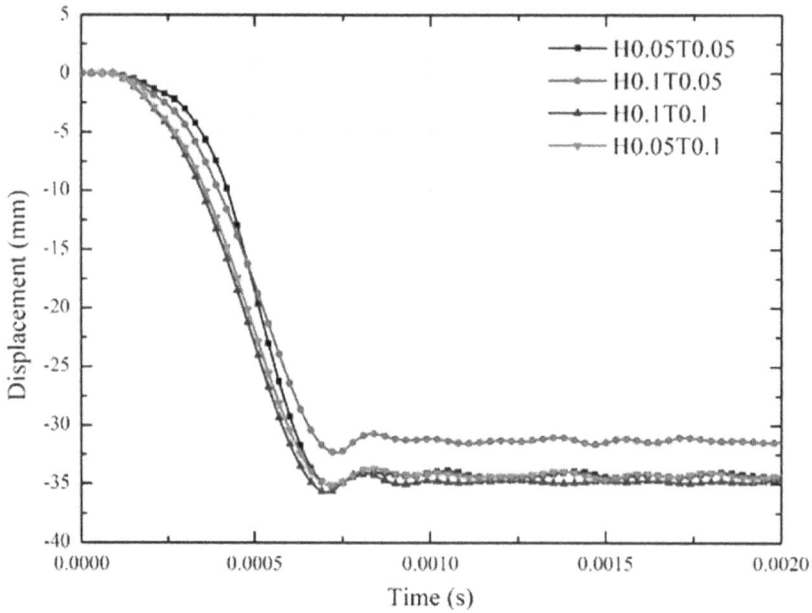

FIGURE 32.6 Displacement-time curve for sandwich panel with different cell wall thicknesses of core and reinforcement.

32.3.3 EFFECT OF CORE DEPTH AND REINFORCEMENT

A parametric study is performed to investigate the effect of depth of core and reinforcement on the behavior of the sandwich structure in resisting blast load. Hence, configuration with varying depths of honeycomb core and reinforcement (i.e., H50T50, H50T25, and H25T25) is considered for the present analysis. Here, H represents the honeycomb core, T represents tube reinforcement, and the number represents the depth in mm of the respective parameter (e.g., H50T25 represents a honeycomb core with 50 mm depth and tube reinforcement with 25 mm depth). In the case of configurations H50T50 and H25T25, the depth of both core and reinforcement is equal and they are in contact with the facesheet. Thereby, both will deform simultaneously under incident blast pressure. The maximum backsheet deflection of H50T50, H50T25, and H25T25 is 32.38 mm, 32.48 mm, and 39.13 mm, respectively (refer to Figure 32.7). Here, in the case of H50T25, initially the cell walls of the honeycomb deform, and when the facesheet reaches the tube reinforcement, the tube and honeycomb deform together. Hence, only 50% depth of reinforcement resists blast load. Therefore, a lesser reduction in backsheet deformation is observed in H50T25 as compared to H50T50.

Also, in the case of H25T25, the least resistance to deformation is obtained compared to the other two configurations (i.e., H50T25 and H50T50). By providing the core with a higher depth between two facesheets, the energy absorption characteristics increase due to the availability of more distance for crushing of the core.

32.3.4 EFFECT OF CHARGE WEIGHT

To evaluate the effect of charge weight on a reinforced honeycomb sandwich structure, the sandwich structure is exposed to 0.5 kg, 1.0 kg, and 1.5 kg TNT charge weight. The scaled distances of the respective charge weights as per Equation (32.4) are 0.38 m/kg$^{1/3}$, 0.3 m/kg$^{1/3}$, and 0.26 m/kg$^{1/3}$. The incident pressure time curve obtained from these three explosives is presented in Figure 32.8. The impulses imparted due to blast load resulting from 0.5 kg, 1.0 kg, and 1.5 kg are

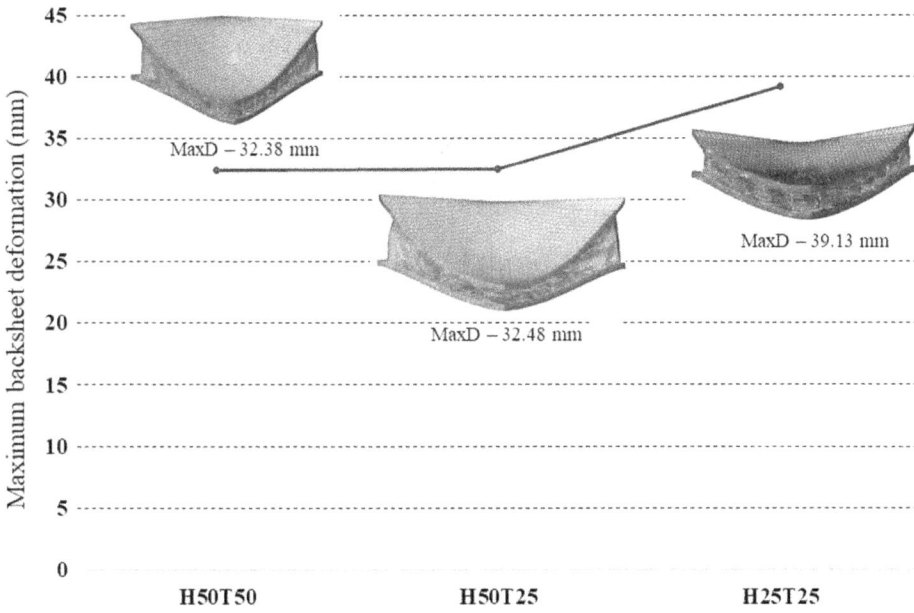

FIGURE 32.7   Maximum backsheet deformation of sandwich structure with varying depths of core and reinforcement.

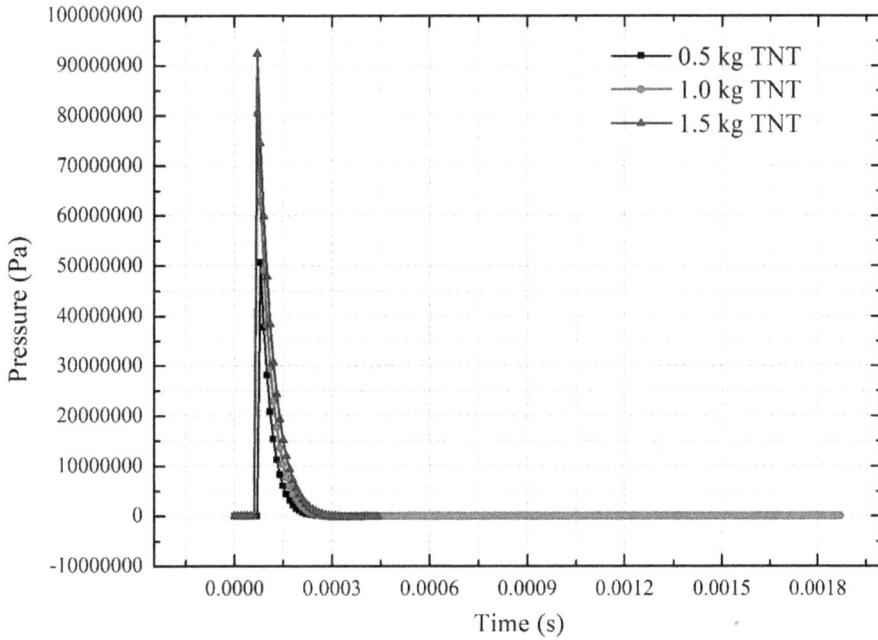

FIGURE 32.8   Pressure versus time curve for varying charge weights.

1,919 Pa.s, 3,511 Pa.s, and 4,508 Pa.s, calculated using the area under the pressure-time curve. The maximum deformation of the reinforced honeycomb sandwich structure is 17.82 mm, 32.38 mm, and 55.91 mm for charge weights of 0.5 kg, 1 kg, and 1.5 kg, respectively, as represented in Figure 32.9. Thus, from this analysis, it can be evidently noted that with an increase in the impulse imparted, the backsheet deformation of the reinforced honeycomb sandwich structure increases. With an increase in charge weight, the energy absorption characteristics of the facesheet and reinforced core also increase.

32.4 CONCLUSION

The blast resistance performance of the reinforced honeycomb sandwich structure is evaluated under blast load. The analysis is then further extended to investigate the effect of varying cell wall thickness and core depth. After performing the finite element analysis, the following conclusions are drawn:

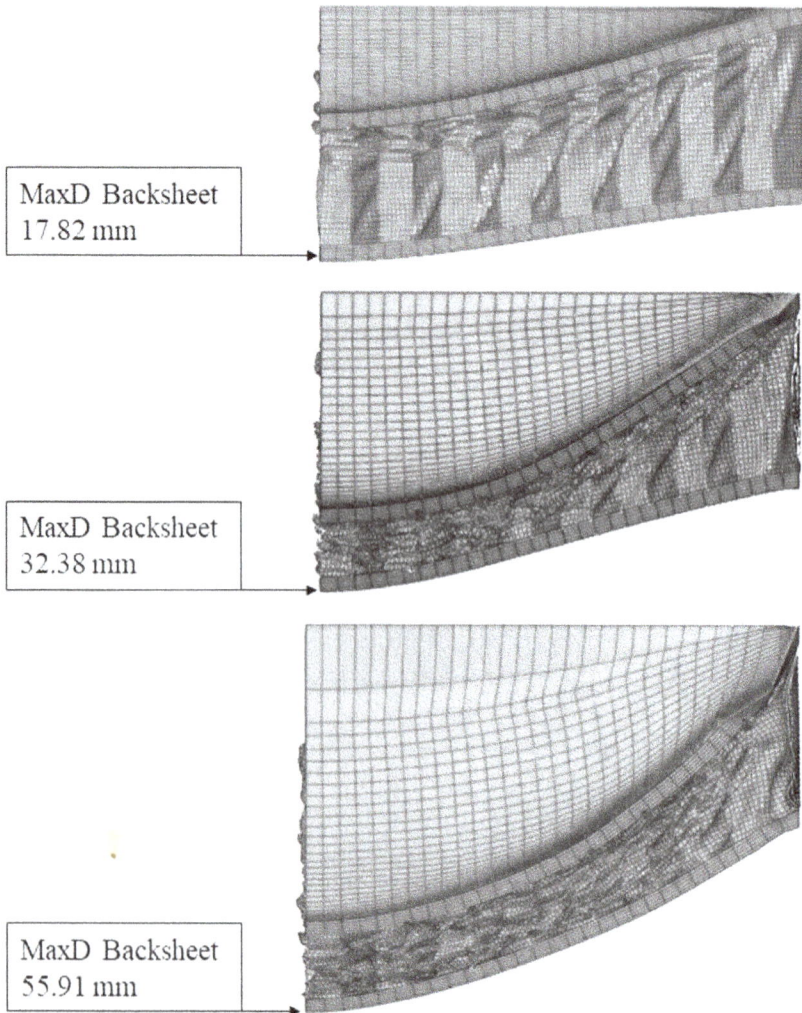

MaxD Backsheet
17.82 mm

MaxD Backsheet
32.38 mm

MaxD Backsheet
55.91 mm

FIGURE 32.9 Maximum deformation modes of sandwich structure with varying charge weights.

1. The tube-reinforced honeycomb showed a reduction in backsheet deformation compared to the conventional honeycomb sandwich structure. Thus, it can be concluded that the provision of reinforcement can be suggested to the designers for getting enhanced blast response mitigation.
2. After analyzing the effect of cell wall thickness, it is observed that reinforced honeycomb with a higher cell wall thickness of the core and lesser thickness of reinforcement promoted blast resistance capacity. Out of several configurations considered in the analysis, the H0.1T0.05 configuration is found to be an effective configuration with core and reinforcement heights being constant and equal to 50 mm.
3. In addition, the current analysis investigated the effect of core depth. As the depth of both core and reinforcement increases, the blast mitigation response also increases with the depth of core and reinforcement due to the availability of space for deformation.
4. The study also extended to considering different charge weights and observed that the deformation of the reinforced honeycomb sandwich structure increases with an increase in charge weights.

REFERENCES

1. R. Alberdi, J. Przywara, K. Khandelwal, Performance evaluation of sandwich panel systems for blast mitigation, *Eng. Struct.* 56 (2013) 2119–2130. doi:10.1016/j.engstruct.2013.08.021.
2. P. Tan, B. Lee, C. Tsangalis, Finite element analysis of sandwich panels subjected to shock tube blast loadings, *J. Sand. Struct. Mater.* 13 (2011) 263–278. doi:10.1177/1099636210375750.
3. Y. Yang, A.S. Fallah, M. Saunders, L.A. Louca, On the dynamic response of sandwich panels with different core set-ups subject to global and local blast loads, *Eng. Struct.* 33 (2011) 2781–2793. doi:10.1016/j.engstruct.2011.06.003.
4. G.N. Nurick, G.S. Langdon, Y. Chi, N. Jacob, Behaviour of sandwich panels subjected to intense air blast – part 1: Experiments, *Compos. Struct.* 91 (2009) 433–441. doi:10.1016/j.compstruct.2009.04.009.
5. T. Wang, Q. Qin, M. Wang, W. Yu, J. Wang, T.J. Wang, Blast response of geometrically asymmetric metal honeycomb sandwich plate: Experimental and theoretical investigations, *Int. J. Impact Eng.* 105 (2017) 24–38. doi:10.1016/j.ijimpeng.2016.10.009.
6. K.P. Dharmasena, H.N.G. Wadley, Z. Xue, and J.W. Hutchinson, Mechanical response of metallic honeycomb sandwich panel structures to high-intensity dynamic loading, *Int. J. Impact Eng.* 35 (2008) 1063–1074. doi:10.1016/j.ijimpeng.2007.06.008.
7. G.S. Langdon, G.N. Nurick, M.Y. Yahya, W.J. Cantwell, The response of honeycomb core sandwich panels, with aluminum and composite face sheets, to blast loading, *J. Sandw. Struct. Mater.* 12 (2010) 733–754. doi:10.1177/1099636210368470.
8. X. Li, P. Zhang, Z. Wang, G. Wu, L. Zhao, Dynamic behavior of aluminum honeycomb sandwich panels under air blast: Experiment and numerical analysis, *Compos. Struct.* 108 (2014) 1001–1008. doi:10.1016/j.compstruct.2013.10.034.
9. M. Stanczak, T. Fras, L. Blanc, P. Pawlowski, A. Rusinek, Blast-induced compression of a thin-walled aluminum honeycomb structure-experiment and modeling, *Metals* 9 (2019) 12. doi:10.3390/met9121350.
10. D. Karagiozova, G. Nurick, G. Langdon, Behaviour of sandwich panels subject to intense air blasts – part 2: Numerical simulation, *Compos. Struct.* 91 (2009) 442–450. doi:10.1016/j.compstruct.2009.04.010.
11. J. Liu, Z. Wang, D. Hui, Blast resistance and parametric study of sandwich structure consisting of honeycomb core filled with circular metallic tubes, *Compos. Part B Eng.* 145 (2018) 261–269. doi:10.1016/j.compositesb.2018.03.005.
12. M.S.H. Fatt, L. Palla, Analytical modeling of composite sandwich panels under blast loads, *J. Sandw. Struct. Mater.* 11 (2009) 357–380. doi:10.1177/1099636209104515.
13. S. Li, X. Li, Z. Wang, G. Wu, G. Lu, L. Zhao, Sandwich panels with layered graded aluminum honeycomb cores under blast loading, *Compos. Struct.* 173 (2017) 242–254. doi:10.1016/j.compstruct.2017.04.037.
14. S. Li, X. Li, Z. Wang, G. Wu, G. Lu, L. Zhao, Finite element analysis of sandwich panels with stepwise graded aluminum honeycomb cores under blast loading, *Compos. Part A Appl. Sci. Manuf.* 80 (2016) 1–12. doi:10.1016/j.compositesa.2015.09.025.
15. J. Li, S. Shi, W. Luo, Q. Wang, Study on explosion-resistance of biomimetic layered honeycomb structure, *Adv. Civ. Eng.* 2020 (2020) 1–15. doi:10.1155/2020/5356145.

16. S. Nayak, A.K. Singh, A.D. Belegundu, C. Yen, Process for design optimization of honeycomb core sandwich panels for blast load mitigation, *Struct. Multidiscip. Optim.* 47 (2013) 749–763. doi:10.1007/s00158-012-0845-x.

17. M.D. Goel, V.A. Matsagar, A.K. Gupta, Dynamic response of stiffened plates under air blast, *Int. J. Prot. Struct.* 2 (2011) 139–156. doi:10.1260/2041-4196.2.1.139.

18. M.D. Goel, V.A. Matsagar, S. Marburg, A.K. Gupta, Comparative performance of stiffened sandwich foam panels under impulsive loading, *J. Perform. Constr. Facil.* 27 (2013) 540–549. doi:10.1061/(ASCE)CF.1943-5509.0000340.

19. M.D. Goel, V.A. Matsagar, A.K. Gupta, Blast resistance of stiffened sandwich panels with aluminum cenosphere syntactic foam, *Int. J. Impact Eng.* 77 (2015) 134–146. doi:10.1016/j.ijimpeng.2014.11.017.

20. C.N. Kingery, G. Bulmash, *Air blast parameters from TNT spherical air burst and hemispherical burst.* Technical report ARBRL-TR-02555, CRC Press, Taylor & Francis group, 1984.

21. K.P. Dharmasena, H.N.G. Wadley, K. Williams, Z. Xue, J.W. Hutchinson, Response of metallic pyramidal lattice core sandwich panels to high intensity impulsive loading in air, *Int. J. Impact Eng.* 38 (2011) 275–289. doi:10.1016/j.ijimpeng.2010.10.00.

Index

For Product Safety Concerns and Information please contact our EU
representative GPSR@taylorandfrancis.com
Taylor & Francis Verlag GmbH, Kaufingerstraße 24, 80331 München, Germany

www.ingramcontent.com/pod-product-compliance
Lightning Source LLC
Chambersburg PA
CBHW080902220326
41598CB00034B/5453

* 9 7 8 1 0 3 2 4 0 2 9 4 9 *